solution
manual

TIME SERIES FORECASTING

TIME SERIES FORECASTING

Unified Concepts and Computer Implementation

Second Edition

BRUCE L. BOWERMAN
RICHARD T. O'CONNELL
MIAMI UNIVERSITY, OHIO

 *Duxbury Press*

Boston

PWS PUBLISHERS

Prindle, Weber & Schmidt • ❧ • Duxbury Press • ♠ • PWS Engineering • ◮ • Breton Publishers • ❀
20 Park Plaza • Boston, Massachusetts 02116

The first edition was published as TIME SERIES AND FORECASTING: AN APPLIED APPROACH.

PWS Publishers is a division of Wadsworth, Inc.

Library of Congress Cataloging-in-Publication Data

Bowerman, Bruce L.
    Time series forecasting.

    Bibliography: p.
    Includes index.
    1. Time-series analysis.   2. Forecasting—
Statistical methods.   I. O'Connell, Richard T.
II. Title.
QA280.B664   1987   003.2   86–25316
ISBN 0–87150–070–1

Printed in the United States of America

87 88 89 90 91 — 10 9 8 7 6 5 4 3 2

Sponsoring Editor: Michael Payne
Editorial Assistant: Dee Hart
Production Coordinator: Ellie Connolly
Production: Editing, Design & Production, Inc.
Interior Design: Ellie Connolly
Cover Design: Julie Gecha
Typesetting: The Alden Press
Cover Printing: New England Book Components, Inc.
Printing and Binding: Halliday Lithograph

To our wives and children:
Drena and Jean
Michael, Christopher, Bradley, Asa, and Nicole

Preface

Time Series Forecasting is designed as a textbook for applied courses in time series forecasting and as a reference book for practitioners who must forecast real-world time series. It is appropriate for advanced (junior and senior level) undergraduates and graduate students in business, engineering, and the sciences (including mathematics, statistics, operations research, and computer science). The required mathematical and statistical background for this book is college algebra and basic statistics.

The objective of this book is to present a concise, applied, and easy to understand discussion of time series forecasting techniques. This is accomplished by placing emphasis on two main themes. First, we use the Box-Jenkins model building and forecasting methodology to unify our presentation of most of the forecasting techniques covered in this book. Second, emphasis is placed on computer implementation of the techniques presented by using SAS.

The text consists of eight chapters. Chapter 1 serves as an introduction. In Chapters 2 and 3 we take a "first principles" approach to introducing the Box-Jenkins methodology. We discuss the general nature of nonseasonal and seasonal Box-Jenkins models, the tentative identification of an appropriate model, and elementary concepts pertaining to the estimation of model parameters and forecasting future time series values. Chapter 4 continues our study of the Box-Jenkins methodology by covering more advanced concepts pertaining to parameter estimation and by presenting various diagnostics that are used to check the adequacy of a tentatively identified model. Chapter 5 discusses combining regression models utilizing deterministic functions of time with Box-Jenkins models to forecast time series. Then, in Chapter 6 we present exponential smoothing forecasting techniques. We also examine the relationships between exponential smoothing, time series regression, and the Box-Jenkins methodology. We will see that the relationships between these techniques provide a framework for a cohesive, unified view of time series forecasting.

The techniques covered in Chapters 2 through 6 are univariate forecasting techniques—techniques that forecast future values of a time series solely on the basis of the past values of the time series. In Chapters 7 and 8 we discuss several multivariate forecasting techniques—techniques that forecast future values of a time series on the basis of past values of the time series and on the basis of values of one or more time series different from but related to the time series to be predicted. Specifically, in Chapters 7 and 8 we will discuss Box-Jenkins transfer function models and classical regression analysis.

We would like to point out that, while some time series books begin with a review of classical regression analysis, we begin instead with Box-Jenkins univariate forecasting techniques because we believe that this methodology provides a powerful framework for thoroughly presenting both the wide range of univariate forecasting techniques and multivariate forecasting techniques. However, for those instructors who wish to begin their course with classical regression analysis, Chapter 8 presents a concise and complete discussion of this subject. Furthermore, the regression chapter is self-contained and can be read independently.

Obviously, implementation of the forecasting methods covered in this book requires a computer. We emphasize the use of SAS, one of the most popular statistical packages in use today. In addition to using SAS output throughout the book, we also include output from a forecasting package called TSERIES. This package was written by (and can be obtained from) Professor William Q. Meeker of the Department of Statistics at Iowa State University. We feel that SAS and TSERIES are representative of most forecasting packages that are currently available. However, because some readers may not have access to SAS or TSERIES, we discuss the specifics of using these packages in separate sections that can be omitted without loss of continuity.

Many people have contributed to this book. We would like to thank David A. Dickey of North Carolina State University for many helpful discussions and for contributing the Farmers' Bureau Co-op gas bill data set analyzed in Chapters 4 and 5. We would also like to thank William Q. Meeker for contributing and analyzing the Travelers' Rest, Inc. data set included in Chapters 3, 4, and 5. We would further like to thank the many reviewers of the book including Joel Fingerman, Roosevelt University; Jack Narayan, Syracuse University; David J. Pack, Miami University, Ohio; Leonard Presby, William Paterson State College of New Jersey; Samuel D. Ramenofsky, Loyola University; Stanley R. Schultz, Cleveland State University; Mack C. Shelley, II, Iowa State University of Science and Technology; and Ronald Tracy, Oakland University. In addition, we wish to thank our editor, Michael Payne, our production coordinator, Ellie Connolly, and our project editor, Rick Batlan of Editing, Design & Production, Inc., as well as the other fine professionals at Duxbury Press. Finally, our wives, Drena and Jean, deserve special thanks for their patience, encouragement, and understanding during the writing process.

Bruce L. Bowerman
Richard T. O'Connell
Oxford, Ohio

Contents

TIME SERIES FORECASTING

AN INTRODUCTION TO FORECASTING

1.1 INTRODUCTION

This chapter introduces the topic of *forecasting*. We will begin in Section 1.2 by discussing *time series data*, the type of data that we will use in this book to make forecasts. Then, in Section 1.3, we explain the general natures of different kinds of forecasting methods. Both *qualitative* and *quantitative* methods are considered. Section 1.4 discusses the fact that forecasts of future time series values are not likely to be perfectly accurate and explains how to measure *forecast errors*. In Section 1.5 we present some important factors that must be considered when choosing a forecasting method. Section 1.6 introduces the *Box-Jenkins forecasting methodology*, which is one of the main topics of this book. That section also presents an overview of the topic coverage for the entire book. We will conclude this chapter with Section 1.7, which briefly introduces the *computer packages* that we will use to implement the forecasting techniques presented in this book.

1.2 FORECASTING AND TIME SERIES

This book is about forecasting and some of the statistical techniques that can be used to produce forecasts. We will begin by making the following definition.

> Predictions of future events and conditions are called *forecasts*, and the act of making such predictions is called *forecasting*.

Forecasting is very important in many types of organizations since predictions of future events must be incorporated into the decision-making process. The government of a country must be able to forecast such things as air quality, water quality, unemployment rate, inflation rate, and welfare payments in order to formulate its policies. A university must be able to forecast student enrollment in order to make decisions concerning faculty resources and housing availability. The university might also wish to forecast daily mean temperature so that it can plan its fuel purchases for the coming months. A local school board must be able to forecast the number of school children of elementary school age who will be living in their school district years in the future in order to decide whether a new school should be built. Any organization must be able to make forecasts in order to make intelligent decisions.

In particular, business firms require forecasts of many events and conditions in all phases of their operations. Some examples of situations in which business forecasts are required are given below.

In marketing departments, reliable forecasts of demand must be available so that sales strategies can be planned. For example, total demand for products must be forecasted in order to plan total promotional effort. Besides this, demand in various market regions and among various consumer groups must be predicted in order to plan effective advertising strategies.

In finance, interest rates must be predicted so that new capital acquisitions can be planned and financed. Financial planners must also forecast receipts and expenditures in order to predict cash flows and maintain company liquidity.

In personnel management, forecasts of the number of workers required in different job categories are required in order to plan job recruiting and training programs. In addition, personnel managers need predictions of the supply of labor in various areas and of the amount of absenteeism and the rate of labor turnover to be expected.

In production scheduling, predictions of demand for each product line are needed. Such predictions are made for specific time periods, for example, for specific weeks and months. These forecasts allow the firm to plan production schedules and inventory maintenance. Forecasts of demand for individual products can be translated into forecasts of raw material requirements so that purchases can be planned. The planning of resource purchases also requires predictions about resource availabilities and prices.

Process control requires forecasts of the behavior of the process in the future. For example, an industrial process may begin to produce increasing numbers of defective items as the process operates over time. If the behavior of this process can be predicted accurately, it will be possible to determine when the process should be shut down and overhauled so that the number of defective items produced can be minimized.

Strategic management requires forecasts of general economic conditions, price and cost changes, technological change, market growth, and the like in order to plan the long-term future of the company. For example, such forecasts might be used to determine whether investment in new plant and equipment will be needed in the future.

In forecasting events that will occur in the future, a forecaster must rely on information concerning events that have occurred in the past. That is, in order to prepare a forecast, the forecaster must analyze past data and must base the forecast on the results of this analysis. Forecasters use past data in the following way. First, *the forecaster analyzes this data in order to identify a pattern* that can be used to describe it. Then *this pattern is extrapolated, or extended, into the future in order to prepare a forecast.* This basic strategy is employed in most forecasting techniques and *rests on the assumption that the pattern that has been identified will continue* in the future. A forecasting technique cannot be expected to give good predictions unless this assumption is valid. If the data pattern that has been identified does not persist in the future, the forecasting technique being used will likely produce inaccurate predictions. A forecaster should not be surprised by such a situation, but rather must try to anticipate when such a change in pattern will take place so that appropriate changes in the forecasting system can be made before the predictions become too inaccurate.

In this book we will use time series data to prepare forecasts.

> A *time series* is a chronological sequence of observations on a particular variable.

As an example, the data in Table 1.1 is a time series which gives the quarterly total value of time deposits held by the Baarth National Bank during 1987 and 1988. Notice that the value of time deposits was observed at equally spaced time points (quarterly). Equally spaced time points are used in most time series studies, although they are not always used. Business time series often consist of yearly, quarterly, or monthly observations, but any other time period may be used. There are many, many examples of time series data; some of these are listed below.

Unit sales of a product over time
Total dollar sales for a company over time
Number of unemployed over time
Unemployment rate over time
Production of a product over time
Air or water quality over time
Inventory level for a product over time
Population of a city over time
Daily mean temperature over time

TABLE 1.1   *Time Series Data: Quarterly Values of Time Deposits*

Year	Quarter	Value of Time Deposits (in millions of dollars)
1987	1	35.3
	2	37.6
	3	38.1
	4	39.5
1988	1	37.9
	2	39.9
	3	40.1
	4	41.2

Time series data are often examined in hopes of discovering a historical pattern that can be exploited in the preparation of a forecast. In order to identify this pattern, it is often convenient to think of a time series as consisting of several components.

The components of a time series are:

1. Trend
2. Cycle
3. Seasonal variations
4. Irregular fluctuations

We will consider each of these components in turn.

Trend refers to the upward or downward movement that characterizes a time series over a period of time. Thus, trend reflects the long-run growth or decline in the time series.

Trend movements can represent a variety of factors. For example, long-run movements in the sales of a particular industry might be determined by one, some, or all of the factors listed below.

1. Technological change in the industry
2. Changes in consumer tastes
3. Increases in per capita income
4. Increases in total population
5. Market growth
6. Inflation or deflation (price changes)

Cycle refers to recurring up and down movements around trend levels. These fluctuations can have a duration of anywhere from two to ten years or even longer measured from peak to peak or trough to trough.

One of the common cyclical fluctuations found in time series data is the "business cycle." The business cycle is represented by fluctuations in the time series caused by recurrent periods of prosperity and recession. Economists have identified several phases in the business cycle. A period of *expansion* in economic or business activity (boom) ends at the *peak* or upper turning point of the business cycle. This peak is followed by a period of *contraction* in economic activity (bust) during which economic activity diminishes. This contraction ends at the lower turning point or *trough* of activity and is then followed by a renewed period of expansion or increase in economic activity.

Cyclical fluctuations need not be caused by changes in economic factors, however. For example, cyclical fluctuations in agricultural yields might reflect changes in weather cycles; the cyclical fluctuations in sales of a particular item of clothing might reflect changes in clothing styles, which are determined by the whims of Paris fashion designers who are bored with the current length of hem lines.

Because there is no single explanation for cyclical fluctuations, they vary greatly in both length and magnitude.

Seasonal variations are periodic patterns in a time series that complete themselves within a calendar year and are then repeated on a yearly basis. Seasonal variations are usually caused by factors such as weather and customs. For example, the average monthly temperature clearly is seasonal in nature since it directly measures changes in the weather. Similarly, the number of monthly housing starts might have a seasonal pattern due to changes in the weather. There might be a high level of housing starts in spring and early summer because of good weather in future months. Housing starts might then decline through late summer and fall, reaching a low point during the coldest months of winter, and then increase rapidly again in early spring. Another time series that might contain a seasonal component is the monthly sales volume in a department store. Here seasonal variation might be caused by the observance of various holidays. Thus department store sales volume might reach high points in December and April because of shopping for the Christmas and Easter holidays.

Ordinarily, series of monthly or quarterly data are used to examine seasonal variations. Clearly, one single yearly observation would not reveal variations that occur during the year.

Irregular fluctuations are erratic movements in a time series that follow no recognizable or regular pattern. Such movements represent what is "left over" in a time series after trend, cycle, and seasonal variations have been accounted for. Many irregular fluctuations in time series are caused by "unusual" events that cannot be forecasted —earthquakes, accidents, hurricanes, wars, wildcat strikes, and the like. Irregular fluctuations can also be caused by errors on the part of the time series analyst.

Time series that exhibit trend, seasonal, and cyclical components are illustrated in Figure 1.1. In Figure 1.1a a time series of sales observations that has an essentially straight line or linear trend is plotted. Figure 1.1b portrays a time series of sales observations that contains a seasonal pattern that repeats annually. Figure 1.1c exhibits

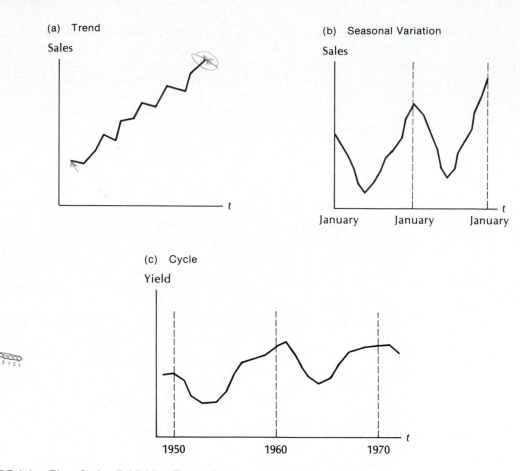

FIGURE 1.1   Time Series Exhibiting Trend, Seasonal, and Cyclical Components

a time series of agricultural yields that is cyclical in nature, repeating a cycle about once every ten years.

It should be pointed out that the time series components we have discussed do not always occur alone; they can occur in any combination or can occur all together. For this reason, *no single best forecasting model exists*. A forecasting model that can be used to forecast a time series characterized by trend alone may not be appropriate in forecasting a time series characterized by a combination of trend and seasonal variations. Thus one of the most important problems to be solved in forecasting is that of *trying to match the appropriate forecasting model to the pattern of the available time series data*. Once an appropriate model has been selected, the methodology usually involves estimating the time series components (model parameters). The estimates are then used to compute a forecast. For example, if a time series is characterized by a combination of trend and seasonal components, the appropriate forecasting technique would first estimate these two components. Forecasts would then be obtained by combining the estimate of the trend component with the estimate of the seasonal component. Again, however, it should be emphasized that the key to this methodology is finding a model that matches the pattern of the historical data that is available.

1.3 FORECASTING METHODS

In Section 1.2 we pointed out that no single best forecasting model exists. In fact, there are many forecasting methods that can be used to predict future events. These methods can be divided into two basic types—*qualitative* methods and *quantitative* methods.

1.3.1 Qualitative Forecasting Methods

Qualitative forecasting methods generally use the opinions of experts to subjectively predict future events. Such methods are often required when historical data either are not available at all or are scarce. For example, consider a situation in which a new product is being introduced. In such a case, no historical sales data for the product are available. In order to forecast sales for the new product, a company must rely on expert opinion, which can be supplied by members of its sales force and market research team. Other situations in which historical data are not available might involve trying to predict if and when new technologies will be discovered and adopted. Qualitative forecasting techniques are also used to predict changes in historical data patterns. Since the use of historical data to predict future events is based on the assumption that the pattern of the historical data will persist, changes in the data pattern cannot be predicted on the basis of historical data. Thus qualitative methods are often used to predict such changes.

We will briefly describe several commonly used qualitative forecasting techniques. The first of these techniques involves *subjective curve fitting*. Consider a firm that is introducing a new product and wishes to forecast sales of this new product over the next several years so that it can estimate the productive capacity needed to produce the product. In predicting sales of a new product, it is often convenient to consider what is known as the "product life cycle." This life cycle is usually thought of as consisting of several stages. During the first stage (growth), sales of the product start slowly, then increase rapidly, and then continue to increase at a slower rate. During the next stage (maturity), sales of the product stabilize, increasing slowly, reaching a plateau, and then decreasing slowly. During the last stage (decline), sales of the product decline at an increasing rate. This product life cycle is illustrated in Figure 1.2. In forecasting sales of the product during the growth stage, the company might use the expert opinion of its sales and marketing personnel to subjectively construct an S-curve, as illustrated in Figure 1.3. Such an S-curve could then be used to forecast sales during this stage. In constructing this S-curve, the company must use its experience with other products and all its knowledge concerning the new product in order to predict how long it will take for the rapid increase in sales to begin, how long this rapid growth will continue, and when sales of the product will begin to stabilize. Note that the construction of this curve

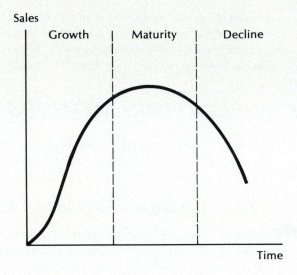

FIGURE 1.2   *Product Life Cycle*

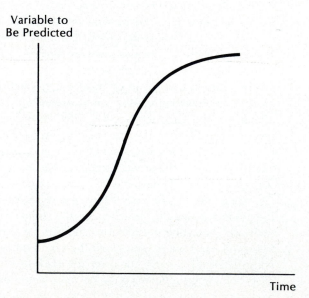

FIGURE 1.3   *S-Curve*

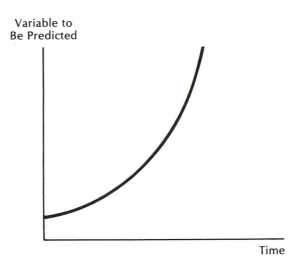

FIGURE 1.4   *Exponential Curve*

is done subjectively since there will be little or no sales data available for the new product. Estimating such a curve is an example of subjective curve fitting. Of course, one of the biggest problems in using this technique is deciding on the form of the curve to be used. In a product life cycle situation the use of an S-curve may be appropriate. But many other functional forms can be used. For example, an exponential curve, as illustrated in Figure 1.4, might be appropriate in some situations. Other situations might call for the use of a logarithmic curve. Thus, the forecaster must first subjectively determine the form of the curve to be used. The subjective construction of such curves is very difficult and requires a great deal of expertise and judgment.

Another common qualitative forecasting method is called the *Delphi Method*. This technique, which was developed by the RAND Corporation, involves using a panel of experts to produce predictions concerning a specific question, such as when a new development will occur in a particular field. The use of the Delphi Method assumes that the panel members are recognized experts in the field of interest, and it also assumes that the combined knowledge of the panel members will produce predictions at least as good as those that would be produced by any one member. When a panel of experts is called upon to make predictions, a panel discussion might seem to be appropriate. But such discussions are often dominated by one individual or by a small group of individuals. Also, the decisions made in panel discussions can be influenced by various kinds of social pressure. The Delphi Method attempts to avoid these problems by keeping the panel members physically separated. Each participant is asked to respond to a series of questionnaires and to return the completed questionnaire to a panel coordinator. After the first questionnaire is completed, subsequent questionnaires are accompanied by information concerning the opinions of the group as a whole. Thus, the participants can review their predictions relative to the group response. It is hoped that after several rounds of questionnaires the group response will converge on a consensus that can be used as a forecast. It should be noted, however, that the Delphi Method does not require that a consensus be reached. Instead, the method allows for justified differences of opinion rather than attempting to produce unanimity. We will not present a more detailed discussion of the Delphi Method here. The interested reader is referred to Brown [1968] or Dalkey [1967].

A third qualitative forecasting technique concerns the use of time independent *technological comparisons*. This method is often used in predicting technological change. The method involves predicting changes in one area by monitoring changes that take place in another area. That is, the forecaster tries to determine a pattern of change in one area, often called a *primary trend*, which he or she believes will result in new developments being made in some other area. A forecast of developments in the second area can then be made by monitoring developments in the first area. For example, consider the problem of trying to forecast when a new metal alloy of very high tensile strength will be used commercially. Suppose that the forecaster determines that metallurgical advances made in industry are related to metallurgical advances in the space program. Then, by following metallurgical advances made in the space program, the forecaster can predict when similar advances will take place in industry. Thus the development of high-tensile-strength alloys in the space program would allow the forecaster to predict when such alloys will be available for commercial use. This type of forecasting poses two basic problems. First, the forecaster must identify a primary trend that will reliably predict events in the area of interest. Second, the forecaster must use his expertise to determine the precise relationship between the primary trend and the events to be forecast. Once these determinations have been made, forecasts in the area of interest can be made by monitoring the primary trend. For a further discussion of this technique the reader is referred to Gerstenfeld [1971].

The qualitative forecasting techniques we have discussed—subjective curve fitting, the Delphi Method, and time independent technological comparisons—represent only some of the subjective forecasting methods available. There are many other subjective methods used to generate predictions of future events—these include the cross-impact method, the relevance tree method, and the morphological research method. The interested reader is referred to, respectively, Gordon and Hayward [1968], Sigford and Parvin [1965], and Zwicky [1962].

1.3.2   Quantitative Forecasting Methods

The rest of this book will be devoted to a discussion of *quantitative forecasting techniques*. These techniques involve the analysis of historical data in an attempt to predict future values of a variable of interest. Quantitative forecasting models can be grouped into two kinds—*univariate* models and *causal* models.

One common type of quantitative forecasting method is called a *univariate* model. Such a model predicts future values of a time series *solely on the basis of the past values of the time series*. When a univariate model is used, historical data are analyzed in an attempt to identify a data pattern. Then, assuming that it will continue in the future, this data pattern is extrapolated in order to produce forecasts. Univariate forecasting models are, therefore, most useful when conditions are expected to remain the same; they are not very useful in forecasting the impact of changes in management policies. For example, while a univariate model can be used to predict sales if a firm expects to continue using its present marketing strategy, such a model would not be useful in predicting the changes in sales that might result from a price increase, increased advertising expenditures, or a new advertising campaign.

The use of *causal forecasting models* involves the identification of other variables that are related to the variable to be predicted. Once these related variables have been identified, a statistical model that describes the relationship between these variables and the variable to be forecast is developed. The statistical relationship derived is then used to forecast the variable of interest. For example, the sales of a product might be related to the price of the product, advertising expenditures to promote the product, competitors' prices charged for similar products, and so on. In such a case, sales would be referred to as the *dependent variable*, while the other variables are referred to as the *independent variables*. The forecaster's job is to statistically estimate the functional relationship between sales and the independent variables. Having determined this relationship, the forecaster would use predicted future values of the independent variables (price of the product, advertising expenditures, competitors' prices, etc.) to predict future values of sales (the dependent variable).

In the business world, causal models are advantageous because they allow management to evaluate the impact of various alternative policies. For example, management might wish to predict how various prices structures and levels of advertising expenditures will affect sales. A causal model relating these variables could be used here. However, causal models have several disadvantages. First, they are quite difficult to develop. Also, they require historical data on all the variables included in the model. Besides this, the ability to predict the dependent variable depends on the ability of the forecaster to accurately predict future values of the independent variables. Despite these disadvantages, causal models are often used.

Before proceeding further, we summarize our discussion of forecasting methods. Quantitative forecasting methods are used when historical data are available: univariate models predict future values of the variable of interest solely on the basis of the historical pattern of that variable, assuming that the historical pattern will continue; causal models predict future values of the variable of interest based on the relationship between that variable and other variables. Qualitative forecasting techniques are used when historical data are scarce or not available at all and depend on the opinions of experts who subjectively predict future events. In actual practice most forecasting systems employ both quantitative and qualitative methods. For example, quantitative methods are used when the existing data pattern is expected to persist, while qualitative methods are used to predict when the existing data pattern might change. Thus, forecasts generated by quantitative methods are almost always subjectively evaluated by management. This evaluation may result in a modification of the forecast based on the manager's "expert opinion".

1.4 ERRORS IN FORECASTING

Unfortunately, all forecasting situations involve some degree of uncertainty. We recognize this fact by including an irregular component in the description of a time series. The presence of this irregular component, which represents unexplained or unpredictable fluctuations in the data, means that some error in forecasting must be expected. If the effect of the irregular component is substantial, our ability to forecast accurately will be limited. If, however, the effect of the irregular component is small, determination of the appropriate trend, seasonal, or cyclical patterns should allow us to forecast with more accuracy.

The irregular component is not the only source of errors in forecasting, however. The accuracy with which we can predict each of the other components of a time series also influences the magnitude of error in our forecasts. Since these components cannot be perfectly predicted in a practical situation, the errors in forecasting represent the combined effects of the irregular component and the accuracy with which the forecasting technique can predict trend, seasonal, or cyclical patterns. Hence, large forecasting errors may indicate that the irregular component is so large that no forecasting technique will produce accurate forecasts, or they may indicate that the forecasting technique being used is not capable of accurately predicting the trend, seasonal, or cyclical components and, therefore, that the technique being used is inappropriate.

1.4.1   Types of Forecasts

The fact that forecasting techniques often produce predictions that are somewhat in error has a bearing on the form of the forecasts we require. In this book we will consider two types of forecasts: (1) the *point forecast* and (2) the *prediction interval forecast*. Whereas a point forecast is a single number that represents our best prediction (or guess) of the actual value of the variable being forecasted, a prediction interval forecast is an interval (or range) of numbers that is calculated so that we are very confident (for example, 95% confident) that the actual value will be contained in the interval. For example, suppose that the Olympia Paper Company, Inc. produces Absorbent Paper Towels and for the past 120 weeks has recorded weekly sales of those towels. In Chapter 2 we will present and forecast this time series. We will find that a point forecast of sales in week 121 is 258,899 rolls, and that a 95% prediction interval forecast of sales in week 121 is [224,621 rolls, 293,177 rolls]. The prediction interval says that the Olympia Paper Company is 95% confident the sales of Absorbent Paper Towels in week 121 will be no less than 224,621 rolls and no more than 293,177 rolls. The company might use the fact that it is quite sure that sales will be no more then 293,177 rolls to determine the level of inventory it should carry for that week. It might also use the fact the sales are not likely to be less than 224,621 rolls to help determine the minimum amount of revenue generated by this product next week.

1.4.2 Measuring Forecast Errors

We now consider the problem of measuring forecasting errors. Denote the actual value of the variable of interest in time period t as y_t. Then, if we denote the predicted value of y_t as $\hat{y}_t$, we can subtract the predicted value of y_t from the actual value y_t to obtain the forecast error e_t. That is:

> The forecast error for a particular forecast $\hat{y}_t$ is
> $$e_t = y_t - \hat{y}_t$$

In Section 1.2 we noted the importance of matching a forecasting technique to the pattern of data characterizing a time series. An examination of forecast errors over time can often indicate whether the forecasting technique being used does or does not match this pattern. For example, if a forecasting technique is accurately forecasting the trend, seasonal, or cyclical components that are present in a time series, the forecast errors should reflect only the irregular component. In such a case, the forecast errors should appear purely random. Figure 1.5a illustrates forecast errors that indicate that the forecasting technique being used appropriately accounts for the trend, seasonal, or cyclical components present in the time series being forecast. Sometimes, when the forecasting technique does not match the data pattern, the forecasting errors will exhibit a pattern over time. In Figure 1.5b, the forecast errors show an upward trend, which indicates that the forecasting methodology being used does not account for an upward trend in the time series. The forecast errors in Figure 1.5c indicate that a seasonal pattern in the data is not accounted for, and the forecast errors in Figure 1.5d indicate that a cyclical pattern in the data is not being accounted for. Patterns of forecast errors such as those illustrated in Figures 1.5b, c, and d indicate that the forecasting technique being employed is not appropriate, that is, it does not match the pattern of the time series data.

If the forecasting errors over time indicate that the forecasting methodology is appropriate (random distribution of errors), it is important to *measure the magnitude of the errors* so that we can determine whether accurate forecasting is possible. In order to do this, one might consider the sum of all forecast errors over time. That is, one might calculate

$$\sum_{t=1}^{n} (y_t - \hat{y}_t)$$

which reads "the summation of the differences between the predicted ($\hat{y}_t$) and actual (y_t) values from time period $t = 1$ through time period $t = n$, where n is the total number of observed time periods." However, this quantity is not used because, if the errors display a random pattern, some errors will be positive while other errors will be

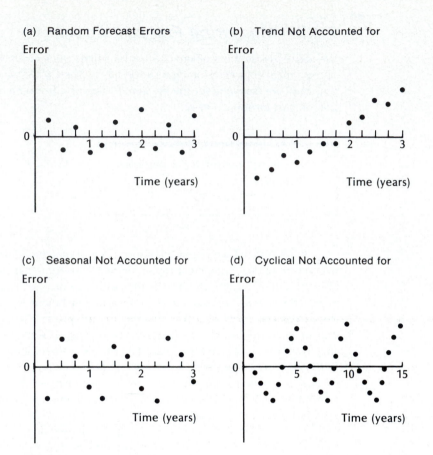

FIGURE 1.5   *Plots of Forecast Errors*

negative, and the sum of the forecast errors will be near zero regardless of the size of the errors. That is, the positive and negative errors, no matter how large or small, will cancel each other out.

One way to remedy this problem is to consider the absolute values of the forecasting errors. These absolute values are called the *absolute deviations*. That is,

$$\text{Absolute Deviation} \; = \; |e_t| \; = \; |y_t - \hat{y}_t|$$

Given the absolute deviations, we can then define a measure known as the *mean absolute deviation* (MAD). This measure is simply the average of the absolute deviations

TABLE 1.2   *Computation of Mean Absolute Deviation*

Actual Value y_t	Predicted Value $\hat{y}_t$	Error $e_t = y_t - \hat{y}_t$	Absolute Deviation $|e_t| = |y_t - \hat{y}_t|$
25	22	3	3
28	30	-2	2
30	29	1	1

$$\sum_{t=1}^{3} |e_t| = 6$$

$$\text{Mean Absolute Deviation} = \frac{\sum_{t=1}^{3} |e_t|}{3} = \frac{6}{3} = 2$$

for all forecasts. That is,

$$\text{Mean Absolute Deviation (MAD)} = \frac{\sum_{t=1}^{n} |e_t|}{n} = \frac{\sum_{t=1}^{n} |y_t - \hat{y}_t|}{n}$$

An example of the calculations involved in computing the MAD is given in Table 1.2. This measure can be used to determine the magnitude of the forecast errors generated by the forecasting methodology being used.

Another way to prevent positive and negative forecast errors from cancelling each other out is to square the forecast errors. These squares are called the *squared errors*:

$$\text{Squared Error} = (e_t)^2 = (y_t - \hat{y}_t)^2$$

Given the squared errors, we can define the *mean squared error* (MSE). This measure is simply the average of the squared errors for all forecasts. That is

$$\text{Mean Squared Error (MSE)} = \frac{\sum_{t=1}^{n} (e_t)^2}{n} = \frac{\sum_{t=1}^{n} (y_t - \hat{y}_t)^2}{n}$$

TABLE 1.3   *Computation of the Mean Squared Error*

Actual Value y_t	Predicted Value $\hat{y}_t$	Error $e_t = y_t - \hat{y}_t$	Squared Error $(e_t)^2 = (y_t - \hat{y}_t)^2$
25	22	3	9
28	30	-2	4
30	29	1	1

$$\sum_{t=1}^{3} (e_t)^2 = 14$$

$$\text{Mean Squared Error} = \frac{\sum_{t=1}^{3}(e_t)^2}{3} = \frac{14}{3} = 4.67$$

An example of the calculations involved in computing the MSE is given in Table 1.3.

Having shown that both the MAD and the MSE can be used to measure the magnitude of forecast errors, we now wish to discuss how these two measures differ. The basic difference between these measures is that *the MSE, unlike the MAD, penalizes a forecasting technique much more for large errors than for small errors*. For example, an error of 2 produces a squared error of 4 while an error of 4 (an error twice as large) produces a squared error of 16 (a squared error four times as large). So, when using the MSE, the forecaster would prefer several smaller forecast errors to one large error. This situation is illustrated in Table 1.4, where we consider two different sets of forecasts generated by methods A and B. Forecasting method A has produced predictions yielding moderate forecast errors, while forecasting method B has produced predictions yielding two small errors along with one large error. Notice that forecasting method A has the larger MAD, while forecasting method B has the larger MSE. This is because, in the calculation of the MSE, forecasting method B is heavily penalized for its large error in forecasting the actual value 67.

Measures such as the MAD and MSE can be used in two different ways. First they can be used to aid in the process of selecting a forecasting model. Suppose that we are trying to choose among several forecasting models in an attempt to determine which of them will likely produce the most accurate predictions of future values of some variable of interest. A common strategy used in making such a selection involves the simulation of historical data. In the simulation process we pretend that we do not know the values of the historical data. Then we use each of the forecasting models to produce "predictions" of the historical data. We next compare these "predictions" with the actual values of the historical data and measure their accuracy using the MAD or MSE. Now, in order to choose the technique we will use in actual forecasting, we compare the performance of the various techniques in "forecasting" the historical data to see which of them provided the most accurate simulated "predictions."

Second, the MAD or MSE can be used to monitor a forecasting system in order

TABLE 1.4 *Comparisons of the Errors Produced by Two Different Forecasting Methods*

	Actual y_t	Predicted $\hat{y}_t$	Error	Absolute Deviation	Squared Error
Forecasting	60	57	+3	3	9
Method A	64	61	+3	3	9
	67	70	−3	3	9
				9	27

$$\text{Mean Absolute Deviation} = \frac{9}{3} = 3$$

$$\text{Mean Squared Error} = \frac{27}{3} = 9$$

	Actual y_t	Predicted $\hat{y}_t$	Error	Absolute Deviation	Squared Error
Forecasting	60	59	+1	1	1
Method B	64	65	−1	1	1
	67	73	−6	6	36
				8	38

$$\text{Mean Absolute Deviation} = \frac{8}{3} = 2.67$$

$$\text{Mean Squared Error} = \frac{38}{3} = 12.67$$

to detect when something has "gone wrong" with the system. For example, in Section 1.2, we stated that forecasts cannot be expected to be accurate unless the historical data pattern that has been identified continues in the future. Consider a situation in which a data pattern that has persisted for an extended period of time suddenly changes. Any forecasting method we might have been using to forecast the variable of interest might now be expected to become inaccurate because of this change. In such a situation, we would like to discover the change in pattern as quickly as possible before forecasts become very inaccurate. This can be done by using measures that incorporate the MAD or MSE to monitor the forecast errors and to "signal" us when these errors become "too large." We will return to this problem in Chapter 6.

1.5 CHOOSING A FORECASTING TECHNIQUE

In choosing a forecasting technique, the forecaster must consider the following factors.

1. The forecast form desired

2. The time frame
3. The pattern of data
4. The cost of forecasting
5. The accuracy desired
6. The availability of data
7. The ease of operation and understanding

The first factor to be considered in choosing a forecasting method is the form in which the forecast is desired. We have discussed the difference between a point forecast and a prediction interval forecast. In some situations a point forecast may be sufficient; in other situations a prediction interval forecast may be required. As we will see in later sections, the form of the forecast can influence the choice of a forecasting method because some techniques yield theoretically correct prediction intervals while others do not.

The second factor that can influence the choice of a forecasting method is the *time frame* of the forecasting situation. Forecasts are generated for points in time that may be a number of days, weeks, months, quarters, or years in the future. This length of time is called the time frame or time horizon. The length of the time frame is usually categorized as follows.

Immediate: less than one month
Short term: one to three months
Medium: more than three months to less than two years
Long term: two years or more

In general, the length of the time frame will influence the choice of the forecasting technique. Typically, a longer time frame makes accurate forecasting more difficult, with qualitative forecasting techniques becoming more useful as the time frame lengthens.

As discussed in Section 1.2, the pattern of data must also be considered when choosing a forecasting model. The components present (trend, cycle, seasonal, or some combination of these) will help determine the model that will be used. Thus, it is extremely important to identify the existing data pattern.

When choosing a forecasting technique, several costs are relevant. First, the cost of developing the model must be considered. We will see in later chapters that the development of a forecasting model requires that a set of procedures be followed. The complexity, and hence the cost, of these procedures vary from technique to technique. Second, the cost of storing the necessary data must be considered. Some forecasting methods require the storage of a relatively small amount of data, while other methods require the storage of large amounts of data. Last, the cost of the actual operation of the forecasting technique is obviously very important. Some forecasting methods are operationally simple, while others are very complex. The degree of complexity can have a definite influence on the total cost of forecasting.

Another very important factor that has a bearing on the choice of a forecasting technique is the desired accuracy of the forecast. In some situations, a forecast that is in error by as much as 20% may be acceptable: in other situations, a forecast that is in error by 1% might be disastrous. The accuracy that can be obtained using any particular forecasting method is always an important consideration.

We have pointed out that historical data on the variable of interest are used when quantitative forecasting methods are employed. The availability of this information is a factor that may determine the forecasting method to be used. Since various forecasting methods require different amounts of historical data, the quantity of data available is important. Beyond this, the accuracy and the timeliness of the data that are available must be examined, since the use of inaccurate or outdated historical data will obviously yield inaccurate predictions. If the needed historical data are not available, special data-collection procedures may be necessary.

Last, the ease with which the forecasting method is operated and understood is important. Managers are held responsible for the decisions they make and if they are to be expected to base their decisions on predictions, they must be able to understand the techniques used to obtain these predictions. A manager simply will not have confidence in the predictions obtained from a forecasting technique he or she does not understand, and if the manager does not have confidence in these predictions, they will not be used in the decision-making process. Thus, the manager's understanding of the forecasting system is of crucial importance.

Choosing the forecasting method to be used in a particular situation involves finding a technique that balances the factors just discussed. It is obvious that the "best" forecasting method for a given situation is not always the "most accurate." Instead, *the forecasting method that should be used is one that meets the needs of the situation at the least cost and inconvenience.*

Suppose that a company wants to predict sales one month in advance. To accomplish this task, the firm develops a complicated forecasting system and finds that the mean absolute deviation for their technique is 2,000 units. The firm must then determine whether the cost and inconvenience of this forecasting system is justified. In order to make this decision, the mean absolute deviation of 2,000 units must be placed in perspective. If monthly sales average 5,000 units, then the mean absolute deviation of 2,000 units is very large and the forecasts will be quite inaccurate. In such a case, using such a complex forecasting system is probably not justified. If, however, monthly sales average 40,000 units, then the forecasts will be quite accurate. But the knowledge that the forecasting method produces accurate forecasts is not enough to tell the firm whether the method is *appropriate*. Suppose that, in the past, monthly sales have differed from 40,000 units by no more than 3,000 units. Then the firm could forecast sales of 40,000 units each month and predict sales nearly as well as with the complex forecasting system, and with much less cost and effort. Thus, if a forecast that is accurate within 3,000 units is adequate, the company should simply forecast sales of 40,000 units each month rather than use the complicated forecasting system; if greater accuracy is required, the more complicated forecasting system would be justified. Thus, the cost and ease of using a simple forecasting method must be balanced against the greater accuracy but higher cost of a more complex forecasting technique.

1.6 THE BOX-JENKINS METHODOLOGY AND AN OVERVIEW

One of the main topics of this book is the *Box-Jenkins forecasting methodology*. This methodology, developed by G. E. P. Box and G. M. Jenkins, consists of four basic steps. The first step, called the *tentative identification step*, involves tentatively identifying a model. This is done by employing a *sample autocorrelation function* and a *sample partial autocorrelation function*. Once a model has been tentatively identified, we estimate the model parameters in the second step. This is called the *estimation step*. The third step is called the *diagnostic checking step*. Here we use diagnostic checks to see whether or not the model we have tentatively identified and estimated is adequate. If the model proves to be inadequate, it must be modified and improved. The diagnostic methods employed will help us to decide how the model can be improved. When a final model is determined, we use the model to forecast future time series values. This fourth step is called the *forecasting step*. Notice that *the Box-Jenkins methodology is an iterative procedure*. If a tentatively identified model is found to be inadequate, we return to the tentative identification step and identify a new (hopefully improved) model. We then estimate the parameters of the new model and check the adequacy of the new model. This cycle of tentative identification, estimation, and diagnostic checking is repeated until we find an adequate final model. The final model is then used to compute forecasts of future time series values.

The Box-Jenkins methodology can be used to forecast either discrete data or continuous data. However, the data must be measured at equally spaced, discrete time intervals (for example, hourly, daily, weekly, or monthly). Moreover, the Box-Jenkins methodology can be used to forecast both nonseasonal and seasonal data.

When we talk about Box-Jenkins models, we are talking about a family of models. That is, there is not a single Box-Jenkins model; rather, there are many different Box-Jenkins models. These models can be grouped into three basic classes—autoregressive models, moving average models, and mixed autoregressive–moving average models. For this reason, Box-Jenkins models are often called ARIMA models (Autoregressive Integrated Moving Average). We will discuss all of these models in this book. As stated previously, in order to use the Box-Jenkins methodology, the forecaster must identify the particular model (among the many possible models) to use in a specific forecasting situation.

Chapters 2, 3, and 4 present a detailed explanation of *univariate Box-Jenkins forecasting models*. Here we use the Box-Jenkins methodology to predict future values of a time series *solely on the basis of the past values of the time series*. Chapter 2 introduces *nonseasonal* Box-Jenkins models and their tentative identification. In that chapter we discuss the concept of *stationarity*, define the *sample autocorrelation function* (*SAC*) and the *sample partial autocorrelation function* (*SPAC*), and present guidelines that can be used to tentatively identify nonseasonal Box-Jenkins models. We also present an example of nonseasonal modeling and forecasting. In Chapter 3 we discuss guidelines that can be used to tentatively identify *seasonal* Box-Jenkins models. We will also further discuss the nature of Box-Jenkins *forecasting*. Chapter 4 gives an in-depth discussion of the *estimation* and *diagnostic checking* steps.

There are many computer forecasting packages that can be used to effectively implement Box-Jenkins univariate forecasting models. We explain how to use two of these packages—SAS and TSERIES—at the conclusion of Chapter 4.

Because it is easy to implement univariate Box-Jenkins models by using a computer, and because these models have proven to provide accurate forecasts in short-term forecasting applications, we recommend the use of these models when sufficient historical data are available. Here, experience indicates that at least fifty historical observations are necessary to build a univariate Box-Jenkins model (for seasonal data, more data is sometimes necessary). However, Box-Jenkins models are not the only useful forecasting models. In some situations, especially when the amount of available historical data is limited, other kinds of forecasting models have been found to be useful. In Chapters 5 and 6 we present some of these models.

Chapter 5 concerns *time series regression models*. These models are useful when the parameters describing the trend, seasonal, or cyclical components of a time series are *deterministic*—that is, the parameters are not changing over time. Included in our discussion are *trend models* (no trend, linear trend, and quadratic trend), *dummy variable models*, *growth curve models*, and *time series decomposition*. Chapter 6 deals with *exponential smoothing models*. These models are useful when the parameters describing a time series are *stochastic*—that is, the parameters are changing over time (although these models have also been used with deterministic parameters). We include presentations of *simple exponential smoothing*, *double exponential smoothing*, and *Winters' Method* (which is an exponential smoothing approach to handling seasonal data). We also discuss *adaptive control procedures*. These procedures are used to monitor an exponential smoothing model and signal when the forecasting errors are becoming "too large." In addition, we explain the relationships between exponential smoothing, the Box-Jenkins methodology, and time series regression. For example, we will find that, although the development of exponential smoothing models predates the advent of the Box-Jenkins methodology, recent research indicates that almost any exponential smoothing model is equivalent to (or essentially equivalent to) a Box-Jenkins model.

When implementing the methods described in Chapters 5 and 6, we use the same four step procedure—tentative identification, estimation, diagnostic checking, and forecasting—that is employed in the Box-Jenkins methodology. However, we will see that, while Box and Jenkins have developed a systematic procedure that uses the sample autocorrelation function and the sample partial autocorrelation function to identify a model, the identification of time series regression models and exponential smoothing models usually involves graphical analysis and a great deal of intuition (identification is much more of an *ad hoc* process).

Chapter 7 discusses *transfer function models*, which are *Box-Jenkins causal models*. That is, a transfer function model predicts future values of a time series on the basis of past values of the time series and on the basis of values of one or more other time series related to the time series to be predicted. For example, we might predict future sales of a product on the basis of past sales of the product and on the basis of advertising expenditures for the product. Transfer function models can be more effective than univariate models if one or more time series can be found that are "closely related" to the time series that we wish to forecast. Box and Jenkins have developed a very precise modeling procedure for building transfer function models. Chapter 7 explains how to use this procedure in situations where we are relating exactly one time series to the time series being predicted. Included is a discussion of how to use SAS to build transfer function models. Although there is a Box-Jenkins procedure for building transfer function models that relate more than one time series to the time series to be forecasted, the procedure is complicated and will not be presented in this book.

When we wish to relate several time series to the time series to be predicted, an alternative approach is to use *classical regression analysis*. This technique is explained in detail in Chapter 8. We include discussions of regression models and their use in estimation and prediction, model building, the assumptions behind regression analysis, residual analysis (used to check the validity of the assumptions), and the use of dummy variables. This chapter is concluded by showing how to implement classical regression analysis on a computer by using SAS.

1.7 COMPUTER PACKAGES: SAS AND TSERIES

In this book we will discuss how to implement forecasting techniques by using two computer packages: SAS (Statistical Analysis System) and TSERIES. Briefly, SAS is a computer package that can be used to analyze data. Among other things (such as data storage and retrieval, report writing, and file handling), SAS can be used to perform a wide variety of statistical analyses, including time series forecasting. TSERIES is a Box-Jenkins forecasting package that was written by (and can be obtained from) Professor William Q. Meeker of the Department of Statistics at Iowa State University. Although SAS is more universally available, we also use TSERIES because TSERIES output includes a useful *t*-statistic that is not given by SAS. Furthermore, the differences between SAS and TSERIES are typical of those between many forecasting packages. While we limit our discussion of computer packages to SAS and TSERIES, other popular forecasting packages produce output that is very similar to SAS and TSERIES output. It should be noted that, whereas both SAS and TSERIES can be used to build Box-Jenkins univariate models, only SAS can be used to build transfer function models. For those readers who do not have access to either SAS or TSERIES, we have presented the specifics of their use in separate sections that can be omitted without loss of continuity.

EXERCISES

1.1   Discuss the need for forecasting.
1.2   Discuss what a time series is, and discuss the meaning of trend effects, seasonal variations, cyclical variations, and irregular effects.
1.3   Discuss the difference between univariate models and transfer function models.
1.4   Discuss the difference between point forecasts and prediction interval forecasts.

1.5 Table 1 gives the forecast errors produced using three different sales forecasting models.
   a. Graph these forecast errors for each model.
   b. Which error patterns appear to be random?
   c. Decide whether models A, B, and C adequately fit the data pattern characterizing sales. Explain your conclusions.

TABLE 1

Year	Quarter	Model A Forecast Error	Model B Forecast Error	Model C Forecast Error
1982	1	+ 25	+ 15	− 20
	2	+ 12	+ 6	+ 6
	3	+ 7	− 10	+ 15
	4	+ 5	− 3	− 10
1983	1	+ 3	+ 12	+ 8
	2	0	+ 4	− 5
	3	− 4	− 7	+ 7
	4	− 11	− 1	− 8
1984	1	− 17	+ 9	+ 3
	2	− 21	+ 7	+ 10
	3	− 28	− 12	− 12
	4	− 34	− 5	+ 4
1985	1	− 21	+ 17	− 7
	2	− 13	+ 3	+ 9
	3	− 7	− 9	+ 19
	4	− 2	− 3	− 7
1986	1	+ 5	+ 13	+ 16
	2	+ 9	+ 5	− 6
	3	+ 15	− 10	− 9
	4	+ 19	− 6	+ 5

1.6 Table 2 presents predicted monthly sales and actual monthly sales for a company over the first six months of 1986.
   a. Calculate the forecast error for each month.
   b. Calculate the MAD.
   c. Calculate the MSE.

TABLE 2

	Actual Sales	Predicted Sales
January 1986	270	265
February 1986	263	268
March 1986	275	269
April 1986	262	267
May 1986	250	245
June 1986	278	275

1.7   Given the following sales forecasting models, determine whether these forecasting models are univariate models or causal models. Note that Sales(t) denotes sales in time t.

    a.   Sales($t + 1$) = .8[Forecasted Sales(t)] + .2[Sales(t)]

    b.   Sales($t + 1$) = 500 + 2.5(Advertising Expenditure in Year t) + 5(Number of Customer Calls by Salesmen in Year t)

    c.   Sales($t + 1$) = $\dfrac{1}{t} \sum\limits_{i=1}^{t}$ Sales(i)

1.8   Table 3 contains the predicted and actual per capita income for a certain area of the United States.

    a.   Calculate the forecast error for each year.

    b.   Plot these forecast errors against time.

    c.   Do you think the forecasting method that was used adequately fits the data pattern present? Why or why not?

    d.   Calculate the MAD.

    e.   Calculate the MSE.

TABLE 3

Year	Per Capita Income ($)	Predicted Per Capita Income ($)
1979	3074	3292
1980	3135	3250
1981	3206	3230
1982	3267	3255
1983	3310	3266
1984	3362	3283
1985	3418	3300
1986	3500	3337

1.9   Table 4 contains actual yearly sales for a company along with the predictions of yearly sales generated using two different forecasting methods.

    a.   Calculate the MAD for both forecasting methods.

    b.   Calculate the MSE for both forecasting methods.

    c.   Explain why these measures of accuracy yield different results in this case.

TABLE 4

Year	Actual Sales (in millions)	Method A Predicted Sales	Method B Predicted Sales
1982	8.0	9.0	9.5
1983	12.0	11.5	10.5
1984	14.0	14.0	12.0
1985	16.0	16.5	13.0
1986	10.0	19.0	15.0

NONSEASONAL BOX-JENKINS MODELS AND THEIR TENTATIVE IDENTIFICATION

2.1 INTRODUCTION

The Box-Jenkins methodology consists of a four-step iterative procedure.

Step 1: Tentative identification: historical data is used to tentatively identify an appropriate Box-Jenkins model.

Step 2: Estimation: historical data is used to estimate the parameters of the tentatively identified model.

Step 3: Diagnostic checking: various diagnostics are used to check the adequacy of the tentatively identified model, and if need be, to suggest an improved model, which is then regarded as a new tentatively identified model.

Step 4: Forecasting: once a final model is obtained, it is used to forecast future time series values.

In this chapter we will discuss the nature of nonseasonal Box-Jenkins models, the tentative identification of an appropriate nonseasonal model (Step 1), elementary concepts pertaining to the estimation of model parameters (Step 2), and forecasting future time series values (Step 4). We will delay a discussion of diagnostic checking (Step 3) until Chapter 4, where we will also discuss more advanced concepts pertaining to the estimation of model parameters.

Classical Box-Jenkins forecasting models describe what we refer to as *stationary time series*. We begin this chapter by defining (in Section 2.2) *stationary* and *nonstationary time series*, and by discussing how *differencing* can often be used to transform a nonstationary time series into a stationary time series. In Section 2.3 we introduce the *sample autocorrelation function* (the *SAC*) and the *sample partial autocorrelation function* (the *SPAC*). We will see how to characterize the *behavior* of these functions, and we will discuss how to use the behavior of the SAC in order to decide whether a time series is nonstationary or stationary. Section 2.4 presents a detailed example of (1) using the behavior of the SAC and the SPAC to tentatively identify a Box-Jenkins model and (2) using the model to forecast future time series values. Then, in Section 2.5, we fully discuss tentative identification of nonseasonal Box-Jenkins models. Here we introduce the *general nonseasonal autoregressive–moving average model* and present five guidelines that show how to use the behavior of the SAC and SPAC to tentatively identify models describing nonseasonal time series. We also discuss some of the most common and most useful special cases of the general nonseasonal model.

2.2 *STATIONARY TIME SERIES AND NONSTATIONARY TIME SERIES*

Classical Box-Jenkins models describe *stationary* time series. Thus, in order to tentatively identify a Box-Jenkins model, we must first determine whether or not the time series we wish to forecast is stationary. If it is not, we must transform the time series into a series of stationary time series values. Intuitively, a time series is *stationary* if the statistical properties (for example, the mean and the variance) of the time series are essentially constant through time. If we have observed n values $y_1, y_2, \ldots, y_n$ of a time series, we can use a plot of these values (against time) to help us determine whether the time series is stationary. If the n values seem to fluctuate with constant variation around a constant mean μ, then it is reasonable to believe that the time series is stationary (in Section 2.3 we will utilize more sophisticated methods to help us determine whether a time series is stationary). If the n values do not fluctuate around a constant mean or do not fluctuate with constant variation, then it is reasonable to believe that the time series is *nonstationary*. As we will illustrate in the following example, if a plot of n time series values $y_1, y_2, \ldots, y_n$ indicates that these values are nonstationary, we can sometimes transform the nonstationary time series values into stationary time series values by taking the *first differences* of the nonstationary time series values. That is,

The *first differences* of the time series values $y_1, y_2, \ldots, y_n$ are

$$z_t = y_t - y_{t-1}$$
where
$$t = 2, \ldots, n$$

Below we list the original values and the first differenced values of a time series.

Original Values	First Differences
y_1	
y_2	$z_2 = y_2 - y_1$
y_3	$z_3 = y_3 - y_2$
.	.
.	.
.	.
y_{n-1}	.
y_n	$z_n = y_n - y_{n-1}$

EXAMPLE 2.1   The Olympia Paper Company, Inc., makes Absorbent Paper Towels. The company would like to develop a prediction model that can be used to give point forecasts and prediction interval forecasts of weekly sales over 100,000 rolls, in units of 10,000 rolls, of Absorbent Paper Towels. With a reliable model, The Olympia Paper Company, Inc., can more effectively plan its production schedule, plan its budget, and estimate requirements for producing and storing this product. For the past 120 weeks the company has recorded weekly sales over 100,000 rolls, in units of 10,000 rolls, of Absorbent Paper Towels. The 120 sales figures, $y_1, y_2, \ldots, y_{120}$, are given in Table 2.1 and are plotted in Figure 2.1a. It should be noticed from Figure 2.1a that the original values of the time series do not seem to fluctuate around a constant mean, and hence it would seem that these values are nonstationary. The first differences $z_2, z_3, \ldots, z_{120}$ of the original values $y_1, y_2, \ldots, y_{120}$ are given in Table 2.2 and are plotted in Figure 2.1b. Since Figure 2.1b indicates that the first differences fluctuate with constant variation around a constant mean, it would seem that these first differences are stationary.

Although taking first differences will sometimes transform nonstationary time series values into stationary time series values, we sometimes need to use other forms of differencing to produce stationary time series values. For example, if the original time series values $y_1, y_2, \ldots, y_n$ are nonstationary, and the first differences of the original time series values

$$z_2 = y_2 - y_1$$
$$z_3 = y_3 - y_2$$
$$\vdots$$
$$z_n = y_n - y_{n-1}$$

are nonstationary, then we can sometimes produce stationary time series values by taking the *second differences* (the first differences of the first differences) of the original

TABLE 2.1   Weekly Sales over 100,000 Rolls of Absorbent Paper Towels (in units of 10,000 rolls)

t	y_t	t	y_t	t	y_t	t	y_t
1	15.0000	31	10.7752	61	-1.3173	91	10.5502
2	14.4064	32	10.1129	62	-0.6021	92	11.4741
3	14.9383	33	9.9330	63	0.1400	93	11.5568
4	16.0374	34	11.7435	64	1.4030	94	11.7986
5	15.6320	35	12.2590	65	1.9280	95	11.8867
6	14.3975	36	12.5009	66	3.5626	96	11.2951
7	13.8959	37	11.5378	67	1.9615	97	12.7847
8	14.0765	38	9.6649	68	4.8463	98	13.9435
9	16.3750	39	10.1043	69	6.5454	99	13.6859
10	16.5342	40	10.3452	70	8.0141	100	14.1136
11	16.3839	41	9.2835	71	7.9746	101	13.8949
12	17.1006	42	7.7219	72	8.4959	102	14.2853
13	17.7876	43	6.8300	73	8.4539	103	16.3867
14	17.7354	44	8.2046	74	8.7114	104	17.0884
15	17.0010	45	8.5289	75	7.3780	105	15.8861
16	17.7485	46	8.8733	76	8.1905	106	14.8227
17	18.1888	47	8.7948	77	9.9720	107	15.9479
18	18.5997	48	8.1577	78	9.6930	108	15.0982
19	17.5859	49	7.9128	79	9.4506	109	13.8770
20	15.7389	50	8.7978	80	11.2088	110	14.2746
21	13.6971	51	9.0775	81	11.4986	111	15.1682
22	15.0059	52	9.3234	82	13.2778	112	15.3818
23	16.2574	53	10.4739	83	13.5910	113	14.1863
24	14.3506	54	10.6943	84	13.4297	114	13.9996
25	11.9515	55	9.8367	85	13.3125	115	15.2463
26	12.0328	56	8.1803	86	12.7445	116	17.0179
27	11.2142	57	7.2509	87	11.7979	117	17.2929
28	11.7023	58	5.0814	88	11.7319	118	16.6366
29	12.5905	59	1.8313	89	11.6523	119	15.3410
30	12.1991	60	-0.9127	90	11.3718	120	15.6453

time series values. Therefore,

The *second differences* of the time series values $y_1, y_2, \ldots, y_n$ are

$$z_t = (y_t - y_{t-1}) - (y_{t-1} - y_{t-2})$$

$$= y_t - 2y_{t-1} + y_{t-2} \quad \text{for} \quad t = 3, 4, \ldots, n$$

TABLE 2.2 First Differences of the Observations in Table 2.1

t	$z_t = y_t - y_{t-1}$	t	$z_t = y_t - y_{t-1}$	t	$z_t = y_t - y_{t-1}$	t	$z_t = y_t - y_{t-1}$
2	$-.5936$	32	$-.6623$	62	.7152	92	.9238
3	.5319	33	$-.1798$	63	.7421	93	.08268
4	1.099	34	1.810	64	1.263	94	.2418
5	$-.4054$	35	.5154	65	.5249	95	.08809
6	-1.235	36	.2419	66	1.635	96	$-.5916$
7	$-.5015$	37	$-.9631$	67	-1.601	97	1.490
8	.1805	38	-1.873	68	2.885	98	1.159
9	2.298	39	.4395	69	1.699	99	$-.2576$
10	.1593	40	.2409	70	1.469	100	.4277
11	$-.1503$	41	-1.062	71	$-.03953$	101	$-.2186$
12	.7167	42	-1.562	72	.5213	102	.3903
13	.6871	43	$-.8918$	73	$-.04202$	103	2.101
14	$-.05226$	44	1.375	74	.2575	104	.7016
15	$-.7344$	45	.3243	75	-1.333	105	-1.202
16	.7475	46	.3444	76	.8124	106	-1.063
17	.4403	47	$-.07841$	77	1.782	107	1.125
18	.4109	48	$-.6371$	78	$-.2790$	108	$-.8497$
19	-1.014	49	$-.2449$	79	$-.2424$	109	-1.221
20	-1.847	50	.8850	80	1.758	110	.3976
21	-2.042	51	.2797	81	.2898	111	.8936
22	1.309	52	.2459	82	1.779	112	.2136
23	1.251	53	1.150	83	.3132	113	-1.195
24	-1.907	54	.2204	84	$-.1613$	114	$-.1867$
25	-2.399	55	$-.8575$	85	$-.1173$	115	1.247
26	.08132	56	-1.656	86	$-.5680$	116	1.772
27	$-.8186$	57	$-.9294$	87	$-.9465$	117	.2750
28	.4881	58	-2.170	88	$-.06604$	118	$-.6564$
29	.8882	59	-3.250	89	$-.07964$	119	-1.296
30	$-.3914$	60	-2.744	90	$-.2804$	120	.3043
31	-1.424	61	$-.4046$	91	$-.8216$		

That is, the second differences of the original time series values $y_1, y_2, \ldots, y_n$ are

$$z_3 = y_3 - 2y_2 + y_1$$

$$z_4 = y_4 - 2y_3 + y_2$$

$$\vdots$$

$$z_n = y_n - 2y_{n-1} + y_{n-2}$$

Henceforth, we will denote the values of the time series we are currently working with

(a) Original Values

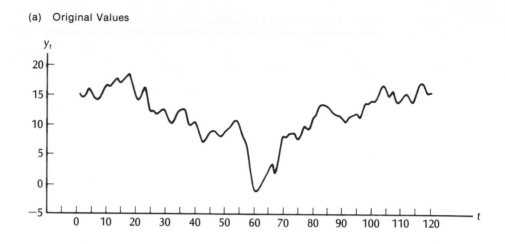

(b) First Differences

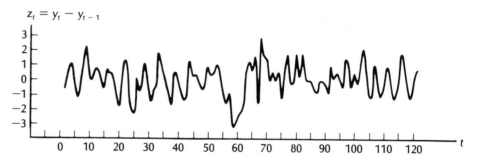

FIGURE 2.1   *Original Values of and First Differences of Weekly Absorbent Paper Towel Sales*

by the symbols $z_b, z_{b+1}, \ldots, z_n$. We will refer to the values $z_b, z_{b+1}, \ldots, z_n$ as the "working series." Note that we do not write the first value of the working series as z_1, because the values $z_b, z_{b+1}, \ldots, z_n$ might be obtained by differencing the nonstationary time series values $y_1, y_2, \ldots, y_n$. For example, if the values $z_b, z_{b+1}, \ldots, z_n$ are obtained by using the transformation $z_t = y_t - y_{t-1}$, then, as illustrated above, $z_b = z_2 = y_2 - y_1$, in which case $b = 2$. As another example, if the values $z_b, z_{b+1}, \ldots, z_n$ are obtained by using the transformation $z_t = y_t - 2y_{t-1} + y_{t-2}$, then, as illustrated above, $z_b = z_3 = y_3 - 2y_2 + y_1$, in which case $b = 3$.

Experience indicates that, if the original time series values $y_1, y_2, \ldots, y_n$ are *nonstationary* and *nonseasonal*, then using the first differencing transformation

$$z_t = y_t - y_{t-1}$$

or the second differencing transformation

$$z_t = (y_t - y_{t-1}) - (y_{t-1} - y_{t-2})$$
$$= y_t - 2y_{t-1} + y_{t-2}$$

will usually produce stationary time series values. If the original time series values y_1, $y_2, \ldots, y_n$ are *nonstationary* and *seasonal*, then more complex transformations may be needed to provide stationary time series values. In Chapter 3 we will discuss a general stationarity transformation. First, however, we will (in the next two sections) discuss the sample autocorrelation function and the sample partial autocorrelation function and how to use these functions to tentatively identify models describing nonseasonal time series.

2.3   THE SAMPLE AUTOCORRELATION AND PARTIAL AUTOCORRELATION FUNCTIONS: THE SAC AND SPAC

Box-Jenkins forecasting models are tentatively identified by examining the behavior of the *sample autocorrelation function (SAC)* and the *sample partial autocorrelation function (SPAC)* for the values of a *stationary time series* $z_b, z_{b+1}, \ldots, z_n$. Here $z_b, z_{b+1}, \ldots, z_n$ may be original time series values or may be transformed time series values. We first consider the SAC.

2.3.1   The Sample Autocorrelation Function (SAC)

Consider the working series of time series values $z_b, z_{b+1}, \ldots, z_n$. The *sample autocorrelation at lag k*, denoted by r_k, is

$$r_k = \frac{\sum_{t=b}^{n-k} (z_t - \bar{z})(z_{t+k} - \bar{z})}{\sum_{t=b}^{n} (z_t - \bar{z})^2}$$

where

$$\bar{z} = \frac{\sum_{t=b}^{n} z_t}{(n - b + 1)}$$

This quantity measures the linear relationship between time series observations separated by a lag of k time units. It can be proven that r_k will always be between -1 and 1. A value of r_k close to 1 indicates that observations separated by a lag of k time units have a strong tendency to move together in a linear fashion with a positive slope, while a value of r_k close to -1 indicates that observations separated by a lag of k time units have a strong tendency to move together in a linear fashion with a negative slope.

To see this, suppose that we have 11 time series values $z_1, z_2, \ldots, z_{11}$, and consider (for example)

$$r_3 = \frac{\sum_{t=1}^{11-3} (z_t - \bar{z})(z_{t+3} - \bar{z})}{\sum_{t=1}^{11} (z_t - \bar{z})^2}$$

$$= \frac{(z_1 - \bar{z})(z_4 - \bar{z}) + (z_2 - \bar{z})(z_5 - \bar{z}) + \cdots + (z_8 - \bar{z})(z_{11} - \bar{z})}{(z_1 - \bar{z})^2 + (z_2 - \bar{z})^2 + \cdots + (z_{11} - \bar{z})^2}$$

where

$$\bar{z} = \frac{\sum_{t=1}^{11} z_t}{11}$$

is the average of the 11 time series values.

To see that r_3 measures the linear relationship between time series observations separated by a lag of 3 time units, first note that since the denominator of r_3

$$\sum_{t=1}^{11} (z_t - \bar{z})^2$$

is positive, the algebraic sign of r_3 is determined by the numerator of r_3

$$\sum_{t=1}^{8} (z_t - \bar{z})(z_{t+3} - \bar{z})$$

A strongly positive value of r_3 results when

$$\sum_{t=1}^{8} (z_t - \bar{z})(z_{t+3} - \bar{z})$$

is strongly positive, which means that most of the products

$$(z_1 - \bar{z})(z_4 - \bar{z}) \qquad (z_2 - \bar{z})(z_5 - \bar{z}) \qquad \ldots \qquad (z_8 - \bar{z})(z_{11} - \bar{z})$$

are positive. Said equivalently, most of the values of

$$(z_t - \bar{z})(z_{t+3} - \bar{z})$$

are positive, which means that, for most of the (z_t, z_{t+3}) combinations:

1. When $(z_t - \bar{z})$ is positive, then $(z_{t+3} - \bar{z})$ is positive, which implies that, *when z_t is greater than $\bar{z}$, then z_{t+3} is greater than $\bar{z}$.* This says that the (z_t, z_{t+3}) combination is, as illustrated in Figure 2.2, in Quadrant I of a coordinate system centered at $(\bar{z}, \bar{z})$.
2. When $(z_t - \bar{z})$ is negative, then $(z_{t+3} - \bar{z})$ is negative, which implies that *when z_t is less than $\bar{z}$, then z_{t+3} is less than $\bar{z}$.* This says that the (z_t, z_{t+3}) combination is, as illustrated in Figure 2.2, in Quadrant III of a coordinate system centered at $(\bar{z}, \bar{z})$.

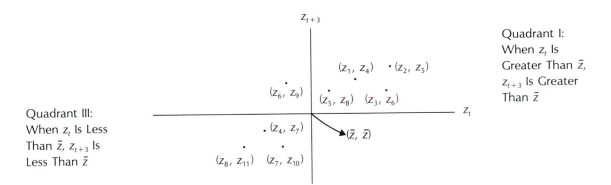

FIGURE 2.2 r_3 *is Strongly Positive: Time Series Observations Separated by a Lag of Three Time Units Have a Strong Tendency to Move Together in a Linear Fashion with a Positive Slope*

In summary, when r_3 is strongly positive, then most of the (z_t, z_{t+3}) combinations are in Quadrants I or III, which implies, as illustrated in Figure 2.2, that time series observations separated by a lag of 3 time units have a strong tendency to move together in a linear fashion with a positive slope. A strongly negative value of r_3 results when

$$\sum_{t=1}^{8} (z_t - \bar{z})(z_{t+3} - \bar{z})$$

is strongly negative, which means that most of the products

$$(z_1 - \bar{z})(z_4 - \bar{z}) \qquad (z_2 - \bar{z})(z_5 - \bar{z}) \qquad \ldots \qquad (z_8 - \bar{z})(z_{11} - \bar{z})$$

are negative. Said equivalently, most of the values of

$$(z_t - \bar{z})(z_{t+3} - \bar{z})$$

are negative, which means that, for most of the (z_t, z_{t+3}) combinations:

1. When $(z_t - \bar{z})$ is positive, then $(z_{t+3} - \bar{z})$ is negative, which implies that, *when z_t is greater than $\bar{z}$, then z_{t+3} is less than $\bar{z}$.* This says that the (z_t, z_{t+3}) combination is, as illustrated in Figure 2.3, in Quadrant IV of a coordinate system centered at $(\bar{z}, \bar{z})$.
2. When $(z_t - \bar{z})$ is negative, then $(z_{t+3} - \bar{z})$ is positive, which implies that, *when z_t is less than $\bar{z}$, then z_{t+3} is greater than $\bar{z}$.* This says that the (z_t, z_{t+3}) combination is, as illustrated in Figure 2.3, in Quadrant II of a coordinate system centered at $(\bar{z}, \bar{z})$.

In summary, when r_3 is strongly negative, then most of the (z_t, z_{t+3}) combinations are in Quadrants II or IV, which implies, as illustrated in Figure 2.3, that time series observations separated by a lag of 3 time units have a tendency to move together in a linear fashion with a negative slope.

In the following box we repeat the formula for r_k and present (for future reference) formulas for the *standard error of r_k, s_{r_k},* and the t_{r_k}-*statistic* related to r_k. In Sections 2.3

Quadrant II:
When z_t Is Less
Than $\bar{z}$, z_{t+3} Is
Greater Than $\bar{z}$

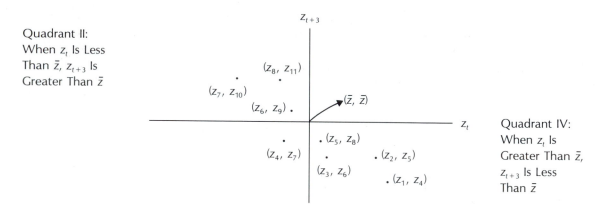

Quadrant IV:
When z_t Is
Greater Than $\bar{z}$,
z_{t+3} Is Less
Than $\bar{z}$

FIGURE 2.3 r_3 is Strongly Negative: Time Series Observations Separated by a Lag of Three Time Units Have a Strong Tendency to Move Together in a Linear Fashion with a Negative Slope

and 2.4 we will see how to use s_{r_k} and t_{r_k} to help us to tentatively identify a Box-Jenkins model. In this box we also define the *sample autocorrelation function (SAC)*.

For the working series $z_b, z_{b+1}, \ldots, z_n$:

1.   The *sample autocorrelation at lag k* is

$$r_k = \frac{\sum_{t=b}^{n-k} (z_t - \bar{z})(z_{t+k} - \bar{z})}{\sum_{t=b}^{n} (z_t - \bar{z})^2}$$

where

$$\bar{z} = \frac{\sum_{t=b}^{n} z_t}{n - b + 1}$$

2.   The *standard error of r_k* is

$$s_{r_k} = \frac{\left(1 + 2\sum_{j=1}^{k-1} r_j^2\right)^{1/2}}{(n - b + 1)^{1/2}}$$

3.   The *t_{r_k}-statistic* is

$$t_{r_k} = \frac{r_k}{s_{r_k}}$$

4.   The sample *autocorrelation function (SAC)* is a listing, or graph, of the sample autocorrelations at lags $k = 1, 2, \ldots$.

Noting that we will henceforth refer to the sample autocorrelation function as the *SAC*, we now present an example.

EXAMPLE 2.2

In this example we consider the original values of Absorbent Paper Towel sales y_1, $y_2, \ldots, y_{120}$ given in Table 2.1. The SAC for this working series is obtained by calculating the sample autocorrelation at lag k for lags $k = 1, 2, 3, \ldots$. As an example of the calculations involved, note that the mean of the 120 original time series values is

$$\bar{y} = \frac{\sum\limits_{t=1}^{120} y_t}{120} = \frac{15.0000 + 14.4064 + \cdots + 15.6453}{120} = 11.58$$

and note that r_3 is calculated as shown below, where $z_b = y_1$, $z_{b+1} = y_2, \ldots,$ $z_n = y_{120}$ and $\bar{z} = \bar{y}$.

$$r_3 = \frac{\sum\limits_{t=b}^{n-k} (z_t - \bar{z})(z_{t+k} - \bar{z})}{\sum\limits_{t=b}^{n} (z_t - \bar{z})^2}$$

$$= \frac{\sum\limits_{t=1}^{120-3} (z_t - \bar{z})(z_{t+3} - \bar{z})}{\sum\limits_{t=1}^{120} (z_t - \bar{z})^2}$$

$$= \frac{(z_1 - \bar{z})(z_4 - \bar{z}) + (z_2 - \bar{z})(z_5 - \bar{z}) + \cdots + (z_{117} - \bar{z})(z_{120} - \bar{z})}{(z_1 - \bar{z})^2 + (z_2 - \bar{z})^2 + \cdots + (z_{120} - \bar{z})^2}$$

$$= [(15.0000 - 11.58)(16.0374 - 11.58)$$
$$+ (14.4064 - 11.58)(15.6320 - 11.58) + \cdots$$
$$+ (17.2929 - 11.58)(15.6453 - 11.58)]$$
$$\div [(15.0000 - 11.58)^2 + (14.4064 - 11.58)^2 + \cdots$$
$$+ (15.6453 - 11.58)^2]$$

$$= .853$$

In Figure 2.4 we present the TSERIES output of the SAC for the original values of the Absorbent Paper Towel sales $y_1, y_2, \ldots, y_{120}$. Here the r_k values are listed under the heading "VALUE" and are also plotted for lags $k = 1, 2, \ldots, 24$.

Furthermore, noting that the TSERIES output of the SAC also gives the standard errors of the sample autocorrelations (that is, the s_{r_k} values) and the related t-statistics (that is, the t_{r_k} values), we calculate s_{r_3} and t_{r_3} as follows.

FIGURE 2.4 *The TSERIES Output of the SAC for the Original Values of Absorbent Paper Towel Sales*

	VALUE[a]	S.E.[b]	T VALUE[c]	−1.0	0.0	1.0
				+ — — — — — — — — — — — — — — — — — +		
1	0.963	0.091	10.545	I	***************************	I
2	0.907	0.154	5.885	I	************************	I
3	0.853	0.194	4.406	I	***********************	I
4	0.801	0.223	3.594	I	*********************	I
5	0.743	0.246	3.024	I	********************	I
6	0.684	0.264	2.596	I	******************	I
7	0.628	0.278	2.257	I	*****************	I
8	0.579	0.290	1.998	I	****************	I
9	0.532	0.299	1.777	I	**************	I
10	0.495	0.307	1.614	I	*************	I
11	0.469	0.313	1.495	I	*************	I
12	0.444	0.319	1.392	I	************	I
13	0.416	0.324	1.282	I	***********	I
14	0.383	0.329	1.166	I	**********	I
15	0.353	0.333	1.061	I	**********	I
16	0.323	0.336	0.962	I	*********	I
17	0.290	0.338	0.859	I	********	I
18	0.251	0.340	0.737	I	*******	I
19	0.203	0.342	0.593	I	******	I
20	0.155	0.343	0.451	I	*****	I
21	0.112	0.343	0.326	I	***	I
22	0.072	0.344	0.211	I	**	I
23	0.033	0.344	0.096	I	*	I
24	0.002	0.344	0.004	I	*	I
				+ — — — — — — — — — — — — — — — — — +		

[a] r_k [b] s_{r_k} [c] t_{r_k}

$$s_{r_3} = \left(1 + 2 \sum_{j=1}^{k-1} r_j^2\right)^{1/2} \Big/ (n - b + 1)^{1/2}$$

$$= \left(1 + 2 \sum_{j=1}^{3-1} r_j^2\right)^{1/2} \Big/ (120 - 1 + 1)^{1/2}$$

$$= (1 + 2[r_1^2 + r_2^2])^{1/2}/(120)^{1/2}$$

$$= (1 + 2[(.963)^2 + (.907)^2])^{1/2}/(120)^{1/2}$$

$$= .194$$

$$t_{r_3} = \frac{r_3}{s_{r_3}} = \frac{.853}{.194} = 4.406$$

One reason we present TSERIES output in this book is that TSERIES calculates values of the *t*-statistic related to r_k. SAS, although giving information similar to these *t*-statistics, does not (at the time of this writing) calculate t_{r_k} values.

In order to employ the Box-Jenkins methodology, we must examine and attempt to classify what we refer to as *"the behavior of the SAC."* The SAC for a nonseasonal time series can display a variety of different behaviors.

First, the SAC for a nonseasonal time series can *cut off*. To see what we mean by this, we say that a *spike at lag k* exists in the SAC if r_k, the sample autocorrelation at lag k, is statistically large. We can judge whether a spike at lag k exists in the SAC by looking at the *t*-statistic related to r_k. When looking at the SAC for a nonseasonal time series, it is most important to consider "low lags". Here, a low lag is considered to be lag $k = 1$, lag $k = 2$, and possibly lag $k = 3$. For low lags, r_k is considered to be statistically large if the absolute value of

$$t_{r_k} = \frac{r_k}{s_{r_k}}$$

is *greater than 1.6*. Here, the choice of the "critical value" 1.6 is somewhat arbitrary. This choice (and other "critical values" given in this book) is motivated by experience with Box-Jenkins modeling—in most cases, we follow the choices of Pankratz [1983]. For nonseasonal time series, the identification of spikes at higher lags ($k > 3$) is not as crucial. Therefore, for lags greater than $k = 3$, a spike is considered to exist in the SAC if the absolute value of

$$t_{r_k} = \frac{r_k}{s_{r_k}}$$

is *greater than 2*. Next, we say that the SAC *cuts off* after lag k if there are no spikes at lags greater than k in the SAC. For example, if

$$|t_{r_1}| = 3.671$$

$$|t_{r_2}| = 1.873$$

$$|t_{r_3}| = 0.517$$

and

$$|t_{r_k}| < 2$$

for all lags $k > 3$, then the SAC would cut off after lag 2. A SAC that cuts off is illustrated in Figure 2.5.

Second, we will say that the SAC *dies down* if this function does not cut off but rather decreases in a "steady fashion". As illustrated in Figure 2.6, the SAC can die down in

1. a damped exponential fashion (with no oscillation or with oscillation),
2. a damped sine-wave fashion, or
3. a fashion dominated by either one of or a combination of both (1) and (2) above.

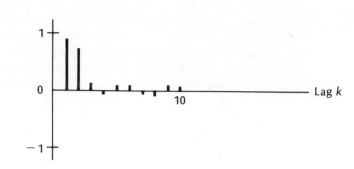

FIGURE 2.5   *A SAC That Cuts Off*

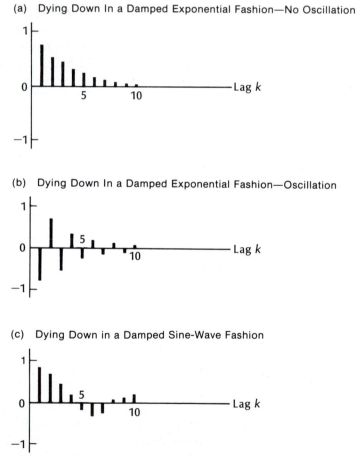

FIGURE 2.6   *Different Dying Down Patterns*

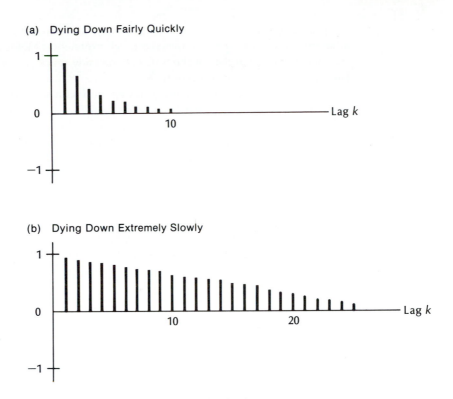

FIGURE 2.7   *Dying Down Fairly Quickly versus Extremely Slowly*

Furthermore, the SAC can die down fairly quickly (as shown in Figure 2.7a) or can die down extremely slowly (as shown in Figure 2.7b).

2.3.2 Using the SAC to Find a Stationary Time Series

The SAC can be used to help us find a working series of stationary time series values z_b, $z_{b+1}, \ldots, z_n$. This can be done because we can relate the behavior of the SAC to stationarity. In general, it can be shown that, for *nonseasonal data*:

1. If the SAC of the time series values $z_b, z_{b+1}, \ldots, z_n$ either *cuts off fairly quickly* or *dies down fairly quickly*, then the time series values should be considered *stationary*.
2. If the SAC of the time series values $z_b, z_{b+1}, \ldots, z_n$ *dies down extremely slowly*, then the time series values should be considered *nonstationary*.

The precise meanings of the terms "fairly quickly" and "extremely slowly" are somewhat arbitrary and can best be determined through experience. Moreover, experience shows that, for nonseasonal data, if the SAC cuts off fairly quickly, it will often do so after a lag k which is less than or equal to 2.

We can now use the following strategy to find a stationary time series. We first compute the SAC for the original time series values $y_1, y_2, \ldots, y_n$. If the SAC either cuts off fairly quickly or dies down fairly quickly, then the original time series values should be considered stationary. If the SAC dies down extremely slowly, the original time series values should be considered nonstationary. In such a case, data transformation is necessary (we would generally begin by trying first differencing). We then compute the SAC for the transformed data. If the SAC for the transformed data either cuts off fairly quickly or dies down fairly quickly, the transformed data should be considered stationary. If the SAC for the transformed data dies down extremely slowly, the transformed data should be considered nonstationary. In such a case, further data transformation (for example, second differencing) is necessary. For nonseasonal data, first or second differencing will generally produce stationary time series values.

EXAMPLE 2.3

Again consider the SAC of the original values of Absorbent Paper Towel sales (shown in Figure 2.4). We can use the SAC for this working series to determine whether the original values of Absorbent Paper Towel sales are stationary. To do this, we note that the SAC in Figure 2.4 can (through experience with Box-Jenkins modeling) be interpreted as dying down extremely slowly (notice the very slow, reasonably steady decrease in the r_k values). Since the SAC for these original values dies down extremely slowly, and since the plot of the original values of Absorbent Paper Towel sales in Figure 2.1a indicates that these original values do not seem to fluctuate around a constant mean, we will conclude that these original values are nonstationary.

This conclusion says that we must use data transformation in order to obtain stationary time series values $z_b, z_{b+1}, \ldots, z_n$. We will attempt to determine whether using the transformation

$$z_t = y_t - y_{t-1} \quad \text{(first differencing)}$$

produces stationary time series values

$$z_b = z_2 = y_2 - y_1 = 14.4064 - 15.0000 = -.5936$$

$$z_{b+1} = z_3 = y_3 - y_2 = 14.9383 - 14.4064 = .5319$$

$$\vdots$$

$$z_n = z_{120} = y_{120} - y_{119} = 15.6453 - 15.3410 = .3043$$

Recalling that all of these first differences are given in Table 2.2, the SAC for this working series is obtained by calculating the sample autocorrelation at lag k for lags $k = 1, 2, 3, \ldots$. As an example of the calculations involved, note that the mean of the 119 first differences is

$$\bar{z} = \frac{\sum\limits_{t=b}^{n} z_t}{n - b + 1} = \frac{\sum\limits_{t=2}^{120} z_t}{120 - 2 + 1} = \frac{-.5936 + .5319 + \cdots + .3043}{119}$$

$$= .005423$$

and note that r_3 is calculated as shown below, where $z_b = z_2$, $z_{b+1} = z_3, \ldots,$ $z_n = z_{120}$

$$r_3 = \frac{\sum\limits_{t=b}^{n-k} (z_t - \bar{z})(z_{t+k} - \bar{z})}{\sum\limits_{t=b}^{n} (z_t - \bar{z})^2}$$

$$= \frac{\sum\limits_{t=2}^{120-3} (z_t - \bar{z})(z_{t+3} - \bar{z})}{\sum\limits_{t=2}^{120} (z_t - \bar{z})^2}$$

$$= \frac{(z_2 - \bar{z})(z_5 - \bar{z}) + \cdots + (z_{117} - \bar{z})(z_{120} - \bar{z})}{(z_2 - \bar{z})^2 + \cdots + (z_{120} - \bar{z})^2}$$

$$= [(-.5936 - .005423)(-.4054 - .005423) + \cdots$$

$$+ (.2750 - .005423)(.3043 - .005423)]$$

$$\div [(-.5936 - .005423)^2 + \cdots + (.3043 - .005423)^2]$$

$$= -.072$$

In Figure 2.8 we present the TSERIES output of the SAC for the first differences of the Absorbent Paper Towel sales $z_2, z_3, \ldots, z_{120}$. Also given are the values of the standard error of r_k and the t-statistic related to r_k. As an example, we calculate:

$$s_{r_3} = \left(1 + 2 \sum\limits_{j=1}^{k-1} r_j^2\right)^{1/2} \bigg/ (n - b + 1)^{1/2}$$

$$= \left(1 + 2 \sum\limits_{j=1}^{3-1} r_j^2\right)^{1/2} \bigg/ (120 - 2 + 1)^{1/2}$$

$$= (1 + 2[r_1^2 + r_2^2])^{1/2}/(119)^{1/2}$$

$$= (1 + 2[(.307)^2 + (-.065)^2])^{1/2}/(119)^{1/2}$$

$$= .100$$

$$t_{r_3} = \frac{r_3}{s_{r_3}} = \frac{-.072}{.100} = -.715$$

FIGURE 2.8   The TSERIES Output of the SAC for the First Differences of Absorbent Paper Towel Sales

	VALUE[a]	S.E.[b]	T VALUE[c]	−1.0	0.0	1.0
				+ − − − − − − − − − − − − − − − − +		
1	0.307	0.092	3.345	I	********	I
2	−0.065	0.100	−0.648	I	**	I
3	−0.072	0.100	−0.715	I	**	I
4	0.105	0.101	1.039	I	***	I
5	0.084	0.102	0.828	I	***	I
6	0.023	0.102	0.223	I	*	I
7	−0.133	0.102	−1.297	I	****	I
8	−0.119	0.104	−1.148	I	***	I
9	−0.174	0.105	−1.659	I	*****	I
10	−0.118	0.107	−1.103	I	***	I
11	−0.052	0.108	−0.477	I	**	I
12	0.021	0.109	0.193	I	*	I
13	0.041	0.109	0.381	I	**	I
14	0.019	0.109	0.178	I	*	I
15	−0.061	0.109	−0.557	I	**	I
16	−0.002	0.109	−0.022	I	*	I
17	0.128	0.109	1.172	I	****	I
18	0.215	0.110	1.949	I	******	I
19	0.056	0.114	0.491	I	**	I
20	−0.045	0.114	−0.398	I	**	I
21	−0.070	0.114	−0.612	I	**	I
22	−0.035	0.114	−0.307	I	*	I
23	−0.052	0.115	−0.451	I	**	I
24	−0.038	0.115	−0.331	I	*	I
				+ − − − − − − − − − − − − − − − − +		

[a] r_k [b] s_{r_k} [c] t_{r_k}

Looking at the SAC for these first differences, we see that the SAC has a spike at lag 1 and cuts off after lag 1 because

$$|t_{r_1}| = |3.345| = 3.345 \quad \text{is greater than} \quad 1.6,$$

$$|t_{r_2}| = |-0.648| = 0.648 \quad \text{is less than} \quad 1.6,$$

$$|t_{r_3}| = |-0.715| = 0.715 \quad \text{is less than} \quad 1.6, \quad \text{and}$$

$$|t_{r_k}| < 2 \quad \text{for all} \quad k > 3$$

Therefore, since the SAC of the first differences of Absorbent Paper Towel sales has a spike at lag 1 and cuts off after lag 1 (that is, cuts off "quickly") and since the plot of these first differences in Figure 2.1b indicates that these first differences do seem to fluctuate around a constant mean, we will assume that these first differences (that is, the time series values produced by using the transformation $z_t = y_t - y_{t-1}$) are stationary.

2.3.3   The Sample Partial Autocorrelation Function (SPAC)

We now present formulas for the *sample partial autocorrelation at lag k*, r_{kk}, the standard error of r_{kk}, and the related *t*-statistic [see (3) below], and we define the *sample partial autocorrelation function* (which we will henceforth refer to as the SPAC).

1. The *sample partial autocorrelation at lag k* is

$$r_{kk} = \begin{cases} r_1 & \text{if } k = 1 \\[2em] \dfrac{r_k - \sum\limits_{j=1}^{k-1} r_{k-1,j} r_{k-j}}{1 - \sum\limits_{j=1}^{k-1} r_{k-1,j} r_j} & \text{if } k = 2, 3, \ldots \end{cases}$$

   where

$$r_{kj} = r_{k-1,j} - r_{kk} r_{k-1,k-j} \quad \text{for } j = 1, 2, \ldots, k - 1$$

2. The *standard error of* r_{kk} is

$$s_{r_{kk}} = 1/(n - b + 1)^{1/2}$$

3. The $t_{r_{kk}}$-*statistic* is

$$t_{r_{kk}} = \frac{r_{kk}}{s_{r_{kk}}}$$

4. The *sample partial autocorrelation function* is a listing, or graph, of the sample partial autocorrelations at lags $k = 1, 2, \ldots$.

It is beyond the scope of this text to give a precise interpretation of the sample partial autocorrelation at lag k. However, this quantity may intuitively be thought of as the sample autocorrelation of time series observations separated by a lag of k time units *with the effects of the intervening observations eliminated.*

Again, in order to employ the Box-Jenkins methodology, we must examine and attempt to classify the behavior of the SPAC. The SPAC, like the SAC, can display a variety of different behaviors. First, the SPAC for a nonseasonal time series can *cut off.* To see what we mean by this, we say that a *spike at lag k* exists in the SPAC if r_{kk}, the sample partial autocorrelation at lag k, is statistically large. We can judge whether a spike at lag k exists in the SPAC by looking at the *t*-statistic related to r_{kk}. Here, we consider a spike at lag k to exist in the SPAC if the absolute value of

$$t_{r_{kk}} = \frac{r_{kk}}{s_{r_{kk}}}$$

is *greater than* 2. Moreover, we say that the SPAC *cuts off* after lag k if there are no spikes at lags greater than k in the SPAC. For nonseasonal data, experience shows that if the SPAC cuts off, it will generally do so after a lag which is less than or equal to 2. Second, we will say that the SPAC *dies down* if this function does not cut off but rather decreases in a "steady fashion". As illustrated in Figure 2.6, the SPAC can die down in (1) a damped exponential fashion (with no oscillation or with oscillation), (2) a damped sine-wave fashion, or (3) a fashion dominated by either one of or a combination of both (1) and (2) above. Furthermore, the SPAC can die down fairly quickly (as shown in Figure 2.7a) or can die down extremely slowly (as shown in Figure 2.7b). We will see how the behavior of the SPAC (as well as the behavior of the SAC) helps us to identify Box-Jenkins models in future sections of this book.

EXAMPLE 2.4   Again consider the first differences of the Absorbent Paper Towel sales given in Table 2.2. The SPAC for these first differences is obtained by calculating the sample partial autocorrelation at lag k for lags $k = 1, 2, 3, \ldots$. The sample partial autocorrelations r_{11}, r_{22}, r_{33}, and r_{44} are computed as follows. Since

$$r_{kk} = \begin{cases} r_1 & \text{if } k = 1 \\[2em] \dfrac{r_k - \sum\limits_{j=1}^{k-1} r_{k-1,j} r_{k-j}}{1 - \sum\limits_{j=1}^{k-1} r_{k-1,j} r_j} & \text{if } k = 2, 3, \ldots \end{cases}$$

where

$$r_{kj} = r_{k-1,j} - r_{kk} r_{k-1,k-j} \qquad \text{for } j = 1, 2, \ldots, k - 1$$

it follows that

$$r_{11} = r_1 = .307 \approx .31$$

$$r_{22} = \frac{r_2 - \sum\limits_{j=1}^{2-1} r_{2-1,j} r_{2-j}}{1 - \sum\limits_{j=1}^{2-1} r_{2-1,j} r_j}$$

$$= \frac{r_2 - r_{11} r_1}{1 - r_{11} r_1} = \frac{-.06 - (.31)(.31)}{1 - (.31)(.31)} = -.18$$

$$r_{21} = r_{11} - r_{22} r_{11} = .31 - (-.18)(.31) = .37$$

$$r_{33} = \frac{r_3 - \sum_{j=1}^{3-1} r_{3-1,j} r_{3-j}}{1 - \sum_{j=1}^{3-1} r_{3-1,j} r_j} = \frac{r_3 - (r_{21} r_2 + r_{22} r_1)}{1 - (r_{21} r_1 + r_{22} r_2)}$$

$$= \frac{-.07 - [(.37)(-.06) + (-.18)(.31)]}{1 - [(.37)(.31) + (-.18)(-.06)]} = .01$$

$$r_{31} = r_{21} - r_{33} r_{22} = .37 - (.01)(-.18) = .37$$

$$r_{32} = r_{22} - r_{33} r_{21} = -.18 - (.01)(.37) = -.18$$

$$r_{44} = \frac{r_4 - \sum_{j=1}^{4-1} r_{4-1,j} r_{4-j}}{1 - \sum_{j=1}^{4-1} r_{4-1,j} r_j} = \frac{r_4 - (r_{31} r_3 + r_{32} r_2 + r_{33} r_1)}{1 - (r_{31} r_1 + r_{32} r_2 + r_{33} r_3)}$$

$$= \frac{.10 - [(.37)(-.07) + (-.18)(-.06) + (.01)(.31)]}{1 - [(.37)(.31) + (-.18)(-.06) + (.01)(-.07)]} = .13$$

In Figure 2.9 we present the TSERIES output of the SPAC for the first differences of the Absorbent Paper Towel sales. Note that the output includes the $s_{r_{kk}}$ and $t_{r_{kk}}$ values. As an example, we calculate (for $k = 1, 2, \ldots$) as follows.

$$s_{r_{kk}} = \frac{1}{(n - b + 1)^{1/2}} = \frac{1}{(120 - 2 + 1)^{1/2}} = .092 \qquad \text{for } k = 1, 2, \ldots$$

$$t_{r_{22}} = \frac{r_{22}}{s_{r_{22}}} = \frac{-.18}{.092} = -1.956$$

Note that the calculations we have demonstrated here do not give precisely the same results as those presented in the TSERIES output of Figure 2.9 because the computer calculations were carried out to more decimal places. Note also that, while TSERIES does compute the t-statistic related to r_{kk}, SAS (at the time of this writing) does not (although it does give information similar to these t-statistics).

We now attempt to describe the behavior of the SPAC in Figure 2.9. This SPAC has a spike at lag 1 and might be interpreted as cutting off after lag 1 because

$$|t_{r_{11}}| = |3.345| = 3.345$$

is greater than 2, and because

$$|t_{r_{kk}}| < 2 \qquad \text{for all } k > 1$$

However, because each of

$$|t_{r_{22}}| = 1.912$$

$$|t_{r_{44}}| = 1.455$$

FIGURE 2.9   *The TSERIES Output of the SPAC for the First Differences of Absorbent Paper Towel Sales*

	VALUE[a]	S.E.[b]	T VALUE[c]	−1.0	0.0	1.0
				+ — — — — — — —	— — — — — — — —	— — +
1	0.307	0.092	3.345	I	********	I
2	−0.175	0.092	−1.912	I	*****	I
3	0.006	0.092	0.067	I	*	I
4	0.133	0.092	1.455	I	****	I
5	−0.010	0.092	−0.104	I	*	I
6	0.020	0.092	0.217	I	*	I
7	−0.140	0.092	−1.523	I	****	I
8	−0.041	0.092	−0.445	I	**	I
9	−0.178	0.092	−1.937	I	*****	I
10	−0.053	0.092	−0.575	I	**	I
11	−0.011	0.092	−0.119	I	*	I
12	0.032	0.092	0.350	I	*	I
13	0.072	0.092	0.787	I	**	I
14	0.010	0.092	0.108	I	*	I
15	−0.058	0.092	−0.630	I	**	I
16	0.002	0.092	0.020	I	*	I
17	0.079	0.092	0.857	I	***	I
18	0.113	0.092	1.235	I	***	I
19	−0.046	0.092	−0.501	I	**	I
20	0.000	0.092	0.001	I	*	I
21	−0.051	0.092	−0.560	I	**	I
22	−0.060	0.092	−0.658	I	**	I
23	−0.062	0.092	−0.681	I	**	I
24	−0.014	0.092	−0.154	I	*	I
				+ — — — — — — — — —	— — — — — — —	— — +

[a] r_{kk} [b] $s_{r_{kk}}$ [c] $t_{r_{kk}}$

and

$$|t_{r_{77}}| \;=\; 1.523$$

is fairly large, the cutoff is not very abrupt. Specifically, comparing the SAC of Figure 2.8 with the SPAC of Figure 2.9, we can say that (particularly at low lags) the SAC seems to cut off more abruptly than the SPAC. Since the cutoff in this SPAC is not very definite, we might also conclude that (particularly at low lags) this SPAC dies down in a fashion "dominated" by damped exponential decay (with oscillation). In the next example we will see how to use the behaviors of the SAC of Figure 2.8 and the SPAC of Figure 2.9 to tentatively identify a model describing the first differences of the Absorbent Paper Towel sales.

2.4 AN EXAMPLE OF NONSEASONAL MODELING AND FORECASTING

Once we have transformed the original time series $y_1, y_2, \ldots, y_n$ into stationary time series values $z_b, z_{b+1}, \ldots, z_n$, we use the SAC and SPAC to identify a Box-Jenkins model describing the stationary time series values. In this section we will consider the model

$$z_t = a_t - \theta_1 a_{t-1}$$

which is called a *nonseasonal moving average model of order 1*. Here, θ_1 is an unknown *parameter* that must be estimated from sample data. Moreover, a_t is called a *random shock* corresponding to time period t, and a_{t-1} is called a *random shock* corresponding to time period $t - 1$. To better understand the nature of the random shocks, suppose (for example) that the stationary time series values

$$z_2 = -.5936$$

$$z_3 = .5319$$

$$\cdot$$
$$\cdot$$
$$\cdot$$

$$z_{120} = .3043$$

in Table 2.2 are described by the model

$$z_t = a_t - \theta_1 a_{t-1}$$

This implies that

$$z_2 = a_2 - \theta_1 a_1$$

$$z_3 = a_3 - \theta_1 a_2$$

$$\cdot$$
$$\cdot$$
$$\cdot$$

$$z_{120} = a_{120} - \theta_1 a_{119}$$

Here, each random shock a_t is a value that is assumed to have been randomly selected from a *normal distribution* that has *mean zero* and a *variance* that is the *same* for each and every time period t. Moreover, the random shocks $a_1, a_2, a_3, \ldots$ are assumed to be *statistically independent* of each other. We will determine how to check the validity of these assumptions as we proceed through this book. For now, consider (for example) how the random shocks $a_1, a_2,$ and a_3 corresponding to time periods 1, 2, and 3 determine the time series values z_2 and z_3. Although we have observed the time series values $z_2 = -.5936$ and $z_3 = .5319$, we cannot know the true values of the random shocks $a_1, a_2,$ and a_3. However, for the purposes of illustration, assume that (a supernatural power knows that) the true value of the parameter θ_1 is $-.3489$ and the true values of the random shocks $a_1, a_2,$ and a_3—which are assumed to have been randomly and independently selected from three normal distributions having mean zero and a

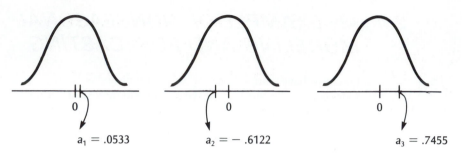

$$a_1 = .0533 \qquad a_2 = -.6122 \qquad a_3 = .7455$$

FIGURE 2.10   *The Random Shocks a_1, a_2, and a_3*

constant variance—are (see Figure 2.10)

$$a_1 = .0533 \qquad a_2 = -.6122 \qquad \text{and} \qquad a_3 = .7455$$

It follows that these random shocks have determined the values $z_2 = -.5936$ and $z_3 = .5319$ through the equations

$$z_2 = a_2 - \theta_1 a_1 = -.6122 - (-.3489)(.0533) = -.5936$$

and

$$z_3 = a_3 - \theta_1 a_2 = .7455 - (-.3489)(-.6122) = .5319$$

It is not intuitively obvious why the nonseasonal moving average model of order 1

$$z_t = a_t - \theta_1 a_{t-1}$$

would adequately represent a stationary time series. However, this and other Box-Jenkins models provide effective representations of many real-world time series. Although there are many Box-Jenkins models, each such model is characterized by its *theoretical autocorrelation function* (TAC) and its *theoretical partial autocorrelation function* (TPAC). We denote the theoretical autocorrelation of a model at lag k by the symbol ϱ_k, and we denote the theoretical partial autocorrelation of the model at lag k by the symbol ϱ_{kk}.

The theoretical autocorrelation of a model at lag k measures the linear relationship between time series values (determined by the model) separated by a lag of k time units. It can be shown that ϱ_k will always be between -1 and 1. A value of ϱ_k close to 1 indicates that time series values separated by a lag of k time units have a strong tendency to move together in a linear fashion with a positive slope, while a value of ϱ_k close to -1 indicates that time series values separated by a lag of k time units have a strong tendency to move together in a linear fashion with a negative slope. It is beyond the scope of this text to give a precise interpretation of the theoretical partial auto-correlation of a model at lag k. Suffice it to say that ϱ_{kk} may intuitively be thought of as the autocorrelation of time series values separated by a lag of k time units *with the effects of the intervening time series values eliminated*. We now define the TAC and the TPAC.

> 1. The *theoretical autocorrelation function (TAC)* of a model is a listing of the theoretical autocorrelations $\varrho_1, \varrho_2, \ldots$.
> 2. The *theoretical partial autocorrelation function (TPAC)* of a model is a listing of the theoretical partial autocorrelations $\varrho_{11}, \varrho_{22}, \ldots$.

As an example, it can be proven that for the nonseasonal moving average model of order 1

$$z_t = a_t - \theta_1 a_{t-1}$$

1. The *TAC* has a nonzero autocorrelation at lag 1 and has zero autocorrelations at all lags after lag 1 (that is, *cuts off after lag 1*). Specifically,

$$\varrho_1 = \frac{-\theta_1}{1 + \theta_1^2}$$

$$\varrho_k = 0 \qquad \text{for } k > 1$$

2. The *TPAC dies down* (that is, the values $\varrho_{11}, \varrho_{22}, \ldots$ decrease) in a fashion dominated by damped exponential decay.

As another example, consider the *nonseasonal moving average model of order 2*

$$z_t = a_t - \theta_1 a_{t-1} - \theta_2 a_{t-2}$$

where θ_1 and θ_2 are unknown parameters and where a_t, a_{t-1}, and a_{t-2} are random shocks corresponding to time periods $t, t-1$, and $t-2$. It can be proven that for this model

1. The *TAC* has nonzero autocorrelations at lags 1 and 2 and has zero auto-correlations at all lags after lag 2 (that is, *cuts off after lag 2*). Specifically,

$$\varrho_1 = \frac{-\theta_1(1 - \theta_1)}{1 + \theta_1^2 + \theta_2^2}$$

$$\varrho_2 = \frac{-\theta_2}{1 + \theta_1^2 + \theta_2^2}$$

$$\varrho_k = 0 \qquad \text{for } k > 2$$

and

2. The *TPAC dies down* (that is, the values $\varrho_{11}, \varrho_{22}, \ldots$ decrease) according to a mixture of damped exponentials and/or damped sine waves.

Now, noting that r_k and r_{kk}, *the sample autocorrelation* and *sample partial autocorrelation* of the time series values $z_b, z_{b+1}, \ldots, z_n$ at lag k, are the point estimates of ϱ_k and ϱ_{kk}, it follows that *the SAC and SPAC are the estimates of the TAC and TPAC*. Therefore, if the behavior of the SAC and SPAC of the time series values $z_b, z_{b+1}, \ldots, z_n$ is consistent with (that is, similar to) the behavior of the TAC and TPAC of a particular model, then it is reasonable to conclude that the particular model has generated the time series values. We illustrate this idea in the following example.

EXAMPLE 2.5

Part 1:   Tentative Identification of a Model

We concluded in Example 2.3 that, for the Absorbent Paper Towel sales time series in Table 2.1, the transformation

$$z_t = y_t - y_{t-1}$$

produces stationary time series values. Recall that the SAC and SPAC of these stationary first differences are given in Figures 2.8 and 2.9, and recall that we have previously concluded that the SAC has a spike at lag 1 and cuts off after lag 1, and that the SPAC either (1) has a spike at lag 1 and cuts off after lag 1 or (2) dies down fairly quickly in a fashion dominated by damped exponential decay. In general, for nonseasonal data, a SAC and SPAC can both die down fairly quickly or can both (appear to) cut off fairly quickly. However, in most situations, if one of these functions cuts off fairly quickly, the other will die down fairly quickly. Therefore, if both of these functions can be interpreted as cutting off, it is useful to decide which of these functions cuts off more abruptly (this usually means that the other function is dying down).

In Example 2.4 we noted that the SAC of Figure 2.8 seems to cut off more abruptly (especially at low lags) than the SPAC of Figure 2.9. It follows that it is reasonable to conclude that the SAC of Figure 2.8 has a spike at lag 1 and cuts off after lag 1; since both the SAC and SPAC seldom cut off simultaneously it is also reasonable to conclude that the SPAC dies down fairly quickly in a fashion dominated by damped exponential decay. Since this behavior of the SAC and SPAC is similar to the behavior of the TAC and TPAC of the nonseasonal moving average model of order 1

$$z_t = a_t - \theta_1 a_{t-1}$$

it follows that this model is a reasonable tentative model describing $z_t = y_t - y_{t-1}$, and thus that a reasonable tentative model describing y_t is

$$y_t - y_{t-1} = a_t - \theta_1 a_{t-1}$$

or

$$y_t = y_{t-1} + a_t - \theta_1 a_{t-1}$$

Part 2:   The Least Squares Point Estimate of the Model Parameter θ_1

Having tentatively identified a model describing the Absorbent Paper Towel sales time series, we must now estimate the value of the model parameter θ_1. Using the data of Table 2.1 and the TSERIES computer package, the *least squares point estimate* of θ_1 in the above model can be calculated to be $\hat{\theta}_1 = -.3545$. In order to explain what we mean by the term "least squares", suppose that we have observed the time series values $y_1, y_2, \ldots, y_{t-1}$ and that we wish to predict the "future time series value"

$$y_t = y_{t-1} + a_t - \theta_1 a_{t-1}$$

The above Box-Jenkins model (and, in fact, any Box-Jenkins model) can be expressed as the sum of a "predictable part p_t" and a "future random shock a_t". That is,

$$y_t = p_t + a_t$$

For example, for the model

$$y_t = y_{t-1} + a_t - \theta_1 a_{t-1} = y_{t-1} - \theta_1 a_{t-1} + a_t$$

it follows that

$$p_t = y_{t-1} - \theta_1 a_{t-1}$$

Since the future random shock a_t has mean zero (that is, has been randomly selected from a normal distribution having mean zero), and since we have no other information about a_t, the *point prediction* $\hat{a}_t$ of the *future random shock a_t is zero*. Therefore, if $\hat{p}_t$ denotes the point prediction of p_t, it follows that the point prediction of

$$y_t = p_t + a_t$$

is

$$\hat{y}_t = \hat{p}_t + \hat{a}_t = \hat{p}_t + 0 = \hat{p}_t$$

Therefore, since

$$a_t = y_t - p_t$$

it follows that when we observe y_t and thus a_t becomes a past random shock, then the *point prediction of the past random shock a_t* is

$$\hat{a}_t = y_t - \hat{p}_t$$
$$= y_t - \hat{y}_t$$

which is called the tth *residual* and is the difference between y_t and the point prediction $\hat{y}_t$. If we assume that we have observed the time series values $y_1, y_2, \ldots, y_{t-1}$, then the point prediction of

$$y_t = y_{t-1} + a_t - \theta_1 a_{t-1}$$

is

$$\hat{y}_t = y_{t-1} + \hat{a}_t - \hat{\theta}_1 \hat{a}_{t-1}$$

where

1. the point prediction $\hat{a}_t$ of the future random shock a_t is zero
2. the point prediction $\hat{a}_{t-1}$ of the past random shock a_{t-1} is the $(t-1)^{st}$ residual $(y_{t-1} - \hat{y}_{t-1})$ if we can calculate $\hat{y}_{t-1}$, and is zero if we cannot calculate $\hat{y}_{t-1}$.

We now calculate (when possible) the first three residuals for the Absorbent Paper Towel sales time series of Table 2.1. For $t = 1$:

$$\hat{y}_1 = y_0 + \hat{a}_1 + .3545\hat{a}_0$$

Since we have not observed y_0, we cannot calculate $\hat{y}_1$, and thus we cannot calculate $(y_1 - \hat{y}_1)$, the first residual. For $t = 2$:

$$\hat{y}_2 = y_1 + \hat{a}_2 + .3545\hat{a}_1$$

$$= 15 + 0 + .3545(0)$$

where the point prediction $\hat{a}_1$ of a_1 is zero, because we cannot calculate $(y_1 - \hat{y}_1)$.

$$y_2 - \hat{y}_2 = 14.4064 - 15$$

$$= -.5936$$

For $t = 3$:

$$\hat{y}_3 = y_2 + \hat{a}_3 + .3545\hat{a}_2$$

$$= 14.4064 + 0 + .3545(y_2 - \hat{y}_2)$$

$$= 14.4064 + .3545(-.5936)$$

$$= 14.1960$$

$$y_3 - \hat{y}_3 = 14.9383 - 14.1960$$

$$= .7423$$

The remaining residuals for periods $t = 4, 5, \ldots, 120$ may be calculated in a similar manner. For future reference it should be noted that $\hat{y}_{120}$ can be calculated to be 14.9553, which implies that

$$y_{120} - \hat{y}_{120} = 15.6453 - 14.9553$$

$$= .6900$$

When we say that $\hat{\theta}_1 = -.3545$ is the *least squares point estimate* of θ_1, we mean that $\hat{\theta}_1 = -.3545$ minimizes the quantity

$$SSE = \sum_{t=2}^{120} (y_t - \hat{y}_t)^2$$

That is, $\hat{\theta}_1 = -.3545$ makes the sum of squared differences between the observed time series values and the predictions (given by the prediction equation $\hat{y}_t = y_{t-1} + \hat{a}_t + .3545\hat{a}_{t-1}$) of these time series values smaller than the value of SSE that would be obtained by using any other value of $\hat{\theta}_1$.

Part 3: *Using the Model to Forecast Future Time Series Values*

Having estimated the parameter θ_1, the next step is to use various diagnostics to check the adequacy of the tentative model

$$y_t = y_{t-1} + a_t - \theta_1 a_{t-1}$$

We will do this in Chapter 4 and find that this model is a reasonable final model, which implies that we should use this model to forecast future time series values.

In order to compute forecasts of future Absorbent Paper Towel sales, we recall that the point prediction of

$$y_t = y_{t-1} + a_t - \theta_1 a_{t-1}$$

is

$$\hat{y}_t = y_{t-1} + \hat{a}_t - \hat{\theta}_1 \hat{a}_{t-1}$$

where

1. the point prediction $\hat{a}_t$ of the future random shock a_t is zero, and
2. the point prediction $\hat{a}_{t-1}$ of the past random shock a_{t-1} is the $(t-1)^{st}$ residual $(y_{t-1} - \hat{y}_{t-1})$ if we can calculate $\hat{y}_{t-1}$, and is zero if we cannot calculate $\hat{y}_{t-1}$.

Since we have observed 120 values of the Absorbent Paper Towel sales time series, we say that we are at "time origin 120". Since the equation

$$y_t = y_{t-1} + a_t - \theta_1 a_{t-1}$$

implies that

$$y_{121} = y_{120} + a_{121} - \theta_1 a_{120}$$

it follows that a point forecast of y_{121} made at time origin 120 is

$$
\begin{aligned}
\hat{y}_{121}(120) &= y_{120} + \hat{a}_{121} - \hat{\theta}_1 \hat{a}_{120} \\
&= y_{120} + 0 - (-.3545)(y_{120} - \hat{y}_{120}) \\
&= 15.6453 + .3545(.6900) \\
&= 15.8899
\end{aligned}
$$

Furthermore, it can be shown that a 95% prediction interval forecast of y_{121} is

$$[15.8899 - 2.0362, \quad 15.8899 + 2.0362] \qquad \text{or} \qquad [13.8537, \quad 17.9261]$$

To explain the meaning of "95% confidence," note that if the actual observation at time 121 is $y_{121} = 16.1099$, then this future observation is contained in the 95% prediction interval [13.8537, 17.9261], which is based on the first 120 time series observations. For this reason we say that the prediction interval–future value combination

$$\{[13.8537, \quad 17.9261]; \quad y_{121} = 16.1099\}$$

is *successful*. Another realization of the Absorbent Paper Towel sales time series (that is, another time series generated by the model $y_t = y_{t-1} + a_t - \theta_1 a_{t-1}$ and a different sequence of random shocks) would yield a different 95% prediction interval based on the first 120 time series observations and a different future observation (y_{121}). If another realization yielded the 95% prediction interval [12.7654, 16.8211] and the future observation $y_{121} = 16.9712$, then the combination

$$\{[12.7654, 16.8211]; y_{121} = 16.9712\}$$

would be *unsuccessful*, because the future observation $y_{121} = 16.9712$ is not contained in the 95% prediction interval [12.7654, 16.8211]. Since there are an infinite number of possible realizations of the Absorbent Paper Towel sales time series, there is an infinite population of possible prediction interval–future value combinations (where the prediction intervals are based on 95% confidence). The interpretation of "95% confidence" here is that 95% of the prediction interval–future value combinations in this population are successful, while 5% of the prediction interval–future value combinations in this population are unsuccessful. Therefore, before we calculate the 95% prediction interval based on the first 120 observations and before we subsequently observe y_{121}, we can be 95% confident that we will be successful (that is, obtain a 95% prediction interval such that y_{121} is contained in our interval), since 95% of the prediction interval–future value combinations in the population of all such combinations are successful, and since we know that we will obtain one prediction interval–future value combination in this population.

Continuing, since

$$y_{122} = y_{121} + a_{122} - \theta_1 a_{121}$$

it follows that a point forecast of y_{122} made at time origin 120 is

$$\hat{y}_{122}(120) = \hat{y}_{121}(120) + \hat{a}_{122} - \hat{\theta}_1 \hat{a}_{121}$$
$$= 15.8899 + 0 - (-.3545)(0)$$
$$= 15.8899$$

Note that, in making the above point forecast, we have used the previously calculated point forecast $\hat{y}_{121}(120) = 15.8899$ because we have not observed y_{121}, and we have set the point prediction $\hat{a}_{121}$ of a_{121} equal to zero because, since we have not observed y_{121}, we cannot calculate the residual $y_{121} - \hat{y}_{121}(120)$. Furthermore, it can be shown that a 95% prediction interval forecast of y_{122} is

$$[15.8899 - 3.4278, 15.8899 + 3.4278] \quad \text{or} \quad [12.4621, 19.3177]$$

In general, since

$$y_{120+\tau} = y_{120+\tau-1} + a_{120+\tau} - \theta_1 a_{120+\tau-1}$$

it follows that a point forecast of $y_{120+\tau}$ made at time origin 120 is, for $\tau \geqslant 2$

$$\hat{y}_{120+\tau}(120) = \hat{y}_{120+\tau-1}(120) + \hat{a}_{120+\tau} - \hat{\theta}_1\hat{a}_{120+\tau-1}$$

$$= 15.8899 + 0 - (-.3545)(0)$$

$$= 15.8899$$

and the larger τ is, the wider is the 95% prediction interval forecast of $y_{120+\tau}$.

Assume now that the actual observation at time 121 is $y_{121} = 16.1099$. Then, although the only way to update the point estimate of θ_1 is to refit the entire model, we can obtain a new point forecast of y_{122} by using the current point estimate of θ_1. Since the one-period-ahead forecast error (or residual) is

$$y_{121} - \hat{y}_{121}(120) = 16.1099 - 15.8899 = .2200$$

and since

$$y_{122} = y_{121} + a_{122} - \theta_1 a_{121}$$

the point forecast of y_{122} made at time origin 121 is

$$\hat{y}_{122}(121) = y_{121} + \hat{a}_{122} - \hat{\theta}_1\hat{a}_{121}$$

$$= y_{121} + 0 - (-.3545)(y_{121} - \hat{y}_{121}(120))$$

$$= 16.1099 + .3545(.2200)$$

$$= 16.1879$$

Moreover, since

$$y_{123} = y_{122} + a_{123} - \theta_1 a_{122}$$

a point forecast of y_{123} made at time origin 121 is

$$\hat{y}_{123}(121) = \hat{y}_{122}(121) + \hat{a}_{123} - \hat{\theta}_1\hat{a}_{122}$$

$$= 16.1879 + 0 - (-.3545)(0)$$

$$= 16.1879$$

2.5 TENTATIVE IDENTIFICATION OF NONSEASONAL BOX-JENKINS MODELS

2.5.1 The General Nonseasonal Autoregressive–Moving Average Model

The nonseasonal moving average models of orders 1 and 2

$$z_t = a_t - \theta_1 a_{t-1} \quad \text{and} \quad z_t = a_t - \theta_1 a_{t-1} - \theta_2 a_{t-2}$$

are only two of many nonseasonal Box-Jenkins models.

> The *general nonseasonal (Box-Jenkins) autoregressive–moving average model of order* (p, q) is
>
> $$z_t = \delta + \phi_1 z_{t-1} + \phi_2 z_{t-2} + \cdots + \phi_p z_{t-p}$$
> $$+ a_t - \theta_1 a_{t-1} - \theta_2 a_{t-2} - \cdots - \theta_q a_{t-q}$$

This model

1. Utilizes a *constant term* δ (the meaning of which will be explained later).
2. Is called *autoregressive* because it expresses the current time series value z_t as a function of past time series values $z_{t-1}, z_{t-2}, \ldots, z_{t-p}$ (note that $\phi_1, \phi_2, \ldots, \phi_p$ are unknown parameters relating z_t to $z_{t-1}, z_{t-2}, \ldots, z_{t-p}$).
3. Is called *moving average* because, in addition to using the current random shock a_t (which is used by all Box-Jenkins models), it uses the past random shocks $a_{t-1}, a_{t-2}, \ldots, a_{t-q}$ (note that $\theta_1, \theta_2, \ldots, \theta_q$ are unknown parameters relating z_t to $a_{t-1}, a_{t-2}, \ldots, a_{t-q}$). Here, each random shock a_t is a value that is assumed to have been randomly selected from a *normal distribution* that has *mean zero* and a *variance* that is the *same* for each and every time period t. Furthermore, the random shocks $a_1, a_2, a_3, \ldots$ are assumed to be *statistically independent*.

In order to write the above general model in a more compact form, we need to become familiar with the symbol B, which is called the *backshift operator* and which shifts the subscript of a time series observation backward in time by one period. That is,

$$By_t = y_{t-1}; \quad \text{for example,} \quad By_{50} = y_{49}$$

Next, the symbol B^k, which intuitively represents B raised to a power equal to k, shifts the subscript of a time series observation backward in time by k periods. That is,

$$B^k y_t = y_{t-k}; \quad \text{for example,} \quad B^{12} y_{50} = y_{38}$$

Although we will subsequently perform various algebraic manipulations with the backshift operator B, we will not discuss the theory justifying these manipulations. However, it can be shown that the manipulations are legitimate.

Using the backshift operator B, we can express the nonseasonal autoregressive–moving average model of order (p, q)

$$z_t = \delta + \phi_1 z_{t-1} + \phi_2 z_{t-2} + \cdots + \phi_p z_{t-p}$$
$$+ a_t - \theta_1 a_{t-1} - \theta_2 a_{t-2} - \cdots - \theta_q a_{t-q}$$

as follows:

$$z_t - \phi_1 z_{t-1} - \phi_2 z_{t-2} - \cdots - \phi_p z_{t-p} = \delta + a_t - \theta_1 a_{t-1} - \theta_2 a_{t-2}$$
$$- \cdots - \theta_q a_{t-q}$$

or

$$z_t - \phi_1 B z_t - \phi_2 B^2 z_t - \cdots - \phi_p B^p z_t = \delta + a_t - \theta_1 B a_t - \theta_2 B^2 a_t$$
$$- \cdots - \theta_q B^q a_t$$

or

$$(1 - \phi_1 B - \phi_2 B^2 - \cdots - \phi_p B^p)z_t = \delta + (1 - \theta_1 B - \theta_2 B^2 - \cdots - \theta_q B^q)a_t$$

or

$$\phi_p(B)z_t = \delta + \theta_q(B)a_t$$

Here:

1. $\phi_p(B) = (1 - \phi_1 B - \phi_2 B^2 - \cdots - \phi_p B^p)$

    and is called the *nonseasonal autoregressive operator of order p.*

2. $\theta_q(B) = (1 - \theta_1 B - \theta_2 B^2 - \cdots - \theta_q B^q)$

    and is called the *nonseasonal moving average operator of order q.*

Each specific nonseasonal Box-Jenkins model is a special case of the general nonseasonal autoregressive–moving average model of order (p, q)

$$\phi_p(B)z_t = \delta + \theta_q(B)a_t$$

Therefore, in order to tentatively identify a nonseasonal model describing a set of stationary time series values $z_b, z_{b+1}, \ldots, z_n$, we must tentatively determine

1. Whether the constant term δ should be included in the model.
2. Which of the operators $\theta_q(B)$ and $\phi_p(B)$ should be included in the model. That is, in a specific application, should we employ the nonseasonal moving average operator of order q, the nonseasonal autoregressive operator of order p, both operators, or neither operator? (If we determine that an operator should not be included, we set this operator equal to 1.)
3. The order of each operator that is included in the model (that is, the value of p if $\phi_p(B)$ is to be included, and the value of q if $\theta_q(B)$ is to be included).

2.5.2   The Constant Term δ

We will first discuss determining whether the constant term δ should be included in the general nonseasonal model. To do this, we should note that it can be proven that

$$\delta = \mu\phi_p(B)$$

where μ is the true mean of (all possible realizations) of the stationary time series under consideration. For example, then,

1. If we determine that $\phi_p(B)$ should not be included and $\theta_q(B)$ should be included in the general nonseasonal model

    $$\phi_p(B)z_t = \delta + \theta_q(B)a_t$$

then this model becomes the (so-called) *nonseasonal moving average model of order q*:

$$z_t = \delta + \theta_q(B)a_t$$

$$= \delta + (1 - \theta_1 B - \theta_2 B^2 - \cdots - \theta_q B^q)\, a_t$$

$$= \delta + a_t - \theta_1 a_{t-1} - \theta_2 a_{t-2} - \cdots - \theta_q a_{t-q}$$

in which case (since $\phi_p(B) = 1$)

$$\delta = \mu\phi_p(B) = \mu$$

2. If we determine that $\phi_p(B)$ should be included and $\theta_q(B)$ should not be included in the general nonseasonal model

$$\phi_p(B)z_t = \delta + \theta_q(B)a_t$$

then this model becomes the (so-called) *nonseasonal autoregressive model of order p*

$$\phi_p(B)z_t = \delta + a_t$$

$$(1 - \phi_1 B - \phi_2 B^2 - \cdots - \phi_p B^p)\, z_t = \delta + a_t$$

$$z_t - \phi_1 z_{t-1} - \phi_2 z_{t-2} - \cdots - \phi_p z_{t-p} = \delta + a_t$$

or

$$z_t = \delta + \phi_1 z_{t-1} + \phi_2 z_{t-2} + \cdots + \phi_p z_{t-p} + a_t$$

in which case

$$\delta = \mu\phi_p(B)$$

$$= \mu(1 - \phi_1 B - \phi_2 B^2 - \cdots - \phi_p B^p)$$

$$= \mu - \phi_1 B\mu - \phi_2 B^2 \mu - \cdots - \phi_p B^p \mu$$

$$= \mu - \phi_1 \mu - \phi_2 \mu - \cdots - \phi_p \mu$$

$$= \mu(1 - \phi_1 - \phi_2 - \cdots - \phi_p)$$

3. If we determine that both $\phi_p(B)$ and $\theta_q(B)$ should be included in the general nonseasonal model

$$\phi_p(B)z_t = \delta + \theta_q(B)a_t$$

then

$$\delta = \mu\phi_p(B)$$

$$= \mu(1 - \phi_1 B - \phi_2 B^2 - \cdots - \phi_p B^p)$$

$$= \mu(1 - \phi_1 - \phi_2 - \cdots - \phi_p)$$

Now, since the sample mean of the stationary time series values $z_b, z_{b+1}, \ldots, z_n$

$$\bar{z} = \frac{\sum\limits_{t=b}^{n} z_t}{n - b + 1}$$

is one possible point estimate of μ, the true mean of (all possible realizations) of the stationary time series, it follows that, if $\bar{z}$ is statistically different from zero, then it is reasonable to assume that μ does not equal zero and, therefore, to assume that δ does not equal zero. Thus, in such a case it is reasonable to include the constant time δ in the model. On the other hand, if $\bar{z}$ is not statistically different from zero, it is reasonable to assume that μ is equal to (or nearly equal to) zero. In such a case, we can therefore assume that δ is equal to (or nearly equal to) zero, and we do not include the constant term δ in the model. One rough rule of thumb is to decide that $\bar{z}$ is statistically different from zero if the absolute value of

$$\frac{\bar{z}}{s_z/\sqrt{n - b + 1}} \qquad \text{where} \qquad s_z = \left(\frac{\sum\limits_{t=b}^{n} (z_t - \bar{z})^2}{(n - b + 1) - 1} \right)^{1/2}$$

is greater than 2 (this would lead us to include the constant term δ in the general nonseasonal model). Here, the use of this rough rule of thumb yields only "approximately correct" results because the denominator $s_z/\sqrt{n - b + 1}$ in the above statistic is only very approximate (see pp. 193–195 of Box and Jenkins [1976] for a discussion of the theoretically correct denominator). We will discuss a better way to decide whether to include the constant term δ in a Box-Jenkins model in Section 4.3.

If the stationary time series values $z_b, z_{b+1}, \ldots, z_n$ are the original time series values $y_1, y_2, \ldots, y_n$, then the assumption that μ equals zero implies that these original time series values are fluctuating around a zero mean, whereas the assumption that μ does not equal zero implies that these original values are fluctuating around a non-zero mean. If the stationary time series values $z_b, z_{b+1}, \ldots, z_n$ are *differences* of the original time series values $y_1, y_2, \ldots, y_n$, then assuming that μ equals zero is equivalent to assuming that there is no *deterministic trend* (or drift) in the original time series values, whereas assuming that μ does not equal zero is equivalent to assuming that there is a deterministic trend in those original values. Here, the term *deterministic trend* refers to a tendency for the original time series values to *move persistently in a particular direction*. If the constant term

$$\delta = \mu \phi_p(B)$$

is positive, then the deterministic trend is upward, whereas if δ is negative, then the deterministic trend is downward. If a time series does not exhibit a deterministic trend, then any trend (that is, failure of the time series to exhibit an affinity for a central value) is *stochastic*.

EXAMPLE 2.6

We have previously concluded that, for the Absorbent Paper Towel sales time series in Table 2.1, the nonseasonal moving average model of order 1

$$z_t = a_t - \theta_1 a_{t-1}$$
$$= a_t - \theta_1 B a_t$$
$$= (1 - \theta_1 B) a_t$$
$$= \theta_1(B) a_t \qquad \text{where} \qquad \theta_1(B) = (1 - \theta_1 B)$$

is a reasonable tentative model describing the stationary first differences produced by the transformation

$$z_t = y_t - y_{t-1}$$

If we insert a constant term δ to form the model

$$z_t = \delta + \theta_1(B) a_t$$

where $z_t = y_t - y_{t-1}$ and where $\delta = \mu \phi_p(B) = \mu$ (since $\phi_p(B) = 1$), it follows that this model implies that

$$y_t - y_{t-1} = \delta + \theta_1(B) a_t$$

or, equivalently, that

$$y_t = \delta + y_{t-1} + \theta_1(B) a_t$$

This last equation says that y_t will move persistently upward by an amount δ per time period if δ is greater than 0, or y_t will move persistently downward by an amount δ per time period if δ is less than 0. Since δ is a constant, the trend defined by δ is called deterministic.

To determine whether δ should indeed be included in the above model, we calculate the mean and standard deviation of the stationary time series values $z_2 = y_2 - y_1, z_3 = y_3 - y_2, \ldots, z_{120} = y_{120} - y_{119}$ in Table 2.2 to be

$$\bar{z} = \frac{\sum_{t=b}^{n} z_t}{n - b + 1} = \frac{\sum_{t=2}^{120} z_t}{120 - 2 + 1} = .005423$$

and

$$s_z = \left(\frac{\sum_{t=b}^{n} (z_t - \bar{z})^2}{(n - b + 1) - 1} \right)^{1/2} = \left(\frac{\sum_{t=2}^{120} (z_t - \bar{z})^2}{(120 - 2 + 1) - 1} \right)^{1/2} = 1.104$$

and then note that since the absolute value of

$$\frac{\bar{z}}{s_z / \sqrt{n - b + 1}} = \frac{.005423}{1.104 / \sqrt{120 - 2 + 1}} = .0536$$

is less than 2, we conclude that $\bar{z}$ is not statistically different from zero, which implies that we should not include $\delta\ (=\mu)$ in the model.

The fact that we have decided to not include δ in the model

$$z_t = a_t - \theta_1 a_{t-1}$$

implies that this model assumes that the original time series values, which are plotted in Figure 2.1, do not exhibit a deterministic trend. Therefore, since Figure 2.1a shows that the original time series values do not exhibit an affinity for a central value, it follows that the trend exhibited by these nonstationary time series values is stochastic. Moreover, since examination of Figures 2.1a and b indicates that using the transformation

$$z_t = y_t - y_{t-1}$$

has reduced the nonstationary time series values $y_1, y_2, \ldots, y_{120}$ to stationary time series values $z_2 = y_2 - y_1$, $z_3 = y_3 - y_2, \ldots, z_{120} = y_{120} - y_{119}$, we can say that using this transformation has modeled the stochastic trend.

2.5.3 Guidelines for Determining Operators in the General Nonseasonal Model

As stated previously, the general nonseasonal model

$$\phi_p(B)z_t = \delta + \theta_q(B)a_t$$

employs the nonseasonal moving average operator of order q

$$\theta_q(B) = (1 - \theta_1 B - \theta_2 B^2 - \cdots - \theta_q B^q)$$

and the nonseasonal autoregressive operator of order p

$$\phi_p(B) = (1 - \phi_1 B - \phi_2 B^2 - \cdots - \phi_p B^p)$$

Table 2.3 presents five guidelines that show how to use the behaviors of the SAC and the SPAC to determine which (if any) of the operators $\theta_q(B)$ and $\phi_p(B)$ should be included in the general nonseasonal model in a specific application.

Guideline 1: Nonseasonal Moving Average Models

To begin our discussion of these guidelines, recall that if we insert (only) the operator $\theta_q(B)$ into the general nonseasonal model, we obtain the nonseasonal moving average model of order q

$$\begin{aligned}
z_t &= \delta + \theta_q(B)a_t \\
&= \delta + (1 - \theta_1 B - \theta_2 B^2 - \cdots - \theta_q B^q)\, a_t \\
&= \delta + a_t - \theta_1 a_{t-1} - \theta_2 a_{t-2} - \cdots - \theta_q a_{t-q}
\end{aligned}$$

It can be proved that for this model

1. The *TAC* has nonzero autocorrelations at lags 1, 2, . . . , q and has zero autocorrelations at all lags after q (that is, *cuts off* after lag q). Said equivalently,

$$\varrho_k \neq 0 \qquad \text{for } k = 1, 2, \ldots, q$$

$$\varrho_k = 0 \qquad \text{for } k > q \qquad \text{and}$$

2. The *TPAC dies down* (that is, the values $\varrho_{11}, \varrho_{22}, \ldots$ decrease in a steady fashion).

Therefore, if for the time series values $z_b, z_{b+1}, \ldots, z_n$

   i.   The *SAC* has spikes at lags 1, 2, . . . , q and *cuts off* after lag q, and
   ii. The *SPAC dies down*, then
   iii. we should tentatively conclude that the time series values are described by the nonseasonal moving average model of order q

$$z_t = \delta + \theta_q(B)a_t$$

In other words, if the SAC and SPAC display the behavior of (i) and (ii) above, then we should insert (only) the nonseasonal moving average operator of order q

$$\theta_q(B) = (1 - \theta_1 B - \theta_2 B^2 - \cdots - \theta_q B^q)$$

into the general nonseasonal model

$$\phi_p(B)z_t = \delta + \theta_q(B)a_t$$

to form the nonseasonal moving average model of order q

$$z_t = \delta + \theta_q(B)a_t$$

This is stated as Guideline 1 of Table 2.3.

For example, it can be proven that for the nonseasonal moving average model of order 1

$$z_t = \delta + \theta_1(B)a_t$$
$$= \delta + (1 - \theta_1 B)a_t$$
$$= \delta + a_t - \theta_1 a_{t-1}$$

1. The *TAC* has a nonzero autocorrelation at lag 1 and has zero autocorrelations at all lags after lag 1 (that is, *cuts off* after lag 1). Specifically,

$$\varrho_1 = \frac{-\theta_1}{1 + \theta_1^2}$$

and

$$\varrho_k = 0 \qquad \text{for } k > 1$$

2. The *TPAC dies down* (that is, the values $\varrho_{11}, \varrho_{22}, \ldots$ decrease) in a fashion dominated by damped exponential decay.

TABLE 2.3 *Guidelines for Choosing Nonseasonal Operators*

Guideline	Behavior of SAC and SPAC	Nonseasonal Operator(s) to be Used and Determination of the Form of the Operator(s)
1	SAC has spikes at lags 1, 2, . . . , q and *cuts off* after lag q, and the SPAC *dies down*	$\theta_q(B) = (1 - \theta_1 B - \theta_2 B^2 - \cdots - \theta_q B^q)$ = nonseasonal moving average operator of order q. Form determined by spikes in SAC. For example: 1. If SAC has a spike at lag 1 and cuts off after lag 1, and the SPAC dies down in a fashion dominated by damped exponential decay, use the nonseasonal moving average operator of order 1: $$\theta_1(B) = (1 - \theta_1 B)$$ 2. If SAC has spikes at lags 1 and 2 and cuts off after lag 2, and the SPAC dies down according to a mixture of damped exponentials and/or damped sine waves, use the nonseasonal moving average operator of order 2: $$\theta_2(B) = (1 - \theta_1 B - \theta_2 B^2)$$
2	SAC *dies down*, and SPAC has spikes at lags 1, 2, . . . , p and *cuts off* after lag p	$\phi_p(B) = (1 - \phi_1 B - \phi_2 B^2 - \cdots - \phi_p B^p)$ = nonseasonal autoregressive operator of order p. Form determined by spikes in SPAC. For example: 1. If SPAC has a spike at lag 1 and cuts off after lag 1, and the SAC dies down in a damped exponential fashion, use the nonseasonal autoregressive operator of order 1: $$\phi_1(B) = (1 - \phi_1 B)$$ 2. If SPAC has spikes at lags 1 and 2 and cuts off after lag 2, and the SAC dies down according to a mixture of damped exponentials and/or damped sine waves, use the nonseasonal autoregressive operator of order 2: $$\phi_2(B) = (1 - \phi_1 B - \theta_2 B^2)$$

(Table 2.3 continued on p. 64.)

Therefore, referring to Guideline 1 of Table 2.3, if the SAC has a spike at lag 1 and cuts off after lag 1 and the SPAC dies down, then we should insert the nonseasonal moving average operator of order 1

$$\theta_1(B) = (1 - \theta_1 B)$$

into the general nonseasonal model

$$\phi_p(B)z_t = \delta + \theta_q(B)a_t$$

TABLE 2.3   Continued

Guideline	Behavior of SAC and SPAC	Nonseasonal Operator(s) to be Used and Determination of the Form of the Operator(s)
3	SAC has spikes at lags 1, 2, . . . , q and *cuts off* after lag q, and the SPAC has spikes at lags 1, 2, . . . , p and *cuts off* after lag p	$\theta_q(B)$ or $\phi_p(B)$ If $\theta_q(B)$, form determined as discussed in Guideline 1. If $\phi_p(B)$, form determined as discussed in Guideline 2. If SAC cuts off more abruptly than SPAC, use $\theta_q(B)$. If SPAC cuts off more abruptly than SAC, use $\phi_p(B)$. If both the SAC and the SPAC appear to cut off equally abruptly, 1. use $\theta_q(B)$ and not $\phi_p(B)$ in a model and 2. use $\phi_p(B)$ and not $\theta_q(B)$ in a model. Then, choose the operator that yields the best model. Often $\theta_q(B)$ yields the best model. Therefore, one might first consider a model using $\theta_q(B)$.
4	SAC contains small sample autocorrelations (that is, has no spikes) at all lags and SPAC contains small sample autocorrelations (that is, has no spikes) at all lags.	No nonseasonal operator
5	SAC *dies down*, and SPAC *dies down*	Both $\theta_q(B)$ and $\phi_p(B)$ Simple forms of these operators are usually sufficient. For example, if the SAC dies down in a damped exponential fashion, and the SPAC dies down in a fashion dominated by damped exponential decay, it is appropriate to use $\theta_1(B) = (1 - \theta_1 B)$ and $\phi_1(B) = (1 - \phi_1 B)$

to form the nonseasonal moving average model of order 1

$$z_t = \delta + \theta_1(B)a_t$$

As another example, it can be proven that for the nonseasonal moving average model of order 2

$$z_t = \delta + \theta_2(B)a_t$$
$$= \delta + (1 - \theta_1 B - \theta_2 B^2)a_t$$
$$= \delta + a_t - \theta_1 a_{t-1} - \theta_2 a_{t-2}$$

1. The *TAC* has nonzero autocorrelations at lags 1 and 2 and has zero auto-correlations at all lags after lag 2 (that is, *cuts off* after lag 2). Specifically,

$$\varrho_1 = \frac{-\theta_1(1 - \theta_1)}{1 + \theta_1^2 + \theta_2^2}$$

$$\varrho_2 = \frac{-\theta_2}{1 + \theta_1^2 + \theta_2^2}$$

$$\varrho_k = 0 \qquad \text{for } k > 2$$

and

2. The *TPAC dies down* (that is, the values $\varrho_{11}, \varrho_{22}, \dots$ decrease) according to a mixture of damped exponentials and/or damped sine waves.

Therefore, referring to Guideline 1 of Table 2.3, if the SAC has spikes at lags 1 and 2 and cuts off after lag 2 and the SPAC dies down, then we should insert the nonseasonal moving average operator of order 2

$$\theta_2(B) = (1 - \theta_1 B - \theta_2 B^2)$$

into the general nonseasonal model

$$\phi_p(B)z_t = \delta + \theta_q(B)a_t$$

to form the nonseasonal moving average model of order 2

$$z_t = \delta + \theta_2(B)a_t$$

EXAMPLE 2.7

We concluded in Example 2.3 that, for the Absorbent Paper Towel sales time series in Table 2.1, the transformation

$$z_t = y_t - y_{t-1}$$

produces stationary time series values. Recall that the SAC and SPAC of these stationary time series values are given in Figures 2.8 and 2.9. Recall also we concluded that the SAC has a spike at lag 1 and cuts off after lag 1 and that the SPAC dies down fairly quickly in a fashion dominated by damped exponential decay. It follows, by Guideline 1 in Table 2.3, that we should use the nonseasonal moving average operator of order 1

$$\theta_1(B) = (1 - \theta_1 B)$$

Inserting this operator into the general nonseasonal model

$$\phi_p(B)z_t = \delta + \theta_q(B)a_t$$

it follows that a reasonable tentative model describing $z_t = y_t - y_{t-1}$ is

$$z_t = (1 - \theta_1 B)a_t$$

and thus that a reasonable tentative model describing y_t is

$$y_t - y_{t-1} = (1 - \theta_1 B)a_t$$

or

$$y_t = y_{t-1} + a_t - \theta_1 a_{t-1}$$

Note that we have not included the constant term $\delta = \mu\phi_p(B) = \mu$ in the above model, because we concluded in Example 2.6 that $\bar{z} = .005423$ is not statistically different from zero. Also, recall that we have used this model in Example 2.5 to forecast future values of the Absorbent Paper Towel sales time series.

Guideline 2: Nonseasonal Autoregressive Models

To continue our discussion of the guidelines in Table 2.3, next recall that if we insert (only) the operator $\phi_p(B)$ into the general nonseasonal model

$$\phi_p(B)z_t = \delta + \theta_q(B)a_t$$

we obtain the nonseasonal autoregressive model of order p

$$\phi_p(B)z_t = \delta + a_t$$

or

$$(1 - \phi_1 B - \phi_2 B^2 - \cdots - \phi_p B^p)\, z_t = \delta + a_t$$

or

$$z_t = \delta + \phi_1 z_{t-1} + \phi_2 z_{t-2} + \cdots + \phi_p z_{t-p} + a_t$$

It can be proved that for this model

1. The *TAC dies down* (that is, the values $\varrho_1, \varrho_2, \ldots$ decrease in a steady fashion) and
2. The *TPAC* has nonzero partial autocorrelations at lags 1, 2, . . . , p and has zero partial autocorrelations at all lags after lag p (that is, *cuts off* after lag p). Said equivalently,

$$\varrho_{kk} \neq 0 \quad \text{for } k = 1, 2, \ldots, p$$

$$\varrho_{kk} = 0 \quad \text{for } k > p$$

Therefore, if for the time series values $z_b, z_{b+1}, \ldots, z_n$

i.   The *SAC dies down* and
ii. The *SPAC* has spikes at lags 1, 2, . . . , p and *cuts off* after lag p

we should tentatively conclude that the time series values are described by the nonseasonal autoregressive model of order p

$$\phi_p(B)z_t = \delta + a_t$$

In other words, if the SAC and SPAC display the behavior of (i) and (ii) above, then we should insert (only) the nonseasonal autoregressive operator of order p

$$\phi_p(B) = (1 - \phi_1 B - \phi_2 B^2 - \cdots - \phi_p B^p)$$

into the general nonseasonal model

$$\phi_p(B)z_t \; = \; \delta + \theta_q(B)a_t$$

to form the nonseasonal autoregressive model of order p

$$\phi_p(B)z_t \; = \; \delta + a_t$$

This is stated as Guideline 2 of Table 2.3.

For example, it can be proven that for the nonseasonal autoregressive model of order 1

$$\phi_1(B)z_t \; = \; \delta + a_t$$

or

$$(1 - \phi_1 B)z_t \; = \; \delta + a_t$$

or

$$z_t \; = \; \delta + \phi_1 z_{t-1} + a_t$$

1. The *TAC dies down* in a damped exponential fashion. Specifically,

$$\varrho_k \; = \; (\phi_1)^k \qquad \text{for } k \geqslant 1$$

2. The *TPAC* has a nonzero partial autocorrelation at lag 1 and zero partial autocorrelations at all lags after lag 1 (that is, *cuts off* after lag 1).

Therefore, referring to Guideline 2 of Table 2.3, if the SAC dies down and the SPAC has a spike at lag 1 and cuts off after lag 1, then we should insert the nonseasonal autoregressive operator of order 1

$$\phi_1(B) \; = \; (1 - \phi_1 B)$$

into the general nonseasonal model

$$\phi_p(B)z_t \; = \; \delta + \theta_q(B)a_t$$

to form the nonseasonal autoregressive model of order 1

$$\phi_1(B)z_t \; = \; \delta + a_t$$

As another example, it can be proven that for the nonseasonal autoregressive model of order 2

$$\phi_2(B)z_t \; = \; \delta + a_t$$

or

$$(1 - \phi_1 B - \phi_2 B^2)z_t \; = \; \delta + a_t$$

or

$$z_t \; = \; \delta + \phi_1 z_{t-1} + \phi_2 z_{t-2} + a_t$$

1. The *TAC dies down* according to a mixture of damped exponentials and/or damped sine waves. Specifically,

$$\varrho_1 = \frac{\phi_1}{1 - \phi_2}$$

$$\varrho_2 = \frac{\phi_1^2}{1 - \phi_2} + \phi_2$$

$$\varrho_k = \phi_1\varrho_{k-1} + \phi_2\varrho_{k-2} \quad \text{for } k \geqslant 3$$

2. The *TPAC* has nonzero partial autocorrelations at lags 1 and 2 and zero partial autocorrelations at all lags after lag 2 (that is, *cuts off* after lag 2).

Therefore, referring to Guideline 2 of Table 2.3, if the SAC dies down and the SPAC has spikes at lags 1 and 2 and cuts off after lag 2, then we should insert the nonseasonal autoregressive operator of order 2

$$\phi_2(B) = (1 - \phi_1 B - \phi_2 B^2)$$

into the general nonseasonal model

$$\phi_p(B)z_t = \delta + \theta_q(B)a_t$$

to form the nonseasonal autoregressive model of order 2

$$\phi_2(B)z_t = \delta + a_t$$

EXAMPLE 2.8

Chemo, Inc., produces Chemical Product XB-77-5, a product that must have a rather precisely controlled viscosity. In order to develop a control scheme for their production process, Chemo, Inc., needs to develop a forecasting model that will give point forecasts and prediction interval forecasts of the daily viscosity readings of Chemical Product XB-77-5. For the past 95 days, Chemo, Inc., has recorded the daily readings of the viscosity of Chemical Product XB-77-5. The 95 daily readings, $y_1, y_2, \ldots, y_{95}$, are given in Table 2.4 and are plotted in Figure 2.11. It should be noticed from Figure 2.11 that the original values of the time series seem to fluctuate around a constant mean, and therefore seem to be stationary. To more precisely determine whether the original time series values are stationary or nonstationary, consider Figure 2.12, which presents the SAC and SPAC for these original time series values. Since the SAC dies down fairly quickly in a damped sine wave fashion, we conclude that the original time series values are stationary. Noting that the SPAC has spikes at lags 1 and 2 (since the absolute values of the *t*-statistics related to r_{11} and r_{22} are greater than 2) and cuts off after lag 2, it follows by Guideline 2 in Table 2.3 that we should use the nonseasonal autoregressive operator of order 2

$$\phi_2(B) = (1 - \phi_1 B - \phi_2 B^2)$$

TABLE 2.4   *Daily Readings of the Viscosity of Chemical Product XB-77-5*

t	y_t	t	y_t	t	y_t
1	25.0000	33	34.4337	65	32.2754
2	27.0000	34	35.4844	66	33.2214
3	33.5142	35	33.2381	67	34.5786
4	35.4962	36	36.1684	68	32.3448
5	36.9029	37	34.4116	69	31.5316
6	37.8359	38	33.7668	70	37.8044
7	34.2654	39	33.4246	71	36.0536
8	31.8978	40	33.5719	72	35.7297
9	33.7567	41	35.9222	73	36.7991
10	36.6298	42	33.2125	74	34.9502
11	36.3518	43	37.1668	75	33.5246
12	40.0762	44	35.8138	76	35.1012
13	38.0928	45	33.6847	77	35.9774
14	34.5412	46	33.2761	78	38.0977
15	34.8567	47	38.8163	79	33.4598
16	34.5316	48	42.0838	80	32.9278
17	32.3851	49	40.0069	81	36.5121
18	32.6058	50	33.4514	82	37.4243
19	34.8913	51	30.8413	83	35.1550
20	38.2418	52	30.0655	84	34.4797
21	36.8926	53	37.0544	85	33.2898
22	33.8942	54	39.0982	86	33.9252
23	34.1710	55	37.9075	87	36.1036
24	35.4268	56	36.2393	88	36.7351
25	38.5831	57	34.9535	89	35.4576
26	34.6184	58	33.2061	90	37.5924
27	33.9741	59	34.4261	91	34.4895
28	30.2072	60	37.4511	92	39.1692
29	30.5429	61	37.3335	93	35.8242
30	34.8686	62	38.4679	94	32.3875
31	35.8892	63	33.0976	95	31.2846
32	35.2035	64	32.9285		

$$\bar{y} = 34.93$$

Inserting this operator into the general nonseasonal model

$$\phi_p(B)z_t = \delta + \theta_q(B)a_t$$

it follows that a reasonable tentative model describing $z_t = y_t$ is

$$(1 - \phi_1 B - \phi_2 B^2)y_t = \delta + a_t$$

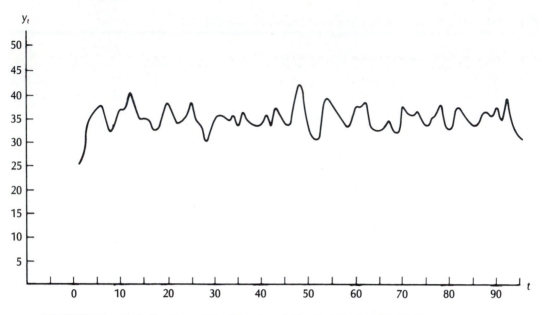

FIGURE 2.11   Daily Readings of the Viscosity of Chemical Product XB-77-5

or

$$y_t = \delta + \phi_1 y_{t-1} + \phi_2 y_{t-2} + a_t$$

Note that we have included $\delta = \mu\phi_2(B) = \mu(1 - \phi_1 B - \phi_2 B^2) = \mu(1 - \phi_1 - \phi_2)$ in the above model because

$$\bar{z} = \bar{y} = \frac{\sum_{t=1}^{95} y_t}{95} = \frac{25.0000 + 27.0000 + \cdots + 31.2846}{95} = 34.93$$

is (obviously) statistically different from zero. Since diagnostic checks (to be presented in Chapter 4) indicate that the above tentative model is a reasonable final model, we should use this model to forecast future daily chemical viscosities.

The least squares point estimates of the parameters δ, ϕ_1, and ϕ_2 in the viscosity model

$$y_t = \delta + \phi_1 y_{t-1} + \phi_2 y_{t-2} + a_t$$

can be computer-calculated to be $\hat{\delta} = 26.34$, $\hat{\phi}_1 = .6857$, and $\hat{\phi}_2 = -.4395$. Therefore, since we have observed 95 values of the chemical viscosity time series, it follows that point forecasts of

$$y_{96} = \delta + \phi_1 y_{95} + \phi_2 y_{94} + a_{96}$$

$$y_{97} = \delta + \phi_1 y_{96} + \phi_2 y_{95} + a_{97}$$

FIGURE 2.12 TSERIES Output of the SAC and SPAC for the Original Values of Chemical Viscosity

(a) The SAC

	VALUE	S.E.	T VALUE	−1.0	0.0	1.0
				+ − − − − − − − − − −	− − − − − − − − −	− − − +
1	0.437	0.103	4.260	I	************	I
2	−0.113	0.121	−0.939	I	***	I
3	−0.342	0.122	−2.809	I	*********	I
4	−0.247	0.131	−1.879	I	*******	I
5	0.001	0.136	0.010	I	*	I
6	0.191	0.136	1.402	I	*****	I
7	0.116	0.139	0.834	I	****	I
8	−0.048	0.140	−0.342	I	**	I
9	−0.131	0.140	−0.937	I	****	I
10	−0.109	0.142	−0.774	I	***	I
11	−0.029	0.142	−0.205	I	*	I
12	0.021	0.142	0.144	I	*	I
13	0.036	0.142	0.254	I	*	I
14	0.025	0.143	0.177	I	*	I
15	−0.019	0.143	−0.131	I	*	I
16	0.005	0.143	0.039	I	*	I
17	0.009	0.143	0.064	I	*	I
18	−0.076	0.143	−0.532	I	**	I
19	−0.145	0.143	−1.010	I	****	I
20	−0.126	0.145	−0.871	I	****	I
21	−0.049	0.146	−0.338	I	**	I
22	0.035	0.146	0.243	I	*	I
23	0.048	0.146	0.331	I	**	I
24	−0.026	0.146	−0.179	I	*	I
				+ − − − − − − −	− − − − −	− − − +

(Part (b), The SPAC follows on p. 72.)

and

$$y_{98} = \delta + \phi_1 y_{97} + \phi_2 y_{96} + a_{98}$$

made at time origin 95 are

$$\hat{y}_{96}(95) = 26.34 + .6857 y_{95} - .4395 y_{94} + \hat{a}_{96}$$

$$= 26.34 + .6857(31.2846) - .4395(32.3875) + 0$$

$$= 33.5576$$

$$\hat{y}_{97}(95) = 26.34 + .6857\hat{y}_{96}(95) - .4395 y_{95} + \hat{a}_{97}$$

$$= 26.34 + .6857(33.5576) - .4395(31.2846) + 0$$

$$= 35.6008$$

FIGURE 2.12   *Continued*

(b)   The SPAC

	VALUE	S.E.	T VALUE	−1.0	0.0	1.0
				+ — — — — — — — — — — — — — — +		
1	0.437	0.103	4.260	I	************	I
2	−0.376	0.103	−3.666	I	**********	I
3	−0.158	0.103	−1.536	I	****	I
4	−0.036	0.103	−0.353	I	*	I
5	0.051	0.103	0.495	I	**	I
6	0.072	0.103	0.700	I	**	I
7	−0.088	0.103	−0.853	I	***	I
8	−0.034	0.103	−0.333	I	*	I
9	−0.016	0.103	−0.157	I	*	I
10	−0.034	0.103	−0.330	I	*	I
11	−0.029	0.103	−0.283	I	*	I
12	−0.048	0.103	−0.464	I	**	I
13	0.008	0.103	0.073	I	*	I
14	0.007	0.103	0.073	I	*	I
15	−0.039	0.103	−0.376	I	*	I
16	0.057	0.103	0.554	I	**	I
17	−0.027	0.103	−0.266	I	*	I
18	−0.118	0.103	−1.147	I	***	I
19	−0.092	0.103	−0.898	I	***	I
20	−0.063	0.103	−0.619	I	**	I
21	−0.045	0.103	−0.439	I	**	I
22	−0.040	0.103	−0.389	I	*	I
23	−0.056	0.103	−0.545	I	**	I
24	−0.068	0.103	−0.665	I	**	I
				+ — — — — — — — — — — — — — — +		

and

$$\hat{y}_{98}(95) = 26.34 + .6857\hat{y}_{97}(95) - .4395\hat{y}_{96(95)} + \hat{a}_{98}$$

$$= 26.34 + .6857(35.6008) - .4395(33.5576) + 0$$

$$= 36.0029$$

Moreover, it can be shown that 95% prediction intervals for y_{96}, y_{97}, and y_{98} are as follows.

for y_{96}:   [33.5576 ± 4.1568]   =   [29.4008,   37.7144]

for y_{97}:   [35.6008 ± 5.0401]   =   [30.5607,   40.6409]

for y_{98}:   [36.0029 ± 5.0417]   =   [30.9612,   41.0446]

Guideline 3: Moving Average or Autoregressive Operators

We next discuss Guideline 3 in Table 2.3. There is no theoretical Box-Jenkins model for which

1. The *TAC* has nonzero autocorrelations at lags 1, 2, . . . , q and has zero autocorrelations at all lags after q (that is, *cuts off* after lag q) and
2. The *TPAC* has nonzero partial autocorrelations at lags 1, 2, . . . , p and has zero partial autocorrelations at all lags after p (that is, *cuts off* after lag p).

However (in practice), sometimes for the time series values $z_b, z_{b+1}, \ldots, z_n$

1. The *SAC* has spikes at lags 1, 2, . . . , q and *cuts off* after lag q and
2. The *SPAC* has spikes at lags 1, 2, . . . , p and *cuts off* after lag p.

If this occurs, experience indicates that (as stated in Example 2.4) we should attempt to determine which of the SAC or SPAC is cutting off more abruptly. If the SAC is cutting off more abruptly, then we should use Guideline 1 in Table 2.3 and insert the nonseasonal moving average operator of order q, $\theta_q(B)$, into the general nonseasonal model to form the nonseasonal moving average model of order q:

$$z_t = \delta + \theta_q(B)a_t$$

If the SPAC is cutting off more abruptly, then we should use Guideline 2 in Table 2.3 and insert the nonseasonal autoregressive operator of order p, $\phi_p(B)$, into the general nonseasonal model to form the nonseasonal autoregressive model of order p:

$$\phi_p(B)z_t = \delta + a_t$$

If the SAC and SPAC appear to cut off equally abruptly, then we should consider both of the above models and use the techniques of Chapter 4 to select the "best" model. Some practitioners feel that any time the SAC has spikes at lags 1, 2, . . . , q and cuts off after lag q, a moving average model is appropriate. That is, these practitioners feel that if the SAC cuts off, the behavior of the SPAC is irrelevant. However, these authors have found that, while the nonseasonal moving average model of order q is often the best model to use when both the SAC and SPAC cut off equally abruptly, occasionally the nonseasonal autoregressive model of order p is a superior model. We, therefore, suggest that both models be considered.

Guideline 4: No Nonseasonal Operator

Next, it can be proven that for the model

$$z_t = \delta + a_t$$

the *TAC has zero autocorrelations* at all lags and the *TPAC has zero partial autocorrelations* at all lags. Therefore, as stated in Guideline 4 of Table 2.3, if the SAC has no spikes at all lags and the SPAC has no spikes at all lags, we should insert neither $\theta_q(B)$ nor $\phi_p(B)$ into the general nonseasonal model

$$\phi_p(B)z_t = \delta + \theta_q(B)a_t$$

and thus form the model

$$z_t = \delta + a_t$$

Guideline 5: Nonseasonal Mixed Autoregressive–Moving Average Models

To complete our discussion of the guidelines in Table 2.3, it can be proven that for the nonseasonal autoregressive–moving average model of order (p, q)

$$\phi_p(B)z_t = \delta + \theta_q(B)a_t$$

which uses *both* of the operators $\phi_p(B)$ and $\theta_q(B)$, the *TAC dies down* and the *TPAC dies down*. Therefore, if, for the time series values $z_b, z_{b+1}, \ldots, z_n$, the SAC dies down and the SPAC dies down, we should use both of the operators $\phi_p(B)$ and $\theta_q(B)$ and thus form the model

$$\phi_p(B)z_t = \delta + \theta_q(B)a_t$$

Usually, when using the above model, simple forms of the operators $\phi_p(B)$ and $\theta_q(B)$ are used. For example, it can be proven that for the nonseasonal autoregressive–moving average model of order (1, 1)

$$\phi_1(B)z_t = \delta + \theta_1(B)a_t$$

or

$$(1 - \phi_1 B)z_t = \delta + (1 - \theta_1 B)a_t$$

or

$$z_t = \delta + \phi_1 z_{t-1} + a_t - \theta_1 a_{t-1}$$

1. The TAC dies down in a damped exponential fashion. Specifically,

$$\varrho_1 = \frac{(1 - \phi_1\theta_1)(\phi_1 - \theta_1)}{1 + \theta_1^2 - 2\theta_1\phi_1}$$

$$\varrho_k = \phi_1\varrho_{k-1} \qquad \text{for } k \geqslant 2$$

   and

2. The TPAC dies down in a fashion dominated by damped exponential decay.

Therefore, if for the stationary time series values $z_b, z_{b+1}, \ldots, z_n$

    i. The SAC dies down in a damped exponential fashion and
    ii. The SPAC dies down in a fashion dominated by damped exponential decay,

then we should insert the nonseasonal autoregressive operator of order 1

$$\phi_1(B) = (1 - \phi_1 B)$$

and the nonseasonal moving average operator of order 1

$$\theta_1(B) = (1 - \theta_1 B)$$

TABLE 2.5   *Specific Nonseasonal Models*

Model	Theoretical Partial Autocorrelation Function (TPAC)	Theoretical Autocorrelation Function (TAC)
First-order moving average $z_t = \delta + (1 - \theta_1 B)a_t$	Dies down in a fashion dominated by damped exponential decay	Cuts off after lag 1; specifically: $\varrho_1 = \dfrac{-\theta_1}{1 + \theta_1^2}$ $\varrho_k = 0 \quad$ for $k > 1$
Second-order moving average $z_t = \delta + (1 - \theta_1 B - \theta_2 B^2)a_t$	Dies down according to a mixture of damped exponentials and/or damped sine waves	Cuts off after lag 2; specifically: $\varrho_1 = \dfrac{-\theta_1(1 - \theta_1)}{1 + \theta_1^2 + \theta_2^2}$ $\varrho_2 = \dfrac{-\theta_2}{1 + \theta_1^2 + \theta_2^2}$ $\varrho_k = 0 \quad$ for $k > 2$
First-order autoregressive $(1 - \phi_1 B)z_t = \delta + a_t$	Cuts off after lag 1	Dies down in a damped exponential fashion; specifically: $\varrho_k = (\phi_1)^k \quad$ for $k \geqslant 1$
Second-order autoregressive $(1 - \phi_1 B - \phi_2 B^2)z_t = \delta + a_t$	Cuts off after lag 2	Dies down according to a mixture of damped exponentials and/or damped sine waves; specifically: $\varrho_1 = \dfrac{\phi_1}{1 - \phi_2}$ $\varrho_2 = \dfrac{\phi_1^2}{1 - \phi_2} + \phi_2$ $\varrho_k = \phi_1 \varrho_{k-1} + \phi_2 \varrho_{k-2}$ for $k \geqslant 3$
Mixed autoregressive–moving average of order (1, 1) $(1 - \phi_1 B)z_t = \delta + (1 - \theta_1 B)a_t$	Dies down in a fashion dominated by damped exponential decay	Dies down in a damped exponential fashion; specifically: $\varrho_1 = \dfrac{(1 - \phi_1 \theta_1)(\phi_1 - \theta_1)}{1 + \theta_1^2 - 2\theta_1 \phi_1}$ $\varrho_k = \phi_1 \varrho_{k-1} \quad$ for $k \geqslant 2$

into the general nonseasonal model

$$\phi_p(B)z_t \ = \ \delta \ + \ \theta_q(B)a_t$$

to form the nonseasonal autoregressive–moving average model of order (1, 1)

$$\phi_1(B)z_t \ = \ \delta \ + \ \theta_1(B)a_t$$

In Table 2.5 we summarize the behaviors of the TAC and TPAC for five of the previously discussed and most useful nonseasonal Box-Jenkins models.

EXERCISES

2.1   Consider Table 1, which gives 90 values $y_1, y_2, \ldots, y_{90}$ of weekly sales, in units of 1000 tubes, of Ultra Shine Toothpaste. The SAC of these values is presented in Figure 1. Next, consider Table 2, which gives the 89 first differences produced by the transformation

$$z_t \ = \ y_t \ - \ y_{t-1}$$

FIGURE 1   TSERIES Output of the SAC for the Original Toothpaste Sales Values

	VALUE	S.E.	T VALUE	−1.0	0.0	1.0
1	0.968	0.105	9.188	I	**************************	I
2	0.937	0.179	5.239	I	************************	I
3	0.904	0.227	3.986	I	***********************	I
4	0.872	0.264	3.304	I	**********************	I
5	0.839	0.294	2.853	I	*********************	I
6	0.807	0.320	2.524	I	********************	I
7	0.774	0.341	2.267	I	********************	I
8	0.741	0.360	2.057	I	*******************	I
9	0.709	0.377	1.879	I	******************	I
10	0.676	0.392	1.726	I	*****************	I
11	0.643	0.404	1.590	I	****************	I
12	0.610	0.415	1.467	I	***************	I
13	0.577	0.425	1.355	I	***************	I
14	0.544	0.434	1.253	I	***************	I
15	0.512	0.441	1.159	I	**************	I
16	0.479	0.448	1.070	I	*************	I
17	0.448	0.454	0.988	I	************	I
18	0.417	0.459	0.910	I	***********	I
19	0.387	0.463	0.836	I	***********	I
20	0.357	0.466	0.765	I	**********	I
21	0.327	0.469	0.697	I	*********	I
22	0.298	0.472	0.631	I	********	I
23	0.269	0.474	0.568	I	*******	I
24	0.241	0.476	0.506	I	*******	I

TABLE 1   *Weekly Sales of Ultra-Shine Toothpaste (in units of 1000 tubes)*

t	y_t	t	y_t	t	y_t
1	235.000	31	551.925	61	846.962
2	239.000	32	557.929	62	853.830
3	244.090	33	564.285	63	860.840
4	252.731	34	572.164	64	871.075
5	264.377	35	582.926	65	877.792
6	277.934	36	595.295	66	881.143
7	286.687	37	607.028	67	884.226
8	295.629	38	617.541	68	890.208
9	310.444	39	622.941	69	894.966
10	325.112	40	633.436	70	901.288
11	336.291	41	647.371	71	913.138
12	344.459	42	658.230	72	922.511
13	355.399	43	670.777	73	930.786
14	367.691	44	685.457	74	941.306
15	384.003	45	690.992	75	950.305
16	398.042	46	693.557	76	952.373
17	412.969	47	700.675	77	960.042
18	422.901	48	712.710	78	968.100
19	434.960	49	726.513	79	972.477
20	445.853	50	736.429	80	977.408
21	455.929	51	743.203	81	977.602
22	465.584	52	751.227	82	979.505
23	477.894	53	764.265	83	982.934
24	491.408	54	777.852	84	985.833
25	507.712	55	791.070	85	991.350
26	517.237	56	805.844	86	996.291
27	524.349	57	815.122	87	1003.100
28	532.104	58	822.905	88	1010.320
29	538.097	59	830.663	89	1018.420
30	544.948	60	839.600	90	1029.480

$$\bar{y} = 674.3$$

The TSERIES output of the SAC and SPAC of these first differences is given in Figure 2, and the mean and standard deviation of these first differences can be calculated to be $\bar{z} = 8.927$ and $s_z = 3.638$.

a. Using the values of r_k in Figure 2, hand calculate s_{r_3} and t_{r_3}.

b. Consider Figure 2. Hand calculate $s_{r_{22}}$ and, using r_{22}, hand calculate $t_{r_{22}}$.

c. Use the above information to rationalize the tentative model

$$(1 - \phi_1 B)z_t = \delta + a_t$$

where

$$z_t = y_t - y_{t-1}$$

TABLE 2   First Differences of the Observations in Table 1

t	$z_t = y_t - y_{t-1}$	t	$z_t = y_t - y_{t-1}$	t	$z_t = y_t - y_{t-1}$
2	4.000	31	6.977	61	7.362
3	5.090	32	6.004	62	6.868
4	8.641	33	6.356	63	7.010
5	11.65	34	7.879	64	10.235
6	13.56	35	10.76	65	6.717
7	8.753	36	12.37	66	3.351
8	8.942	37	11.73	67	3.083
9	14.81	38	10.51	68	5.982
10	14.67	39	5.400	69	4.758
11	11.18	40	10.49	70	6.322
12	8.168	41	13.94	71	11.85
13	10.94	42	10.86	72	9.373
14	12.29	43	12.55	73	8.275
15	16.31	44	14.68	74	10.52
16	14.04	45	5.535	75	8.999
17	14.93	46	2.565	76	2.068
18	9.932	47	7.118	77	7.669
19	12.06	48	12.04	78	8.058
20	10.89	49	13.80	79	4.347
21	10.08	50	9.916	80	4.961
22	9.655	51	6.774	81	.1938
23	12.31	52	8.024	82	1.903
24	13.51	53	13.04	83	3.429
25	16.30	54	13.59	84	2.899
26	9.525	55	13.22	85	5.517
27	7.112	56	14.77	86	4.941
28	7.755	57	9.278	87	6.809
29	5.993	58	7.783	88	7.220
30	6.851	59	7.758	89	8.100
		60	8.937	90	11.06

$$\bar{z} = 8.927$$

d.   The least squares point estimates of the parameters δ and ϕ_1 in the model specified in (c) can be calculated to be $\hat{\delta} = 3.043$ and $\hat{\phi}_1 = .6591$.

1.   Calculate the residual $y_3 - \hat{y}_3$.
2.   Calculate $\hat{y}_{91}(90)$, $\hat{y}_{92}(90)$, and $\hat{y}_{93}(90)$, which are point forecasts of y_{91}, y_{92}, and y_{93} made at time origin 90.
3.   95% prediction interval forecasts of y_{91}, y_{92}, and y_{93} made at time origin 90 can be shown to be of the form

    $[— \pm 5.42]$

    $[— \pm 10.5]$

FIGURE 2 *TSERIES Output of the SAC and SPAC for the Toothpaste Sales Values Obtained by Using the Transformation*
$z_t = y_t - y_{t-1}$

(a) The SAC

	VALUE	S.E.	T VALUE	−1.0	0.0	1.0
				+ − — — — — — — — — — — — — — +		
1	0.643	0.106	6.064	I	****************	I
2	0.321	0.143	2.242	I	*********	I
3	0.246	0.151	1.625	I	*******	I
4	0.238	0.156	1.527	I	*******	I
5	0.256	0.160	1.602	I	*******	I
6	0.262	0.164	1.595	I	*******	I
7	0.168	0.169	0.997	I	*****	I
8	0.090	0.171	0.526	I	***	I
9	0.041	0.171	0.242	I	**	I
10	0.042	0.171	0.247	I	**	I
11	0.045	0.171	0.262	I	**	I
12	0.068	0.171	0.398	I	**	I
13	0.051	0.172	0.298	I	**	I
14	0.037	0.172	0.218	I	*	I
15	0.123	0.172	0.715	I	****	I
16	0.139	0.173	0.801	I	****	I
17	0.084	0.174	0.482	I	***	I
18	0.142	0.175	0.814	I	****	I
19	0.177	0.176	1.003	I	*****	I
20	0.069	0.178	0.386	I	**	I
21	−0.024	0.178	−0.137	I	*	I
22	−0.109	0.178	−0.613	I	***	I
23	−0.114	0.179	−0.636	I	***	I
24	−0.048	0.180	−0.267	I	**	I
				+ — — — — — — — — — — — — — +		

(Part (b), The SPAC follows on p. 80.)

and

$$[— \pm 15.45]$$

Calculate these prediction interval forecasts.

2.2 Consider Example 2.5. Assume that the actual observation at time 122 is $y_{122} = 15.9265$. Calculate $\hat{y}_{123}(122)$ and $\hat{y}_{124}(122)$, which are point forecasts of y_{123} and y_{124} made at time origin 122.

2.3 Assume that for the transformed values

$$z_t = (1 - B)^2 y_t$$

the SAC has spikes at lags 1 and 2 and cuts off after lag 2, and the SPAC dies down. Specify a tentative model describing z_t and then algebraically expand this model to express y_t as a function of a_t, past y_t values, and past a_t values.

FIGURE 2　Continued

(b)　The SPAC

```
        VALUE    S.E.    T VALUE   -1.0                        0.0                       1.0
                                    +  — — — — — — — —  — — — — — — — —  +
 1      0.643    0.106    6.064     I                          ****************           I
 2     -0.157    0.106   -1.478     I                      ****                           I
 3      0.188    0.106    1.771     I                        *****                        I
 4      0.040    0.106    0.379     I                          **                         I
 5      0.122    0.106    1.154     I                        ****                         I
 6      0.057    0.106    0.538     I                          **                         I
 7     -0.089    0.106   -0.841     I                       ***                           I
 8      0.006    0.106    0.054     I                          *                          I
 9     -0.066    0.106   -0.618     I                        **                           I
10      0.033    0.106    0.310     I                          *                          I
11     -0.030    0.106   -0.282     I                          *                          I
12      0.066    0.106    0.624     I                          **                         I
13     -0.031    0.106   -0.296     I                          *                          I
14      0.040    0.106    0.375     I                          **                         I
15      0.170    0.106    1.605     I                        *****                        I
16     -0.058    0.106   -0.544     I                        **                           I
17      0.002    0.106    0.017     I                          *                          I
18      0.145    0.106    1.367     I                        ****                         I
19     -0.006    0.106   -0.057     I                          *                          I
20     -0.159    0.106   -1.497     I                      ****                           I
21     -0.076    0.106   -0.718     I                        **                           I
22     -0.156    0.106   -1.472     I                      ****                           I
23      0.017    0.106    0.165     I                          *                          I
24      0.017    0.106    0.164     I                          *                          I
                                    +  — — — — — — — — — — — — — — — — —  +
```

SEASONAL BOX-JENKINS MODELS AND THEIR TENTATIVE IDENTIFICATION

3.1 INTRODUCTION

In this chapter we will discuss seasonal Box-Jenkins models, the tentative identification of an appropriate seasonal model, and the basic nature of forecasts of future time series values. We will begin in Section 3.2 by presenting a general stationarity transformation and by discussing the analysis of the SAC and SPAC at the seasonal level. Then, in Section 3.3 we will discuss using the SAC and SPAC to tentatively identify models describing seasonal time series. Here we will present several helpful guidelines to aid in the tentative identification process. We also consider several examples of tentative identification. We complete this chapter with Section 3.4, which explains the nature of the forecasts provided by Box-Jenkins models.

3.2 A GENERAL STATIONARITY TRANSFORMATION AND THE ANALYSIS OF THE SAC AND SPAC AT THE SEASONAL LEVEL

We will begin this section by illustrating in the following example that, if the variability of a time series increases as time advances (which implies that the time series is nonstationary with respect to its variance), then we can sometimes stabilize the variance of the time series by using a pre-differencing transformation (such as taking the natural logarithms of the time series values).

EXAMPLE 3.1

Traveler's Rest Inc. operates four hotels in Central City. The analysts in the operating division of the corporation were asked to develop a model that could be used to obtain short-term forecasts (up to 1 year) of the number of occupied rooms in the hotels. These forecasts were needed by various personnel to assist in decision making with regard to hiring additional help during the summer months, ordering materials that have long delivery lead times, budgeting of local advertising expenditures, etc.

The available historical data consisted of the number of occupied rooms during each day for the 15 years from 1973 to 1987. Because it was desired to obtain monthly forecasts, these data were reduced to monthly averages by dividing each monthly total by the number of days in the month. The monthly room averages for 1973 to 1986 are denoted by $y_1, y_2, \ldots, y_{168}$, are given in Table 3.1, and are plotted in Figure 3.1.

At the outset, it was decided to perform all analyses with the data from 1973 to 1986 so that forecasts for 1987 could be used as a check on the validity of the model. Figure 3.1 shows that the monthly room averages follow a strong trend and that they have a seasonal pattern with one major and several minor peaks during the year. It also appears that the amount of seasonal variation is increasing with the level of the time series, indicating that the use of a pre-differencing transformation might be warranted. The natural logarithms of the room averages are denoted by $y_1^*, y_2^*, \ldots, y_{168}^*$ and are plotted in Figure 3.2. Examination of Figures 3.1 and 3.2 indicates that the log transformation has equalized the amount of seasonal variation over the range of the data. Since, however, Figure 3.2 indicates that the natural logarithms do not fluctuate around a constant mean (since they exhibit a strong trend and seasonal variation), the natural logarithms are nonstationary. We will therefore now discuss a general stationary transformation that can be used to transform nonseasonal or seasonal nonstationary time series values into stationary time series values.

To discuss a general stationarity transformation, we will define the *nonseasonal operator* ∇ to be

$$\nabla = 1 - B$$

TABLE 3.1   Monthly Hotel Room Averages for 1973–1986

t	y_t	t	y_t	t	y_t	t	y_t
1	501.	43	785.	85	645.	127	1067.
2	488.	44	830.	86	602.	128	1038.
3	504.	45	645.	87	601.	129	812.
4	578.	46	643.	88	709.	130	790.
5	545.	47	551.	89	706.	131	692.
6	632.	48	606.	90	817.	132	782.
7	728.	49	585.	91	930	133	758
8	725.	50	553.	92	983	134	709.
9	585.	51	576.	93	745.	135	715
10	542.	52	665.	94	735.	136	788.
11	480.	53	656.	95	620.	137	794.
12	530.	54	720.	96	698.	138	893.
13	518.	55	826.	97	665.	139	1046.
14	489.	56	838.	98	626.	140	1075.
15	528.	57	652.	99	649.	141	812.
16	599.	58	661.	100	740.	142	822.
17	572.	59	584.	101	729.	143	714.
18	659.	60	644.	102	824.	144	802.
19	739.	61	623.	103	937.	145	748.
20	758.	62	553.	104	994.	146	731.
21	602.	63	599.	105	781.	147	748.
22	587.	64	657.	106	759	148	827.
23	497.	65	680.	107	643.	149	788.
24	558.	66	759.	108	728.	150	937.
25	555.	67	878.	109	691.	151	1076.
26	523.	68	881.	110	649.	152	1125.
27	532.	69	705.	111	656.	153	840.
28	623.	70	684.	112	735.	154	864.
29	598.	71	577.	113	748.	155	717.
30	683.	72	656.	114	837.	156	813.
31	774.	73	645.	115	995.	157	811.
32	780.	74	593.	116	1040.	158	732.
33	609.	75	617.	117	809.	159	745.
34	604.	76	686.	118	793.	160	844.
35	531.	77	679.	119	692.	161	833.
36	592.	78	773.	120	763.	162	935.
37	578.	79	906.	121	723.	163	1110.
38	543.	80	934.	122	655.	164	1124.
39	565.	81	713.	123	658.	165	868.
40	648.	82	710.	124	761.	166	860.
41	615.	83	600.	125	768.	167	762.
42	697.	84	676.	126	885.	168	877.

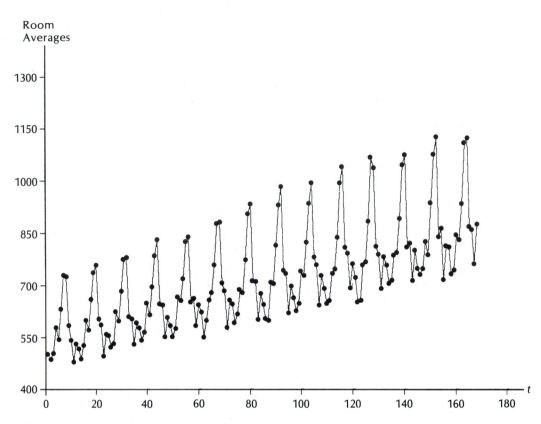

FIGURE 3.1   Monthly Hotel Room Averages, 1973–1986

and we will define the *seasonal operator* ∇_L to be

$$\nabla_L = 1 - B^L$$

where L is the number of seasons in a year ($L = 4$ for quarterly data and $L = 12$ for monthly data). Moreover, we will let y_t^* represent an appropriate pre-differencing transformation (for example, $y_t^* = \ln y_t$ if we need to take the natural logarithms of the original time series values and $y_t^* = y_t$ if we do not need a pre-differencing transformation). Then

The *general stationarity transformation* is

$$z_t = \nabla_L^D \nabla^d y_t^*$$
$$= (1 - B^L)^D (1 - B)^d y_t^*$$

where d is the *degree of non-seasonal differencing* used, and D is the *degree of seasonal differencing* used.

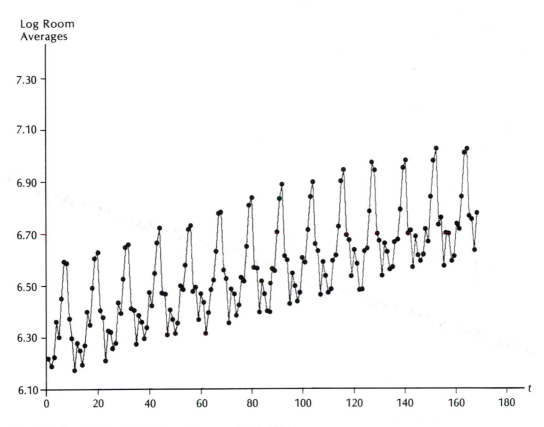

FIGURE 3.2   *Monthly Log Hotel Room Averages, 1973–1986*

If a time series possesses no seasonal variation, then experience with Box-Jenkins modeling indicates that setting $D = 0$ and $d = 0, 1,$ or 2 will usually produce stationary time series values. Setting $D = 0$ implies that

$$z_t = \nabla_L^D \nabla^d y_t^*$$

$$= \nabla_L^0 \nabla^d y_t^*$$

$$= (1 - B^L)^0 (1 - B)^d y_t^*$$

$$= (1 - B)^d y_t^*$$

which means that we are using no seasonal differencing. Setting $d = 0$ implies that

$$z_t = (1 - B)^d y_t^*$$

$$= (1 - B)^0 y_t^*$$

$$= y_t^*$$

which means that, as illustrated in the first column of Table 3.2, the transformed values

TABLE 3.2 Special Cases of the General Stationarity Transformation $z_t = \nabla_L^D \nabla^d y_t^*$

1) $z_t = \nabla_L^0 \nabla^0 y_t^*$
$= (1 - B^L)^0 (1 - B)^0 y_t^*$
$= y_t^*$

$z_1 = y_1^*$
$z_2 = y_2^*$
$\cdot$
$\cdot$
$\cdot$
$z_n = y_n^*$

2) $z_t = \nabla_L^0 \nabla^1 y_t^*$
$= (1 - B^L)^0 (1 - B)^1 y_t^*$
$= y_t^* - y_{t-1}^*$

$z_2 = y_2^* - y_1^*$
$z_3 = y_3^* - y_2^*$
$\cdot$
$\cdot$
$\cdot$
$z_n = y_n^* - y_{n-1}^*$

3) $z_t = \nabla_L^0 \nabla^2 y_t^*$
$= (1 - B^L)^0 (1 - B)^2 y_t^*$
$= y_t^* - 2y_{t-1}^* + y_{t-2}^*$

$z_3 = y_3^* - 2y_2^* + y_1^*$
$z_4 = y_4^* - 2y_3^* + y_2^*$
$\cdot$
$\cdot$
$\cdot$
$z_n = y_n^* - 2y_{n-1}^* + y_{n-2}^*$

4) $z_t = \nabla_L^1 \nabla^0 y_t^*$
$= (1 - B^L)^1 (1 - B)^0 y_t^*$
$= y_t^* - y_{t-L}^*$

$z_{L+1} = y_{L+1}^* - y_1^*$
$z_{L+2} = y_{L+2}^* - y_2^*$
$\cdot$
$\cdot$
$\cdot$
$z_n = y_n^* - y_{n-L}^*$

5) $z_t = \nabla_L^1 \nabla^1 y_t^*$
$= (1 - B^L)^1 (1 - B)^1 y_t^*$
$= y_t^* - y_{t-1}^* - y_{t-L}^* + y_{t-L-1}^*$

$z_{L+2} = y_{L+2}^* - y_{L+1}^* - y_2^* + y_1^*$
$z_{L+3} = y_{L+3}^* - y_{L+2}^* - y_3^* + y_2^*$
$\cdot$
$\cdot$
$\cdot$
$z_n = y_n^* - y_{n-1}^* - y_{n-L}^* + y_{n-L-1}^*$

$z_1, z_2, \ldots, z_n$ are the original pre-differenced values $y_1^*, y_2^*, \ldots, y_n^*$. Setting $d = 1$ implies that

$$
\begin{aligned}
z_t &= (1 - B)^d y_t^* \\
&= (1 - B)^1 y_t^* \\
&= (1 - B) y_t^* \\
&= y_t^* - B y_t^* \\
&= y_t^* - y_{t-1}^*
\end{aligned}
$$

which means that the transformed values are the "first regular differenced" values given in the second column of Table 3.2. Setting $d = 2$ implies that

$$
\begin{aligned}
z_t &= (1 - B)^d y_t^* \\
&= (1 - B)^2 y_t^* \\
&= (1 - 2B + B^2) y_t^* \\
&= y_t^* - 2B y_t^* + B^2 y_t^* \\
&= y_t^* - 2 y_{t-1}^* + y_{t-2}^*
\end{aligned}
$$

which means that the transformed values are the "second regular differenced" values given in the third column of Table 3.2. Therefore, to summarize, if a time series possesses no seasonal variation, then one of the transformations illustrated in columns 1, 2, and 3 of Table 3.2 will usually produce stationary time series values.

If a time series possesses seasonal variation, then experience with Box-Jenkins modeling indicates that setting $D = 0$ or 1 and $d = 0$ or 1 will usually produce stationary time series values. As illustrated above, setting $D = 0$ and $d = 0$ or 1 yields the transformations producing the values given in columns 1 and 2 of Table 3.2. Setting $D = 1$ and $d = 0$ implies that

$$
\begin{aligned}
z_t &= (1 - B^L)^D (1 - B)^d y_t^* \\
&= (1 - B^L)^1 (1 - B)^0 y_t^* \\
&= (1 - B^L) y_t^* \\
&= y_t^* - B^L y_t^* \\
&= y_t^* - y_{t-L}^*
\end{aligned}
$$

which means that the transformed values are the "first seasonal differenced" values given in the fourth column of Table 3.2. Setting $D = 1$ and $d = 1$ implies that

$$
\begin{aligned}
z_t &= (1 - B^L)^D (1 - B)^d y_t^* \\
&= (1 - B^L)^1 (1 - B)^1 y_t^* \\
&= (1 - B - B^L + B^{L+1}) y_t^*
\end{aligned}
$$

$$= y_t^* - By_t^* - B^L y_t^* + B^{L+1} y_t^*$$

$$= y_t^* - y_{t-1}^* - y_{t-L}^* + y_{t-L-1}^*$$

which means that the transformed values are the "first regular differenced and first seasonal differenced" values given in the fifth column of Table 3.2. Therefore, to summarize, if a time series possesses seasonal variation, then one of the transformations illustrated in columns 1, 2, 4, and 5 of Table 3.2 will usually produce stationary time series values.

EXAMPLE 3.2

In Example 3.1 we stated that since Figure 3.2 indicates that the natural logarithms ($y_t^* = \ln y_t$) of the monthly hotel room averages do not fluctuate around a constant mean (since they exhibit a strong trend and seasonal variation), the natural logarithms are nonstationary. We will show in the next example that using the transformation

$$z_t = y_t^* - y_{t-12}^* = \ln y_t - \ln y_{t-12}$$

produces stationary time series values

$$z_{13} = \ln y_{13} - \ln y_1 = \ln 518 - \ln 501 = .0333691$$

$$z_{14} = \ln y_{14} - \ln y_2 = \ln 489 - \ln 488 = .0020471$$

$$z_{15} = \ln y_{15} - \ln y_3 = \ln 528 - \ln 504 = .04652$$

$$z_{16} = \ln y_{16} - \ln y_4 = \ln 599 - \ln 578 = .0356877$$

$$z_{17} = \ln y_{17} - \ln y_5 = \ln 572 - \ln 545 = .0483532$$

$$z_{18} = \ln y_{18} - \ln y_6 = \ln 659 - \ln 632 = .0418341$$

$$z_{19} = \ln y_{19} - \ln y_7 = \ln 739 - \ln 728 = .0149969$$

$$z_{20} = \ln y_{20} - \ln y_8 = \ln 758 - \ln 725 = .0445117$$

$$z_{21} = \ln y_{21} - \ln y_9 = \ln 602 - \ln 585 = .0286456$$

$$z_{22} = \ln y_{22} - \ln y_{10} = \ln 587 - \ln 542 = .0797588$$

$$z_{23} = \ln y_{23} - \ln y_{11} = \ln 497 - \ln 480 = .0348039$$

$$z_{24} = \ln y_{24} - \ln y_{12} = \ln 558 - \ln 530 = .051482$$

.

.

.

$$z_{167} = \ln y_{167} - \ln y_{155} = \ln 762 - \ln 717 = .0608707$$

$$z_{168} = \ln y_{168} - \ln y_{156} = \ln 877 - \ln 813 = .0757759$$

Noting that these values seem (intuitively) to fluctuate around a constant mean, we can say, in summary, that taking the natural logarithms of the hotel room averages has made the logarithms fluctuate with constant variation, and taking the seasonal differences of the natural logarithms has caused the differences to fluctuate around a constant mean.

In general, how do we determine whether a particular stationarity transformation

$$z_t = (1 - B^L)^D (1 - B)^d y_t^*$$

has transformed original time series values $y_1, y_2, \ldots, y_n$ possessing seasonal variation into stationary time series values $z_b, z_{b+1}, \ldots, z_n$? To do this, we examine the behavior of the SAC of the values $z_b, z_{b+1}, \ldots, z_n$ at the *nonseasonal level* and at the *seasonal level*. We (somewhat arbitrarily) define the behavior of the SAC (or SPAC) at the nonseasonal level to be the behavior of this function at lags 1 through $(L - 3)$. For monthly data $(L = 12)$, this is the behavior at lags 1 through 9. For quarterly data $(L = 4)$, this is the behavior at lag 1. However, for quarterly data, we sometimes consider lag 2 and possibly lag 3 to be part of the nonseasonal level. The behaviors displayed by the SAC and SPAC at the nonseasonal level are similar to those described in Chapter 2. Specifically, we define the lags 1, 2, and possibly 3 to be low nonseasonal lags in the SAC, and we say that a spike exists at a low nonseasonal lag k in the SAC if

$$|t_{r_k}| > 1.6$$

Moreover, we define the lags, $4, 5, \ldots, L - 3$ to be higher nonseasonal lags in the SAC, and we say that a spike exists at a higher nonseasonal lag k in the SAC if

$$|t_{r_k}| > 2$$

Furthermore, we say that a spike exists at a nonseasonal lag k $(1, 2, \ldots, L - 3)$ in the SPAC if

$$|t_{r_{kk}}| > 2$$

Next, we define the behavior of the SAC (or SPAC) at the *seasonal level* to be the behavior of this function at lags equal to (or nearly equal to) $L, 2L, 3L$, and $4L$. Furthermore,

1. We define the lags $L, 2L, 3L$, and $4L$ to be *exact seasonal lags*, and we say that a spike exists at an *exact seasonal lag k in the SAC* if

   $$|t_{r_k}| > 1.25$$

   Moreover, we say that a spike exists at an *exact seasonal lag k in the SPAC* if

   $$|t_{r_{kk}}| > 2$$

2. We define the lags $L - 2, L - 1, L + 1, L + 2, 2L - 2, 2L - 1, 2L + 1,$ $2L + 2, 3L - 2, 3L - 1, 3L + 1, 3L + 2, 4L - 2, 4L - 1, 4L + 1,$ and $4L + 2$ to be *near seasonal lags*, and we say that a spike exists at a *near seasonal lag k in the SAC* if

   $$|t_{r_k}| > 1.6$$

Moreover, we say that a spike exists at a *near seasonal lag k in the SPAC* if

$$|t_{r_{kk}}| > 2$$

We say that *the SAC (or SPAC) cuts off after lag k at the seasonal level if there are no spikes at exact seasonal lags or near seasonal lags greater than lag k in this function.* Furthermore, we say that *the SAC (or SPAC) dies down at the seasonal level if this function does not cut off but rather decreases in a steady fashion at the seasonal level.* In general, it can be shown that, if the SAC of the time series values $z_b, z_{b+1}, \ldots, z_n$

1. cuts off fairly quickly or dies down fairly quickly at the *nonseasonal level,* and
2. cuts of fairly quickly or dies down fairly quickly at the *seasonal level*

then these values should be considered *stationary.* Otherwise, these values should be considered nonstationary. Here, if the SAC (or SPAC) cuts off fairly quickly at the nonseasonal level, it will often do so after a lag k which is less than or equal to 2. Moreover, if the SAC (or SPAC) cuts off fairly quickly at the seasonal level, it will often do so after a lag which is less than or equal to $L + 2$.

EXAMPLE 3.3

Reconsider Example 3.1, and recall that the original monthly hotel room averages are given in Table 3.1. Also, recall that, since the plot in Figure 3.1 indicates that the amount of seasonal variation is increasing with the level of this time series, it is appropriate to use the pre-differencing transformation

$$y_t^* = \ln y_t$$

The TSERIES output of the SAC and SPAC obtained by using the transformation

1. $y_t^* = \ln y_t$ is presented in Figure 3.3
2. $z_t = y_t^* - y_{t-1}^*$ is presented in Figure 3.4
3. $z_t = y_t^* - y_{t-12}^*$ is presented in Figure 3.5
4. $z_t = y_t^* - y_{t-1}^* - y_{t-12}^* + y_{t-13}^*$ is presented in Figure 3.6

Examining Figures 3.3 and 3.4 we see that

1. Since the SAC in Figure 3.3 dies down extremely slowly both at the nonseasonal level and at the seasonal level (that is, at lags equal to, or nearly equal to, $L = 12$, $2L = 24$, and $3L = 36$), it follows that the values obtained by using the transformation $y_t^* = \ln y_t$ should be considered nonstationary.
2. Since the SAC in Figure 3.4 dies down slowly at the nonseasonal level and extremely slowly at the seasonal level, the values obtained by using the transformation $z_t = y_t^* - y_{t-1}^*$ should also be considered non-stationary.

We next examine Figure 3.5, which presents the SAC and SPAC of the values obtained by using the transformation $z_t = y_t^* - y_{t-12}^*$. At the nonseasonal level, the SAC has spikes at lags 1, 3, and 5, and the SPAC also has spikes at lags 1, 3, and 5. Although it is somewhat difficult to classify one of these functions as cutting off and one of these functions as dying down at the nonseasonal level, since the spikes at lags 1, 3, and 5 in the SAC might be part of a sinusoidal dying down pattern (see Figure 3.5(a)), while the spikes at lags 1, 3, and 5 in the SPAC appear to be more isolated (see Figure 3.5(b)), we will conclude that, *at the nonseasonal level*, the SPAC has spikes at lags 1, 3, and 5 and cuts off after lag 5 and the SAC dies down fairly quickly (in a fashion dominated by sine wave decay). Next, examining the SAC and SPAC in Figure 3.5 at the *seasonal level* (that is, at lags equal to, or nearly equal to, $L = 12$, $2L = 24$, and $3L = 36$), we conclude that

1. The SAC has a spike at lag 12 (since

   $$|t_{r_{12}}| = 3.511$$

   is greater than 1.25) and cuts off after lag 12 (since there are no spikes at lags equal to, or nearly equal to, 24 and 36), and

2. The SPAC dies down fairly quickly, since the spikes in this function at lags 12, 25 (≈ 24), and 37 (≈ 36) are of decreasing size. Specifically, note that

   $$|t_{r_{12,12}}| = 4.057$$
   $$|t_{r_{25,25}}| = 2.192$$

   and

   $$|t_{r_{37,37}}| = 2.027$$

In the next example we will use the above conclusions to help us identify the time series model that describes the values obtained by using the transformation $z_t = y_t^* - y_{t-12}^*$. For now, it suffices to say that, since the SAC in Figure 3.5a dies down fairly quickly at the nonseasonal level and cuts off fairly quickly (after lag 12) at the seasonal level, the values obtained by using the transformation

$$z_t = y_t^* - y_{t-12}^*$$

should be considered stationary.

Finally, examining Figure 3.6, which presents the SAC of the values obtained by using the transformation

$$z_t = y_t^* - y_{t-1}^* - y_{t-12}^* + y_{t-13}^*$$

we see that this SAC might be interpreted as dying down fairly quickly at the nonseasonal level and cutting off fairly quickly (after lag 18) at the seasonal level. However, the above described dying down behavior at the nonseasonal level and cutting off behavior at the seasonal level do not appear to be as "quick" as the dying down behavior at the nonseasonal level and the cutting off behavior at the seasonal level illustrated in Figure 3.5. This implies that the values obtained by using the transformation

$$z_t = y_t^* - y_{t-1}^* - y_{t-12}^* + y_{t-13}^*$$

FIGURE 3.3   The TSERIES Output of the SAC and SPAC of the Hotel Room Occupancy Values Obtained by Using the
Transformation $y_t^* = \ln y_t$
(a) The SAC

	VALUE	S.E.	T VALUE	−1.0	0.0	1.0
				+ — — — — — — — — — — — —	— — — — —	— — — — — — +
1	0.797	0.077	10.324	I	*********************	I
2	0.600	0.116	5.162	I	****************	I
3	0.391	0.133	2.929	I	***********	I
4	0.271	0.140	1.938	I	********	I
5	0.137	0.143	0.961	I	****	I
6	0.131	0.144	0.913	I	****	I
7	0.134	0.145	0.924	I	****	I
8	0.253	0.145	1.741	I	*******	I
9	0.356	0.148	2.407	I	**********	I
10	0.531	0.153	3.475	I	**************	I
11	0.690	0.164	4.219	I	*****************	I
12	0.854	0.180	4.740	I	**********************	I
13	0.672	0.203	3.314	I	*****************	I
14	0.491	0.216	2.277	I	*************	I
15	0.296	0.222	1.334	I	********	I
16	0.187	0.224	0.832	I	*****	I
17	0.058	0.225	0.259	I	**	I
18	0.047	0.225	0.210	I	**	I
19	0.046	0.226	0.206	I	**	I
20	0.156	0.226	0.690	I	*****	I
21	0.251	0.226	1.108	I	*******	I
22	0.413	0.228	1.814	I	***********	I
23	0.561	0.232	2.416	I	***************	I
24	0.714	0.240	2.971	I	******************	I
25	0.550	0.253	2.178	I	***************	I
26	0.383	0.260	1.475	I	**********	I
27	0.205	0.263	0.781	I	******	I
28	0.100	0.264	0.380	I	***	I
29	−0.020	0.264	−0.075	I	*	I
30	−0.032	0.264	−0.122	I	*	I
31	−0.031	0.264	−0.119	I	*	I
32	0.065	0.264	0.245	I	**	I
33	0.152	0.264	0.575	I	****	I
34	0.302	0.265	1.141	I	********	I
35	0.443	0.267	1.659	I	***********	I
36	0.584	0.271	2.155	I	***************	I
37	0.436	0.279	1.564	I	************	I
38	0.283	0.283	1.002	I	********	I
				+ — — — — — — — — — — — —	— — — — —	— — — — — — +

FIGURE 3.3   Continued

(b) The SPAC

```
                                      −1.0            0.0                1.0
            VALUE     S.E.    T VALUE   + — — — — — — — — — — — — — +
   1        0.797    0.077    10.324    I                 ********************    I
   2       −0.095    0.077    −1.228    I                 ***                     I
   3       −0.157    0.077    −2.040    I                 ****                    I
   4        0.103    0.077     1.335    I                 ***                     I
   5       −0.140    0.077    −1.818    I                 ****                    I
   6        0.229    0.077     2.973    I                 ******                  I
   7        0.017    0.077     0.217    I                 *                       I
   8        0.286    0.077     3.704    I                 ********                I
   9        0.129    0.077     1.668    I                 ****                    I
  10        0.318    0.077     4.120    I                 *********               I
  11        0.398    0.077     5.163    I                 ***********             I
  12        0.466    0.077     6.046    I                 *************           I
  13       −0.653    0.077    −8.464    I     *****************                   I
  14        0.017    0.077     0.217    I                 *                       I
  15        0.030    0.077     0.390    I                 *                       I
  16       −0.073    0.077    −0.941    I                 **                      I
  17       −0.037    0.077    −0.482    I                 *                       I
  18       −0.126    0.077    −1.636    I                 ****                    I
  19       −0.028    0.077    −0.368    I                 *                       I
  20       −0.009    0.077    −0.117    I                 *                       I
  21       −0.014    0.077    −0.188    I                 *                       I
  22        0.076    0.077     0.983    I                 **                      I
  23        0.066    0.077     0.852    I                 **                      I
  24        0.074    0.077     0.959    I                 **                      I
  25       −0.148    0.077    −1.912    I                 ****                    I
  26        0.024    0.077     0.308    I                 *                       I
  27        0.070    0.077     0.913    I                 **                      I
  28       −0.093    0.077    −1.202    I                 ***                     I
  29        0.051    0.077     0.657    I                 **                      I
  30       −0.043    0.077    −0.554    I                 **                      I
  31       −0.004    0.077    −0.056    I                 *                       I
  32       −0.065    0.077    −0.842    I                 **                      I
  33       −0.022    0.077    −0.279    I                 *                       I
  34        0.038    0.077     0.489    I                 *                       I
  35        0.015    0.077     0.190    I                 *                       I
  36       −0.013    0.077    −0.171    I                 *                       I
  37       −0.062    0.077    −0.807    I                 **                      I
  38        0.037    0.077     0.474    I                 *                       I
                                        + — — — — — — — — — — — — — +
```

FIGURE 3.4   The TSERIES Output of the SAC and SPAC of the Hotel Room Occupancy Values Obtained by Using the
Transformation $z_t = y_t^* - y_{t-1}^*$
(a) The SAC

	VALUE	S.E.	T VALUE		−1.0		0.0		1.0
				+	— — — — —	— — —	—	— — — — —	+
1	−0.010	0.077	−0.123	I			*		I
2	0.026	0.077	0.337	I			*		I
3	−0.218	0.077	−2.814	I		******			I
4	0.005	0.081	0.059	I			*		I
5	−0.314	0.081	−3.874	I		********			I
6	0.014	0.088	0.161	I			*		I
7	−0.299	0.088	−3.398	I		********			I
8	0.010	0.094	0.102	I			*		I
9	−0.191	0.094	−2.036	I		*****			I
10	0.027	0.096	0.279	I			*		I
11	−0.007	0.096	−0.075	I			*		I
12	0.904	0.096	9.393	I			************************		I
13	0.001	0.138	0.006	I			*		I
14	0.029	0.138	0.210	I			*		I
15	−0.203	0.138	−1.467	I		******			I
16	0.014	0.140	0.102	I			*		I
17	−0.284	0.140	−2.033	I		********			I
18	0.003	0.143	0.019	I			*		I
19	−0.271	0.143	−1.890	I		*******			I
20	−0.001	0.146	−0.009	I			*		I
21	−0.174	0.146	−1.189	I		*****			I
22	0.016	0.148	0.108	I			*		I
23	0.002	0.148	0.015	I			*		I
24	0.826	0.148	5.594	I			*********************		I
25	0.010	0.173	0.057	I			*		I
26	0.017	0.173	0.097	I			*		I
27	−0.168	0.173	−0.968	I		*****			I
28	0.008	0.174	0.044	I			*		I
29	−0.263	0.174	−1.510	I		*******			I
30	−0.002	0.176	−0.010	I			*		I
31	−0.240	0.176	−1.360	I		******			I
32	−0.012	0.178	−0.069	I			*		I
33	−0.160	0.178	−0.895	I		****			I
34	0.007	0.179	0.039	I			*		I
35	0.006	0.179	0.035	I			*		I
36	0.758	0.179	4.230	I			*******************		I
37	0.012	0.197	0.060	I			*		I
38	0.022	0.197	0.110	I			*		I
				+	— — — — —	— — —	—	— — — — —	+

FIGURE 3.4 Continued

(b) The SPAC

	VALUE	S.E.	T VALUE			
				−1.0	0.0	1.0
				+ — — — — — —	— — — —	— — +
1	−0.010	0.077	−0.123	I	*	I
2	0.026	0.077	0.336	I	*	I
3	−0.218	0.077	−2.812	I	******	I
4	0.001	0.077	0.018	I	*	I
5	−0.318	0.077	−4.114	I	********	I
6	−0.041	0.077	−0.534	I	**	I
7	−0.357	0.077	−4.608	I	*********	I
8	−0.189	0.077	−2.443	I	*****	I
9	−0.381	0.077	−4.918	I	**********	I
10	−0.507	0.077	−6.555	I	*************	I
11	−0.727	0.077	−9.399	I	*******************	I
12	0.560	0.077	7.239	I	***************	I
13	−0.128	0.077	−1.655	I	****	I
14	−0.181	0.077	−2.339	I	*****	I
15	−0.015	0.077	−0.196	I	*	I
16	−0.068	0.077	−0.878	I	**	I
17	0.114	0.077	1.473	I	***	I
18	−0.072	0.077	−0.929	I	**	I
19	0.071	0.077	0.914	I	**	I
20	−0.008	0.077	−0.099	I	*	I
21	−0.025	0.077	−0.320	I	*	I
22	−0.008	0.077	−0.100	I	*	I
23	0.011	0.077	0.138	I	*	I
24	0.086	0.077	1.117	I	***	I
25	−0.009	0.077	−0.112	I	*	I
26	−0.069	0.077	−0.896	I	**	I
27	0.134	0.077	1.737	I	****	I
28	−0.045	0.077	−0.581	I	**	I
29	−0.035	0.077	−0.453	I	*	I
30	0.051	0.077	0.663	I	**	I
31	0.012	0.077	0.151	I	*	I
32	0.035	0.077	0.448	I	*	I
33	−0.052	0.077	−0.675	I	**	I
34	0.032	0.077	0.415	I	*	I
35	−0.008	0.077	−0.103	I	*	I
36	0.009	0.077	0.111	I	*	I
37	−0.028	0.077	−0.366	I	*	I
38	0.077	0.077	0.990	I	**	I
				+ — — — — — —	— — — —	— — +

FIGURE 3.5   The TSERIES Output of the SAC and SPAC of the Hotel Room Occupancy Values Obtained by Using the Transformation $z_t = y_t^* - y_{t-12}^*$
(a) The SAC

	VALUE	S.E.	T VALUE			
				−1.0	0.0	1.0
				+ — — — — — — — — —	— — — — —	— +
1	0.193	0.080	2.415	I	******	I
2	0.024	0.083	0.294	I	*	I
3	−0.244	0.083	−2.941	I	*******	I
4	−0.152	0.088	−1.731	I	****	I
5	−0.212	0.089	−2.376	I	******	I
6	0.092	0.092	0.992	I	***	I
7	0.111	0.093	1.190	I	***	I
8	0.129	0.094	1.373	I	****	I
9	0.062	0.095	0.651	I	**	I
10	0.079	0.095	0.830	I	***	I
11	−0.078	0.096	−0.818	I	**	I
12	−0.337	0.096	−3.511	I	*********	I
13	−0.011	0.103	−0.107	I	*	I
14	0.112	0.103	1.088	I	***	I
15	0.149	0.104	1.433	I	****	I
16	0.102	0.105	0.963	I	***	I
17	0.088	0.106	0.829	I	***	I
18	−0.144	0.107	−1.354	I	****	I
19	0.059	0.108	0.546	I	**	I
20	0.037	0.108	0.339	I	*	I
21	0.087	0.108	0.807	I	***	I
22	−0.055	0.109	−0.511	I	**	I
23	0.010	0.109	0.088	I	*	I
24	−0.051	0.109	−0.472	I	**	I
25	0.060	0.109	0.552	I	**	I
26	−0.041	0.109	−0.374	I	**	I
27	0.096	0.109	0.881	I	***	I
28	−0.015	0.110	−0.139	I	*	I
29	0.007	0.110	0.062	I	*	I
30	0.022	0.110	0.201	I	*	I
31	−0.013	0.110	−0.122	I	*	I
32	−0.123	0.110	−1.122	I	****	I
33	0.028	0.111	0.256	I	*	I
34	0.013	0.111	0.122	I	*	I
35	0.109	0.111	0.988	I	***	I
36	0.047	0.111	0.423	I	**	I
37	0.054	0.112	0.481	I	**	I
38	0.018	0.112	0.159	I	*	I
				+ — — — — — — — — —	— — — — —	— +

FIGURE 3.5   Continued

(b) The SPAC

	VALUE	S.E.	T VALUE		−1.0	0.0	1.0
				+	— — — — — — — —	— — — — — —	— +
1	0.193	0.080	2.415	I		******	I
2	−0.014	0.080	−0.169	I		*	I
3	−0.256	0.080	−3.198	I		*******	I
4	−0.062	0.080	−0.777	I		**	I
5	−0.176	0.080	−2.199	I		*****	I
6	0.119	0.080	1.481	I		****	I
7	0.042	0.080	0.528	I		**	I
8	0.002	0.080	0.029	I		*	I
9	0.056	0.080	0.701	I		**	I
10	0.081	0.080	1.007	I		***	I
11	−0.035	0.080	−0.437	I		*	I
12	−0.325	0.080	−4.057	I		*********	I
13	0.190	0.080	2.373	I		*****	I
14	0.126	0.080	1.572	I		****	I
15	−0.036	0.080	−0.449	I		*	I
16	0.018	0.080	0.221	I		*	I
17	0.009	0.080	0.116	I		*	I
18	−0.038	0.080	−0.471	I		*	I
19	0.227	0.080	2.830	I		******	I
20	0.040	0.080	0.496	I		**	I
21	0.010	0.080	0.122	I		*	I
22	0.018	0.080	0.220	I		*	I
23	−0.035	0.080	−0.436	I		*	I
24	−0.121	0.080	−1.505	I		****	I
25	0.175	0.080	2.192	I		*****	I
26	−0.009	0.080	−0.116	I		*	I
27	0.038	0.080	0.474	I		*	I
28	0.026	0.080	0.324	I		*	I
29	−0.082	0.080	−1.029	I		***	I
30	−0.002	0.080	−0.023	I		*	I
31	0.105	0.080	1.306	I		***	I
32	−0.119	0.080	−1.485	I		***	I
33	0.077	0.080	0.956	I		**	I
34	−0.024	0.080	−0.299	I		*	I
35	0.002	0.080	0.021	I		*	I
36	−0.053	0.080	−0.665	I		**	I
37	0.162	0.080	2.027	I		*****	I
38	0.019	0.080	0.238	I		*	I
				+	— — — — — — — —	— — — — — —	— +

FIGURE 3.6   The TSERIES Output of the SAC and SPAC of the Hotel Room Occupancy Values Obtained by Using the Transformation $z_t = y_t^* - y_{t-1}^* - y_{t-12}^* + y_{t-13}^*$

(a) The SAC

	VALUE	S.E.	T VALUE	−1.0	0.0	1.0
				+ — — — — — — — — — — — +		
1	−0.396	0.080	−4.930	I	**********	I
2	0.084	0.092	0.910	I	***	I
3	−0.239	0.093	−2.583	I	******	I
4	0.107	0.096	1.111	I	***	I
5	−0.239	0.097	−2.458	I	******	I
6	0.190	0.101	1.878	I	*****	I
7	−0.018	0.103	−0.176	I	*	I
8	0.065	0.103	0.629	I	**	I
9	−0.045	0.103	−0.435	I	**	I
10	0.108	0.104	1.043	I	***	I
11	0.038	0.104	0.365	I	*	I
12	−0.346	0.104	−3.311	I	*********	I
13	0.130	0.112	1.168	I	****	I
14	0.040	0.113	0.354	I	**	I
15	0.058	0.113	0.515	I	**	I
16	−0.025	0.113	−0.223	I	*	I
17	0.143	0.113	1.267	I	****	I
18	−0.280	0.114	−2.456	I	********	I
19	0.160	0.118	1.349	I	*****	I
20	−0.064	0.120	−0.534	I	**	I
21	0.122	0.120	1.020	I	****	I
22	−0.126	0.121	−1.042	I	****	I
23	0.093	0.122	0.768	I	***	I
24	−0.121	0.122	−0.988	I	****	I
25	0.131	0.123	1.070	I	****	I
26	−0.152	0.124	−1.230	I	****	I
27	0.169	0.125	1.353	I	*****	I
28	−0.095	0.126	−0.753	I	***	I
29	0.022	0.127	0.174	I	*	I
30	0.022	0.127	0.177	I	*	I
31	0.039	0.127	0.306	I	**	I
32	−0.165	0.127	−1.296	I	*****	I
33	0.089	0.128	0.692	I	***	I
34	−0.068	0.129	−0.531	I	**	I
35	0.110	0.129	0.850	I	***	I
36	−0.036	0.130	−0.277	I	*	I
37	0.035	0.130	0.266	I	*	I
38	0.023	0.130	0.178	I	*	I
				+ — — — — — — — — — — — +		

FIGURE 3.6   Continued

(b) The SPAC

	VALUE	S.E.	T VALUE	−1.0	0.0	1.0
				+ — — — — —	— — — — —	— — — — +
1	−0.396	0.080	−4.930	I	**********	I
2	−0.087	0.080	−1.079	I	***	I
3	−0.284	0.080	−3.532	I	********	I
4	−0.125	0.080	−1.561	I	****	I
5	−0.348	0.080	−4.335	I	*********	I
6	−0.172	0.080	−2.138	I	*****	I
7	−0.102	0.080	−1.264	I	***	I
8	−0.125	0.080	−1.561	I	****	I
9	−0.088	0.080	−1.099	I	***	I
10	0.032	0.080	0.394	I	*	I
11	0.234	0.080	2.911	I	*******	I
12	−0.287	0.080	−3.568	I	********	I
13	−0.130	0.080	−1.617	I	****	I
14	0.032	0.080	0.398	I	*	I
15	−0.030	0.080	−0.378	I	*	I
16	−0.021	0.080	−0.266	I	*	I
17	0.028	0.080	0.347	I	*	I
18	−0.208	0.080	−2.592	I	******	I
19	0.018	0.080	0.221	I	*	I
20	0.029	0.080	0.359	I	*	I
21	0.025	0.080	0.308	I	*	I
22	0.083	0.080	1.031	I	***	I
23	0.099	0.080	1.235	I	***	I
24	−0.173	0.080	−2.150	I	*****	I
25	0.039	0.080	0.480	I	**	I
26	−0.024	0.080	−0.304	I	*	I
27	0.028	0.080	0.345	I	*	I
28	0.101	0.080	1.261	I	***	I
29	−0.001	0.080	−0.014	I	*	I
30	−0.067	0.080	−0.834	I	**	I
31	0.127	0.080	1.580	I	****	I
32	−0.086	0.080	−1.071	I	***	I
33	−0.006	0.080	−0.069	I	*	I
34	−0.031	0.080	−0.385	I	*	I
35	0.028	0.080	0.348	I	*	I
36	−0.163	0.080	−2.026	I	*****	I
37	0.001	0.080	0.016	I	*	I
38	0.052	0.080	0.644	I	**	I
				+ — — — — —	— — — — —	— — — — +

should probably be considered "less stationary" than the values obtained by using the transformation

$$z_t = y_t^* - y_{t-12}^*$$

3.3 TENTATIVE IDENTIFICATION OF A SEASONAL BOX-JENKINS MODEL

3.3.1 The General Box-Jenkins Model

Once we have used the stationarity transformation

$$z_t = (1 - B^L)^D (1 - B)^d y_t^*$$

to transform time series values $y_1, y_2, \ldots, y_n$ possessing seasonal variation into stationary time series values $z_b, z_{b+1}, \ldots, z_n$, we use the SAC and SPAC of the values $z_b, z_{b+1}, \ldots, z_n$ to tentatively identify a Box-Jenkins model describing these values. Each (seasonal or nonseasonal) Box-Jenkins model is a special case (or a slight modification) of *the general Box-Jenkins model of order* (p, P, q, Q), which is

$$\phi_p(B)\,\phi_P(B^L)z_t = \delta + \theta_q(B)\theta_Q(B^L)a_t$$

Here

1. $\phi_p(B) = (1 - \phi_1 B - \phi_2 B^2 - \cdots - \phi_p B^p)$

   and is the (previously discussed) *nonseasonal autoregressive operator of order p.*

2. $\phi_P(B^L) = (1 - \phi_{1,L} B^L - \phi_{2,L} B^{2L} - \cdots - \phi_{P,L} B^{PL})$

   and is called the *seasonal autoregressive operator of order P.*

3. $\theta_q(B) = (1 - \theta_1 B - \theta_2 B^2 - \cdots - \theta_q B^q)$

   and is the (previously discussed) *nonseasonal moving average operator of order q.*

4. $\theta_Q(B^L) = (1 - \theta_{1,L} B^L - \theta_{2,L} B^{2L} - \cdots - \theta_{Q,L} B^{QL})$

   and is called the *seasonal moving average operator of order Q.*

5. $\delta = \mu\phi_p(B)\phi_P(B^L)$

   and is called a *constant term,* where μ is the true mean of the stationary time series being modeled.

6. $\phi_1, \phi_2, \ldots, \phi_p; \phi_{1,L}, \phi_{2,L}, \ldots, \phi_{P,L}; \theta_1, \theta_2, \ldots, \theta_q; \theta_{1,L}, \theta_{2,L}, \ldots, \theta_{Q,L};$ and δ are *unknown parameters* which must be estimated from sample data.
7. $a_t, a_{t-1}, \ldots$ are *random shocks* that are assumed to be statistically independent of each other; each is assumed to have been randomly selected from a *normal distribution* that has *mean zero* and a *variance* that is the *same* for each and every time period t.

3.3.2   Guidelines for Identifying Seasonal Models

Identification of the particular special form of the general Box-Jenkins model of order (p, P, q, Q) that describes a particular stationary time series $z_b, z_{b+1}, \ldots, z_n$ involves determining

1. Whether the constant term δ should be included in the model,
2. Which of the operators $\phi_p(B)$, $\phi_P(B^L)$, $\theta_q(B)$, and $\theta_Q(B^L)$ should be included in the model, and
3. The order of each operator that is included in the model.

Letting

$$\bar{z} = \frac{\sum\limits_{t=b}^{n} z_t}{n - b + 1} \quad \text{and} \quad s_z = \left(\frac{\sum\limits_{t=b}^{n} (z_t - \bar{z})^2}{(n - b + 1) - 1} \right)^{1/2}$$

denote the mean and standard deviation of the time series values $z_b, z_{b+1}, \ldots, z_n$, it follows (as in Chapter 2) that one very approximate procedure is to decide that $\bar{z}$ is statistically different from zero and thus to include the constant term

$$\delta = \mu \phi_p(B) \phi_P(B^L)$$

in the general Box-Jenkins model if the absolute value of

$$\frac{\bar{z}}{s_z / \sqrt{n - b + 1}}$$

is greater than 2. Moreover, in order to determine which of the operators $\phi_p(B)$, $\phi_P(B^L)$, $\theta_q(B)$, and $\theta_Q(B^L)$ should be included in the general Box-Jenkins model, we

1. Use the behaviors of the SAC and SPAC of the values $z_b, z_{b+1}, \ldots, z_n$ at the *nonseasonal level* to determine which (if any) of the nonseasonal moving average operator of order q

$$\theta_q(B) = (1 - \theta_1 B - \theta_2 B^2 - \cdots - \theta_q B^q)$$

and the nonseasonal autoregressive operator of order p

$$\phi_p(B) = (1 - \phi_1 B - \phi_2 B^2 - \cdots - \phi_p B^p)$$

should be utilized.

TABLE 3.3 Guidelines For Choosing Seasonal Operators

Guideline	Seasonal Behavior of SAC and SPAC	Seasonal Operator(s) to be Used and Determination of the Form of the Operator(s)
6	SAC has spikes at lags L, $2L$, . . . , QL and *cuts off* after lag QL, and the SPAC *dies down*.	$\theta_Q(B^L) = (1 - \theta_{1,L}B^L - \theta_{2,L}B^{2L} - \cdots - \theta_{Q,L}B^{QL})$ $\quad = $ seasonal moving average operator of order Q Form determined by spikes in SAC at seasonal level. For example, if SAC has a spike at lag L and cuts off after lag L, use the seasonal moving average operator of order 1 $\theta_1(B^L) = (1 - \theta_{1,L}B^L)$
7	SAC *dies down*, and SPAC has spikes at lags L, $2L$, . . . , PL and *cuts off* after lag PL	$\phi_P(B^L) = (1 - \phi_{1,L}B^L - \phi_{2,L}B^{2L} - \cdots - \phi_{P,L}B^{PL})$ $\quad = $ seasonal autoregressive operator of order P Form determined by spikes in SPAC at seasonal level. For example, if SPAC has a spike at lag L and cuts off after lag L, use the seasonal autoregressive operator of order 1 $\phi_1(B^L) = (1 - \phi_{1,L}B^L)$
8	SAC has spikes at lags L, $2L$, . . . , QL and *cuts off* after lag QL, and the SPAC has spikes at lags L, $2L$, . . . , PL and *cuts off* after lag PL	$\theta_Q(B^L) \quad$ or $\quad \phi_P(B^L)$ If $\theta_Q(B^L)$, form determined as discussed in Guideline 6. If $\phi_P(B^L)$, form determined as discussed in Guideline 7. If SAC cuts off more abruptly at seasonal level than SPAC, use $\theta_Q(B^L)$. If SPAC cuts off more abruptly at seasonal level than SAC, use $\phi_P(B^L)$. If both the SAC and the SPAC appear to cut off equally abruptly at the seasonal level, 1. Use $\theta_Q(B^L)$ and not $\phi_P(B^L)$ in a model and 2. Use $\phi_P(B^L)$ and not $\theta_Q(B^L)$ in a model Then, choose the operator that yields the best model. Often $\theta_Q(B^L)$ yields the best model. Therefore, one might "first" consider a model using $\theta_Q(B^L)$.
9	SAC contains small sample auto-correlations (that is, has no spikes) at all seasonal lags and SPAC contains small sample partial autocorrelations (that is, has no spikes) at all seasonal lags	No seasonal operator
10	SAC *dies down* fairly quickly at the seasonal level, and SPAC *dies down* fairly quickly at the seasonal level	Both $\theta_Q(B^L)$ and $\phi_P(B^L)$ Simple forms of these operators are usually sufficient. For example, it might be appropriate to use $\theta_1(B^L) = (1 - \theta_{1,L}B^L)$ and $\phi_1(B^L) = (1 - \phi_{1,L}B^L)$

2.   Use the behaviors of the SAC and SPAC of the values z_b, z_{b+1}, . . . , z_n at the *seasonal level* to determine which (if any) of the seasonal moving average operator of order Q

$$\theta_Q(B^L) = (1 - \theta_{1,L}B^L - \theta_{2,L}B^{2L} - \cdots - \theta_{Q,L}B^{QL})$$

and the seasonal autoregressive operator of order P

$$\phi_P(B^L) = (1 - \phi_{1,L}B^L - \phi_{2,L}B^{2L} - \cdots - \phi_{P,L}B^{PL})$$

should be utilized.

In this regard, recall that Table 2.3 presents five guidelines describing nonseasonal behaviors of the SAC and SPAC and the nonseasonal operator(s) that should be used if each of these nonseasonal behaviors is observed. Moreover, Table 3.3 presents five guidelines describing seasonal behaviors of the SAC and SPAC and the seasonal operator(s) that should be used if each of these seasonal behaviors is observed. It should be noted that, whereas the seasonal behaviors described in Guidelines 6 through 9 occur frequently in practice, the seasonal behavior described in Guideline 10 occurs rarely in practice. For this reason, we *rarely* use *both* the seasonal moving average operator of order Q *and* the seasonal autoregressive operator of order P in the same model.

Once we have tentatively identified an appropriate special form of the general Box-Jenkins model

$$\phi_p(B)\phi_P(B^L)z_t = \delta + \theta_q(B)\theta_Q(B^L)a_t$$

we insert the appropriate stationarity transformation

$$z_t = \nabla_L^D \nabla^d y_t^*$$

into this model to obtain the model

$$\phi_p(B)\phi_P(B^L)\nabla_L^D \nabla^d y_t^* = \delta + \theta_q(B)\theta_Q(B^L)a_t$$

3.3.3   Examples

We will now present three examples of using Guidelines 6 through 10 in Table 3.3 to help us to tentatively identify Box-Jenkins seasonal models, and then we will discuss the rationale behind Guidelines 6 through 10.

EXAMPLE 3.4   We concluded in Example 3.3 that, for the monthly hotel room averages in Table 3.1, the transformation

$$z_t = y_t^* - y_{t-12}^*$$

where

$$y_t^* = \ln y_t$$

produces stationary time series values. As previously discussed, the SAC and SPAC of these stationary time series values are presented in Figure 3.5.

1. At the *nonseasonal level*, the SAC dies down fairly quickly, and the SPAC has spikes at lags 1, 3, and 5 and cuts off after lag 5. It follows, by Guideline 2 in Table 2.3, that we should use some form of $\phi_p(B)$, the nonseasonal autoregressive operator order of p. Since the SPAC has spikes at lags 1, 3, and 5 and cuts off after lag 5 at the nonseasonal level, it might be appropriate to use one of the forms

   Form 1: $(1 - \phi_1 B - \phi_3 B^3 - \phi_5 B^5)$

   Form 2: $(1 - \phi_1 B - \phi_3 B^3)$

   Form 3: $(1 - \phi_1 B - \phi_2 B^2 - \phi_3 B^3)$

   of $\phi_p(B)$, the nonseasonal autoregressive operator. While Form 1 is directly suggested by the spikes at lags 1, 3, and 5, it follows, since use of the term $\phi_5 B^5$ would lead to an unusual model, that Forms 2 and 3 are also reasonable.

2. At the *seasonal level*, the SAC has a spike at lag 12 and cuts off after lag 12, and the SPAC dies down fairly quickly. It follows, by Guideline 6 in Table 3.3, that we should use some form of $\theta_Q(B^L)$, the seasonal moving average operator of order Q. Since the SAC has a spike at lag 12 and cuts off after lag 12, it seems appropriate (from Guideline 6) to use the seasonal moving average operator of order 1

$$\theta_1(B^{12}) = (1 - \theta_{1,12} B^{12})$$

By inserting each of the nonseasonal operators

$$(1 - \phi_1 B - \phi_3 B^3 - \phi_5 B^5)$$

$$(1 - \phi_1 B - \phi_3 B^3)$$

and

$$(1 - \phi_1 B - \phi_2 B^2 - \phi_3 B^3)$$

and the seasonal operator

$$(1 - \theta_{1,12} B^{12})$$

into the general model

$$\phi_p(B)\phi_P(B^L)z_t = \delta + \theta_q(B)\theta_Q(B^L)a_t$$

we are led to consider the tentative models

$$(1 - \phi_1 B - \phi_3 B^3 - \phi_5 B^5)z_t = \delta + (1 - \theta_{1,12} B^{12})a_t$$

$$(1 - \phi_1 B - \phi_3 B^3)z_t = \delta + (1 - \theta_{1,12} B^{12})a_t$$

and

$$(1 - \phi_1 B - \phi_2 B^2 - \phi_3 B^3)z_t = \delta + (1 - \theta_{1,12} B^{12})a_t$$

Note that we have included

$$\delta = \mu \phi_p(B)\phi_P(B^L)$$

in the above models because the mean and standard deviation of the values

$$z_{13} = y^*_{13} - y^*_1$$

$$z_{14} = y^*_{14} - y^*_2$$

.

.

.

$$z_{168} = y^*_{168} - y^*_{156}$$

produced by the transformation

$$z_t = y^*_t - y^*_{t-12}$$

are

$$\bar{z} = \frac{\sum\limits_{t=b}^{n} z_t}{n - b + 1} = \frac{\sum\limits_{t=13}^{168} z_t}{168 - 13 + 1} = .033$$

and

$$s_z = \left(\frac{\sum\limits_{t=b}^{n} (z_t - \bar{z})^2}{(n - b + 1) - 1} \right)^{1/2} = \left(\frac{\sum\limits_{t=13}^{168} (z_t - \bar{z})^2}{(168 - 13 + 1) - 1} \right)^{1/2} = .02319$$

which imply that the absolute value of

$$\frac{\bar{z}}{s_z / \sqrt{n - b + 1}} = \frac{.033}{.02319 / \sqrt{168 - 13 + 1}} = 17.7951$$

is greater than 2, which implies that $\bar{z}$ is statistically different from zero. Further analysis to be carried out in Chapter 4 indicates that the last of the above models is probably the most appropriate model. Since

$$z_t = y^*_t - y^*_{t-12}$$

where

$$y^*_t = \ln y_t$$

it follows that

$$(1 - \phi_1 B - \phi_2 B^2 - \phi_3 B^3)(y^*_t - y^*_{t-12}) = \delta + (1 - \theta_{1,12} B^{12})a_t$$

or

$$y_t^* - \phi_1 B y_t^* - \phi_2 B^2 y_t^* - \phi_3 B^3 y_t^* - y_{t-12}^* + \phi_1 B y_{t-12}^* + \phi_2 B^2 y_{t-12}^* + \phi_3 B^3 y_{t-12}^*$$
$$= \delta + a_t - \theta_{1,12} B^{12} a_t$$

or

$$y_t^* = \delta + y_{t-12}^* + \phi_1(y_{t-1}^* - y_{t-13}^*) + \phi_2(y_{t-2}^* - y_{t-14}^*) + \phi_3(y_{t-3}^* - y_{t-15}^*)$$
$$- \theta_{1,12} a_{t-12} + a_t$$

adequately describes $y_t^* = \ln y_t$.

The least squares point estimates of δ, ϕ_1, ϕ_2, ϕ_3, and $\theta_{1,12}$ can be calculated to be $\hat{\delta} = .0258$, $\hat{\phi}_1 = .2922$, $\hat{\phi}_2 = .1674$, $\hat{\phi}_3 = -.2408$ and $\hat{\theta}_{1,12} = .5917$. Since we have analyzed 168 values of the monthly hotel room averages time series, we will assume that, at time origin 168, we wish to forecast future values of the time series. Since

$$y_t^* = \delta + y_{t-12}^* + \phi_1(y_{t-1}^* - y_{t-13}^*) + \phi_2(y_{t-2}^* - y_{t-14}^*)$$
$$+ \phi_3(y_{t-3}^* - y_{t-15}^*) - \theta_{1,12} a_{t-12} + a_t$$

it follows that a point forecast of y_{169}^* made at time origin 168 is

$$\hat{y}_{169}^*(168) = \hat{\delta} + y_{157}^* + \hat{\phi}_1(y_{168}^* - y_{156}^*) + \hat{\phi}_2(y_{167}^* - y_{155}^*)$$
$$+ \hat{\phi}_3(y_{166}^* - y_{154}^*) - \hat{\theta}_{1,12} \hat{a}_{157} + \hat{a}_{169}$$
$$= .0258 + y_{157}^* + .2922(y_{168}^* - y_{156}^*) + .1674(y_{167}^* - y_{155}^*)$$
$$- .2408(y_{166}^* - y_{154}^*) - .5917[y_{157}^* - \hat{y}_{157}^*] + 0$$

In the above equation, y_{168}^*, y_{167}^*, y_{166}^*, y_{157}^*, y_{156}^*, y_{155}^*, and y_{154}^* are the natural logarithms of the time series values y_{168}, y_{167}, y_{166}, y_{157}, y_{156}, y_{155}, and y_{154}, and $[y_{157}^* - \hat{y}_{157}^*]$ is the 157th residual. When the values of these quantities are inserted into the above equation, we find that $\hat{y}_{169}^*(168)$ equals 6.7325468. Moreover, it can be shown that a 95% prediction interval forecast of y_{169}^* is

[6.6949048, 6.7701872]

Since $y_{169}^* = \ln y_{169}$, which implies that

$$y_{169} = e^{y_{169}^*}$$

it follows, since

$$\hat{y}_{169}^*(168) = 6.7325468$$

is a point forecast of y_{169}^*, that

$$e^{\hat{y}_{169}^*(168)} = e^{6.7325468} = 839.282$$

is a point forecast of y_{169}. Moreover, it follows, since

[6.6949048, 6.7701872]

TABLE 3.4 TSERIES Output of Forecasts of Monthly Hotel Room Averages for Periods 169 to 192 (1987 and 1988)

FORECASTS IN TERMS OF THE ORIGINAL SERIES

PERIOD	FORECAST	95 PERCENT LIMITS	
		LOWER	UPPER
169	839.282	808.277	871.475
170	772.404	742.700	803.295
171	778.176	747.400	810.219
172	873.158	838.421	909.333
173	860.308	826.026	896.013
174	984.442	945.052	1025.472
175	1156.817	1110.529	1205.034
176	1184.384	1136.991	1233.750
177	905.888	869.634	943.655
178	905.797	869.544	943.560
179	783.736	752.368	816.411
180	890.941	855.283	928.086
181	858.496	821.849	896.777
182	794.571	760.484	830.185
183	805.146	770.474	841.378
184	904.309	865.334	945.038
185	890.905	852.499	931.040
186	1018.169	974.252	1064.065
187	1195.705	1144.131	1249.604
188	1223.750	1170.966	1278.914
189	936.085	895.707	978.282
190	936.099	895.720	978.297
191	810.066	775.124	846.583
192	920.907	881.184	962.421

is a 95% prediction interval forecast of y^*_{169}, that

$$[e^{6.6949048}, e^{6.7701872}] = [808.277, 871.475]$$

is a 95% prediction interval forecast of y_{169}.

In Table 3.4 we summarize the point forecasts and prediction interval forecasts given by TSERIES of the hotel room averages in periods 169 through 192. Recall that, although Traveler's Rest observed the monthly hotel room averages from 1973 to 1987, we have developed our model by using the data from 1973 to 1986, so that we could use the data observed in 1987 to check the validity of the forecasts generated by the model based on the data from 1973 to 1986. The prediction interval forecasts for months 169 to 180 (1987) are graphed in Figure 3.7, along with the actual hotel room averages for these months. Since 11 out of 12 prediction intervals contain the actual hotel room averages, and since manage-

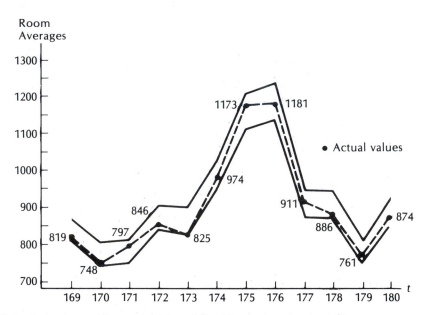

FIGURE 3.7   95% Prediction Interval Forecasts of Actual Hotel Room Averages for 1987

ment of Traveler's Rest thought that the prediction intervals were short enough to be useful for planning purposes, Traveler's Rest decided to use the model

$$y_t^* = \delta + y_{t-12}^* + \phi_1(y_{t-1}^* - y_{t-13}^*) + \phi_2(y_{t-2}^* - y_{t-14}^*) + \phi_3(y_{t-3}^* - y_{t-15}^*)$$
$$- \theta_{1,12}a_{t-12} + a_t$$

to predict monthly hotel room averages for months 181 to 192 (1988). To do this, Traveler's Rest first used all of the previously observed data (that is, used the monthly hotel room averages observed from 1973 to 1987) to calculate the point estimates

$$\hat{\delta} = .0225 \quad \hat{\phi}_1 = .3100 \quad \hat{\phi}_2 = .1407 \quad \hat{\phi}_3 = -.1412$$

and $\hat{\theta}_{1,12} = .7707$

of the parameters in the above model. Then, Traveler's Rest used these point estimates and the above model to obtain the desired forecasts

In order to understand the nature of the deterministic trend implied by the constant term δ in the model

$$(1 - \phi_1 B - \phi_2 B^2 - \phi_3 B^3)z_t = \delta + (1 - \theta_{1,12}B^{12})a_t$$

note that this model implies that

$$\delta = \mu\phi_p(B)\phi_P(B^L) = \mu(1 - \phi_1 B - \phi_2 B^2 - \phi_3 B^3)$$

Therefore, we can write the model

$$(1 - \phi_1 B - \phi_2 B^2 - \phi_3 B^3)z_t = \delta + (1 - \theta_{1,12}B^{12})a_t$$

where

$$z_t = y_t^* - y_{t-12}^*$$

as

$$(1 - \phi_1 B - \phi_2 B^2 - \phi_3 B^3)(y_t^* - y_{t-12}^*) = \mu(1 - \phi_1 B - \phi_2 B^2 - \phi_3 B^3)$$
$$+ (1 - \theta_{1,12} B^{12})a_t$$

or

$$y_t^* - y_{t-12}^* = \mu + \frac{(1 - \theta_{1,12} B^{12})}{(1 - \phi_1 B - \phi_2 B^2 - \phi_3 B^3)} a_t$$

or

$$\ln y_t - \ln y_{t-12} = \mu + \frac{(1 - \theta_{1,12} B^{12})}{(1 - \phi_1 B - \phi_2 B^2 - \phi_3 B^3)} a_t$$

or

$$\ln \frac{y_t}{y_{t-12}} = \mu + \frac{(1 - \theta_{1,12} B^{12})}{(1 - \phi_1 B - \phi_2 B^2 - \phi_3 B^3)} a_t$$

or

$$\frac{y_t}{y_{t-12}} = (e^\mu)(e^{\frac{(1 - \theta_{1,12} B^{12})}{(1 - \phi_1 B - \phi_2 B^2 - \phi_3 B^3)} a_t})$$

Since the point estimate of μ is $\bar{z} = .033$, and since $e^{\bar{z}} = e^{.033} = 1.034$, we estimate that the deterministic trend resulting from including the constant term δ in the model implies that, roughly, the forecast of the monthly hotel room average for a particular month increases by 3.4% over the forecast for the same month of the previous year. Therefore, since we have analyzed the natural logarithms of the hotel room averages, the deterministic trend is exponential (implying a percentage growth) rather than linear. Since management at Traveler's Rest believed that the modest growth rate (3.4%) would continue into 1987 they used the above model (with the constant term δ included) to forecast the monthly hotel room averages for 1987. However, if management had believed that this growth rate would not continue into 1987, then it would have been appropriate for them to remove the constant term from the model.

To conclude this example, we present in Figure 3.8 the SAS output of the SAC and SPAC of the hotel room occupancy values obtained by using the transformation

$$z_t = y_t^* - y_{t-12}^*$$

where

$$y_t^* = \ln y_t$$

FIGURE 3.8 SAS Output of the SAC and SPAC of the Hotel Room Occupancy Values Obtained by Using the Transformation $z_t = y_t^* - y_{t-12}^*$
(a) The SAC

AUTOCORRELATIONS

LAG	COVARIANCE	CORRELATION	STD
0	.000534177	1.00000	0
1	.000103267	0.19332	0.0800641
2	.000013019	0.02437	0.0830023
3	−.00013047	−0.24424	0.0830482
4	−8.094E−05	−0.15153	0.0875315
5	−.00011321	−0.21194	0.0891972
6	.000048928	0.09160	0.0923689
7	.000059097	0.11063	0.0929494
8	.000068788	0.12877	0.0937896
9	.000032999	0.06178	0.0949162
10	.000042221	0.07904	0.0951736
11	−4.177E−05	−0.07820	0.0955935
12	−.00018003	−0.33702	0.0960027
13	−5.932E−06	−0.01110	0.103309
14	.000060051	0.11242	0.103317
15	.000079669	0.14914	0.104098
16	.000054249	0.10156	0.105459
17	.000047002	0.08799	0.106084
18	−7.707E−05	−0.14428	0.10655
19	.000031468	0.05891	0.107795
20	.000019557	0.03661	0.108002
21	.000046615	0.08726	0.108081
22	−2.960E−05	−0.05541	0.108532
23	.000005134	0.00961	0.108713
24	−2.741E−05	−0.05132	0.108718

"." MARKS TWO STANDARD ERRORS

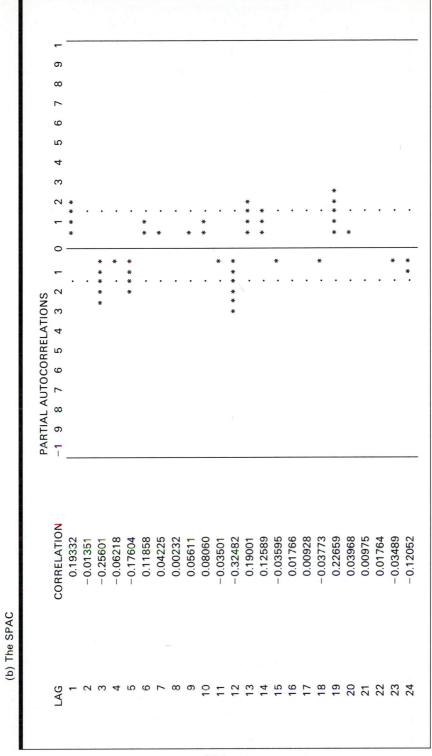

FIGURE 3.8 Continued

(b) The SPAC

LAG	CORRELATION
1	0.19332
2	-0.01351
3	-0.25601
4	-0.06218
5	-0.17604
6	0.11858
7	0.04225
8	0.00232
9	0.05611
10	0.08060
11	-0.03501
12	-0.32482
13	0.19001
14	0.12589
15	-0.03595
16	0.01766
17	0.00928
18	-0.03773
19	0.22659
20	0.03968
21	0.00975
22	0.01764
23	-0.03489
24	-0.12052

PARTIAL AUTOCORRELATIONS

Note that, whereas the TSERIES output in Figure 3.5 of the SAC and SPAC gives the values of

$$r_k, \; s_{r_k}, \; t_{r_k}, \; r_{kk}, \; s_{r_{kk}}, \quad \text{and} \quad t_{r_{kk}}$$

the SAS output only gives the values of

$$r_k, \; s_{r_k}, \quad \text{and} \quad r_{kk}$$

However, the SAS output presents "bands" (represented by the dots); if the plotted value (represented by the asterisks) of r_k (or r_{kk}) exceeds the appropriate dot, then the absolute value of

$$t_{r_k} \;=\; \frac{r_k}{s_{r_k}} \quad \left(\text{or } t_{r_{kk}} \;=\; \frac{r_{kk}}{s_{r_{kk}}} \right)$$

is greater than 2. Both SAS and TSERIES lead us to tentatively identify the same model.

EXAMPLE 3.5

We consider an example discussed by Box and Jenkins]1976], who analyzed monthly passenger totals (measured in thousands of passengers) in international air travel from 1949 to 1959. This time series is presented in Table B1 (see Appendix B). Box and Jenkins produced stationary time series values by using the stationarity transformation

$$z_t \;=\; (1 - B^{12})(1 - B)y_t^* \;=\; y_t^* - y_{t-1}^* - y_{t-12}^* + y_{t-13}^*$$

where

$$y_t^* \;=\; \ln y_t$$

The SAS output of the SAC and SPAC of these stationary time series values is presented in Figure 3.9.

1. At the *nonseasonal level*, the SAC has a spike at lag 1 and (with the exception of a spike at lag 3) cuts off after lag 1, and the SPAC dies down fairly quickly. It follows, by Guideline 1 in Table 2.3, that we should use the nonseasonal moving average operator of order 1

$$\theta_1(B) \;=\; (1 - \theta_1 B)$$

2. At the *seasonal level*, the SAC has a spike at lag 12 and cuts off after lag 12, and the SPAC has a spike at lag 12 and either cuts off after lag 12 or dies down. It follows, by either Guideline 6 or Guideline 8 in Table 3.3, that we should use the seasonal moving average operator of order 1

$$\theta_1(B^{12}) \;=\; (1 - \theta_{1,12} B^{12})$$

By inserting

$$\theta_1(B) \;=\; (1 - \theta_1 B)$$

and

$$\theta_1(B^{12}) \;=\; (1 \,-\, \theta_{1,12}B^{12})$$

into the general model

$$\phi_p(B)\phi_P(B^{12})z_t \;=\; \delta \,+\, \theta_q(B)\theta_Q(B^{12})a_t$$

we are led to try the model

$$z_t \;=\; (1 \,-\, \theta_1 B)\,(1 \,-\, \theta_{1,12}B^{12})a_t$$

where we exclude δ (which assumes that there is no deterministic trend) because $\bar{z}$ can be shown to be not statistically different from zero. It follows that

$$(1 \,-\, B^{12})(1 \,-\, B)y_t^* \;=\; (1 \,-\, \theta_1 B)(1 \,-\, \theta_{1,12}B^{12})a_t$$

or

$$(1 \,-\, B \,-\, B^{12} \,+\, B^{13})y_t^* \;=\; (1 \,-\, \theta_1 B \,-\, \theta_{1,12}B^{12} \,+\, \theta_1\theta_{1,12}B^{13})a_t$$

or

$$y_t^* \,-\, y_{t-1}^* \,-\, y_{t-12}^* \,+\, y_{t-13}^* \;=\; a_t \,-\, \theta_1 a_{t-1} \,-\, \theta_{1,12}a_{t-12} \,+\, \theta_1\theta_{1,12}a_{t-13}$$

or

$$y_t^* \;=\; y_{t-1}^* \,+\, y_{t-12}^* \,-\, y_{t-13}^* \,+\, a_t \,-\, \theta_1 a_{t-1} \,-\, \theta_{1,12}a_{t-12} \,+\, \theta_1\theta_{1,12}a_{t-13}$$

is a reasonable tentative model.

Further analysis to be carried out in Chapter 4 indicates that the above model is a reasonable final model. The point forecasts and 95% prediction interval forecasts of the logged monthly passenger totals for 1960 are given in Table 3.5. To obtain forecasts of the actual monthly passenger totals, we must exponentiate the forecasts in Table 3.5. For example, the point forecast and 95% prediction interval forecast of the passenger total in period 133 (January 1960) are, respectively,

$$e^{6.0373} \;=\; 418.76085 \quad (418{,}761 \text{ passengers})$$

and

$$[e^{5.9659}, \quad e^{6.1087}] \;=\; [389.90378, \quad 449.75366]$$

or

$$[389{,}904 \text{ passengers, } 449{,}754 \text{ passengers}]$$

We next note that our procedure for identifying a seasonal Box-Jenkins model involves examining the SAC and SPAC at the nonseasonal level and then examining them at the seasonal level. Therefore, our procedure implicitly suggests separating the behaviors of the SAC and SPAC at the nonseasonal level from their behaviors at the seasonal level. However, it can be shown that for many seasonal Box-Jenkins models involving both nonseasonal and seasonal operators, the behaviors of the TAC and TPAC at the nonseasonal level are related to their behaviors at the seasonal level. Therefore,

FIGURE 3.9 SAS Output of the SAC and SPAC of the Monthly Passenger Totals Obtained by Using the Transformation $z_t = y_t^* - y_{t-1}^* - y_{t-12}^* + y_{t-13}^*$ where $y_t^* = \ln y_t$.
(a) The SAC

AUTOCORRELATIONS

LAG	COVARIANCE	CORRELATION	-1 9 8 7 6 5 4 3 2 1 0 1 2 3 4 5 6 7 8 9 1	STD
0	0.00201141	1.00000	| *	0
1	-0.0063782	-0.31710	* * * * * * |	0.0916698
2	.000219144	0.10895	| * *	0.100466
3	-0.0004343	-0.21592	* * * * |	0.101454
4	.000089982	0.04474	| *	0.105244
5	.000062465	0.03106	| *	0.105404
6	.000085971	0.04274	| *	0.105481
7	-0.0001303	-0.06478	* |	0.105626
8	.000025101	0.01248	|	0.10596
9	.000326642	0.16239	| * * *	0.105972
10	-.00010417	-0.05179	* |	0.108043
11	.000148605	0.07388	| *	0.108251
12	-.00082537	-0.41035	* * * * * * * * |	0.108674
13	.00032666	0.16240	| * * *	0.120996
14	-.00010719	-0.05329	* |	0.122814
15	.000293742	0.14604	| * * *	0.123008
16	-0.0002927	-0.14552	* * * |	0.124457
17	.000186388	0.09267	| * *	0.125879
18	-6.018E-05	-0.02992	* |	0.12645
19	.000108789	0.05409	| *	0.12651
20	-.00028365	-0.14102	* * * |	0.126704
21	.000072268	0.03593	| *	0.128016
22	-.00016215	-0.08061	* * |	0.128101
23	.00043818	0.21667	| * * * *	0.128527
24	-6.069E-05	-0.03017	|	0.13156

'.' MARKS TWO STANDARD ERRORS

114

FIGURE 3.9 Continued

(b) The SPAC

PARTIAL AUTOCORRELATIONS

LAG	CORRELATION
1	-0.31710
2	0.00934
3	-0.19873
4	-0.09252
5	0.03361
6	0.03035
7	-0.06064
8	-0.00849
9	0.21218
10	0.04408
11	0.07212
12	-0.35474
13	-0.08283
14	-0.02168
15	-0.01661
16	-0.13096
17	0.05560
18	0.07705
19	-0.00278
20	-0.14277
21	0.11704
22	-0.04138
23	0.18134
24	-0.10472

TABLE 3.5   Forecasts of Logged Monthly Passenger Totals for 1960

			SAS	
FORECASTS FOR VARIABLE LY				
OBS	FORECAST	STD ERROR	LOWER 95%	UPPER 95%
— — — —	FORECAST BEGINS	— — —		
133	6.0373	0.0364	5.9659	6.1087
134	5.9907	0.0436	5.9051	6.0762
135	6.1462	0.0498	6.0486	6.2439
136	6.1201	0.0553	6.0118	6.2285
137	6.1564	0.0603	6.0383	6.2746
138	6.3020	0.0649	6.1748	6.4292
139	6.4276	0.0692	6.2920	6.5632
140	6.4379	0.0733	6.2943	6.5814
141	6.2646	0.0771	6.1136	6.4157
142	6.1338	0.0807	5.9755	6.2920
143	6.0048	0.0842	5.8397	6.1699
144	6.1129	0.0876	5.9412	6.2845

sometimes the behaviors of the SAC and SPAC at the nonseasonal level are related to their behaviors at the seasonal level. In spite of this, experience has shown that our procedure (and guidelines) yield tentative seasonal Box-Jenkins models that can be shown (using the diagnostic checking procedures presented in Chapter 4) to be good final models or models that require little modification in order to form a good final model. Moreover, we now present an example in which we utilize a relationship between the nonseasonal and seasonal behaviors of the SAC and SPAC to tentatively identify a seasonal Box-Jenkins model.

EXAMPLE 3.6

In this example we consider data given by Nelson [1973]. Nelson analyzed monthly U.S. auto registration data from 1947 to 1968 and produced stationary time series values by using the stationarity transformation

$$z_t = y_t^* - y_{t-1}^* - y_{t-12}^* + y_{t-13}^* \quad \text{where} \quad y_t^* = \ln y_t$$

The SAC and SPAC of these stationary time series values are presented in Figure 3.10.

1. At the *nonseasonal level*, the SAC has spikes at (possibly) lag 1 and lag 2 and cuts off after lag 2, and the SPAC has spikes at (possibly) lag 1 and lag 2 and cuts off after lag 2. It follows, by Guideline 3 in Table 2.3, that we should use either the nonseasonal moving average operator of order 2

$$\theta_2(B) = (1 - \theta_1 B - \theta_2 B^2)$$

(a)   The SAC

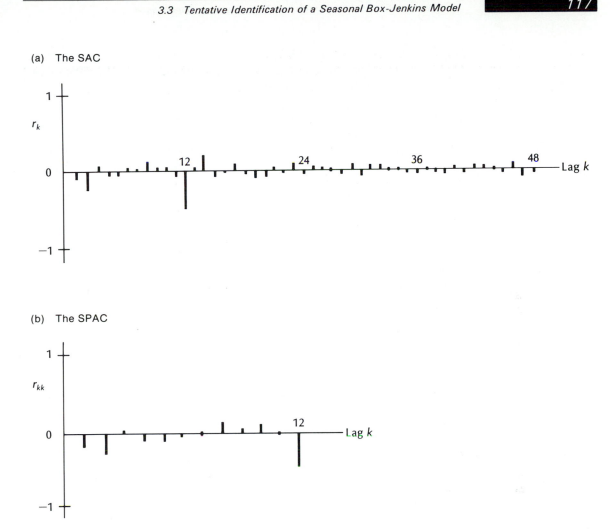

(b)   The SPAC

FIGURE 3.10   *The SAC and SPAC of the Automobile Registration Values Obtained by Using the Transformation*
$$z_t = y_t^* - y_{t-1}^* - y_{t-12}^* + y_{t-13}^*$$

or the nonseasonal autoregressive operator of order 2

$$\phi_2(B) = (1 - \phi_1 B - \phi_2 B^2)$$

Since both the SAC and SPAC appear to cut off equally abruptly at the nonseasonal level, we will first try using (by Guideline 3 in Table 2.3) the nonseasonal moving average operator of order 2

$$\theta_2(B) = (1 - \theta_1 B - \theta_2 B^2)$$

2.   At the *seasonal level*, the SAC has spikes at lags 12 and 14 and cuts off abruptly after lag 14 ($=L + 2 = 12 + 2$), and the SPAC (which is not

shown past lag 12) has a spike at lag 12. It follows, by either Guideline 6 or Guideline 8 in Table 3.3, that we should use the seasonal moving average operator of order 1

$$\theta_1(B^{12}) \;=\; (1 - \theta_{1,12}B^{12})$$

Note that we have not used a term (such as $\theta_{1,14}B^{14}$) explaining the spike at lag 14 in the SAC because we are assuming that the spike at lag 2 in the SAC at the nonseasonal level and the spike at lag 12 in the SAC at the seasonal level combine to explain the spike at lag 14.

By inserting

$$\theta_2(B) \;=\; (1 - \theta_1 B - \theta_2 B^2)$$

and

$$\theta_1(B^{12}) \;=\; (1 - \theta_{1,12}B^{12})$$

into the general model

$$\phi_p(B)\phi_P(B^{12})z_t \;=\; \delta + \theta_q(B)\theta_Q(B^{12})a_t$$

we are led to try the model

$$z_t \;=\; (1 - \theta_1 B - \theta_2 B^2)(1 - \theta_{1,12}B^{12})a_t$$

where we exclude δ because $\bar{z}$ is not statistically different from zero. On considering the above model, note that, intuitively speaking, the terms $\theta_2 B^2$ and $\theta_{1,12}B^{12}$ *together* intuitively explain the spike at lag 14 in the SAC. It follows that

$$y_t^* - y_{t-1}^* - y_{t-12}^* + y_{t-13}^* \;=\; (1 - \theta_1 B - \theta_2 B^2)(1 - \theta_{1,12}B^{12})a_t$$

where

$$y_t^* \;=\; \ln y_t$$

is a reasonable tentative model, and further analysis (see Nelson [1973]) indicates that the above model is an appropriate final model.

3.3.4   Rationale for Guidelines 6 through 10

The rationale for Guidelines 6 through 10 is based on various theoretical facts concerning seasonal Box-Jenkins models. Specifically, if we insert the seasonal moving average operator of order 1

$$\theta_1(B^L) \;=\; (1 - \theta_{1,L}B^L)$$

into the general model

$$\phi_P(B)\phi_P(B^L)z_t \;=\; \delta + \theta_q(B)\theta_Q(B^L)a_t$$

we obtain the model

$$z_t \;=\; \delta + (1 - \theta_{1,L}B^L)a_t$$

which is called the *seasonal moving average model of order 1*. It can be shown that for this model the TAC has a nonzero autocorrelation at lag L and zero autocorrelations elsewhere and the TPAC dies down at the seasonal lags $L, 2L, 3L, 4L, \ldots$. This is why we say in Guideline 6 of Table 3.3) that if, for the time series values $z_b, z_{b+1}, \ldots, z_n$, the SAC has a spike at lag L and cuts off after lag L and the SPAC dies down at the seasonal level, then we should use the seasonal moving average operator of order 1

$$\theta_1(B^L) = (1 - \theta_{1,L}B^L)$$

Next, if we insert the seasonal autoregressive operator of order 1

$$\phi_1(B^L) = (1 - \phi_{1,L}B^L)$$

into the general model

$$\phi_p(B)\phi_P(B^L)z_t = \delta + \theta_q(B)\theta_Q(B^L)a_t$$

we obtain the model

$$(1 - \phi_{1,L}B^L)z_t = \delta + a_t$$

which is called the *seasonal autoregressive model of order 1*. It can be shown that for this model the TPAC has a nonzero partial autocorrelation at lag L and zero partial autocorrelations elsewhere and the TAC dies down at the seasonal lags $L, 2L, 3L, 4L, \ldots$. Therefore, Guideline 7 of Table 3.3 states that if, for the time series values $z_b, z_{b+1}, \ldots, z_n$, the SPAC has a spike at lag L and cuts off after lag L and the SAC dies down at the seasonal level, then we should use the seasonal autoregressive operator of order 1

$$\phi_1(B^L) = (1 - \phi_{1,L}B^L)$$

The rationale for Guidelines 8, 9, and 10 in Table 3.3 is analagous to the previously discussed rationale for Guidelines 3, 4, and 5 in Table 2.3.

The general Box-Jenkins model

$$\phi_p(B)\phi_P(B^L)z_t = \delta + \theta_q(B)\theta_Q(B^L)a_t$$

is a *multiplicative model*, because it multiplies the two autoregressive operators $\phi_p(B)$ and $\phi_P(B^L)$, and because it multiplies the two moving average operators $\theta_q(B)$ and $\theta_Q(B^L)$. Experience with Box-Jenkins modeling indicates that multiplicative models are usually the most appropriate models describing seasonal time series. However, it is sometimes appropriate to modify a multiplicative model to form an *additive model*. For example, we might modify the multiplicative model

$$z_t = (1 - \theta_1 B)(1 - \theta_{1,12}B^{12})a_t$$
$$= (1 - \theta_1 B - \theta_{1,12}B^{12} + \theta_1\theta_{1,12}B^{13})a_t$$

to form the additive model

$$z_t = (1 - \theta_1 B - \theta_{12}B^{12})a_t$$

or to form the additive model

$$z_t = (1 - \theta_1 B - \theta_{12}B^{12} - \theta_{13}B^{13})a_t$$

where this second additive model differs from the above multiplicative model in that the parameter multiplied by B^{13} in the multiplicative model is restricted to equal the product of θ_1 and $\theta_{1,12}$, whereas there is no restriction on the parameter multiplied by B^{13} in the additive model. Although there is a slight difference between the TAC and TPAC of a multiplicative model and the TAC and TPAC of an additive model formed by modifying the multiplicative model, the difference is seldom enough to recognize by using the SAC and SPAC. Therefore, we would advise the novice modeler to use multiplicative models in the tentative identification stage and to remember (in the event that a tentatively identified model should prove to be inadequate) that in some situations it can be useful to modify a multiplicative model to form an additive model. In Chapter 4 we will present two examples illustrating how to use various model building techniques to choose between a multiplicative model and an additive model.

3.4 THE BASIC NATURE OF BOX-JENKINS FORECASTING

We next discuss the nature of the forecasts provided by Box-Jenkins models. Considering the general Box-Jenkins model

$$\phi_p(B)\phi_P(B^L)(1 - B^L)^D(1 - B)^d y_t^* = \delta + \theta_q(B)\theta_Q(B^L)a_t$$

it can be shown that the constant term δ and the autoregressive and differencing operators

$$\phi_p(B)\phi_P(B^L)(1 - B^L)^D(1 - B)^d$$

determine the "basic nature" of the forecasts provided by this model, and it can be shown that the moving average operators

$$\theta_q(B)\theta_Q(B^L)$$

determine how previous random shocks (or residuals, which are the predictions of previous random shocks) modify the basic nature of the forecasts. We have previously shown that, if the stationary time series values $z_b, z_{b+1}, \ldots, z_n$ given by the transformation

$$z_t = (1 - B^L)^D(1 - B)^d y_t^*$$

are differences of the original time series values (either $d > 0$ or $D > 0$), then including the constant term δ in the general Box-Jenkins model implies that forecasts are made by assuming that there is a deterministic trend in the data that will continue into the future. To illustrate how the autoregressive and differencing operators determine the basic nature of the forecasts, consider the following model, which has been shown to describe many seasonal monthly time series (such as the monthly passenger totals series in Example 3.5)

$$(1 - B^{12})(1 - B)y_t^* = (1 - \theta_1 B)(1 - \theta_{1,12}B^{12})a_t$$

or

$$(1 - B^1 - B^{12} + B^{13})y_t^* = (1 - \theta_1 B - \theta_{1,12} B^{12} + \theta_1 \theta_{1,12} B^{13})a_t$$

or

$$y_t^* - y_{t-1}^* - y_{t-12}^* + y_{t-13}^* = a_t - \theta_1 a_{t-1} - \theta_{1,12} a_{t-12} + \theta_1 \theta_{1,12} a_{t-13}$$

or

$$y_t^* = \underbrace{y_{t-12}^* + (y_{t-1}^* - y_{t-13}^*)}_{\substack{\text{portion of the model determined} \\ \text{by autoregressive and differencing} \\ \text{operators}}} \underbrace{- \theta_1 a_{t-1} - \theta_{1,12} a_{t-12} + \theta_1 \theta_{1,12} a_{t-13} + a_t}_{\substack{\text{portion of the model determined by} \\ \text{moving average operators}}}$$

Note that the portion of the model determined by the autoregressive and differencing operators says that the basic nature of the forecast of y_t^* equals

1. y_{t-12}^*, the value of the time series in the same month of the previous year, plus
2. $(y_{t-1}^* - y_{t-13}^*)$, which is a stochastic trend component.

If y_{t-1}^* is greater than y_{t-13}^*, then there is an indication of an upward trend from one month of one year to the same month of the next year, and thus the basic nature of the forecast of y_t^* equals y_{t-12}^* plus a positive trend component $(y_{t-1}^* - y_{t-13}^*)$. If y_{t-1}^* is less than y_{t-13}^*, then there is an indication of a downward trend from one month of one year to the same month of the next year, and thus the basic nature of the forecast of y_t^* equals y_{t-12}^* plus a negative trend component $(y_{t-1}^* - y_{t-13}^*)$.

As another example, consider the following model, which describes, as we saw previously, the natural logarithms of the monthly hotel room averages.

$$(1 - \phi_1 B - \phi_2 B^2 - \phi_3 B^3)(1 - B^{12})y_t^* = \delta + (1 - \theta_{1,12} B^{12})a_t$$

or

$$y_t^* = \underbrace{\delta + y_{t-12}^* + \phi_1(y_{t-1}^* - y_{t-13}^*)}_{} $$
$$\underbrace{+ \phi_2(y_{t-2}^* - y_{t-14}^*) + \phi_3(y_{t-3}^* - y_{t-15}^*)}_{\substack{\text{portion of the model determined} \\ \text{by the constant term } \delta \text{ and the} \\ \text{autogressive and differencing operators}}} \underbrace{- \theta_{1,12} a_{t-12} + a_t}_{\substack{\text{portion of the model} \\ \text{determined by moving} \\ \text{average operators}}}$$

Note that the portion of the model determined by the constant term δ and the autoregressive and differencing operators says that the basic nature of the forecast of y_t^* equals

1. y_{t-12}^*, the value of the time series in the same month of the previous year, plus
2. a deterministic trend component δ, which was discussed in Example 3.4, plus
3. $\phi_1(y_{t-1}^* - y_{t-13}^*) + \phi_2(y_{t-2}^* - y_{t-14}^*) + \phi_3(y_{t-3}^* - y_{t-15}^*)$

    which is a linear combination of the stochastic trend components $(y_{t-1}^* - y_{t-13}^*)$, $(y_{t-2}^* - y_{t-14}^*)$, and $(y_{t-3}^* - y_{t-15}^*)$.

Interpretations of the basic natures of the forecasts provided by other Box-Jenkins models can be made in a similar fashion.

EXERCISES

3.1.   Miller and Wichern [1977] reported 178 monthly values $y_1, y_2, \ldots, y_{178}$ of the number of people in Wisconsin employed in "Trade" from 1961 to 1975 (all observations are given in units of 1000 employees). This time series is presented in Table B2 (see Appendix B). In Figures 1, 2, and 3 we present the TSERIES output of the SAC of the values produced by each of the transformations $z_t = y_t$, $z_t = y_t - y_{t-1}$, and $z_t = y_t - y_{t-12}$. In Figure 4 we present the TSERIES output of the SAC and SPAC of the values produced by the transformation $z_t = y_t - y_{t-1} - y_{t-12} + y_{t-13}$. It can also be shown that the mean and standard deviation of the values produced by the transformation $z_t = y_t - y_{t-1} - y_{t-12} + y_{t-13}$ are $\bar{z} = .08727$ and $s_z = 1.443$.

a.   Consider Figure 1 which shows the SAC of the values produced by the transformation $z_t = y_t$.
1.   Describe the behavior of the SAC at the nonseasonal level.
2.   Describe the behavior of the SAC at the seasonal level.
3.   Should the values produced by the transformation $z_t = y_t$ be considered stationary? Explain your answer.

b.   Consider Figure 2 which shows the SAC of the values produced by the transformation $z_t = y_t - y_{t-1}$.
1.   Describe the behavior of the SAC at the nonseasonal level.
2.   Describe the behavior of the SAC at the seasonal level.
3.   Should the values produced by the transformation $z_t = y_t - y_{t-1}$ be considered stationary? Explain your answer.

c.   Consider Figure 3 which shows the SAC of the values produced by the transformation $z_t = y_t - y_{t-12}$.
1.   Describe the behavior of the SAC at the nonseasonal level.
2.   Describe the behavior of the SAC at the seasonal level.
3.   Should the values produced by the transformation $z_t = y_t - y_{t-12}$ be considered stationary? Explain your answer.

d.   Consider Figure 4 which shows the SAC of the values produced by the transformation $z_t = y_t - y_{t-1} - y_{t-12} + y_{t-13}$.
1.   Describe the behavior of the SAC at the nonseasonal level. Justify your answer using t-statistics.
2.   Describe the behavior of the SAC at the seasonal level. Justify your answer using t-statistics.
3.   Should the values produced by the transformation

$$z_t = y_t - y_{t-1} - y_{t-12} + y_{t-13}$$

be considered stationary? Explain your answer.
4.   Describe the behavior of the SPAC at the nonseasonal level. Justify your answer using t-statistics.
5.   Describe the behavior of the SPAC at the seasonal level. Justify your answer using t-statistics.

FIGURE 1 TSERIES Output of the SAC for the Original Trade Values

	VALUE	S.E.	T VALUE
1	0.971	0.075	12.956
2	0.941	0.127	7.391
3	0.915	0.162	5.660
4	0.899	0.189	4.764
5	0.884	0.211	4.186
6	0.871	0.231	3.769
7	0.854	0.249	3.431
8	0.838	0.265	3.165
9	0.826	0.279	2.955
10	0.819	0.293	2.799
11	0.813	0.305	2.660
12	0.807	0.317	2.544
13	0.777	0.329	2.364
14	0.746	0.339	2.203
15	0.720	0.348	2.070
16	0.702	0.356	1.971
17	0.687	0.364	1.888
18	0.673	0.371	1.813
19	0.655	0.378	1.734
20	0.640	0.384	1.665
21	0.627	0.390	1.607
22	0.620	0.396	1.566
23	0.611	0.401	1.523
24	0.604	0.406	1.485

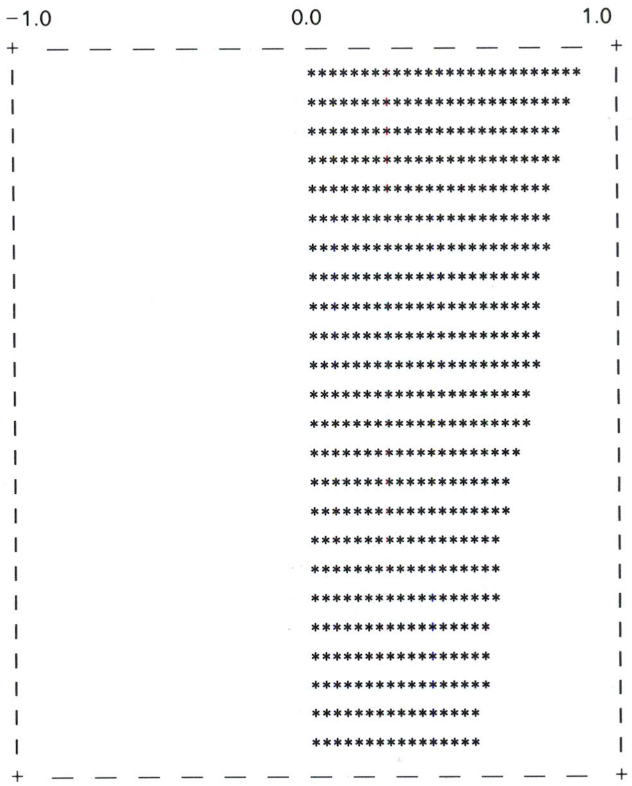

123

FIGURE 2 TSERIES Output of the SAC for the Trade Values Obtained by Using the Transformation $z_t = y_t - y_{t-1}$

	VALUE	S.E.	T VALUE
1	0.026	0.075	0.343
2	-0.194	0.075	-2.584
3	-0.315	0.078	-4.044
4	-0.126	0.085	-1.481
5	0.014	0.086	0.161
6	0.224	0.086	2.606
7	0.007	0.089	0.083
8	-0.107	0.089	-1.204
9	-0.304	0.090	-3.381
10	-0.179	0.096	-1.870
11	0.041	0.097	0.422
12	0.902	0.098	9.244
13	0.015	0.137	0.113
14	-0.186	0.137	-1.363
15	-0.294	0.138	-2.128
16	-0.120	0.142	-0.847
17	0.017	0.142	0.119
18	0.205	0.142	1.444
19	0.010	0.144	0.066
20	-0.096	0.144	-0.664
21	-0.275	0.144	-1.909
22	-0.159	0.147	-1.083
23	0.032	0.148	0.219
24	0.815	0.148	5.497

FIGURE 3 TSERIES Output of the SAC for the Trade Values Obtained by Using the Transformation $z_t = y_t - y_{t-12}$

	VALUE	S.E.	T VALUE
1	0.940	0.078	12.110
2	0.868	0.129	6.724
3	0.798	0.160	4.971
4	0.734	0.183	4.014
5	0.680	0.200	3.403
6	0.624	0.213	2.929
7	0.564	0.224	2.520
8	0.507	0.232	2.182
9	0.444	0.239	1.858
10	0.376	0.244	1.542
11	0.309	0.247	1.251
12	0.240	0.250	0.963
13	0.200	0.251	0.798
14	0.164	0.252	0.652
15	0.126	0.253	0.500
16	0.090	0.253	0.354
17	0.042	0.253	0.167
18	-0.004	0.253	-0.016
19	-0.037	0.253	-0.147
20	-0.065	0.253	-0.256
21	-0.090	0.253	-0.356
22	-0.117	0.254	-0.463
23	-0.146	0.254	-0.577
24	-0.178	0.254	-0.701

FIGURE 4 TSERIES Output of the SAC and SPAC of the Trade Values Obtained by Using the Transformation $z_t = y_t - y_{t-1} - y_{t-12} + y_{t-13}$
(a) The SAC

	VALUE	S.E.	T VALUE			
				−1.0	0.0	1.0
1	0.184	0.078	2.361		*****	
2	−0.027	0.080	−0.342		*	
3	−0.010	0.080	−0.129		*	
4	−0.141	0.081	−1.745		****	
5	−0.002	0.082	−0.026		*	
6	0.097	0.082	1.180		***	
7	0.020	0.083	0.242		*	
8	0.090	0.083	1.093		***	
9	0.096	0.083	1.147		***	
10	0.000	0.084	0.003		*	
11	−0.070	0.084	−0.834		**	
12	−0.277	0.084	−3.282		********	
13	−0.097	0.090	−1.079		***	
14	0.027	0.090	0.304		*	
15	0.003	0.090	0.033		*	
16	0.139	0.090	1.539		****	
17	0.024	0.092	0.263		*	
18	−0.101	0.092	−1.107		***	
19	−0.096	0.092	−1.037		***	
20	−0.055	0.093	−0.592		**	
21	−0.019	0.093	−0.205		*	
22	−0.019	0.093	−0.206		*	
23	0.051	0.093	0.550		**	
24	−0.080	0.093	−0.855		**	

126

FIGURE 4 *Continued*

(b) The SPAC

	VALUE	S.E.	T VALUE				
				-1.0	0.0	1.0	
1	0.184	0.078	2.361		*****		
2	-0.063	0.078	-0.814		**		
3	0.007	0.078	0.089		*		
4	-0.147	0.078	-1.889		****		
5	0.056	0.078	0.719		**		
6	0.077	0.078	0.988		***		
7	-0.010	0.078	-0.124		*		
8	0.081	0.078	1.041		***		
9	0.071	0.078	0.910		**		
10	0.001	0.078	0.014		*		
11	-0.074	0.078	-0.945		**		
12	-0.259	0.078	-3.333		********		
13	0.016	0.078	0.200		*		
14	0.006	0.078	0.079		*		
15	-0.030	0.078	-0.391		*		
16	0.092	0.078	1.180		***		
17	-0.019	0.078	-0.244		*		
18	-0.044	0.078	-0.559		**		
19	-0.081	0.078	-1.035		***		
20	0.031	0.078	0.399		*		
21	0.037	0.078	0.474		*		
22	-0.067	0.078	-0.864		**		
23	0.035	0.078	0.447		*		
24	-0.190	0.078	-2.440		*****		

e. Consider the values produced by the transformation

$$z_t = y_t - y_{t-1} - y_{t-12} + y_{t-13}$$

Should $\bar{z}$ be considered to be statistically different from zero? Justify your answer.

f. Use any information available to rationalize the tentative model

$$z_t = (1 - \theta_1 B)(1 - \theta_{1,12} B^{12})a_t$$

where

$$z_t = y_t - y_{t-1} - y_{t-12} + y_{t-13}$$

g. Explain which guidelines (in Tables 2.3 and 3.3) you used to identify the above tentative model.

h. Explain why the constant term δ is not included in the above model.

i. Algebraically expand the tentative model given in part (f), and express y_t as a function of past time series observations and past and present random shocks.

j. The least squares point estimates of θ_1 and $\theta_{1,12}$ are $\hat{\theta}_1 = -.1728$ and $\hat{\theta}_{1,12} = .4588$. Moreover, using the model specified in (f) and these least squares point estimates, we can calculate (at time origin 178) point forecasts and prediction interval forecasts of $y_{179}, y_{180}, \ldots, y_{190}$. The TSERIES output of these forecasts is given below:

| | FORECASTS IN TERMS OF THE ORIGINAL SERIES | | |
| | | 95 PERCENT LIMITS | |
PERIOD	FORECAST	LOWER	UPPER
179	404.235	401.637	406.833
180	408.915	404.911	412.919
181	387.031	382.000	392.062
182	382.970	377.088	388.852
183	385.959	379.335	392.583
184	394.111	386.820	401.402
185	402.554	394.652	410.456
186	407.294	398.825	415.762
187	407.071	398.071	416.071
188	409.878	400.376	419.380
189	410.552	400.573	420.530
190	413.195	402.762	423.628

1. Demonstrate the calculation of $\hat{y}_{179}(178) = 404.235$
2. Demonstrate the calculation of $\hat{y}_{180}(178) = 408.915$
3. Demonstrate the calculation of $\hat{y}_{181}(178) = 387.031$
4. Explain the basic nature of the forecasts obtained by using the model

$$z_t = (1 - \theta_1 B)(1 - \theta_{1,12} B^{12})a_t$$

where

$$z_t = y_t - y_{t-1} - y_{t-12} + y_{t-13}$$

FIGURE 5 *TSERIES Output of the SAC and SPAC of the Values Obtained by Using the Transformation* $z_t = y_t - y_{t-1}$ *and the Fabricated Metals Employment Data*

a) The SAC

	VALUE	S.E.	T VALUE	−1.0	0.0
1	−0.039	0.075	−0.522	|	*
2	0.123	0.075	1.632	|	****
3	0.001	0.076	0.016	|	*
4	0.060	0.076	0.784	|	**
5	−0.040	0.077	−0.522	|	**
6	−0.025	0.077	−0.329	|	*
7	−0.101	0.077	−1.314	|	***
8	0.069	0.078	0.883	|	**
9	−0.146	0.078	−1.877	|	****
10	0.098	0.079	1.237	|	***
11	−0.036	0.080	−0.453	|	*
12	0.234	0.080	2.914	|	*******
13	−0.057	0.084	−0.682	|	**
14	0.143	0.084	1.694	|	****
15	−0.109	0.086	−1.279	|	***
16	−0.143	0.086	−1.652	|	****
17	−0.122	0.088	−1.393	|	****
18	0.026	0.089	0.289	|	*
19	−0.196	0.089	−2.214	|	*****
20	−0.114	0.091	−1.249	|	***
21	−0.131	0.092	−1.424	|	****
22	0.131	0.093	1.406	|	****
23	−0.073	0.094	−0.773	|	**
24	0.179	0.094	1.896	|	*****

(Continued on page 130.)

3.2 Miller and Wichern [1977] reported 178 monthly values $y_1, y_2, \ldots, y_{178}$ of the number of people in Wisconsin employed in the fabricated metals industry from 1961 to 1975 (all observations are given in units of 1000 employees). This time series is presented in Table B3 (see Appendix B). Figure 5 gives the TSERIES output of the SAC and SPAC for the time series values produced by using the transformation $z_t = y_t - y_{t-1}$.

    a. Describe the behavior of the SAC at the nonseasonal level. Use t-statistics to justify your answer.

    b. Describe the behavior of the SAC at the seasonal level. Use t-statistics to justify your answer.

    c. Should the values produced by the transformation $z_t = y_t - y_{t-1}$ be considered stationary? Explain your answer.

FIGURE 5   *Continued*

b) The SPAC

	VALUE	S.E.	T VALUE		
				−1.0	0.0
				+ − − − − − − − − − − −	
1	−0.039	0.075	−0.522	I	*
2	0.122	0.075	1.617	I	****
3	0.010	0.075	0.139	I	*
4	0.046	0.075	0.614	I	**
5	−0.038	0.075	−0.511	I	*
6	−0.041	0.075	−0.552	I	**
7	−0.097	0.075	−1.291	I	***
8	0.069	0.075	0.919	I	**
9	−0.118	0.075	−1.565	I	***
10	0.084	0.075	1.115	I	***
11	0.003	0.075	0.040	I	*
12	0.215	0.075	2.862	I	******
13	−0.044	0.075	−0.579	I	**
14	0.090	0.075	1.194	I	***
15	−0.120	0.075	−1.593	I	***
16	−0.224	0.075	−2.978	I	******
17	−0.097	0.075	−1.286	I	***
18	0.044	0.075	0.584	I	**
19	−0.108	0.075	−1.435	I	***
20	−0.151	0.075	−2.013	I	****
21	−0.038	0.075	−0.511	I	*
22	0.090	0.075	1.193	I	***
23	−0.015	0.075	−0.194	I	*
24	0.122	0.075	1.624	I	****
				+ − − − − − − − − − − −	

d. Describe the behavior of the SPAC at the nonseasonal level. Use t-statistics to justify your answer.

e. Describe the behavior of the SPAC at the seasonal level. Use t-statistics to justify your answer.

f. Use any of the conclusions that you made in parts (a)–(e) to rationalize the tentative model

$$(1 - \phi_{1,12}B^{12})z_t = a_t$$

Assume that $\bar{z}$ is not statistically different from zero.

g. Explain which guidelines (in Tables 2.3 and 3.3) you used to identify the above tentative model.

h. Algebraically expand the tentative model in (f) and express y_t as a function of past time series observations and of the random shock a_t.

i. Explain the basic nature of the forecasts provided by the model

$$(1 - \phi_{1,12}B^{12})z_t = a_t \qquad \text{where} \qquad z_t = y_t - y_{t-1}$$

3.3 Algebraically expand each of the following models and express y_t as a function of past time series observations and of past and present random shocks.

a. $z_t = (1 - \theta_1 B - \theta_2 B^2)(1 - \theta_{1,12}B^{12})a_t$

where

$$z_t = y_t - y_{t-1} - y_{t-12} + y_{t-13}$$

b. $\phi_2(B^4)z_t = \delta + \theta_2(B)\theta_1(B^4)a_t$

where

$$z_t = y_t - y_{t-4}$$

c. $\phi_2(B)\phi_1(B^{12})z_t = \theta_1(B^{12})a_t$

where

$$z_t = (1 - B^{12})^1(1 - B)^2 y_t$$

d. $z_t = (1 - \theta_6 B^6)(1 - \theta_{1,12}B^{12})a_t$

where

$$z_t = y_t - y_{t-1}$$

e. $z_t = (1 - \theta_6 B^6 - \theta_{12}B^{12})a_t$

where

$$z_t = y_t - y_{t-1}$$

f. $(1 - \phi_1 B - \phi_2 B^2)(1 - \phi_{1,12}B^{12})y_t = \delta + a_t$

3.4. Consider Example 3.4 in which we forecasted the Traveler's Rest monthly hotel room averages using the model

$$y_t^* = \delta + y_{t-12}^* + \phi_1(y_{t-1}^* - y_{t-13}^*) + \phi_2(y_{t-2}^* - y_{t-14}^*) + \phi_3(y_{t-3}^* - y_{t-15}^*)$$
$$- \theta_{1,12}a_{t-12} + a_t$$

a. Using the estimates $\hat{\delta} = .0258$, $\hat{\phi}_1 = .2922$, $\hat{\phi}_2 = .1674$, $\hat{\phi}_3 = -.2408$, and $\hat{\theta}_{1,12} = .5917$ and any other information in this chapter, compute $\hat{y}_{181}^*(180)$.

b. Use your result from part (a) to compute $\hat{y}_{181}(180)$.

ESTIMATION AND DIAGNOSTIC CHECKING FOR BOX-JENKINS MODELS

4.1 INTRODUCTION

This chapter concerns *estimation* and *diagnostic checking*. We begin with Section 4.2, which discusses the *preliminary estimates* that are required by Box-Jenkins computer packages. We will see that our discussion of this topic involves *stationarity* and *invertibility conditions*. Section 4.3 deals with advanced concepts pertaining to the *estimation of model parameters*. Included in the topics covered in this section are estimation using SAS and TSERIES and the use of *t-values* and *prob-values* to judge the importance of model parameters. Section 4.4 covers *diagnostic checking*—checking to see whether or not a tentatively identified model is adequate. We discuss the *Box-Pierce* and *Ljung-Box statistics*, use the techniques of this chapter to choose a model that can be used to forecast the hotel room occupancy data, and explain how to approach *finding an improved model* when a tentatively identified model has been found to be inadequate. Section 4.5 presents several more examples of Box-Jenkins modeling. Finally, we conclude this chapter with Section 4.6, which explains the use of SAS and TSERIES in implementing the Box-Jenkins methodology.

4.2 STATIONARITY AND INVERTIBILITY CONDITIONS, AND PRELIMINARY ESTIMATION

4.2.1 Stationarity and Invertibility Conditions

The Box-Jenkins methodology requires that the model

$$\phi_p(B)\phi_P(B^L)z_t = \delta + \theta_q(B)\theta_Q(B^L)a_t$$

to be used in describing and forecasting a time series be both *stationary* and *invertible*. We have previously discussed the meaning of stationarity. Although we will not formally discuss the meaning of invertibility, we will intuitively discuss it. Noting that a Box-Jenkins model can be used to express z_t as a function of past z-observations (that is, z_{t-1}, z_{t-2}, . . .), it follows that a Box-Jenkins model is *not invertible* if the weights placed on the past z-observations when expressing z_t as a function of these observations do not decline as we move further into the past. However, an invertible Box-Jenkins model implies that these weights do decline—a condition that intuition indicates should hold (since it seems that a recent observation should count more heavily than a more distantly past observation). Each of the conditions of stationarity and invertibility implies that the parameters of the operators $\phi_p(B)$, $\phi_P(B^L)$, $\theta_q(B)$, and $\theta_Q(B^L)$ used in the model under consideration satisfy certain conditions. The stationarity and invertibility conditions on the parameters of the first and second orders of the operators $\phi_p(B)$, $\phi_P(B^L)$, $\theta_q(B)$, and $\theta_Q(B^L)$ are as given in Table 4.1. Using these conditions, we summarize in Table 4.2 the stationarity and invertibility conditions on the parameters of the nonseasonal models summarized in Table 2.5. The stationarity and invertibility conditions on the parameters of the *general* forms of the operators $\phi_p(B)$, $\phi_P(B^L)$, $\theta_q(B)$, and $\theta_Q(B^L)$ are complicated and will not be given here. However, we will say that

1. There are stationarity conditions, but there are no invertibility conditions, on the parameters of any form of each of the autoregressive operators $\phi_p(B)$ and $\phi_P(B^L)$. A necessary (but not sufficient) stationarity condition on the parameters of any form of each of the operators $\phi_p(B)$ and $\phi_P(B^L)$ is that the sum of the values of the parameters in the operator is less than 1.
2. There are invertibility conditions, but there are no stationarity conditions, on the parameters of any form of each of the moving average operators $\theta_q(B)$ and $\theta_Q(B^L)$. A necessary (but not sufficient) invertibility condition on the parameters of any form of each of the operators $\theta_q(B)$ and $\theta_Q(B^L)$ is that the sum of the values of the parameters in the operator is less than 1.

4.2.2 Preliminary Estimates

We discuss the stationarity and invertibility conditions on the parameters of the above operators for two practical reasons. To understand these reasons, note that Box-Jenkins computer packages start with *preliminary point estimates* of the parameters to be

TABLE 4.1   Stationarity and Invertibility Conditions on the Parameters in the First and Second Orders of $\phi_p(B)$, $\phi_P(B^L)$, $\theta_q(B)$, and $\theta_Q(B^L)$

Operator	Stationarity Conditions	Invertibility Conditions
$\theta_1(B) = (1 - \theta_1 B)$	None	$|\theta_1| < 1$
$\theta_2(B) = (1 - \theta_1 B - \theta_2 B^2)$	None	$\theta_1 + \theta_2 < 1$ $\theta_2 - \theta_1 < 1$ $|\theta_2| < 1$
$\theta_1(B^L) = (1 - \theta_{1,L} B^L)$	None	$|\theta_{1,L}| < 1$
$\theta_2(B^L) = (1 - \theta_{1,L} B^L - \theta_{2,L} B^{2L})$	None	$\theta_{1,L} + \theta_{2,L} < 1$ $\theta_{2,L} - \theta_{1,L} < 1$ $|\theta_{2,L}| < 1$
$\phi_1(B) = (1 - \phi_1 B)$	$|\phi_1| < 1$	None
$\phi_2(B) = (1 - \phi_1 B - \phi_2 B^2)$	$\phi_1 + \phi_2 < 1$ $\phi_2 - \phi_1 < 1$ $|\phi_2| < 1$	None
$\phi_1(B^L) = (1 - \phi_{1,L} B^L)$	$|\phi_{1,L}| < 1$	None
$\phi_2(B^L) = (1 - \phi_{1,L} B^L - \phi_{2,L} B^{2L})$	$\phi_{1,L} + \phi_{2,L} < 1$ $\phi_{2,L} - \phi_{1,L} < 1$ $|\phi_{2,L}| < 1$	None

TABLE 4.2   Stationarity and Invertibility Conditions For Some Specific Nonseasonal Models

Model	Stationarity Conditions	Invertibility Conditions
First-order moving average $z_t = \delta + (1 - \theta_1 B)a_t$	None	$|\theta_1| < 1$
Second-order moving average $z_t = \delta + (1 - \theta_1 B - \theta_2 B^2)a_t$	None	$\theta_1 + \theta_2 < 1$ $\theta_2 - \theta_1 < 1$ $|\theta_2| < 1$
First-order autoregressive $(1 - \phi_1 B)z_t = \delta + a_t$	$|\phi_1| < 1$	None
Second-order autoregressive $(1 - \phi_1 B - \phi_2 B^2)z_t = \delta + a_t$	$\phi_1 + \phi_2 < 1$ $\phi_2 - \phi_1 < 1$ $|\phi_2| < 1$	None
Mixed autoregressive–moving average of order (1, 1) $(1 - \phi_1 B)z_t = \delta + (1 - \theta_1 B)a_t$	$|\phi_1| < 1$	$|\theta_1| < 1$

estimated and then apply an iterative search technique to a sum of squares function to obtain final least squares point estimates of the parameters. Some computer packages (for example, SAS and TSERIES) automatically supply default preliminary point estimates, but other computer packages require that the user supply preliminary point estimates as inputs. The first reason, then, for discussing the stationarity and invertibility conditions is that the preliminary point estimates should satisfy these conditions. Moreover, it can be shown that we can obtain preliminary point estimates that do satisfy these conditions if

1. We set the preliminary point estimate of any parameter in any of the operators $\phi_p(B)$, $\phi_P(B^L)$, $\theta_q(B)$, and $\theta_Q(B^L)$ equal to .1.
2. We set the preliminary point estimate of

$$\begin{aligned}
\delta &= \mu\phi_p(B)\phi_P(B^L) \\
&= \mu(1 - \phi_1 B - \phi_2 B^2 - \cdots - \phi_p B^p)(1 - \phi_{1,L}B^L \\
&\quad - \phi_{2,L}B^{2L} - \cdots - \phi_{P,L}B^{PL}) \\
&= \mu(1 - \phi_1 - \phi_2 - \cdots - \phi_p)(1 - \phi_{1,L} - \phi_{2,L} - \cdots - \phi_{P,L})
\end{aligned}$$

equal to

$$\bar{z}(1 - .1 - .1 - \cdots - .1)(1 - .1 - .1 - \cdots - .1)$$

where $\bar{z}$ is the mean of the stationary time series values $z_b, z_{b+1}, \ldots, z_n$.

The second reason for discussing the stationarity and invertibility conditions is that, when we obtain the final least squares point estimates of the parameters in our model, we should verify that these point estimates satisfy the stationarity and invertibility conditions. If they do not, this suggests that the model may not be adequate.

EXAMPLE 4.1

We said in Example 2.7 that the model

$$z_t = (1 - \theta_1 B)a_t$$

where

$$z_t = y_t - y_{t-1}$$

adequately describes the original values of Absorbent Paper Towels sales. By examining Table 4.1, we see that the invertibility condition on the parameter θ_1 in the nonseasonal moving average operator of order 1

$$\theta_1(B) = (1 - \theta_1 B)$$

is

$$|\theta_1| < 1$$

Note that the preliminary point estimate $\hat{\theta}_1 = .1$ satisfies this condition, and note that $\hat{\theta}_1 = -.3545$, which we have said in Example 2.5 is the final least squares point estimate of θ_1, also satisfies this condition.

EXAMPLE 4.2

We said in Example 3.4 that the model

$$(1 - \phi_1 B - \phi_2 B^2 - \phi_3 B^3)z_t = \delta + (1 - \theta_{1,12} B^{12})a_t$$

where

$$z_t = y_t^* - y_{t-12}^* \quad \text{and} \quad y_t^* = \ln y_t$$

adequately describes the original monthly hotel room averages in Table 3.1. A necessary (but not sufficient) stationarity condition on the parameters ϕ_1, ϕ_2, and ϕ_3 in the nonseasonal autoregressive operator of order 3

$$\phi_3(B) = (1 - \phi_1 B - \phi_2 B^2 - \phi_3 B^3)$$

is

$$\phi_1 + \phi_2 + \phi_3 < 1$$

Moreover, the invertibility condition on the parameter $\theta_{1,12}$ in the seasonal moving average operator of order 1

$$\theta_1(B^{12}) = (1 - \theta_{1,12} B^{12})$$

is (from Table 4.1)

$$|\theta_{1,12}| < 1$$

Note that the preliminary point estimates $\hat{\phi}_1 = .1$, $\hat{\phi}_2 = .1$, $\hat{\phi}_3 = .1$, and $\hat{\theta}_{1,12} = .1$ satisfy these stationarity and invertibility conditions, and note that, since $\bar{z} = .033$ is (as stated in Example 3.4) the mean of the stationary time series values $z_{13}, z_{14}, \ldots, z_{168}$, a preliminary point estimate of

$$\delta = \mu\phi_p(B)\phi_P(B^L)$$
$$= \mu(1 - \phi_1 B - \phi_2 B^2 - \phi_3 B^3)$$
$$= \mu(1 - \phi_1 - \phi_2 - \phi_3)$$

is

$$\bar{z}(1 - .1 - .1 - .1) = .033(.7) = .0231$$

Finally, note that

$$\hat{\phi}_1 = .2922 \quad \hat{\phi}_2 = .1674 \quad \hat{\phi}_3 = -.2408 \quad \text{and} \quad \hat{\theta}_{1,12} = .5917$$

(which are the final least squares point estimates of ϕ_1, ϕ_2, ϕ_3, and $\theta_{1,12}$ of Example 3.4) also satisfy the above stationarity and invertibility conditions.

To conclude this section, note that although the easiest way to obtain a preliminary estimate of any parameter in any of the operators $\phi_p(B)$, $\phi_P(B^L)$, $\theta_q(B)$, and $\theta_Q(B^L)$ is to set the preliminary estimate equal to .1, we can sometimes obtain a better preliminary point estimate (that is, a preliminary point estimate that is likely to be nearer the final point estimate that will be obtained). To see this, recall that we expressed the theoretical autocorrelations of each model in Table 2.5 in terms of the parameters of that model. One application of such relationships is in obtaining better preliminary point estimates. For example, since Table 2.5 tells us that for the first order autoregressive model

$$(1 - \phi_1 B)z_t = \delta + a_t$$

we know that

$$\varrho_k = (\phi_1)^k \qquad \text{for } k \geqslant 1$$

which implies that

$$\varrho_1 = \phi_1$$

it follows that an initial point estimate of ϕ_1 is

$$\hat{\phi}_1 = r_1$$

where r_1, the point estimate of ϱ_1, is the sample autocorrelation of the time series values $z_b, z_{b+1}, \ldots, z_n$ at lag 1. However, although better preliminary point estimates can be obtained, current Box-Jenkins computer packages generally do not require them in order to obtain good final point estimates (some older packages did require these better preliminary estimates).

4.3 ESTIMATION

4.3.1 Estimation Using SAS and TSERIES

After a tentative Box-Jenkins model

$$\phi_p(B)\phi_P(B^L)z_t = \delta + \theta_q(B)\theta_Q(B^L)a_t$$

has been identified, we estimate the parameters of the model. Although Box and Jenkins [1976] favor the *maximum likelihood* approach to calculating point estimates, this approach can be somewhat difficult and costly to implement and so they suggest using the *least squares* approach to calculating point estimates. SAS and TSERIES calculate least squares point estimates, and SAS has an option that allows us to calculate maximum likelihood point estimates. It can be shown that, if the random shocks are normally distributed (as we assume that they are), then least squares point estimates are either exactly or very nearly maximum likelihood point estimates.

TSERIES obtains least squares point estimates of the parameters in the operators $\phi_p(B)$, $\phi_P(B^L)$, $\theta_q(B)$, and $\theta_Q(B^L)$ and *separately* obtains a point estimate of the true mean μ of the stationary time series by calculating the mean

$$\bar{z} = \frac{\sum_{t=b}^{n} z_t}{n - b + 1}$$

of the time series values z_b, z_{b+1}, $\ldots$, z_n. It follows that the point estimate of

$$
\begin{aligned}
\delta &= \mu\phi_p(B)\phi_P(B^L) \\
&= \mu(1 - \phi_1 B - \phi_2 B^2 - \cdots - \phi_p B^p)(1 - \phi_{1,L} B^L - \phi_{2,L} B^{2L} - \cdots - \phi_{P,L} B^{PL}) \\
&= \mu(1 - \phi_1 - \phi_2 - \cdots - \phi_p)(1 - \phi_{1,L} - \phi_{2,L} - \cdots - \phi_{P,L})
\end{aligned}
$$

that is provided by TSERIES is

$$
\hat{\delta} = \bar{z}(1 - \hat{\phi}_1 - \hat{\phi}_2 - \cdots - \hat{\phi}_p)(1 - \hat{\phi}_{1,L} - \hat{\phi}_{2,L} - \cdots - \hat{\phi}_{P,L})
$$

where $\hat{\phi}_1$, $\hat{\phi}_2$, $\ldots$, $\hat{\phi}_p$, $\hat{\phi}_{1,L}$, $\hat{\phi}_{2,L}$, $\ldots$, $\hat{\phi}_{P,L}$ are the least squares point estimates of ϕ_1, ϕ_2, $\ldots$, ϕ_p, $\phi_{1,L}$, $\phi_{2,L}$, $\ldots$, $\phi_{P,L}$. In contrast, SAS *simultaneously* obtains least squares point estimates of the parameters in the operators $\phi_p(B)$, $\phi_P(B^L)$, $\theta_q(B)$, and $\theta_Q(B^L)$ and of the true mean μ (here, the least squares point estimate of μ will usually not exactly equal $\bar{z}$). It follows that the point estimate of δ that is provided by SAS is

$$
\hat{\delta} = \hat{\mu}(1 - \hat{\phi}_1 - \hat{\phi}_2 - \cdots - \hat{\phi}_p)(1 - \hat{\phi}_{1,L} - \hat{\phi}_{2,L} - \cdots - \hat{\phi}_{P,L})
$$

where $\hat{\phi}_1$, $\hat{\phi}_2$, $\ldots$, $\hat{\phi}_p$, $\hat{\phi}_{1,L}$, $\hat{\phi}_{2,L}$, $\ldots$, $\hat{\phi}_{P,L}$, and $\hat{\mu}$ are the least squares point estimates of ϕ_1, ϕ_2, $\ldots$, ϕ_p, $\phi_{1,L}$, $\phi_{2,L}$, $\ldots$, $\phi_{P,L}$, and μ.

4.3.2   t-Values and Prob-Values

Associated with the *point estimate* of each parameter in a Box-Jenkins model is its *standard error* and *t-value*. Let θ denote any particular parameter in a Box-Jenkins model, and let $\hat{\theta}$ denote the point estimate of θ and $s_{\hat{\theta}}$ denote the standard error of the point estimate $\hat{\theta}$ (note that SAS and TSERIES calculate $\hat{\theta}$ and $s_{\hat{\theta}}$). Then, the *t*-value associated with $\hat{\theta}$ is calculated by the equation

$$
t_{\hat{\theta}} = \frac{\hat{\theta}}{s_{\hat{\theta}}}
$$

If the absolute value of $t_{\hat{\theta}}$ is "large", then $\hat{\theta}$ is "large". This implies that θ does not equal zero, and thus, that we should reject H_0: $\theta = 0$, which implies that we should include the parameter θ in the Box-Jenkins model. To decide how large $t_{\hat{\theta}}$ must be before we reject H_0: $\theta = 0$ (and thus include θ in the Box-Jenkins model), we consider the errors that can be made in hypothesis testing. A *Type I error* is committed if we reject H_0: $\theta = 0$ when H_0: $\theta = 0$ is true (which means we would include θ in the Box-Jenkins model when θ should not be included). A *Type II error* is committed if we do not reject H_0: $\theta = 0$ when H_0: $\theta = 0$ is false (which means we would not include θ in the Box-Jenkins model when θ should be included). We obviously desire that both the *probability of a Type I error*, which we denote by the symbol α, and the *probability of a Type II error*, which we denote by the symbol β, be *small*. The hypothesis testing procedure that we use in this book assumes that we observe a time series of n values and set α equal to a specified value. Here, we usually choose α to be between .05 and .01, with .05 being the most frequent choice. Note that the lower we set α, the lower is the probability that we will include θ in a Box-Jenkins model when θ should not be included, and thus, if we do reject H_0: $\theta = 0$, the more sure we are that we should include θ. The reason that we usually do not choose α to be lower than .01 is that setting α extremely small often leads to a probability of a Type II error (not including θ when θ should be included) that is unacceptably large. We now state the following result.

Hypothesis Testing Result I

Assume that the Box-Jenkins model under consideration utilizes n_p parameters, and define the $t_{\hat{\theta}}$-statistic to be

$$t_{\hat{\theta}} = \frac{\hat{\theta}}{s_{\hat{\theta}}}$$

Then, we can reject the null hypothesis H_0: $\theta = 0$ in favor of the alternative hypothesis H_1: $\theta \neq 0$ by setting the probability of a Type I error equal to α if and only if either of the following equivalent conditions hold

1. $|t_{\hat{\theta}}| > t_{[\alpha/2]}^{(n-n_p)}$ that is $t_{\hat{\theta}} > t_{[\alpha/2]}^{(n-n_p)}$ or $t_{\hat{\theta}} < -t_{[\alpha/2]}^{(n-n_p)}$

   where the *rejection point*

   $t_{[\alpha/2]}^{(n-n_p)}$

   is the point on the scale of the *t*-distribution having $(n - n_p)$ degrees of freedom so that the area under the curve of this *t*-distribution to the right of $t_{[\alpha/2]}^{(n-n_p)}$ is $\alpha/2$. If $(n - n_p)$ is at least 30, $t_{[\alpha/2]}^{(n-n_p)}$ should be approximated by $z_{[\alpha/2]}$ which is the point on the scale of the standard normal curve so that the area under this curve to the right of $z_{[\alpha/2]}$ is $\alpha/2$.
2. *Prob-value* is less than α, where prob-value is twice the area under the curve of the *t*-distribution having $(n - n_p)$ degrees of freedom to the right of $|t_{\hat{\theta}}|$. If $(n - n_p)$ is at least 30, then the prob-value is approximately found by calculating twice the area under the standard normal curve to the right of $|t_{\hat{\theta}}|$.

We will not discuss the precise rationale for Hypothesis Testing Result I in this book. However, noting that we can look up $z_{[\alpha/2]}$ points in Table A1 (in Appendix A) and we can look up $t_{[\alpha/2]}^{(n-n_p)}$ points in Table A2 (in Appendix A), we fully illustrate this result in the following example.

EXAMPLE 4.3

We concluded in Example 2.7 that the model

$$z_t = (1 - \theta_1 B)a_t$$

where

$$z_t = y_t - y_{t-1}$$

is a reasonable tentative model describing the $n = 120$ original values of Absorbent Paper Towel sales in Table 2.1. Table 4.3 presents the TSERIES output of

TABLE 4.3   TSERIES Output of the Least Squares Estimation of θ_1 in the Model $z_t = (1 - \theta_1 B)a_t$ where $z_t = y_t - y_{t-1}$

(a) The Iterative Search

NO CONSTANT TERM IN THE MODEL
MAX BACKORDER = 1

BEGINING ESTIMATION

ITERATION	SUM OF SQUARES	PARAMETER VALUES
0	0.1539888D+03	0.100000[a]
1	0.1429642D+03	−0.010000
2	0.1349555D+03	−0.120000
3	0.1297019D+03	−0.230000
4	0.1274483D+03	−0.340000
5	0.1274149D+03	−0.356599
6	0.1274141D+03	−0.354081
7	0.1274141D+03	−0.354469
8	0.1274141D+03	−0.354467[b]

RELATIVE CHANGE IN EACH ESTIMATE LESS THAN 0.1000E−03
ESTIMATION COMPLETED

(b) The Final Point Estimate and its Associated *t*-Value

TERM#	TYPE	ORDER	ESTIMATE	STD. ERROR	T VALUE	95% LIMITS LOWER	UPPER
1	REG. MA	1	−0.3545	0.0864	−4.1022	−0.5239	−0.1851

[a] Preliminary point estimate
[b] Final least squares point estimate

1.   the least squares point estimate of θ_1, which is

$$\hat{\theta}_1 = -.3545$$

2.   the standard error of the point estimate $\hat{\theta}_1$, which is

$$s_{\hat{\theta}_1} = .0864$$

and

3.   the *t*-value associated with $\hat{\theta}_1$, which is

$$t_{\hat{\theta}_1} = \frac{\hat{\theta}_1}{s_{\hat{\theta}_1}} = \frac{-.3545}{.0864} = -4.1022$$

(a) The Rejection Points For Testing H_0: $\theta_1 = 0$ versus H_1: $\theta_1 \neq 0$

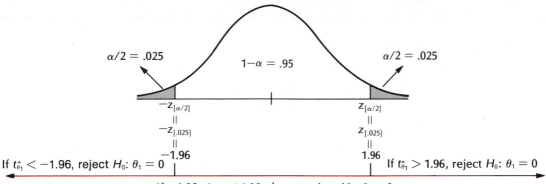

$\alpha/2 = .025$ $1-\alpha = .95$ $\alpha/2 = .025$

$-z_{[\alpha/2]}$ $z_{[\alpha/2]}$
$\parallel$ $\parallel$
$-z_{[.025]}$ $z_{[.025]}$
$\parallel$ $\parallel$
-1.96 1.96

If $t_{\hat{\theta}_1} < -1.96$, reject H_0: $\theta_1 = 0$ If $t_{\hat{\theta}_1} > 1.96$, reject H_0: $\theta_1 = 0$

If $-1.96 \leq t_{\hat{\theta}_1} \leq 1.96$, do not reject H_0: $\theta_1 = 0$

(b) The Prob-Value For Testing H_0: $\theta_1 = 0$ versus H_1: $\theta_1 \neq 0$:
Reject H_0: $\theta_1 = 0$ if Prob-Value is less than α

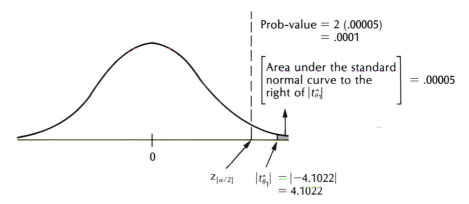

Prob-value $= 2 \, (.00005)$
$= .0001$

$\begin{bmatrix} \text{Area under the standard} \\ \text{normal curve to the} \\ \text{right of } |t_{\hat{\theta}_1}| \end{bmatrix} = .00005$

0

$z_{[\alpha/2]}$ $|t_{\hat{\theta}_1}| = |-4.1022|$
$= 4.1022$

FIGURE 4.1   *Testing H_0: $\theta_1 = 0$ versus H_1: $\theta_1 \neq 0$*

Noting that the above model uses $n_p = 1$ parameter, it follows that if we wish to test H_0: $\theta_1 = 0$ versus H_1: $\theta_1 \neq 0$ by setting α equal to .05, we would use the rejection point

$$t_{[\alpha/2]}^{(n-n_p)} = t_{[.05/2]}^{(120-1)} = t_{[.025]}^{(119)}$$

which (since $n - n_p = 119$ is at least 30) can be approximated by (see Figure 4.1a)

$$z_{[\alpha/2]} = z_{[.05/2]} = z_{[.025]} = 1.96$$

To explain what we mean when we say that α equals .05, note that although the $n = 120$ observed values of the Absorbent Paper Towel sales time series in Table 2.1 yielded the point estimate

$$\hat{\theta}_1 \; = \; -.3545$$

the standard error

$$s_{\hat{\theta}_1} \; = \; .0864$$

and the *t*-value

$$t_{\hat{\theta}_1} \; = \; -4.1022$$

another realization (that is, another $n = 120$ values) of the Absorbent Paper Towel sales time series would yield a somewhat different point estimate, standard error, and *t*-value. It can be proven that, if the null hypothesis $H_0: \theta_1 = 0$ is true, then the population of all possible *t*-values that could be observed (by observing all possible realizations of the Absorbent Paper Towel time series) has (approximately) a *t*-distribution with $(n - n_p)$ degrees of freedom. It follows that setting α equal to .05 means that, if $H_0: \theta_1 = 0$ is true, then, as illustrated in Figure 4.1,

1. .95 (that is, 95%) of all possible *t*-values would be between $-z_{[.025]} = -1.96$ and $z_{[.025]} = 1.96$ and thus would be close enough to zero to cause us to not reject $H_0: \theta_1 = 0$ when $H_0: \theta_1 = 0$ is true (which would be a correct decision).
2. .025 (that is, 2.5%) of all possible *t*-values would be less than $-z_{[.025]} = -1.96$, and .025 (that is, 2.5%) of all possible *t*-values would be greater than $z_{[.025]} = 1.96$, and thus—in total—.05 (that is, 5%) of all possible *t*-values would be far enough away from zero to cause us to reject $H_0: \theta_1 = 0$ when $H_0: \theta_1 = 0$ is true (which would be a Type I error).

Recalling that the *t*-value computed from the $n = 120$ values of the Absorbent Paper Towel time series that we have actually observed is

$$t_{\hat{\theta}_1} \; = \; -4.1022$$

it follows, since the absolute value of this *t*-value is greater than $z_{[.025]} = 1.96$, that we can reject $H_0: \theta_1 = 0$ in favor of $H_1: \theta_1 \neq 0$ by setting α equal to .05.

If we wish to use condition (2) in Hypothesis Testing Result I, then we must first calculate the prob-value, which is twice the area under the standard normal curve (since $n - n_p = 119$ is at least 30) to the right of

$$|t_{\hat{\theta}_1}| \; = \; |-4.1022| \; = \; 4.1022$$

Since Table A1 indicates that the area to the right of 3 is .001, the area to the right of 4.1022 is less than .001, and thus the prob-value is less than .002. More detailed calculation indicates that the area under the standard normal curve to the right of 4.1022 is roughly .00005 (see Figure 3.10b) and thus that the prob-value is roughly 2(.00005) = .0001. Since this prob-value is less than α = .05, it follows by condition (2) in Hypothesis Testing Result I that we can reject H_0: θ_1 = 0 in favor of H_1: $\theta_1 \neq 0$ by setting α equal to .05.

To see the logic behind condition (2) in Hypothesis Testing Result I, note that if, for *any value* of α:

$$\text{prob-value} = 2 \times \begin{bmatrix} \text{area under the standard} \\ \text{normal curve to the right} \\ \text{of } |t_{\hat{\theta}_1}| \end{bmatrix}$$

is less than α, then

$$\begin{bmatrix} \text{area under the standard} \\ \text{normal curve to the right} \\ \text{of } |t_{\hat{\theta}_1}| \end{bmatrix} < \alpha/2$$

This implies (as can be seen by comparing Figure 4.1a with Figure 4.1b) that

$$|t_{\hat{\theta}_1}| > z_{[\alpha/2]}$$

which implies by condition (1) in Hypothesis Testing Result I that we should reject H_0: θ_1 = 0 in favor of H_1: $\theta_1 \neq 0$. For example, since

$$\text{prob-value} = .0001$$

is less than .05, .01, and .0005, we can reject H_0: θ_1 = 0 in favor of H_1: $\theta_1 \neq 0$ by setting α equal to .05, or by setting α equal to .01, or by setting α equal to .0005. Therefore, since we can reject H_0: θ_1 = 0 in favor of H_1: $\theta_1 \neq 0$ by using such low values of α, we are very confident that θ_1 should be included in the time series model.

As illustrated in the above example, since condition (2) in Hypothesis Testing Result I requires calculating the area under the curve of the *t*-distribution having $n - n_p$ degrees of freedom (or the area under the standard normal curve) to the right of $|t_{\hat{\theta}}|$, whereas condition (1) requires only that we calculate the $t_{\hat{\theta}}$-statistic and look up the rejection point $t_{[\alpha/2]}^{(n-n_p)}$ in a *t*-table (or $z_{[\alpha/2]}$ in a normal table), condition (2) is more complicated than condition (1) from a computational standpoint. However, condition (2) has the following advantage over condition (1). If there were several hypothesis testers, all of whom wished to use different values of α (the probability of a Type I error), and if condition (1) were used, each hypothesis tester would have to look up a different rejection point to decide whether to reject H_0 at the hypothesis tester's particular chosen value of α. However, if condition (2) were used, only the prob-value would need to be calculated, and each hypothesis tester would know that if the prob-value is less than the particular chosen value of α, then H_0 should be rejected.

Up to this point we have considered the prob-value as a decision rule for deciding when to reject the null hypothesis H_0: θ = 0 in favor of the alternative hypothesis H_1: $\theta \neq 0$. However, the prob-value has a more important use. Although it is a frequently used convention to set α at .05, this choice, or any other choice, of α is quite arbitrary. This has led many statisticians to use the prob-value to determine the amount of "probabilistic doubt" cast upon $H_0: \theta = 0$ by the $t_{\hat{\theta}}$-statistic. Specifically, the prob-value can be interpreted to be the probability, if $H_0: \theta = 0$ is true, of observing a $t_{\hat{\theta}}$-statistic that is at least as far from zero (negatively or positively), and thus at least as contradictory to $H_0: \theta = 0$, as the $t_{\hat{\theta}}$-statistic that we have actually observed. For example, if we consider Example 4.3, the

$$\text{prob-value} = .0001$$

says that, if $H_0: \theta_1 = 0$ is true, then the t-value we observed

$$t_{\hat{\theta}_1} = -4.1022$$

is so rare that only .0001 (that is, 1 in 10,000) of all possible t-values are at least as far away from zero and thus at least as contradictory to $H_0: \theta_1 = 0$ as the observed t-value. Hence, two conclusions are possible. The first conclusion is that $H_0: \theta_1 = 0$ is true and a sample result has occurred that is so rare that there is only a .0001 chance of observing a sample result more contradictory to $H_0: \theta_1 = 0$. The second conclusion is that $H_0: \theta_1 = 0$ is false and that $H_1: \theta_1 \neq 0$ is true. A reasonable person would probably believe the second conclusion rather than the 1 in 10,000 chance. From this example we also see that the smaller the prob-value is, the more contradictory to $H_0: \theta = 0$ is the observed $t_{\hat{\theta}}$-statistic, and hence the more doubt is cast upon the validity of $H_0: \theta = 0$.

In practice, precise calculation of the prob-value often requires use of a computer package. Unfortunately, at this time neither TSERIES nor the SAS time series package calculates the prob-value associated with the $t_{\hat{\theta}}$-statistic (although, as we will illustrate in Chapter 8, various regression packages in SAS calculate prob-values). Henceforth, since in most Box-Jenkins models $(n - n_p)$ is at least 30, and since we have seen in Example 4.3 that the rejection point for testing $H_0: \theta = 0$ versus $H_1: \theta \neq 0$, when α is set equal to .05, is $z_{[.025]} = 1.96$, which is nearly equal to 2, we will say, as a practical rule, that it is reasonable to include in a model any parameter θ whose absolute $t_{\hat{\theta}}$-statistic is greater than 2. If the absolute t-value of a parameter is not greater than 2, we should seriously consider excluding the parameter from the model, because including it would be likely to produce a model that is not *parsimonious*. Here, a parsimonious model is one that adequately fits the historical data without using any unnecessary parameters. Box and Jenkins [1976] emphasize the importance of obtaining parsimonious models because such models usually produce more accurate forecasts.

Since SAS simultaneously obtains least squares point estimates of the parameters in the operators $\phi_p(B)$, $\phi_P(B^L)$, $\theta_q(B)$, and $\theta_Q(B^L)$ and of the true mean μ, SAS provides standard errors and t-values for all of these point estimates. Therefore, when using SAS, it is reasonable to decide to include the constant term

$$\delta = \mu\phi_p(B)\phi_P(B^L)$$

in a Box-Jenkins model if the absolute t-value associated with the least squares point estimate of μ is at least 2. In contrast, since TSERIES obtains least squares point estimates of the parameters in the operators $\phi_p(B)$, $\phi_P(B^L)$, $\theta_q(B)$, and $\theta_Q(B^L)$ and separately calculates the point estimate $\bar{z}$ of μ, TSERIES provides standard errors and t-values for the least squares point estimates of the parameters in the operators $\phi_p(B)$, $\phi_P(B^L)$, $\theta_q(B)$, and $\theta_Q(B^L)$ but does not provide a standard error and t-value for the point estimate $\bar{z}$ of μ. In this case, one very approximate procedure is (as previously discussed) to decide that $\bar{z}$ is statistically different from zero and thus to include the constant term

$$\delta = \mu\phi_p(B)\phi_P(B^L)$$

in a Box-Jenkins model if the absolute value of

$$\frac{\bar{z}}{s_z/\sqrt{n-b+1}}$$

is greater than 2, where $\bar{z}$ and s_z are the mean and standard deviation of the observed time series values $z_b, z_{b+1}, \ldots, z_n$. The above differences between the estimation procedures in SAS and TSERIES are typical of the differences between the estimation procedures in various Box-Jenkins computer programs.

EXAMPLE 4.4

We concluded in Example 3.3 that, for the monthly hotel room averages in Table 3.1, the transformation

$$z_t = y_t^* - y_{t-12}^* \qquad \text{where } y_t^* = \ln y_t$$

produces stationary time series values. In this example and in Example 4.6 we will consider the following models

Model 1: $(1 - \phi_1 B - \phi_3 B^3 - \phi_5 B^5)z_t = \delta + (1 - \theta_{1,12}B^{12})a_t$

Model 2: $(1 - \phi_1 B - \phi_3 B^3)z_t = \delta + (1 - \theta_{1,12}B^{12})a_t$

Model 3: $(1 - \phi_1 B - \phi_2 B^2 - \phi_3 B^3)z_t = \delta + (1 - \theta_{1,12}B^{12})a_t$

which have been tentatively identified in Example 3.4. We will first consider Model 3. Table 4.4 presents the TSERIES output of

1. the least squares point estimates of the parameters ϕ_1, ϕ_2, ϕ_3, and $\theta_{1,12}$ in Model 3, which are $\hat{\phi}_1 = .2922$, $\hat{\phi}_2 = .1674$, $\hat{\phi}_3 = -.2408$ and $\hat{\theta}_{1,12} = .5917$
2. the standard errors of the least squares point estimates
3. the t-values associated with the least squares point estimates, which are all (in absolute value) greater than 2 and thus imply that it is reasonable to include the parameters ϕ_1, ϕ_2, ϕ_3, and $\theta_{1,12}$ in the above model
4. the mean of the values $z_{13} = y_{13}^* - y_1^*$, $z_{14} = y_{14}^* - y_2^*$, $\ldots$, $z_{168} = y_{168}^* - y_{156}^*$ produced by the transformation $z_t = y_t^* - y_{t-12}^*$, which is $\bar{z} = .033$, and which implies that a point estimate of

$$\delta = \mu\phi_p(B)\phi_P(B^L)$$

$$= \mu(1 - \phi_1 B - \phi_2 B^2 - \phi_3 B^3)$$

$$= \mu(1 - \phi_1 - \phi_2 - \phi_3)$$

TABLE 4.4   *TSERIES Output of Estimation and Diagnostic Checking for the Model $(1 - \phi_1 B - \phi_2 B^2 - \phi_3 B^3)z_t = \delta + (1 - \theta_{1,12}B^{12})a_t$ where $z_t = y_t - y_{t-12}$*

(a)   The Iterative Search

A CONSTANT TERM IS TO BE INCLUDED IN THE MODEL
MAX BACKORDER = 12

BEGINING ESTIMATION

ITERATION	SUM OF SQUARES	PARAMETER VALUES			
0	0.8109055D−01	0.100000	0.100000	0.100000	0.100000
1	0.7091125D−01	0.126299	0.097189	−0.010000	0.208598
2	0.6493448D−01	0.170801	0.097124	−0.120000	0.302699
3	0.6092257D−01	0.201075	0.113110	−0.185393	0.412699
4	0.5884863D−01	0.244349	0.138055	−0.227250	0.522699
5	0.5847978D−01	0.282222	0.160532	−0.235951	0.597602
6	0.5831770D−01	0.292957	0.166753	−0.244674	0.558569
7	0.5800218D−01	0.292182	0.167381	−0.240809	0.591711

UNABLE TO REDUCE SUM OF SQUARES ANY FURTHER
LARGEST CORRECTION IS   0.3314E−01 FOR ESTIMATE NUMBER   4
ESTIMATION COMPLETED

(b)   The Final Point Estimates and Their Associated *t*-Values

TERM #	TYPE	ORDER	ESTIMATE	STD. ERROR	T VALUE	95% LIMITS LOWER	UPPER
1	REG. AR	1	0.2922	0.0796	3.6688	0.1361	0.4483
2	REG. AR	2	0.1674	0.0820	2.0408	0.0066	0.3282
3	REG. AR	3	−0.2408	0.0796	−3.0239	−0.3969	−0.0847
4	SEA. MA	12	0.5917	0.0725	8.1663	0.4497	0.7338

ESTIMATED MEAN OF THE WORKING SERIES = 0.0330
THIS YIELDS A CONSTANT TERM = 0.0258

CORRELATION MATRIX OF THE ESTIMATED PARAMETERS

	1	2	3	4
1	0.100E+01	−0.252E+00	−0.106E+00	0.854E−01
2	−0.252E+00	0.100E+01	−0.264E+00	0.630E−01
3	−0.106E+00	−0.264E+00	0.100E+01	0.612E−01
4	0.854E−01	0.630E−01	0.612E−01	0.100E+01

is

$$\hat{\delta} = \bar{z}(1 - \hat{\phi}_1 - \hat{\phi}_2 - \hat{\phi}_3)$$

$$= .033(1 - .2922 - .1674 - (-.2408))$$

$$= .0258$$

and

5. the *correlation matrix* of the estimated parameters.

Considering the correlation matrix in Table 4.4, we see that, for example, the estimated correlation between $\hat{\phi}_1$ and $\hat{\phi}_2$ is $-.252$, the estimated correlation between $\hat{\phi}_1$ and $\hat{\theta}_{1,12}$ is .0854, and the estimated correlation between $\hat{\phi}_2$ and $\hat{\phi}_3$ is $-.264$ (which, in absolute value, is the largest correlation in the correlation matrix). Although the point estimates of the parameters in a Box-Jenkins model will always be correlated, very high correlations suggest that the point estimates may be of poor quality, in the sense that a slightly different realization of the time series would have yielded considerably different point estimates. As a practical rule, we should consider two point estimates to be highly correlated if the absolute estimated correlation between these point estimates is greater than .9. If this occurs, we should consider using the SAC and SPAC to find an alternative model with less correlated point estimates. Since the highest absolute estimated correlation between the point estimates in Model 3 is .264, we conclude that the point estimates in this model are not very correlated.

4.3.3 Multicollinearity and the Standard Error

Before leaving our discussion of using the $t_{\hat{\theta}}$-statistic and related prob-value to test $H_0: \theta = 0$ versus $H_1: \theta \neq 0$, we should make one final comment. Consider the model analyzed in the previous example

$$(1 - \phi_1 B - \phi_2 B^2 - \phi_3 B^3)(y_t^* - y_{t-12}^*) = \delta + (1 - \theta_{1,12} B^{12})a_t$$

which we have seen in Example 3.4 is equivalent to

$$y_t^* = \delta + \phi_1(y_{t-1}^* - y_{t-13}^*) + \phi_2(y_{t-2}^* - y_{t-14}^*) + \phi_3(y_{t-3}^* - y_{t-15}^*)$$
$$- \theta_{1,12} a_{t-12} + a_t$$

Using the language of a branch of statistics called *regression analysis*, we call y_t^* the *dependent variable* and $(y_{t-1}^* - y_{t-13}^*)$, $(y_{t-2}^* - y_{t-14}^*)$, $(y_{t-3}^* - y_{t-15}^*)$, and a_{t-12} *independent variables*. It can be proven that, for any parameter θ in a time series model, the $t_{\hat{\theta}}$-statistic and related prob-value measure the *additional importance* of the independent variable multiplied by θ over and above the combined importance of the other independent variables in the time series model. For example, in the above model, the $t_{\hat{\phi}_2}$-statistic measures the additional importance of $(y_{t-2}^* - y_{t-14}^*)$ over and above the combined importance of $(y_{t-1}^* - y_{t-13}^*)$, $(y_{t-3}^* - y_{t-15}^*)$, and a_{t-12}. For this reason, *multicollinearity* (which exists when the independent variables are related to each other and thus to some extent contribute redundant information for the description and prediction of the dependent variable) can cause the $t_{\hat{\theta}}$-statistics to make individual independent variables look unimportant when they really are important. This did not happen with the above model (because the absolute $t_{\hat{\theta}}$-statistics in Table 4.4 are all greater than 2), but it can happen with other time series models. Because of multicollinearity, it is useful to have measures of the "overall fit" of a time series model. One such measure is the *standard error*.

The *standard error* is

$$s = \sqrt{\frac{SSE}{n - n_p}} = \sqrt{\frac{\sum_{t=1}^{n} (y_t - \hat{y}_t)^2}{n - n_p}}$$

---

where n is the number of observations in the original time series, n_p is the number of parameters in the model, and the tth residual $(y_t - \hat{y}_t)$ is the difference between the observed time series value y_t and the prediction $\hat{y}_t$ given by the model. The smaller the standard error is, the better is the overall fit of the model. In particular, a model with a smaller standard error often yields shorter (that is, more accurate) prediction interval forecasts. We will use the standard error in future examples.

4.4 DIAGNOSTIC CHECKING

4.4.1 The Box-Pierce and Ljung-Box Statistics

A good way to check the adequacy of an *overall* Box-Jenkins model is to analyze the residuals obtained from the model in the following manner. Just as we can calculate the sample autocorrelation and partial autocorrelation functions of the time series values z_b, $z_{b+1}, \ldots, z_n$, we can calculate such functions for the residuals. We let "RSAC" denote the sample autocorrelation function of the residuals and denote "RSPAC" the sample partial autocorrelation function of the residuals. One way to use the residuals to check the adequacy of the overall model is to examine a statistic that determines whether the first K sample autocorrelations of the residuals, considered together, indicate adequacy of the model. Two such statistics have been suggested. These statistics are summarized below (the choice of K is somewhat arbitrary and will be further discussed below).

---

1. The Box-Pierce statistic is

$$Q = n' \sum_{l=1}^{K} r_l^2(\hat{a})$$

2. The Ljung-Box statistic is

$$Q^* = n'(n' + 2) \sum_{l=1}^{K} (n' - l)^{-1} r_l^2(\hat{a})$$

Here

$n' = n - (d + LD)$ where n is the number of observations in the original time series, L is the number of seasons in a year (if the time series exhibits seasonal variation), and d and D are, respectively, the degrees of nonseasonal and seasonal differencing used to transform the original time series values into stationary time series values (note: some computer programs calculate Q and Q^* by using n instead of n')

$r_l^2(\hat{a})$ is the square of $r_l(\hat{a})$, the sample autocorrelation of the residuals at lag l—that is, the sample autocorrelation of residuals separated by a lag of l time units.

We use both of the above statistics to test the adequacy of the overall model in exactly the same way. However, since theory indicates that Q^* is the better of the two above statistics, we will discuss how to test model adequacy by using Q^*. The modeling process is supposed to account for the relationship between the time series observations. If it does account for these relationships, the residuals should be unrelated, and hence the autocorrelations of the residuals should be small. Thus, Q^* should be small. The larger Q^* is, the larger are the autocorrelations of the residuals and the more related are the residuals. Hence, a large value of Q^* indicates that the model is inadequate. We can reject the adequacy of the model under consideration by setting the probability of a Type I error equal to α if and only if either of the following equivalent conditions hold:

1. Q^* is greater than $\chi^2_{[\alpha]}(K - n_p)$, the point on the scale of the chi-square distribution having $K - n_p$ degrees of freedom such that there is an area of α under the curve of this distribution above this point. Here, n_p is the number of parameters that must be estimated in the model under consideration. Table A3 (in Appendix A) is a table of chi-square points.
2. Prob-value is less than α, where prob-value is the area under the curve of the chi-square distribution having $K - n_p$ degrees of freedom to the right of Q^*.

Frequently, α is chosen to equal .05, but this choice is not sacred. Usually, however, we set α somewhere between .01 and .05. If the prob-value is less than .01, this is very strong evidence that the model is inadequate. If the prob-value is greater than .01 but less than .05, this is fairly strong evidence that the model is inadequate. If the prob-value is greater than .05 [or equivalently, if Q^* is less than $\chi^2_{[.05]}(K - n_p)$], then it is reasonable to conclude that the model is adequate. Moreover, the greater the prob-value is (and thus the smaller Q^* is), the more we believe in the adequacy of the model. The adequacy of the model can be further investigated by examining the individual sample autocorrelations and individual sample partial autocorrelations of the residuals. We suggest that, if Q^* does

FIGURE 4.2   *TSERIES Output of the RSAC and RSPAC for the Model* $z_t = (1 - \theta_1 B) a_t$ *where* $z_t = y_t - y_{t-1}$

(a)   The RSAC

	VALUE	S.E.	T VALUE	−1.0	0.0	1.0
				+ − − − − − −	− − − − − − −	− − − +
1	0.005	0.092	0.053	I	*	I
2	−0.036	0.092	−0.391	I	*	I
3	−0.102	0.092	−1.115	I	***	I
4	0.128	0.093	1.383	I	****	I
5	0.027	0.094	0.292	I	*	I
6	0.061	0.094	0.647	I	**	I
7	−0.142	0.095	−1.496	I	****	I
8	−0.031	0.096	−0.319	I	*	I
9	−0.146	0.096	−1.516	I	****	I
10	−0.058	0.098	−0.589	I	**	I
11	−0.040	0.099	−0.411	I	**	I
12	0.026	0.099	0.259	I	*	I
13	0.019	0.099	0.196	I	*	I
14	0.039	0.099	0.394	I	**	I
15	−0.074	0.099	−0.746	I	**	I
16	−0.002	0.099	−0.023	I	*	I
17	0.068	0.099	0.682	I	**	I
18	0.196	0.100	1.964	I	******	I
19	0.002	0.103	0.022	I	*	I
20	−0.027	0.103	−0.263	I	*	I
21	−0.060	0.103	−0.581	I	**	I
22	−0.001	0.103	−0.006	I	*	I
23	−0.055	0.103	−0.532	I	**	I
24	0.013	0.104	0.122	I	*	I
				+ − − − − − − −	− − − − − −	− − +

BOX-PIERCE TEST CHISQUARE STATISTIC WITH 20 DEGREES OF FREEDOM = 16.116

(*Continued on page 151.*)

not indicate that the adequacy of the overall model should be rejected, then we should only be concerned about individual sample autocorrelations and individual sample partial autocorrelations of the residuals having absolute *t*-values greater than 2. If, however, Q^* indicates that the adequacy of the overall model should be rejected, then we should use a more detailed procedure to identify spikes in the RSAC and RSPAC. Later in this section we will discuss this more detailed procedure and a general procedure for using the RSAC and RSPAC to make necessary improvements in a Box-Jenkins model.

Whereas TSERIES calculates the Box-Pierce statistic

$$Q = n' \sum_{l=1}^{K} r_l^2(\hat{a})$$

FIGURE 4.2 Continued

(b) The RSPAC

	VALUE	S.E.	T VALUE	−1.0	0.0	1.0
				+ − − − − − − − − −	− − − − − − − −	− − − +
1	0.005	0.092	0.053	I	*	I
2	−0.036	0.092	−0.391	I	*	I
3	−0.102	0.092	−1.114	I	***	I
4	0.129	0.092	1.408	I	****	I
5	0.019	0.092	0.204	I	*	I
6	0.060	0.092	0.657	I	**	I
7	−0.119	0.092	−1.299	I	***	I
8	−0.037	0.092	−0.399	I	*	I
9	−0.156	0.092	−1.696	I	****	I
10	−0.103	0.092	−1.125	I	***	I
11	−0.032	0.092	−0.344	I	*	I
12	0.004	0.092	0.045	I	*	I
13	0.063	0.092	0.687	I	**	I
14	0.056	0.092	0.613	I	**	I
15	−0.047	0.092	−0.515	I	**	I
16	−0.029	0.092	−0.314	I	*	I
17	0.030	0.092	0.328	I	*	I
18	0.144	0.092	1.570	I	****	I
19	−0.006	0.092	−0.064	I	*	I
20	−0.004	0.092	−0.044	I	*	I
21	−0.029	0.092	−0.319	I	*	I
22	−0.055	0.092	−0.602	I	**	I
23	−0.079	0.092	−0.861	I	**	I
24	−0.012	0.092	−0.126	I	*	I
				+ − − − − − − − − −	− − − − − − − −	− − − +

for a value of K that makes the number of degrees of freedom $(K - n_p)$ equal to 20, SAS calculates the preferred Ljung-Box statistic

$$Q^* = n'(n' + 2) \sum_{l=1}^{K} (n' - l)^{-1} r_l^2(\hat{a})$$

for values of K equal to 6, 12, 18, 24, and 30.

EXAMPLE 4.5

Figure 4.2 presents the TSERIES output of the RSAC, the RSPAC, and the value of Q, the Box-Pierce statistic, having $K - n_p = 20$ degrees of freedom that result when the model of Example 2.7

$$z_t = (1 - \theta_1 B)a_t$$

where

$$z_t = y_t - y_{t-1}$$

is fit to the $n = 120$ original values of Absorbent Paper Towel sales in Table 2.1. Since $K - n_p = 20$, and since the above model utilizes $n_p = 1$ parameter (θ_1), Q has been calculated by using the first

$$K = 20 + n_p = 20 + 1 = 21$$

sample autocorrelations of the residuals. Therefore, recalling that $d = 1$ is the degree of nonseasonal differencing used to transform the original time series values into stationary time series values, which implies that

$$n' = n - (d + LD)$$

$$= 120 - (1 + L(0))$$

$$= 119$$

it follows that Q has been calculated as shown below:

$$Q = n' \sum_{l=1}^{K} r_l^2(\hat{a})$$

$$= 119 \sum_{l=1}^{21} r_l^2(\hat{a})$$

$$= 119[r_1^2(\hat{a}) + r_2^2(\hat{a}) + \cdots + r_{21}^2(\hat{a})]$$

$$= 119[(.005)^2 + (-.036)^2 + \cdots + (-.06)^2]$$

$$= 16.116$$

Since Table A3 tells us that

$$\chi^2_{[.05]}(K - n_p = 20) = 31.4104$$

and since

$$Q = 16.116 < 31.4104 = \chi^2_{[.05]}(20)$$

we cannot reject the adequacy of the above model by setting α, the probability of a Type I error, equal to .05. Moreover, since all of the absolute t-values corresponding to the individual sample autocorrelations and individual sample partial autocorrelations of the residuals are less than 2, we conclude that the model

$$z_t = (1 - \theta_1 B)a_t$$

where

$$z_t = y_t - y_{t-1}$$

is adequate. In Example 2.5 we have used this model to forecast future Absorbent Paper Towel sales.

4.4.2 Choosing A Model to Forecast the Hotel Room Occupancy Data

EXAMPLE 4.6

Figure 4.3 presents the TSERIES output of the RSAC, RSPAC, and the value of Q, the Box-Pierce statistic, having $K - n_p = 20$ degrees of freedom, that result when Model 3 of Example 4.4

$$(1 - \phi_1 B - \phi_2 B^2 - \phi_3 B^3)z_t = \delta + (1 - \theta_{1,12} B^{12})a_t$$

where

$$z_t = y_t^* - y_{t-12}^* \quad \text{and} \quad y_t^* = \ln y_t$$

is fit to the $n = 168$ monthly hotel room averages in Table 3.1. Although Model 3 uses $n_p = 5$ parameters (ϕ_1, ϕ_2, ϕ_3, $\theta_{1,12}$, and δ), since TSERIES calculates the estimate of δ separately from the estimates of ϕ_1, ϕ_2, ϕ_3, and $\theta_{1,12}$, TSERIES assumes that $n_p = 4$ and thus calculates Q by using the first

$$K = 20 + n_p = 20 + 4 = 24$$

sample autocorrelations of the residuals. Therefore, recalling that $d = 0$ is the degree of nonseasonal differencing and $D = 1$ is the degree of seasonal differencing used to transform the original time series values into stationary time series values, which implies that

$$n' = n - (d + LD)$$

$$= 168 - (0 + 12(1))$$

$$= 156$$

it follows that Q has been calculated as

$$Q = n' \sum_{l=1}^{K} r_l^2(\hat{a})$$

$$= 156 \sum_{l=1}^{24} r_l^2(\hat{a})$$

$$= 26.712$$

Since Table A3 tells us that

$$\chi_{[.05]}^2 (K - n_p = 20) = 31.4104$$

and since

$$Q = 26.712 < 31.4104 = \chi_{[.05]}^2(20)$$

we cannot reject the adequacy of Model 3 by setting α equal to .05. Moreover, since all but one of the absolute t-values corresponding to the sample autocorrelations and sample partial autocorrelations of the residuals in Figure 4.3 are less than 2 (only the absolute t-value corresponding to the sample autocorrelation of

FIGURE 4.3   TSERIES Output of the RSAC and RSPAC for the Model $(1 - \phi_1 B - \phi_2 B^2 - \phi_3 B^3)z_t = \delta + (1 - \theta_{1,12}B^{12})a_t$ where $z_t = y_t^* - y_{t-12}^*$ and $y_t^* = \ln y_t$

(a)   The RSAC

	VALUE	S.E.	T VALUE	−1.0	0.0	1.0
				+ − − − − − − − − − − − − − − − +		
1	0.008	0.080	0.103	I	*	I
2	−0.040	0.080	−0.504	I	**	I
3	0.041	0.080	0.509	I	**	I
4	0.013	0.080	0.157	I	*	I
5	−0.129	0.080	−1.610	I	****	I
6	0.073	0.082	0.892	I	**	I
7	0.081	0.082	0.990	I	***	I
8	0.005	0.083	0.066	I	*	I
9	−0.014	0.083	−0.168	I	*	I
10	0.176	0.083	2.126	I	*****	I
11	0.053	0.085	0.620	I	**	I
12	−0.021	0.085	−0.249	I	*	I
13	0.143	0.085	1.680	I	****	I
14	0.158	0.087	1.817	I	*****	I
15	0.020	0.089	0.231	I	*	I
16	0.048	0.089	0.541	I	**	I
17	0.070	0.089	0.786	I	**	I
18	−0.169	0.089	−1.901	I	*****	I
19	0.103	0.091	1.134	I	***	I
20	0.035	0.092	0.379	I	*	I
21	0.085	0.092	0.923	I	***	I
22	−0.018	0.092	−0.190	I	*	I
23	0.061	0.093	0.656	I	**	I
24	−0.006	0.093	−0.066	I	*	I
				+ − − − − − − − − − − − − − − +		

BOX-PIERCE TEST CHISQUARE STATISTIC WITH 20 DEGREES OF FREEDOM = 26.712

(Continued on page 155.)

the residuals at lag 10 is greater than 2), we conclude that Model 3 is adequate. In Example 3.4 we used this model to forecast future monthly hotel room averages. Next we present

1.   In Table 4.5 the SAS output of the least squares point estimates of the parameters ϕ_1, ϕ_2, ϕ_3, $\theta_{1,12}$, and μ in

   Model 3: $(1 - \phi_1 B - \phi_2 B^2 - \phi_3 B^3)z_t = \delta + (1 - \theta_{1,12}B^{12})a_t$
   where
   $\delta = \mu(1 - \phi_1 - \phi_2 - \phi_3)$

FIGURE 4.3   *Continued*

(b)   The RSPAC

	VALUE	S.E.	T VALUE	-1.0	0.0	1.0
1	0.008	0.080	0.103		*	
2	-0.040	0.080	-0.505		**	
3	0.042	0.080	0.519		**	
4	0.010	0.080	0.128		*	
5	-0.127	0.080	-1.583		****	
6	0.076	0.080	0.952		**	
7	0.070	0.080	0.876		**	
8	0.018	0.080	0.229		*	
9	-0.012	0.080	-0.148		*	
10	0.158	0.080	1.973		*****	
11	0.067	0.080	0.834		**	
12	0.002	0.080	0.019		*	
13	0.137	0.080	1.715		****	
14	0.152	0.080	1.899		****	
15	0.078	0.080	0.979		***	
16	0.055	0.080	0.691		**	
17	0.049	0.080	0.617		**	
18	-0.153	0.080	-1.911		****	
19	0.140	0.080	1.751		****	
20	-0.033	0.080	-0.406		*	
21	0.069	0.080	0.864		**	
22	-0.030	0.080	-0.376		*	
23	-0.022	0.080	-0.278		*	
24	-0.043	0.080	-0.534		**	

which are

$$\hat{\phi}_1 = .2764 \qquad \hat{\phi}_2 = .1603 \qquad \hat{\phi}_3 = -.2338$$

$$\hat{\theta}_{1,12} = .5282 \qquad \text{and} \qquad \hat{\mu} = .0332$$

and which differ slightly from the estimates obtained using TSERIES (see Table 4.4). Note that the SAS output of the estimated correlations of these least squares point estimates is also given in Table 4.5, and note that these least squares point estimates yield the following point estimate of δ:

$$\hat{\delta} = \hat{\mu}(1 - \hat{\phi}_1 - \hat{\phi}_2 - \hat{\phi}_3)$$

$$= .0332 (1 - .2764 - .1603 - (-.2338)) = .0264$$

TABLE 4.5 SAS Output of the Least Squares Estimation of ϕ_1, ϕ_2, ϕ_3, and $\theta_{1,12}$ in the Model $(1 - \phi_1 B - \phi_2 B^2 - \phi_3 B^3)z_t = \delta + (1 - \theta_{1,12}B^{12})a_t$ where $z_t = y_t^* - y_{t-12}^*$

(a)   The Least Squares Point Estimates and their Associated t-values

ARIMA: LEAST SQUARES ESTIMATION				
PARAMETER	ESTIMATE	STD ERROR	T RATIO	LAG
MU	0.0331582	0.00103717	31.97	0
MA1,1	0.528167	0.0744367	7.10	12
AR1,1	0.276362	0.0799611	3.46	1
AR1,2	0.160343	0.0822188	1.95	2
AR1,3	−0.233759	0.0808655	−2.89	3

CONSTANT ESTIMATE   = 0.0264288
VARIANCE ESTIMATE   = .000402545
STD ERROR ESTIMATE = 0.0200635[a]
NUMBER OF RESIDUALS = 156

(b)   The Estimated Correlations of the Least Squares Point Estimates

CORRELATIONS OF THE ESTIMATES					
	MU	MA1,1	AR1,1	AR1,2	AR1,3
MU	1.000	0.133	0.017	0.028	0.003
MA1,1	0.133	1.000	0.060	0.107	0.076
AR1,1	0.017	0.060	1.000	−0.226	−0.114
AR1,2	0.028	0.107	−0.226	1.000	−0.232
AR1,3	0.003	0.076	−0.114	−0.232	1.000

[a] Standard error (s)

Also, note that Table 4.5 gives the SAS output of the standard errors and t-values associated with the least squares point estimates $\hat{\phi}_1$, $\hat{\phi}_2$, $\hat{\phi}_3$, $\hat{\theta}_{1,12}$, and $\hat{\mu}$ (recall that TSERIES does not give a t-value for the point estimate of μ).

2.   In Table 4.6 the SAS output of the RSAC and the values of Q^*, the Ljung-Box statistic, and the related prob-values for $K = 6, 12, 18, 24$, and 30. For example, recalling that $d = 0$ is the degree of nonseasonal differencing and $D = 1$ is the degree of seasonal differencing used to transform the original time series values into stationary time series values, which implies that

$$n' = n - (d + LD)$$

$$= 168 - (0 + 12(1))$$

$$= 156$$

TABLE 4.6   *SAS Output of the RSAC for the Model* $(1 - \phi_1 B - \phi_2 B^2 - \phi_3 B^3)z_t = \delta + (1 - \theta_{1,12}B^{12})a_t$ *where*
$z_t = y_t^* - y_{t-12}^*$

AUTOCORRELATION CHECK OF RESIDUALS

TO LAG	CHI SQUARE	DF	PROB	AUTOCORRELATIONS					
6	2.82	1	0.093	−0.003	−0.014	0.036	0.012	−0.104	0.069
12	10.28	7	0.173	0.108	0.005	0.001	0.177	0.038	−0.015
18	24.70	13	0.025	0.135	0.163	0.026	0.053	0.087	−0.163
24	29.76	19	0.055	0.126	0.030	0.090	0.003	0.054	0.007
30	34.65	25	0.095	0.102	0.021	0.113	−0.045	0.002	0.015

it follows that Q^* has been calculated for $K = 18$ as follows:

$$Q^* = n'(n' + 2) \sum_{l=1}^{K} (n' - l)^{-1} r_l^2(\hat{a})$$

$$= 156(156 + 2) \sum_{l=1}^{18} (156 - l)^{-1} r_l^2(\hat{a})$$

$$= 24.70$$

Moreover, since, for $K = 18$ the prob-value is the area under the curve of the chi-square distribution having $K - n_p = 18 - 5 = 13$ degrees of freedom to the right of $Q^* = 24.70$, and since the SAS output tells us that this prob-value is .025, which is less than .05, we can reject the adequacy of Model 3 by using $K = 18$ and $\alpha = .05$. However, since the prob-values associated with the values of Q^* for $K = 6, 12, 24,$ and 30 are greater than .05 (although, admittedly, some of these prob-values are fairly small), we will not reject the overall adequacy of Model 3. Although we have not presented the SAS output of the RSPAC, in later discussions we will present SAS output of the RSPAC (and SAS output of both the RSAC and RSPAC that are more informative than the output given in Table 4.6).

3.   In Table 4.7 the SAS output of the point forecasts and prediction interval forecasts of logged and actual hotel room occupancies in periods 169–180. Although TSERIES and SAS give somewhat different point estimates of the parameters in Model 3, comparison of Tables 3.4 and 4.7 tells us that both give extremely similar point forecasts and prediction interval forecasts of the hotel room occupancies in periods 169–180.

To complete this example, we note that

1.   Since the absolute *t*-value associated with $\hat{\phi}_2$ in

TABLE 4.7   SAS Output of Point Forecasts and Prediction Interval Forecasts of Logged and Actual Hotel Room Occupancies in Periods 169–180 Made Using the Model $(1 - \phi_1 B - \phi_2 B^2 - \phi_3 B^3)z_t = \delta + (1 - \theta_{1,12}B^{12})a_t$
where $z_t = y_t^* - y_{t-12}^*$

(a)   Forecasts of Logged Hotel Room Occupancies

SAS

FORECASTS FOR VARIABLE LY

OBS	FORECAST	STD ERROR	LOWER 95%	UPPER 95%
- - - - - - - - - - FORECAST BEGINS - - - - - - - - - -				
169	6.7337	0.0201	6.6944	6.7730
170	6.6487	0.0208	6.6079	6.6895
171	6.6562	0.0214	6.6144	6.6981
172	6.7715	0.0215	6.7293	6.8136
173	6.7556	0.0215	6.7134	6.7978
174	6.8901	0.0216	6.8478	6.9325
175	7.0522	0.0216	7.0099	7.0946
176	7.0750	0.0216	7.0326	7.1173
177	6.8064	0.0216	6.7640	6.8488
178	6.8070	0.0216	6.7646	6.8494
179	6.6637	0.0216	6.6213	6.7061
180	6.7936	0.0216	6.7513	6.8360

(b)   Forecasts of Actual Hotel Room Occupancies

SAS

OBS	Y	L95CI	FY	U95CI
169	.	807.84	840.24	873.94
170	.	740.90	771.75	803.89
171	.	745.74	777.61	810.84
172	.	836.60	872.59	910.14
173	.	823.38	858.87	895.88
174	.	941.80	982.54	1025.05
175	.	1107.49	1155.41	1205.39
176	.	1133.01	1182.03	1233.17
177	.	866.14	903.61	942.71
178	.	866.66	904.16	943.28
179	.	750.95	783.44	817.34
180	.	855.14	892.15	930.75

Model 3: $(1 - \phi_1 B - \phi_2 B^2 - \phi_3 B^3)z_t = \delta + (1 - \theta_{1,12} B^{12})a_t$

is nearly equal to 2 (note that the TSERIES output in Table 4.4 indicates that

$$t_{\hat{\phi}_2} = 2.0408$$

and the SAS output in Table 4.5 indicates that

$$t_{\hat{\phi}_2} = 1.95)$$

we should include ϕ_2 in Model 3 and thus choose Model 3 over
Model 2: $(1 - \phi_1 B - \phi_3 B^3)z_t = \delta + (1 - \theta_{1,12} B^{12})a_t$

2. Since the SAS output in Table 4.8 of estimation, diagnostic checking, and forecasting for

Model 1: $(1 - \phi_1 B - \phi_3 B^3 - \phi_5 B^5)z_t = \delta + (1 - \theta_{1,12} B^{12})a_t$

indicates that

a. ϕ_5 should probably not be included in Model 1 (because

$$t_{\hat{\phi}_5} = -1.40$$

is not greater than 2 in absolute value—see Table 4.8a);

b. the standard error for Model 1 ($s = .0202$—see Table 4.8a) is slightly greater than the standard error for Model 3 ($s = .0201$—see Table 4.5a);

c. the prob-values associated with the values of Q^* for $K = 6, 12, 18$, and 24 are smaller for Model 1 (see Table 4.8b) than for Model 3 (see Table 4.6), indicating Model 1 is inferior to Model 3;

d. the 95% prediction intervals for future values of $y_t^* = \ln y_t$ for periods $t = 169$ through 180 are slightly longer for Model 1 (see Figure 4.8d) than for Model 3 (see Table 4.7a);

we will choose Model 3 over Model 1. In summary, then, we choose Model 3 over Model 1 and Model 2.

Next, note that although SAS and TSERIES calculate a 95% prediction interval for y_t, we can calculate a $100(1 - \alpha)\%$ prediction interval for y_t for any value of α by using the equation

$$[\hat{y}_t \pm [t_{[\alpha/2]}^{(n-n_p)}/t_{[.025]}^{(n-n_p)}][(U - L)/2]]$$

or, if $(n - n_p)$ is at least 30, by using the equation

$$[\hat{y}_t \pm [z_{[\alpha/2]}/z_{[.025]}][(U - L)/2]]$$

Here, $\hat{y}_t$ is the point forecast of y_t, and $(U - L)$ is the difference between the upper and lower limits of a 95% prediction interval—[L, U]—for y_t. For example, noting that in

TABLE 4.8 *SAS Output of Estimation, Diagnostic Checking, and Forecasting for Model 1:* $(1 - \phi_1 B - \phi_3 B^3 - \phi_5 B^5)z_t = \delta + (1 - \theta_{1,12}B^{12})a_t$

(a) Estimation

ARIMA: LEAST SQUARES ESTIMATION

ITERATION	SSE	MU	MA1,1	AR1,1	AR1,2	AR1,3	CONSTANT	LAMBDA
0	0.0843462	.0330432	0.1	0.1	0.1	0.1	0.0231303	1.0E−05
1	0.0623326	.0331693	0.383237	0.176277	−.213693	−.150517	0.0394029	1.0E−06
2	0.061585	.0331426	0.431551	0.259346	−.198627	−.125825	0.0353003	1.0E−07
3	0.06152	.0331321	0.456369	0.272472	−.192197	−.117361	0.0343608	1.0E−08
4	0.0615137	.0331144	0.46116	0.278457	−.189557	−.113913	0.0339427	1.0E−09
5	0.0615131	.0331128	0.463486	0.279776	−.188923	−.113042	0.0338476	1.0E−10
6	0.061513	.033111	0.463995	0.280351	−.188674	−.112707	0.0338074	1.0E−11

PARAMETER	ESTIMATE	STD ERROR	T RATIO	LAG
MU	0.033111	.000911575	36.32	0
MA1,1	0.463995	0.0785255	5.91	12
AR1,1	0.280351	0.0790211	3.55	1
AR1,2	−0.188674	0.0802503	−2.35	3
AR1,3	−0.112707	0.0805691	−1.40	5

CONSTANT ESTIMATE = 0.0338074
VARIANCE ESTIMATE = .000407371
STD ERROR ESTIMATE = 0.0201834
NUMBER OF RESIDUALS = 156

CORRELATIONS OF THE ESTIMATES

	MU	MA1,1	AR1,1	AR1,2	AR1,3
MU	1.000	0.132	0.026	0.013	0.008
MA1,1	0.132	1.000	0.115	0.096	0.150
AR1,1	0.026	0.115	1.000	−0.166	0.121
AR1,2	0.013	0.096	−0.166	1.000	−0.153
AR1,3	0.008	0.150	0.121	−0.153	1.000

TABLE 4.8 Continued   (Part (c), the RSPAC, and Part (d), the Forecast, follow on pages 162 and 163.)

(b) The RSAC

AUTOCORRELATION CHECK OF RESIDUALS

TO LAG	CHI SQUARE	DF	PROB	AUTOCORRELATIONS					
6	5.34	1	0.021	-0.015	0.132	0.018	0.050	-0.014	0.111
12	12.66	7	0.081	0.095	0.052	0.020	0.166	0.063	-0.003
18	28.94	13	0.007	0.137	0.175	0.078	0.067	0.116	-0.140
24	38.15	19	0.006	0.169	0.027	0.130	0.008	0.068	-0.013
30	44.87	25	0.009	0.132	0.004	0.127	-0.033	0.027	0.013

AUTOCORRELATION PLOT OF RESIDUALS

LAG	COVARIANCE	CORRELATION	-1 ... 1	STD
0	.000407371	1.00000	|********************|	0
1	-5.908E-06	-0.01450	.	0.0800641
2	0.00005363	0.13165	***	0.0800809
3	7.524E-06	0.01847	.	0.0814564
4	.000020206	0.04960	*	0.0814833
5	-5.661E-06	-0.01390	.	0.0816766
6	.000045417	0.11149	**	0.0816918
7	0.00003856	0.09466	**	0.0826613
8	.000021242	0.05214	*	0.0833533
9	8.322E-06	0.02043	.	0.0835621
10	.000067599	0.16594	***	0.0835941
11	.000025524	0.06266	*	0.0856796
12	-1.353E-06	-0.00332	.	0.0859728
13	.000055648	0.13660	***	0.0859737
14	.000071253	0.17491	***	0.0873539
15	.000031827	0.07813	**	0.0895708
16	.000027385	0.06722	*	0.0900065
17	0.00004728	0.11606	**	0.0903278
18	-.000005703	-0.13999	***	0.0912787
19	.000006871	0.16867	***	0.0926449
20	.000010947	0.02687	*	0.0945928
21	.000052858	0.12975	***	0.0946417
22	3.101E-06	0.00761	.	0.0957753
23	.000027614	0.06779	*	0.0957792
24	-5.231E-06	-0.01284	.	0.0960862

'.' MARKS TWO STANDARD ERRORS

161

TABLE 4.8 Continued

(c) The RSPAC

PARTIAL AUTOCORRELATIONS

LAG	CORRELATION
1	−0.01450
2	0.13147
3	0.02243
4	0.03347
5	−0.01833
6	0.10181
7	0.10316
8	0.02941
9	−0.00578
10	0.15097
11	0.06982
12	−0.05002
13	0.10350
14	0.17922
15	0.06651
16	0.00136
17	0.06915
18	−0.16228
19	0.14044
20	0.00366
21	0.03645
22	0.00002
23	−0.01555
24	−0.06818

TABLE 4.8 *Continued*

(d) Forecasts of Logged Hotel Room Occupancies

SAS

FORECASTS FOR VARIABLE LY

OBS	FORECAST	STD ERROR	LOWER 95%	UPPER 95%
- - - - - -	- - FORECAST BEGINS	- - - - - -		
169	6.7357	0.0202	6.6961	6.7752
170	6.6445	0.0210	6.6034	6.6855
171	6.6565	0.0210	6.6153	6.6977
172	6.7692	0.0213	6.7275	6.8109
173	6.7512	0.0214	6.7093	6.7931
174	6.8852	0.0216	6.8429	6.9276
175	7.0501	0.0216	7.0077	7.0925
176	7.0733	0.0216	7.0309	7.1156
177	6.8053	0.0217	6.7629	6.8477
178	6.8059	0.0217	6.7635	6.8484
179	6.6637	0.0217	6.6213	6.7062
180	6.7947	0.0217	6.7522	6.8372

Table 4.7a a 95% prediction interval for $y^*_{169} = \ln y_{169}$ given by Model 3 is

$$[6.6944, \quad 6.7730] \quad = \quad [L, U]$$

it follows, since

1. $(U - L)/2 \ = \ (6.7730 - 6.6944)/2 \ = \ .0393$

2. $\hat{y}^*_{169} \ = \ 6.7337$ \qquad (see Table 4.7a)

3. $(n - n_p) \ = \ (168 - 5) \ = \ 163$ \qquad is large

that a 99% prediction interval for y^*_{169} is (since $\alpha = .01$ and $z_{[\alpha/2]} = z_{[.005]}$)

$$[\hat{y}^*_{169} \pm (z_{[.005]}/z_{[.025]})((U - L)/2)]$$

$$= \quad [6.7337 \pm (2.58/1.96)(.0393)]$$

$$= \quad [6.7337 \pm .0517]$$

$$= \quad [6.682, \quad 6.7854]$$

and thus that a 99% prediction interval for

$$y_{169} \ = \ e^{y^*_{169}}$$

is

$$[e^{6.682}, e^{6.7854}] \ = \ [797.91, \quad 884.83]$$

4.4.3 *Identifying Improved Models*

In general, if Q^* (or Q) indicates that we should reject the adequacy of a Box-Jenkins model describing a particular time series, we should use the behaviors of the RSAC and RSPAC at the nonseasonal and seasonal levels to (tentatively) identify (by using the guidelines in Tables 2.3 and 3.3) a model describing the residuals. We then combine the model describing the original time series values with the model describing the residuals (obtained from the former model) to arrive at a new model describing the original time series values. We will illustrate this process in Example 4.7. For now, note that when we discussed in Sections 2.4 and 3.3 the use of the guidelines in Tables 2.3 and 3.3 to identify a model describing the original time series values, we suggested various critical absolute *t*-values for identifying spikes at the nonseasonal and seasonal levels of the SAC.

In Table 4.9 we summarize these previously suggested critical absolute *t*-values and suggest somewhat different values for identifying spikes at the nonseasonal and seasonal levels of the RSAC (except for the critical values marked by a *, the suggestions are those of Pankratz [1983]).

Before presenting an example, two comments should be made. First, we recommend using the critical absolute *t*-values in Table 4.9 for identifying spikes in the RSAC only if Q^* (or Q) indicates that we should reject the adequacy of the overall model. If Q^* (or Q) indicates not rejecting the model's adequacy, we should only be concerned about individual sample autocorrelations of the residuals having absolute *t*-values

TABLE 4.9   Critical Absolute t-Values for Identifying Spikes in the SAC and RSAC (* = see text)

Lags	Critical Absolute t-values in SAC	Critical Absolute t-Values in RSAC
Nonseasonal lags		
Low (1, 2, perhaps 3)	1.6	1.25
Other nonseasonal lags (except Half Seasonal Lags)	2.0*	1.6
Seasonal lags		
Exact seasonal (L, $2L$, $3L$, $4L$)	1.25	1.25
Near seasonal ($L - 2$, $L - 1$, $L + 1$, $L + 2$, $2L - 2$, $2L - 1$, $2L + 1$, $2L + 2$, $3L - 2$, $3L - 1$, $3L + 1$, $3L + 2$, $4L - 2$, $4L - 1$, $4L + 1$, $4L + 2$)	1.6	1.25
Half seasonal ($.5L$, $1.5L$, $2.5L$, $3.5L$)	1.6	1.25
Other lags	2.0*	1.6

greater than 2. Second, whether or not Q^* (or Q) indicates that we should reject the adequacy of the overall model, we recommend only being concerned about individual sample *partial* autocorrelations of the residuals having absolute t-values greater than 2.

EXAMPLE 4.7

In the hotel room occupancy example, consider the model

$$z_t = \delta + (1 - \theta_{1,12}B^{12})a_t$$

and note that Figure 4.4 presents the TSERIES output of the RSAC, the RSPAC, and the value of Q having 20 degrees of freedom. Since

$$Q = 51.374 > 31.4104 = \chi^2_{[.05]}(20)$$

we reject the adequacy of the model by setting α equal to .05. This implies that the "random shocks" in the model are really not statistically independent; to denote this fact, we will replace the symbol a_t by η_t and write the above model as

$$z_t = \delta + (1 - \theta_{1,12}B^{12})\eta_t$$

To tentatively identify a model describing η_t (which represents the residuals), note that

1. At the *nonseasonal level*, the RSAC dies down fairly quickly, and the RSPAC has spikes at lags 1 and 3 (that is, the absolute t-values corresponding to the sample partial autocorrelations of the residuals at lags 1 and 3 are greater than 2) and cuts off after lag 3. It follows, by Guideline 2 in Table 2.3, that we should use some form of $\phi_p(B)$, the nonseasonal autoregressive operator of order p. Since the RSPAC has spikes at lags

FIGURE 4.4　TSERIES Output of RSAC and RSPAC for the Model $z_t = \delta + (1 - \theta_{1,12}B^{12})a_t$　　where　$z_t = y_t^* - y_{t-12}^*$

(a)　The RSAC

	VALUE	S.E.	T VALUE	−1.0	0.0	1.0
				+ − − − − − − − − − − − − − − − − − +		
1	0.281	0.080	3.505	I	********	I
2	0.152	0.086	1.763	I	****	I
3	−0.143	0.088	−1.627	I	****	I
4	−0.092	0.089	−1.032	I	***	I
5	−0.170	0.090	−1.887	I	*****	I
6	0.050	0.092	0.549	I	**	I
7	0.069	0.092	0.753	I	**	I
8	0.094	0.092	1.012	I	***	I
9	0.051	0.093	0.547	I	**	I
10	0.146	0.093	1.564	I	****	I
11	0.070	0.095	0.734	I	**	I
12	0.040	0.095	0.419	I	**	I
13	0.158	0.095	1.661	I	*****	I
14	0.191	0.097	1.975	I	*****	I
15	0.124	0.099	1.246	I	****	I
16	0.037	0.100	0.373	I	*	I
17	0.025	0.100	0.248	I	*	I
18	−0.140	0.100	−1.398	I	****	I
19	0.077	0.102	0.761	I	***	I
20	0.038	0.102	0.368	I	*	I
21	0.132	0.102	1.299	I	****	I
22	−0.012	0.103	−0.120	I	*	I
23	0.068	0.103	0.661	I	**	I
24	0.002	0.103	0.022	I	*	I
				+ − − − − − − − − − − − − − − − +		

BOX-PIERCE TEST CHISQUARE STATISTIC WITH 20 DEGREES OF FREEDOM = 51.374

(Continued on page 167.)

1 and 3 and cuts off after lag 3, it might be appropriate to use either

$$(1 - \phi_1 B - \phi_3 B^3)$$

or

$$(1 - \phi_1 B - \phi_2 B^2 - \phi_3 B^3)$$

2. At the *seasonal level*, neither the RSAC nor RSPAC has a spike at lag 12 (although there are spikes in the RSAC and RSPAC at near seasonal lags). It follows, by Guideline 9 in Table 3.3, that it might be appropriate to not use a seasonal operator.

FIGURE 4.4 Continued

(b) The RSPAC

	VALUE	S.E.	T VALUE	−1.0	0.0	1.0
				+ − − − − − − − −	− − − − − − − +	
1	0.281	0.080	3.505	I	********	I
2	0.079	0.080	0.991	I	***	I
3	−0.223	0.080	−2.789	I	******	I
4	−0.013	0.080	−0.157	I	*	I
5	−0.101	0.080	−1.260	I	***	I
6	0.121	0.080	1.509	I	****	I
7	0.052	0.080	0.654	I	**	I
8	−0.015	0.080	−0.182	I	*	I
9	0.030	0.080	0.371	I	*	I
10	0.144	0.080	1.794	I	****	I
11	0.035	0.080	0.440	I	*	I
12	−0.007	0.080	−0.091	I	*	I
13	0.212	0.080	2.645	I	******	I
14	0.149	0.080	1.865	I	****	I
15	0.041	0.080	0.507	I	**	I
16	−0.003	0.080	−0.034	I	*	I
17	0.055	0.080	0.690	I	**	I
18	−0.106	0.080	−1.321	I	***	I
19	0.196	0.080	2.451	I	******	I
20	−0.027	0.080	−0.342	I	*	I
21	0.001	0.080	0.008	I	*	I
22	−0.047	0.080	−0.583	I	**	I
23	−0.010	0.080	−0.131	I	*	I
24	0.006	0.080	0.078	I	*	I
				+ − − − − − − − −	− − − − − − − +	

By inserting each of the forms

$$(1 - \phi_1 B - \phi_3 B^3)$$

and

$$(1 - \phi_1 B - \phi_2 B^2 - \phi_3 B^3)$$

of $\phi_p(B)$ into the general model

$$\phi_p(B)\phi_P(B^L)\eta_t = \delta + \theta_q(B)\theta_Q(B^L)a_t$$

we are led to consider two tentative models describing the residuals

$$(1 - \phi_1 B - \phi_3 B^3)\eta_t = a_t$$

and

$$(1 - \phi_1 B - \phi_2 B^2 - \phi_3 B^3)\eta_t = a_t$$

Combining the model

$$(1 - \phi_1 B - \phi_2 B^2 - \phi_3 B^3)\eta_t = a_t$$

which implies that

$$\eta_t = \frac{a_t}{(1 - \phi_1 B - \phi_2 B^2 - \phi_3 B^3)}$$

with the model

$$z_t = \delta + (1 - \theta_{1,12} B^{12})\eta_t$$

yields the model

$$z_t = \delta + (1 - \theta_{1,12} B^{12}) \left(\frac{a_t}{(1 - \phi_1 B - \phi_2 B^2 - \phi_3 B^3)} \right)$$

or

$$(1 - \phi_1 B - \phi_2 B^2 - \phi_3 B^3)z_t = \delta(1 - \phi_1 B - \phi_2 B^2 - \phi_3 B^3)$$
$$+ (1 - \theta_{1,12} B^{12})a_t$$

or

$$(1 - \phi_1 B - \phi_2 B^2 - \phi_3 B^3)z_t = \delta' + (1 - \theta_{1,12} B^{12})a_t$$

Since

$$\delta' = \delta(1 - \phi_1 B - \phi_2 B^2 - \phi_3 B^3)$$

is a *constant term*, we can rewrite δ' as δ and rewrite the above model as

$$(1 - \phi_1 B - \phi_2 B^2 - \phi_3 B^3)z_t = \delta + (1 - \theta_{1,12} B^{12})a_t$$

Thus, the use of the RSAC and RSPAC resulting from the model

$$z_t = \delta + (1 - \theta_{1,12} B^{12})\eta_t$$

has led to a new model that we determined in Example 4.6 adequately describes the monthly hotel room averages time series. In fact, since the absolute *t*-value corresponding to the sample partial autocorrelation of the residuals at lag 5 (1.26 from Figure 4.4), is less than the absolute *t*-value corresponding to the sample partial autocorrelation of the time series values produced by the transformation $z_t = y_t^* - y_{t-12}^*$ at lag 5 (2.199 from Figure 3.5), utilizing the RSPAC resulting from the model

$$z_t = \delta + (1 - \theta_{1,12} B^{12})\eta_t$$

makes the spike at lag 5 look less important and thus makes the nonseasonal nature of the time series clearer than does using the SPAC of the time series values produced by the transformation $z_t = y_t^* - y_{t-12}^*$.

In general, it has been found that fitting a model describing only the seasonal nature of a time series and then analyzing the RSAC and RSPAC resulting from the model can clarify the nonseasonal nature of the time series.

EXAMPLE 4.8

In Example 3.5 we concluded that the model

$$z_t = (1 - \theta_1 B)(1 - \theta_{1,12} B^{12}) a_t$$

where

$$z_t = y_t^* - y_{t-1}^* - y_{t-12}^* + y_{t-13}^* \quad \text{and} \quad y_t^* = \ln y_t$$

is a reasonable tentative model describing monthly passenger totals (measured in thousands of passengers) in international air travel from 1949 to 1959. The SAS output of estimation and diagnostic checking for this model, which is given in Figure 4.5, indicates that this model is an adequate final model.

4.5 SEVERAL EXAMPLES OF BOX-JENKINS MODELING

We will complete this chapter with three examples. The first two illustrate modifying a multiplicative model to form an additive model and then choosing between them by using the techniques of this chapter. The third example illustrates a difficult Box-Jenkins modeling problem in which we use the concept of invertibility to show that it might be appropriate to use a model other than the Box-Jenkins models of Chapters 2, 3, and 4.

EXAMPLE 4.9

The data for this example is given by Miller and Wichern [1977]. They analyzed 120 monthly values of average weekly total investments per month of large New York City banks. This time series is presented in Table 4B of Appendix B. Stationary time series values were produced by using the stationarity transformation

$$z_t = y_t - y_{t-1}$$

The SAC and SPAC of these stationary values are presented in Figure 4.6.

1. At the *nonseasonal level*, the SAC has a spike at lag 6 and cuts off after lag 6, and the SPAC seems to die down fairly quickly. It follows, by Guideline 1 in Table 2.3 that we should use the form

   $$\theta_q(B) = (1 - \theta_6 B^6)$$

   of the nonseasonal moving average operator.

2. At the *seasonal level*, the SAC has a spike at lag 12 and cuts off after lag

FIGURE 4.5   SAS Output of Estimation and Diagnostic Checking for the Model

$$z_t = (1 - \theta_1 B)(1 - \theta_{1,12} B^{12}) a_t$$

where

$$z_t = y_t^* - y_{t-1}^* - y_{t-12}^* + y_{t-13}^*$$

(a)   Estimation

SAS

ARIMA: PRELIMINARY ESTIMATION

MOVING AVERAGE ESTIMATES

1   0.31710

MOVING AVERAGE ESTIMATES

12   0.41035

WHITE NOISE VARIANCE EST = 0.00172153

ARIMA: LEAST SQUARES ESTIMATION

ITERATION	SSE	MA1,1	MA2,1	LAMBDA
0	0.164104	0.317101	0.410346	1.0E−05
1	0.155327	0.351138	0.630036	1.0E−06
2	0.155312	0.34261	0.629249	1.0E−07
3	0.155311	0.341116	0.629797	1.0E−08
4	0.155311	0.340773	0.630059	1.0E−09

PARAMETER	ESTIMATE	STD ERROR	T RATIO	LAG
MA1,1	0.340773	0.0870011	3.92	1
MA2,1	0.630059	0.0698134	9.02	12

VARIANCE ESTIMATE   = 0.00132745
STD ERROR ESTIMATE   = 0.0364341
NUMBER OF RESIDUALS = 119

CORRELATIONS OF THE ESTIMATES

	MA1,1	MA2,1
MA1,1	1.000	−0.036
MA2,1	−0.036	1.000

(b)   The RSAC

AUTOCORRELATION CHECK OF RESIDUALS

TO LAG	CHI SQUARE	DF	PROB	AUTOCORRELATIONS					
6	5.81	4	0.214	0.008	0.042	−0.158	−0.129	0.036	0.043
12	7.50	10	0.677	−0.027	−0.055	0.077	−0.051	0.022	0.009
18	12.21	16	0.729	0.047	0.040	0.016	−0.165	0.037	−0.033
24	19.70	22	0.602	−0.024	−0.105	−0.048	−0.033	0.188	0.013

FIGURE 4.6 The SAC and SPAC of the Investment Values Obtained Using the Transformation $z_t = y_t - y_{t-1}$

(a) The SAC

	VALUE	S.E.	T VALUE	−1.0	0.0	1.0
				+ − − − − − − − − − −	− − − − − − − −	− − − +
1	−0.060	0.092	−0.656	I	**	I
2	−0.069	0.092	−0.755	I	**	I
3	0.042	0.092	0.450	I	**	I
4	0.012	0.093	0.133	I	*	I
5	−0.056	0.093	−0.609	I	**	I
6	−0.292	0.093	−3.140	I	********	I
7	−0.162	0.100	−1.619	I	*****	I
8	−0.029	0.102	−0.282	I	*	I
9	0.141	0.103	1.373	I	****	I
10	−0.075	0.104	−0.724	I	**	I
11	−0.003	0.105	−0.024	I	*	I
12	0.274	0.105	2.624	I	********	I
13	−0.036	0.111	−0.326	I	*	I
14	0.046	0.111	0.415	I	**	I
15	0.051	0.111	0.456	I	**	I
16	−0.057	0.111	−0.517	I	**	I
17	−0.013	0.111	−0.113	I	*	I
18	−0.214	0.111	−1.925	I	******	I
19	0.052	0.115	0.453	I	**	I
20	−0.080	0.115	−0.696	I	**	I
21	0.081	0.115	0.703	I	***	I
22	−0.122	0.116	−1.054	I	****	I
23	0.002	0.117	0.017	I	*	I
24	0.204	0.117	1.746	I	******	I
				+ − − − − − − − − − −	− − − − − − − −	− − − +

(Part (b), The SPAC, follows on page 172.)

12, and the SPAC seems to die down quickly. It follows, by Guideline 6 in Table 3.3, that we should use the seasonal moving average operator of order 1

$$\theta_1(B^{12}) = (1 - \theta_{1,12}B^{12})$$

By inserting

$$\theta_q(B) = (1 - \theta_6 B^6)$$

and

$$\theta_1(B^{12}) = (1 - \theta_{1,12}B^{12})$$

FIGURE 4.6   Continued

(b)   The SPAC

	VALUE	S.E.	T VALUE	−1.0	0.0	1.0
				+ — — — — — — — —	— — — — — — —	— — +
1	−0.060	0.092	−0.656	I	**	I
2	−0.073	0.092	−0.800	I	**	I
3	0.033	0.092	0.360	I	*	I
4	0.012	0.092	0.131	I	*	I
5	−0.050	0.092	−0.547	I	**	I
6	−0.302	0.092	−3.293	I	********	I
7	−0.236	0.092	−2.571	I	******	I
8	−0.127	0.092	−1.386	I	****	I
9	0.135	0.092	1.472	I	****	I
10	−0.038	0.092	−0.414	I	*	I
11	−0.036	0.092	−0.389	I	*	I
12	0.151	0.092	1.645	I	****	I
13	−0.128	0.092	−1.399	I	****	I
14	0.002	0.092	0.023	I	*	I
15	0.098	0.092	1.064	I	***	I
16	−0.022	0.092	−0.236	I	*	I
17	0.002	0.092	0.021	I	*	I
18	−0.182	0.092	−1.989	I	*****	I
19	0.084	0.092	0.915	I	***	I
20	−0.078	0.092	−0.856	I	**	I
21	0.096	0.092	1.051	I	***	I
22	−0.095	0.092	−1.034	I	***	I
23	−0.077	0.092	−0.836	I	**	I
24	0.024	0.092	0.259	I	*	I
				+ — — — — — — — —	— — — — — — —	— — +

into the general model

$$\phi_p(B)\phi_P(B^{12})z_t = \delta + \theta_q(B)\theta_Q(B^{12})a_t$$

we are led to try the tentative model

$$z_t = (1 - \theta_6 B^6)(1 - \theta_{1,12}B^{12})a_t$$

or the (modified) additive model

$$z_t = (1 - \theta_6 B^6 - \theta_{12}B^{12})a_t$$

Note that we do not use δ in the above models because $\bar{z}$ is not statistically different from zero. If we use the techniques of this chapter to choose the better model (that is, select the model providing the best combination of significant

$t_{\hat{\theta}}$-statistics, a small standard error (s), a small Ljung-Box statistic (Q^*), and short prediction interval forecasts), we conclude that the additive model

$$z_t = (1 - \theta_6 B^6 - \theta_{12} B^{12}) a_t$$

or

$$y_t - y_{t-1} = (1 - \theta_6 B^6 - \theta_{12} B^{12}) a_t$$

is an appropriate final model.

EXAMPLE 4.10

The data for this example is given by Mabert [1976], who analyzed sales of switches by the Jackson Company. Since the Jackson Company divides their year into 13 periods, $L = 13$. Mabert found that the *original* time series values were stationary. The SAC and SPAC of these stationary time series values are presented in Figure 4.7.

1. At the *nonseasonal level*, the SAC dies down fairly quickly, and the SPAC has spikes at lags 1 and 2 and cuts off after lag 2. It follows, by Guideline 2 in Table 2.3, that we should use the nonseasonal autoregressive operator of order 2

    $$\phi_2(B) = (1 - \phi_1 B - \phi_2 B^2)$$

2. At the *seasonal level*, the SAC dies down fairly quickly, and the SPAC (which is shown through lag 13) has spikes at lags 12 and 13. It follows, by Guideline 7 in Table 3.3, that we should use the form

    $$(1 - \phi_{1,12} B^{12})$$

    or possibly the form (recall that $L = 13$)

    $$(1 - \phi_{1,13} B^{13})$$

    of $\phi_P(B^{13})$, the seasonal autoregressive operator.

By inserting

$$\phi_2(B) = (1 - \phi_1 B - \phi_2 B^2)$$

and one of the forms

$$(1 - \phi_{1,12} B^{12})$$

and

$$(1 - \phi_{1,13} B^{13})$$

of $\phi_P(B^{13})$ into the general model

$$\phi_p(B)\phi_P(B^{13}) y_t = \delta + \theta_q(B)\theta_Q(B^{13}) a_t$$

(a)   The SAC

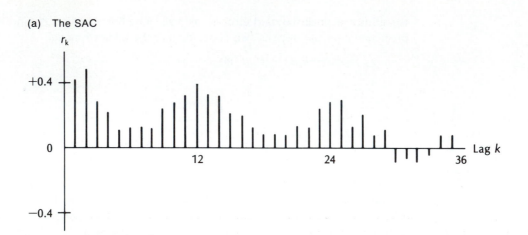

(b)   The SPAC

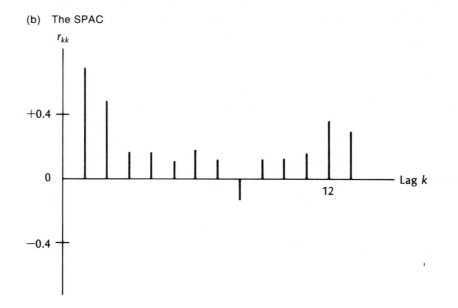

FIGURE 4.7   *The SAC and SPAC of the Switch Sales Original Values*

we are led to try the tentative models

$$(1 - \phi_1 B - \phi_2 B^2)(1 - \phi_{1,12} B^{12}) y_t = \delta + a_t$$

and

$$(1 - \phi_1 B - \phi_2 B^2)(1 - \phi_{1,13} B^{13}) y_t = \delta + a_t$$

and possibly the (modified) additive models

$$(1 - \phi_1 B - \phi_2 B^2 - \phi_{12} B^{12}) y_t = \delta + a_t$$

and

$$(1 - \phi_1 B - \phi_2 B^2 - \phi_{13} B^{13}) y_t = \delta + a_t$$

Here we include δ because $\bar{y}$ is statistically different from zero. Using the techniques of this chapter, we conclude that the additive model

$$(1 - \phi_1 B - \phi_2 B^2 - \phi_{12} B^{12}) y_t = \delta + a_t$$

is an appropriate final model.

EXAMPLE 4.11

Farmers' Bureau Co-op, a small agricultural cooperative, would like to predict its propane gas bills for the four quarters of next year. In order to do this, Farmers' Bureau has compiled its quarterly propane gas bills for the last ten years. These $n = 40$ gas bills, which are given in Table 4.10 and plotted in Figure 4.8, exhibit an upward trend and seasonal variation. The TSERIES output of the SAC obtained by using the transformation

1. $z_t = y_t$ is presented in Figure 4.9.
2. $z_t = y_t - y_{t-1}$ is presented in Figure 4.10.
3. $z_t = y_t - y_{t-4}$ is presented in Figure 4.11a.
4. $z_t = y_t - y_{t-1} - y_{t-4} + y_{t-5}$ is presented in Figure 4.12.

Since both the SAC in Figure 4.9 and the SAC in Figure 4.10 die down extremely slowly at the seasonal level (that is, at lags $L = 4$, $2L = 8$, $3L = 12$, and $4L = 16$), we conclude that (1) the values produced by the transformation $z_t = y_t$ are nonstationary and (2) the values produced by the transformation $z_t = y_t - y_{t-1}$

TABLE 4.10   *Quarterly Propane Gas Bills for Farmers' Bureau Co-op*

Year	Quarter 1	Quarter 2	Quarter 3	Quarter 4
1	$344.39(= y_1)$	$246.63(= y_2)$	$131.53(= y_3)$	$288.87(= y_4)$
2	$313.45(= y_5)$	$189.76(= y_6)$	$179.10(= y_7)$	$221.10(= y_8)$
3	246.84	209.00	51.21	133.89
4	277.01	197.98	50.68	218.08
5	365.10	207.51	54.63	214.09
6	267.00	230.28	230.32	426.41
7	467.06	306.03	253.23	279.46
8	336.56	196.67	152.15	319.67
9	440.00	315.04	216.42	339.78
10	$434.66(= y_{37})$	$399.66(= y_{38})$	$330.80(= y_{39})$	$539.78(= y_{40})$

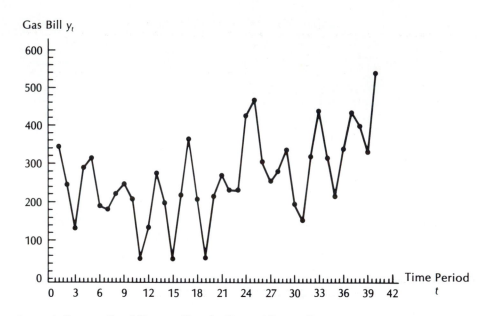

FIGURE 4.8   Quarterly Propane Gas Bill versus Time for Farmers' Bureau Co-op

are nonstationary. Examining the SAC and SPAC in Figure 4.11a of the values produced by the transformation $z_t = y_t - y_{t-4}$, we make the following conclusions:

1. At the *seasonal level*, the SAC has a spike at lag 4 and cuts off after lag 4, and the SPAC dies down (note the steadily decreasing t-values at lags 4, 8, and 12). It follows, by Guideline 6 in Table 3.3, that we should use the seasonal moving average operator of order 1

$$\theta_1(B^4) = (1 - \theta_{1,4}B^4)$$

2. At the *nonseasonal level* (possibly lags 1 and 2), each of the SAC and SPAC has a spike at lag 1. If the SAC is cutting off after lag 1 and the SPAC is dying down, we should, by Guideline 1 in Table 2.3, use the nonseasonal moving average operator of order 1

$$\theta_1(B) = (1 - \theta_1 B)$$

If the SAC is dying down and the SPAC is cutting off after lag 1, we should, by Guideline 2 in Table 2.3, use the nonseasonal autoregressive operator of order 1

$$\phi_1(B) = (1 - \phi_1 B)$$

Since it is difficult to tell which of the above nonseasonal behaviors is occurring, we will try models using each of the above operators. By inserting each of the

FIGURE 4.9   TSERIES Output of the SAC for the Original Gas Bill Values

	VALUE	S.E.	T VALUE	−1.0	0.0	1.0
				+ − − − − − − − − − − −	− − −	− − +
1	0.391	0.158	2.471	I	**********	I
2	−0.055	0.181	−0.304	I	**	I
3	0.269	0.181	1.488	I	********	I
4	0.495	0.191	2.593	I	*************	I
5	0.063	0.221	0.285	I	**	I
6	−0.199	0.221	−0.898	I	*****	I
7	0.157	0.225	0.695	I	*****	I
8	0.393	0.228	1.723	I	**********	I
9	0.049	0.244	0.201	I	**	I
10	−0.227	0.245	−0.928	I	******	I
11	0.069	0.250	0.275	I	**	I
12	0.280	0.250	1.120	I	********	I
13	−0.012	0.258	−0.048	I	*	I
14	−0.251	0.258	−0.971	I	*******	I
15	0.074	0.264	0.279	I	**	I
16	0.237	0.265	0.894	I	*******	I
17	−0.142	0.270	−0.527	I	****	I
18	−0.335	0.272	−1.232	I	*********	I
19	−0.080	0.282	−0.283	I	**	I
20	0.047	0.282	0.166	I	**	I
21	−0.238	0.283	−0.841	I	******	I
22	−0.286	0.288	−0.994	I	********	I
23	−0.043	0.295	−0.146	I	**	I
24	−0.013	0.295	−0.043	I	*	I
				+ − − − − − − − − − − −	− − −	− − +

operators

$$\theta_1(B) = (1 - \theta_1 B)$$

and

$$\phi_1(B) = (1 - \phi_1 B)$$

and the operator

$$\theta_1(B^4) = (1 - \theta_{1,4} B^4)$$

into the general model

$$\phi_p(B)\phi_P(B^L)z_t = \delta + \theta_q(B)\theta_Q(B^L)a_t$$

we are led to consider the tentative models

Model 1: $z_t = (1 - \theta_1 B)(1 - \theta_{1,4} B^4)a_t$

FIGURE 4.10 TSERIES Output of the SAC for the Gas Bill Values Obtained by Using the Transformation $z_t = y_t - y_{t-1}$

	VALUE	S.E.	T VALUE	−1.0	0.0	1.0
1	−0.030	0.160	−0.185		*	
2	−0.756	0.160	−4.715		*******************	
3	0.049	0.234	0.211		**	
4	0.673	0.235	2.868		******************	
5	−0.077	0.280	−0.276		**	
6	−0.641	0.280	−2.285		*****************	
7	0.037	0.316	0.117		*	
8	0.613	0.316	1.940		****************	
9	0.031	0.345	0.091		*	
10	−0.547	0.345	−1.585		***************	
11	−0.010	0.367	−0.028		*	
12	0.489	0.367	1.333		*************	
13	−0.027	0.383	−0.069		*	
14	−0.545	0.383	−1.424		***************	
15	0.039	0.402	0.096		**	
16	0.514	0.403	1.277		**************	
17	−0.049	0.419	−0.117		**	
18	−0.406	0.419	−0.968		************	
19	0.077	0.429	0.178		**	
20	0.412	0.429	0.959		***********	
21	−0.133	0.439	−0.303		****	
22	−0.361	0.441	−0.819		**********	
23	0.087	0.448	0.195		***	
24	0.290	0.448	0.647		********	

and

$$\text{Model 2:} \quad (1 - \phi_1 B)z_t = (1 - \theta_{1,4} B^4)a_t$$

We have not used a constant term δ in these models because it can be shown that $\bar{z}$ is not statistically different from zero. In Figures 4.13 and 4.14 we present the SAS output of estimation, diagnostic checking, and forecasting for Models 1 and 2. Note that, although the prob-values associated with the values of Q^* for $K = 6, 12, 18,$ and 24 are larger for Model 1 than for Model 2, the standard error for Model 2 (which is $s = 60.1324$) is smaller than the standard error for Model 1 (which is $s = 69.9959$). To improve

$$\text{Model 2:} \quad (1 - \phi_1 B)z_t = (1 - \theta_{1,4} B^4)a_t$$

notice that the RSAC resulting from this model has a spike at lag 5, which suggests

FIGURE 4.11 TSERIES Output of the SAC and SPAC for the Gas Bill Values Obtained by Using the Transformation $z_t = y_t - y_{t-4}$

(a) The SAC

	VALUE	S.E.	T VALUE
1	0.597	0.167	3.583
2	0.204	0.218	0.935
3	-0.082	0.223	-0.368
4	-0.378	0.224	-1.685
5	-0.385	0.241	-1.594
6	-0.132	0.258	-0.514
7	0.050	0.260	0.193
8	0.089	0.260	0.341
9	0.134	0.261	0.513
10	0.080	0.263	0.305
11	-0.166	0.263	-0.630
12	-0.262	0.266	-0.984
13	-0.155	0.273	-0.565
14	-0.086	0.276	-0.313
15	0.087	0.276	0.315
16	0.177	0.277	0.640
17	0.147	0.280	0.524
18	0.020	0.282	0.071
19	-0.007	0.282	-0.025
20	0.056	0.282	0.198
21	0.009	0.283	0.031
22	0.053	0.283	0.186
23	0.073	0.283	0.259
24	0.019	0.284	0.067

(Part (b), The SPAC, follows on page 180.)

TABLE 4.11   *Continued*

(b) The SPAC

	VALUE	S.E.	T VALUE	
1	0.597	0.167	3.583	******************
2	-0.237	0.167	-1.424	******
3	-0.150	0.167	-0.900	****
4	-0.345	0.167	-2.070	*********
5	0.067	0.167	0.399	**
6	0.208	0.167	1.249	******
7	-0.012	0.167	-0.074	*
8	-0.207	0.167	-1.240	******
9	0.032	0.167	0.190	*
10	0.065	0.167	0.393	**
11	-0.264	0.167	-1.586	********
12	-0.116	0.167	-0.693	***
13	0.179	0.167	1.071	*****
14	0.049	0.167	0.292	**
15	0.008	0.167	0.048	*
16	-0.252	0.167	-1.510	********
17	0.123	0.167	0.739	****
18	0.036	0.167	0.216	*
19	0.104	0.167	0.622	***
20	0.088	0.167	0.531	***
21	-0.110	0.167	-0.662	***
22	0.054	0.167	0.325	**
23	-0.059	0.167	-0.354	**
24	0.104	0.167	0.625	***

FIGURE 4.12 TSERIES Output of the SAC for the Gas Bill Values Obtained by Using the Transformation
$z_t = y_t - y_{t-1} - y_{t-4} + y_{t-5}$

	VALUE	S.E.	T VALUE	−1.0	0.0	1.0
				+ − − − − − − − − −	− − − − − − − − −	− − +
1	0.043	0.169	0.255	I	**	I
2	−0.109	0.169	−0.642	I	***	I
3	0.058	0.171	0.338	I	**	I
4	−0.440	0.172	−2.561	I	************	I
5	−0.391	0.202	−1.942	I	**********	I
6	0.016	0.222	0.072	I	*	I
7	0.222	0.222	0.998	I	******	I
8	0.065	0.228	0.284	I	**	I
9	0.247	0.229	1.077	I	*******	I
10	0.288	0.236	1.218	I	********	I
11	−0.177	0.246	−0.719	I	*****	I
12	−0.275	0.250	−1.100	I	*******	I
13	−0.099	0.258	−0.384	I	***	I
14	−0.200	0.259	−0.770	I	*****	I
15	0.010	0.264	0.039	I	*	I
16	0.131	0.264	0.497	I	****	I
17	0.187	0.266	0.705	I	*****	I
18	0.013	0.269	0.049	I	*	I
19	−0.020	0.269	−0.075	I	*	I
20	0.083	0.269	0.307	I	***	I
21	−0.167	0.270	−0.617	I	*****	I
22	0.013	0.273	0.048	I	*	I
23	−0.003	0.273	−0.012	I	*	I
24	−0.044	0.273	−0.160	I	**	I
				+ − − − − − − − − − −	− − − − − − − − −	− − +

that we consider

Model 3: $(1 - \phi_1 B)z_t = (1 - \theta_{1,4}B^4 - \theta_{1,5}B^5)a_t$

The SAS output of estimation, diagnostic checking, and forecasting for this model is given in Figure 4.15. Since this figure indicates that Model 3 has a smaller standard error (note that $s = 57.0203$ for Model 3) and yields shorter prediction intervals than Models 1 and 2, we conclude that Model 3 is the best model (note that Models 1, 2, and 3 yield considerably different point forecasts, so the choice between these models is important). However, from Figure 4.15a we see that

$$\hat{\theta}_{1,4} = .971685 \quad \text{and} \quad \hat{\theta}_{1,5} = .305779$$

and that, consequently,

$$\hat{\theta}_{1,4} + \hat{\theta}_{1,5} = 1.277464$$

FIGURE 4.13 SAS Output of Estimation, Diagnostic Checking, and Forecasting for Model $z_t = (1 - \theta_1 B)(1 - \theta_{1,4}B^4)a_t$ where $z_t = y_t - y_{t-4}$

(a) Estimation

ARIMA: LEAST SQUARES ESTIMATION

ITERATION	SSE	MA1,1	MA2,1	LAMBDA
0	176394	-.597198	0.377917	1.0E-05
1	168006	-.818255	0.536974	1.0E-06
2	166673	-.769751	0.551132	1.0E-07
3	166582	-.757075	0.544258	1.0E-08
4	166580	-.755625	0.542217	1.0E-09
5	166580	-.755483	0.541895	1.0E-10

PARAMETER	ESTIMATE	STD ERROR	T RATIO	LAG
MA1,1	-0.755483	0.131314	-5.75	1
MA2,1	0.541895	0.19324	2.80	4

VARIANCE ESTIMATE = 4899.42
STD ERROR ESTIMATE = 69.9959
NUMBER OF RESIDUALS = 36

CORRELATIONS OF THE ESTIMATES

	MA1,1	MA2,1
MA1,1	1.000	0.342
MA2,1	0.342	1.000

FIGURE 4.13 Continued (b) The RSAC

AUTOCORRELATION CHECK OF RESIDUALS

TC LAG	CHI SQUARE	DF	PROB	AUTOCORRELATIONS					
6	7.22	4	0.125	0.127	0.185	0.338	-0.092	0.061	0.039
12	11.01	10	0.357	0.109	0.193	-0.053	0.014	-0.145	0.038
18	15.67	16	0.477	0.089	0.107	0.049	0.103	0.039	0.189
24	18.80	22	0.658	-0.102	-0.134	0.079	-0.030	0.014	0.030

AUTOCORRELATION PLOT OF RESIDUALS

LAG	COVARIANCE	CORRELATION	STD
0	4899.42	1.00000	0
1	905.819	0.18488	0.166667
2	1657.93	0.33839	0.172269
3	623.943	0.12735	0.189838
4	191.254	0.03904	0.192196
5	-450.584	-0.09197	0.192416
6	297.886	0.06080	0.193633
7	947.023	0.19329	0.194163
8	-262.105	-0.05350	0.199437
9	535.193	0.10924	0.199835
10	187.675	0.03831	0.201487
11	69.6682	0.01422	0.201689
12	-708.435	-0.14460	0.201717
13	524.698	0.10709	0.204576
14	240.026	0.04899	0.206127
15	435.218	0.08883	0.20645
16	923.857	0.18856	0.207509
17	503.972	0.10286	0.212216
18	191.835	0.03915	0.213596
19	-658.95	-0.13450	0.213795
20	386.336	0.07885	0.216133
21	-497.717	-0.10159	0.21693
22	147.394	0.03008	0.218248
23	-147.545	-0.03011	0.218363
24	68.6262	0.01401	0.218478

'.' MARKS TWO STANDARD ERRORS

183

(Part (c), the RSPAC, and Part (d), Forecasts, follow on page 184.)

FIGURE 4.13 Continued   (c) The RSPAC

PARTIAL AUTOCORRELATIONS

LAG	CORRELATION
1	0.18488
2	0.31498
3	0.03018
4	-0.09981
5	-0.15799
6	0.12052
7	0.32236
8	-0.17303
9	-0.10813
10	0.05202
11	0.10209
12	-0.15973
13	0.04216
14	0.16599
15	0.14148
16	0.01651
17	-0.13792
18	-0.00353
19	-0.07435
20	0.09505
21	0.00070
22	-0.02416
23	-0.13523
24	-0.04211

(d) Forecasts of y_{41}, y_{42}, y_{43}, and y_{44}

FORECASTS FOR VARIABLE Y

OBS	FORECAST	STD ERROR	LOWER 95%	UPPER 95%
	- - - - - FORECAST BEGINS - - - - -			
41	547.0229	69.9949	409.8338	684.2120
42	328.8408	87.7257	156.9020	500.7796
43	256.5111	87.7257	84.5723	428.4499
44	421.7147	87.7257	249.7759	593.6535

FIGURE 4.14 SAS Output of Estimation, Diagnostic Checking, and Forecasting for Model $(1 - \phi_1 B)z_t = (1 - \theta_{1,4}B^4)a_t$ where $z_t = y_t - y_{t-4}$

(a) Estimation

ARIMA: LEAST SQUARES ESTIMATION

ITERATION	SSE	MA1,1	AR1,1	LAMBDA
0	162599	0.377917	0.597198	1.0E−05
1	128955	0.7931	0.828253	1.0E−06
2	124895	1.01076	0.908404	1.0E−07
3	122965	0.960173	0.872249	1.0E−08
4	122941	0.9553	0.87896	1.0E−09
5	122941	0.954769	0.879566	1.0E−10

PARAMETER	ESTIMATE	STD ERROR	T RATIO	LAG
MA1,1	0.954769	0.0979224	9.75	4
AR1,1	0.879566	0.119063	7.39	1

VARIANCE ESTIMATE   = 3615.9
STD ERROR ESTIMATE   = 60.1324
NUMBER OF RESIDUALS = 36

CORRELATIONS OF THE ESTIMATES

	MA1,1	AR1,1
MA1,1	1.000	0.048
AR1,1	0.048	1.000

(Part (b), The RSAC, Part (c), the RSPAC, and Part (d), the Forecast, follow on pages 186 and 187.)

FIGURE 4.14 *Continued* (b) The RSAC

AUTOCORRELATION CHECK OF RESIDUALS

TO LAG	CHI SQUARE	DF	PROB	AUTOCORRELATIONS					
6	8.11	4	0.088	-0.035	-0.039	0.114	0.034	-0.380	0.150
12	17.90	10	0.057	0.210	-0.044	-0.000	0.263	-0.144	-0.223
18	22.59	16	0.125	0.033	0.036	-0.038	0.198	0.156	0.010
24	26.59	22	0.227	-0.143	0.113	-0.110	0.016	-0.011	0.033

AUTOCORRELATION PLOT OF RESIDUALS

LAG	COVARIANCE	CORRELATION	-1 9 8 7 6 5 4 3 2 1 0 1 2 3 4 5 6 7 8 9 1	STD
0	3615.9	1.00000		0
1	-126.432	-0.03497		0.166667
2	-140.657	-0.03890		0.16687
3	413.537	0.11437		0.167122
4	121.295	0.03354		0.169282
5	-1374.39	-0.38010		0.169467
6	543.893	0.15042		0.19169
7	759.447	0.21003		0.194941
8	-157.93	-0.04368		0.201129
9	-0.887491	-0.00025		0.201392
10	952.154	0.26332		0.201392
11	-518.886	-0.14350		0.210739
12	-808.036	-0.22347		0.213436
13	119.264	0.03298		0.219839
14	129.94	0.03594		0.219977
15	-136.537	-0.03776		0.22014
16	715.471	0.19787		0.22032
17	565.219	0.15631		0.225202
18	37.6372	0.01041		0.228196
19	-516.62	-0.14287		0.228209
20	406.806	0.11250		0.23068
21	-399.273	-0.11042		0.232199
22	59.3715	0.01642		0.233654
23	-39.6466	-0.01096		0.233686
24	120.963	0.03345		0.2337

'.' MARKS TWO STANDARD ERRORS

FIGURE 4.14 Continued   (c)   The RSPAC

PARTIAL AUTOCORRELATIONS

LAG	CORRELATION	-1	9	8	7	6	5	4	3	2	1	0	1	2	3	4	5	6	7	8	9	1
1	-0.03497
2	-0.04017
3	0.11186
4	0.04031
5	-0.37545
6	0.14036
7	0.23127
8	0.03729
9	-0.02885
10	0.09445
11	-0.05374
12	-0.13662
13	-0.06325
14	0.03264
15	0.16094
16	0.12275
17	-0.01984
18	0.06782
19	-0.11900
20	0.10526
21	-0.00688
22	0.09409
23	-0.10388
24	-0.17449

(d)   Forecasts of y_{41}, y_{42}, y_{43}, and y_{44}

FORECASTS FOR VARIABLE Y

OBS	FORECAST	STD ERROR	LOWER 95%	UPPER 95%
- - - - - - - - - - FORECAST BEGINS - - - - - - - - -
41	588.0327	60.1324	470.1756	705.8897
42	460.7349	80.0831	303.7753	617.6945
43	332.2212	92.6146	150.7004	513.7421
44	459.2003	101.2509	260.7527	657.6480

FIGURE 4.15 SAS Output of Estimation, Diagnostic Checking, and Forecasting for Model $(1 - \phi_1 B)z_t = (1 - \theta_{1,4}B^4 - \theta_{1,5}B^5)a_t$ where $z_t = y_t - y_{t-4}$

(a) Estimation

ARIMA: LEAST SQUARES ESTIMATION

ITERATION	SSE	MA1,1	MA1,2	AR1,1	LAMBDA
0	271149	0.1	0.1	0.1	1.0E-05
1	149929	0.554029	-.066417	0.762463	1.0E-06
2	111094	0.977167	0.35516	0.899946	1.0E-07
3	108257	0.930092	0.317221	0.756395	1.0E-08
4	107339	0.96028	0.318824	0.793561	1.0E-09
5	107309	0.962456	0.315573	0.782959	1.0E-10
6	107305	0.964929	0.31375	0.784456	1.0E-11
7	107302	0.966419	0.311938	0.784294	1.0E-12
8	107299	0.967776	0.310381	0.784537	1.0E-13
9	107297	0.968948	0.309	0.784689	1.0E-14
10	107296	0.969981	0.307787	0.784838	1.0E-15
11	107295	0.970888	0.306718	0.784966	1.0E-16
12	107294	0.971685	0.305779	0.785079	1.0E-17

PARAMETER	ESTIMATE	STD ERROR	T RATIO	LAG
MA1,1	0.971685	0.086903	11.18	4
MA1,2	0.305779	0.0863042	3.54	5
AR1,1	0.785079	0.13187	5.95	1

VARIANCE ESTIMATE   = 3251.32
STD ERROR ESTIMATE   = 57.0203
NUMBER OF RESIDUALS = 36

CORRELATIONS OF THE ESTIMATES

	MA1,1	MA1,2	AR1,1
MA1,1	1.000	-0.752	0.229
MA1,2	-0.752	1.000	0.088
AR1,1	0.229	0.088	1.000

FIGURE 4.15 Continued   (b) The RSAC   (Part (c), the RSPAC, and Part (d), the Forecast follow on page 190.)

AUTOCORRELATION CHECK OF RESIDUALS

AUTOCORRELATIONS

TO LAG	CHI SQUARE	DF	PROB						
6	1.68	3	0.642	-0.017	-0.080	0.102	0.037	-0.145	-0.001
12	9.16	9	0.423	0.153	-0.035	0.181	0.126	-0.128	-0.226
18	14.69	15	0.474	0.133	-0.050	-0.010	0.166	0.183	0.015
24	17.73	21	0.666	-0.112	0.109	-0.105	0.002	-0.002	0.001

AUTOCORRELATION PLOT OF RESIDUALS

LAG	COVARIANCE	CORRELATION	STD
0	3251.32	1.00000	0
1	-55.3455	-0.01702	0.166667
2	-258.556	-0.07952	0.166715
3	331.394	0.10193	0.167765
4	119.356	0.03671	0.169477
5	-470.604	-0.14474	0.169697
6	-3.15794	-0.00097	0.173093
7	497.845	0.15312	0.173093
8	-113.851	-0.03502	0.176816
9	587.852	0.18080	0.177008
10	410.709	0.12632	0.182066
11	-415.657	-0.12784	0.184484
12	-736.303	-0.22646	0.186929
13	432.03	0.13288	0.194401
14	-163.605	-0.05032	0.196908
15	-32.8264	-0.01010	0.197264
16	540.918	0.16637	0.197279
17	593.93	0.18267	0.201138
18	49.383	0.01519	0.205695
19	-364.809	-0.11220	0.205726
20	355.796	0.10943	0.207419
21	-341.343	-0.10499	0.209017
22	6.60222	0.00203	0.210476
23	-7.49602	-0.00231	0.210477
24	4.38543	0.00135	0.210478

'.' MARKS TWO STANDARD ERRORS

FIGURE 4.15 *Continued* (c) The RSPAC

PARTIAL AUTOCORRELATIONS

LAG	CORRELATION
1	−0.01702
2	−0.07984
3	0.09977
4	0.03383
5	−0.13032
6	−0.00923
7	0.13159
8	−0.00809
9	0.21705
10	0.09193
11	−0.11913
12	−0.24022
13	0.08515
14	−0.02557
15	0.10911
16	0.10716
17	0.13119
18	0.02972
19	−0.14586
20	0.08014
21	0.02582
22	0.06491
23	−0.09691
24	−0.11348

(d) Forecasts of y_{41}, y_{42}, y_{43}, and y_{44}

FORECASTS FOR VARIABLE Y

OBS	FORECAST	STD ERROR	LOWER 95%	UPPER 95%
	- - - - - FORECAST BEGINS - - - - -			
41	609.4653	57.0203	497.7077	721.2228
42	537.3028	72.4932	395.2190	679.3866
43	472.5770	80.5630	314.6768	630.4773
44	569.6509	85.1567	402.7472	736.5546

Thus, $\hat{\theta}_{1,4}$ and $\hat{\theta}_{1,5}$ do not satisfy one of the invertibility conditions on $\theta_{1,4}$ and $\theta_{1,5}$, which is (from Section 4.2)

$$\theta_{1,4} + \theta_{1,5} < 1$$

Therefore, while it might not be too dangerous to use Model 3 for short term forecasting, we should look for a better model, as described in Chapter 5.

4.6 USING COMPUTER PACKAGES

4.6.1 Using SAS

In this section we first present the SAS programs needed to analyze the hotel room data.

PROGRAM 1   *This SAS program computes the SAC and SPAC for the time series values produced by each of the transformations*

$$z_t = y_t^*, \quad z_t = y_t^* - y_{t-1}^*, \quad z_t = y_t^* - y_{t-12}^*, \quad \text{and} \quad z_t = y_t^* - y_{t-1}^* - y_{t-12}^* + y_{t-13}^*$$

where

$$y_t^* = \ln y_t$$

```
DATA HOTL;    } Defines data file with name HOTL.
INPUT Y;    } Defines a variable with name Y.
LY = LOG(Y);    } Transformation to compute logged data.
CARDS;    } Declares that data follows this statement.
  501 ⎫
  488 ⎬        Time series data. One observation per line.
   ·  ⎪
   ·  ⎬
   ·  ⎪
  877 ⎭
PROC ARIMA DATA=HOTL;    } Declares Box-Jenkins (ARIMA) procedure and data file to be used.
  IDENTIFY VAR=LY;    } Generates SAC and SPAC for logged data z_t = y_t*
  IDENTIFY VAR=LY(1);    } Computes differences and generates SAC and SPAC using z_t = y_t* - y*_{t-1}
  IDENTIFY VAR=LY(12);    } Computes differences and generates SAC and SPAC using z_t = y_t* - y*_{t-12}
  IDENTIFY VAR=LY(1,12);    } Computes differences and generates SAC and SPAC using z_t = y_t* - y*_{t-1} -
                                    y*_{t-12} + y*_{t-13}
```

After examining the output generated by Program 1, we conclude that $z_t = y_t^* - y_{t-12}^*$ is stationary, and one model that we tentatively identify is

$$(1 - \phi_1 B - \phi_2 B^2 - \phi_3 B^3)z_t = \delta + (1 - \theta_{1,12} B^{12})a_t$$

where

$$z_t = y_t^* - y_{t-12}^* \quad \text{and} \quad y_t^* = \ln y_t$$

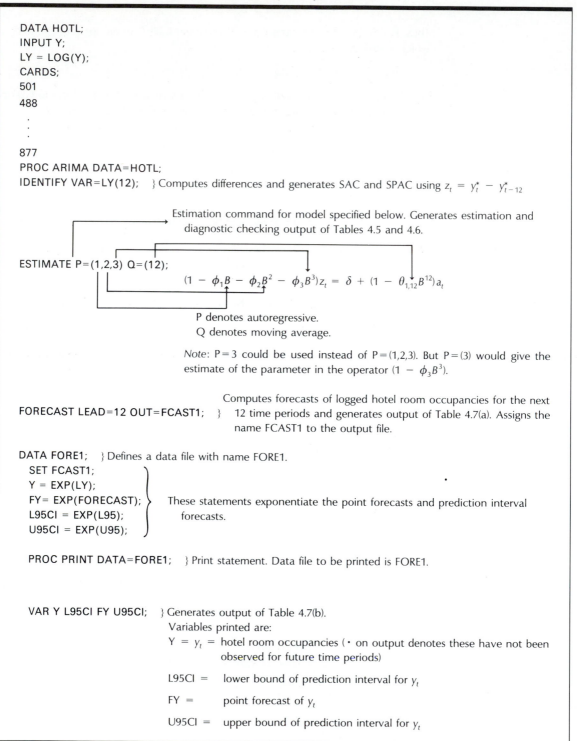

```
DATA HOTL;
INPUT Y;
LY = LOG(Y);
CARDS;
501
488
  .
  .
  .
877
PROC ARIMA DATA=HOTL;
IDENTIFY VAR=LY(12);
```
} Computes differences and generates SAC and SPAC using $z_t = y_t^* - y_{t-12}^*$

Estimation command for model specified below. Generates estimation and diagnostic checking output of Tables 4.5 and 4.6.

```
ESTIMATE P=(1,2,3) Q=(12);
```

$$(1 - \phi_1 B - \phi_2 B^2 - \phi_3 B^3)z_t = \delta + (1 - \theta_{1,12}B^{12})a_t$$

P denotes autoregressive.
Q denotes moving average.

Note: P=3 could be used instead of P=(1,2,3). But P=(3) would give the estimate of the parameter in the operator $(1 - \phi_3 B^3)$.

Computes forecasts of logged hotel room occupancies for the next

```
FORECAST LEAD=12 OUT=FCAST1;
```
} 12 time periods and generates output of Table 4.7(a). Assigns the name FCAST1 to the output file.

```
DATA FORE1;
```
} Defines a data file with name FORE1.
```
  SET FCAST1;
  Y = EXP(LY);
  FY= EXP(FORECAST);
  L95CI = EXP(L95);
  U95CI = EXP(U95);
```
These statements exponentiate the point forecasts and prediction interval forecasts.

```
  PROC PRINT DATA=FORE1;
```
} Print statement. Data file to be printed is FORE1.

```
  VAR Y L95CI FY U95CI;
```
} Generates output of Table 4.7(b).

Variables printed are:

Y = y_t = hotel room occupancies (· on output denotes these have not been observed for future time periods)

L95CI = lower bound of prediction interval for y_t

FY = point forecast of y_t

U95CI = upper bound of prediction interval for y_t

Note 1: ESTIMATE P = (1,2,3) Q = (12) NOCONSTANT;
would exclude δ from the above model.

Note 2: ESTIMATE P = (1,2,3) Q = (12) PRINTALL PLOT;
would be used if we wish to have a more complete SAS output of estimation and diagnostic checking. That is, if we exclude PRINTALL PLOT, we generate the (reduced) SAS output of estimation and diagnostic checking given in Tables 4.5 and 4.6. If we include PRINTALL PLOT, we generate a more complete output, as given in Table 4.8.

Note 3: ESTIMATE P = (1)(6,12) Q = (1,4)(12);
would be appropriate for the model

$$(1 - \phi_1 B)(1 - \phi_{1,6} B^6 - \phi_{1,12} B^{12})z_t$$
$$= \delta + (1 - \theta_1 B - \theta_4 B^4)(1 - \theta_{1,12} B^{12})a_t$$

We next present the SAS programs needed to analyze the Farmers' Bureau Coop gas bill data.

PROGRAM 1 *This SAS program computes the SAC and SPAC for the time series values produced by each of the transformations*

$$z_t = y_t, \quad z_t = y_t - y_{t-1}, \quad z_t = y_t - y_{t-4}, \quad \text{and} \quad z_t = y_t - y_{t-1} - y_{t-4} + y_{t-5}$$

```
DATA GAS;
INPUT Y;
CARDS;
344.39
246.63
    .
    .
    .
539.78
PROC ARIMA DATA=GAS;
IDENTIFY VAR=Y;
IDENTIFY VAR=Y(1);
IDENTIFY VAR=Y(4);
IDENTIFY VAR=Y(1,4);
```

After examining the output generated by Program 1, we conclude that $z_t = y_t - y_{t-4}$ is stationary, and we tentatively identify the models

$$z_t = (1 - \theta_1 B)(1 - \theta_{1,4} B^4)a_t$$

and

$$(1 - \phi_1 B)z_t = (1 - \theta_{1,4} B^4)a_t$$

where

$$z_t = y_t - y_{t-4}$$

PROGRAM 2   *This program fits the above models. It estimates model parameters, computes the RSAC and RSPAC for each model, and forecasts the next four y_t values.*

```
DATA GAS;
INPUT Y;
CARDS;
344.39
246.63
    .
    .
    .
539.78
PROC ARIMA DATA=GAS;
IDENTIFY VAR=Y(4);
ESTIMATE Q=(1)(4) NOCONSTANT
PRINTALL PLOT;
FORECAST LEAD=4;
ESTIMATE P=(1) Q=(4) NOCONSTANT
PRINTALL PLOT;
FORECAST LEAD=4;
```

After examining the RSAC and RSPAC, we consider the improved model

$$(1 - \phi_1 B)z_t = (1 - \theta_{1,4}B^4 - \theta_{1,5}B^5)a_t$$

PROGRAM 3   *This SAS program fits the above model. It estimates model parameters, computes the RSAC and RSPAC, and forecasts the next four y_t values.*

```
DATA GAS;
INPUT Y;
CARDS;
344.39
246.63
    .
    .
    .
539.78
PROC ARIMA DATA=GAS;
IDENTIFY VAR=Y(4);
ESTIMATE P=(1) Q=(4,5) NOCONSTANT
PRINTALL PLOT;
FORECAST LEAD=4;
```

4.6.2 Using TSERIES (Optional)

In this section we present the TSERIES programs needed to analyze the hotel room data.

PROGRAM 1 *This TSERIES program computes the SAC and SPAC for the time series values produced by each of the transformations*

$$z_t = y_t^*, \quad z_t = y_t^* - y_{t-1}^*, \quad z_t = y_t^* - y_{t-12}^*, \quad \text{and} \quad z_t = y_t^* - y_{t-1}^* - y_{t-12}^* + y_{t-13}^*$$

where

$$y_t^* = \ln y_t$$

DATA 168

501.	488.	504.	578.	545.	632.	728.	725.	585.	542.	480.	530.	518.	489.
528.	599.	572.	659.	739.	758.	602.	587.	497.	558.	555.	523.	532.	623.
598.	683.	774.	780.	609.	604.	531.	592.	578.	543.	565.	648.	615.	697.
785.	830.	645.	643.	551.	606.	585.	553.	576.	665.	656.	720.	826.	838.
652.	661.	584.	644.	623.	553.	599.	657.	680.	759.	878.	881.	705.	684.
577.	656.	645.	593.	617.	686.	679.	773.	906.	934.	713.	710.	600.	676.
645.	602.	601.	709.	706.	817.	930.	983.	745.	735.	620.	698.	665.	626.
649.	740.	729.	824.	937.	994.	781.	759.	643.	728.	691.	649.	656.	735.
748.	837.	995.	1040.	809.	793.	692.	763.	723.	655.	658.	761.	768.	
885.	1067.	1038.	812.	790.	692.	782.	758.	709.	715.	788.	794.	893.	
1046.	1075.	812.	822.	714.	802.	748.	731.	748.	827.	788.	937.	1076.	
1125.	840.	864.	717.	813.	811.	732.	745.	844.	833.	935.	1110.	1124.	
868.	860.	762.	877.										

TRAN 0 0 ⎫
IDEN ⎬ Generates output in Figure 3.3.
⎭ TRAN 0 0 takes the natural logarithm of each observation.
IDEN is the identification command (generates differencing, SAC, and SPAC).

TRAN 0 0 ⎫
NRDF 1 ⎬ Generates output in Figure 3.4.
IDEN ⎭ NRDF 1 denotes *regular (nonseasonal)* differencing of order 1.

TRAN 0 0 ⎫ Generates output in Figure 3.5.
NRDF 0 ⎪ NRDF 0 denotes *no regular differencing.*
NSDF 1 ⎬ NSDF 1 denotes *first order seasonal differencing.*
IDEN ⎭ Unless otherwise specified, NSDF assumes L = 12.

TRAN 0 0 ⎫
NRDF 1 ⎪
NSDF 1 ⎬ Generates output in Figure 3.6.
IDEN ⎭

STOP

After examining the output generated by Program 1, we conclude that $z_t = y_t^* - y_{t-12}^*$

is stationary, and one model that we tentatively identify is

$$(1 - \phi_1 B - \phi_2 B^2 - \phi_3 B^3)z_t = \delta + (1 - \theta_{1,12}B^{12})a_t$$

where

$$z_t = y_t^* - y_{t-12}^* \quad \text{and} \quad y_t^* = \ln y_t$$

PROGRAM 2 *This TSERIES program fits the above model. It estimates model parameters, computes the RSAC and RSPAC, and forecasts the next twelve monthly values of $y_t = e^{y_t^*}$*

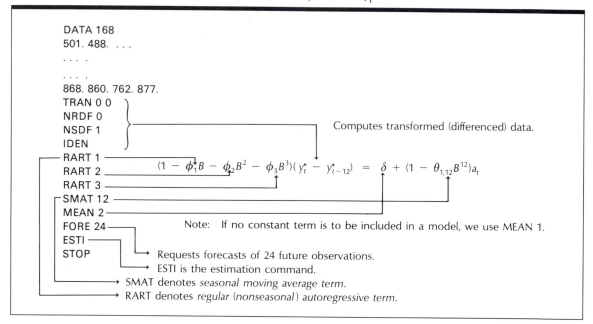

As a second (unrelated) example, if we wished to fit the model

$$(1 - \phi_{1,5}B^5)(y_t - y_{t-4}) = (1 - \theta_1 B)a_t$$

we would employ the following TSERIES program.

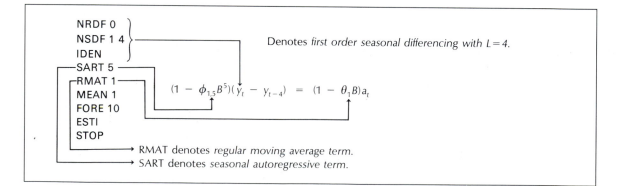

EXERCISES

4.1 Table 1 and Figure 1 present the TSERIES output of estimation and diagnostic checking for the model

$$(1 - \phi_1 B)z_t = \delta + a_t$$

where

$$z_t = y_t - y_{t-1}$$

which is the tentative model describing the Ultra-Shine Toothpaste sales data in Table 1 of the Exercises in Chapter 2.

a. Find and identify the least squares point estimates of the parameters in the above model.

b. Does the least squares point estimate of ϕ_1 satisfy the stationarity and invertibility conditions on ϕ_1? What are these conditions?

c. Using $\hat{\phi}_1$ and $s_{\hat{\phi}_1}$, calculate $t_{\hat{\phi}_1}$.

TABLE 1 *TSERIES Output of the Least Squares Estimation of ϕ_1 in the Model $(1 - \phi_1 B)z_t = \delta + a_t$ where $z_t = y_t - y_{t-1}$*

(a) The Iterative Search

A CONSTANT TERM IS TO BE INCLUDED IN THE MODEL
MAX BACKORDER = 1

BEGINING ESTIMATION

ITERATION	SUM OF SQUARES	PARAMETER VALUES
0	0.1026130D+04	0.100000
1	0.9001845D+03	0.210000
2	0.8017218D+03	0.320000
3	0.7307416D+03	0.430000
4	0.6872441D+03	0.540000
5	0.6712293D+03	0.649999
6	0.6711357D+03	0.658478
7	0.6711353D+03	0.659063
8	0.6711353D+03	0.659104

RELATIVE CHANGE IN EACH ESTIMATE LESS THAN 0.1000E−03
ESTIMATION COMPLETED

(b) The Final Point Estimate and its Associated *t*-Value

						95% LIMITS	
TERM #	TYPE	ORDER	ESTIMATE	STD. ERROR	T VALUE	LOWER	UPPER
1	REG. AR	1	0.6591	0.0804	8.2018	0.5016	0.8166

ESTIMATED MEAN OF THE WORKING SERIES = 8.927
THIS YIELDS A CONSTANT TERM = 3.043

FIGURE 1   TSERIES Output of RSAC and RSPAC for the Model $(1 - \phi_1 B)z_t = \delta + a_t$ where $z_t = y_t - y_{t-1}$

(a)   The RSAC

	VALUE	S.E.	T VALUE	−1.0	0.0	1.0
				+ − − − − − − − − − − − − − − − +		
1	0.093	0.106	0.879	I	***	I
2	−0.211	0.107	−1.972	I	******	I
3	−0.019	0.111	−0.174	I	*	I
4	0.028	0.112	0.255	I	*	I
5	0.052	0.112	0.464	I	**	I
6	0.170	0.112	1.519	I	*****	I
7	0.030	0.115	0.265	I	*	I
8	−0.012	0.115	−0.105	I	*	I
9	−0.056	0.115	−0.486	I	**	I
10	−0.014	0.115	−0.120	I	*	I
11	−0.009	0.115	−0.080	I	*	I
12	0.058	0.115	0.503	I	**	I
13	0.023	0.116	0.196	I	*	I
14	−0.099	0.116	−0.860	I	***	I
15	0.100	0.117	0.854	I	***	I
16	0.097	0.117	0.829	I	***	I
17	−0.101	0.118	−0.856	I	***	I
18	0.054	0.119	0.454	I	**	I
19	0.196	0.120	1.639	I	******	I
20	0.007	0.123	0.053	I	*	I
21	−0.005	0.123	−0.039	I	*	I
22	−0.112	0.123	−0.911	I	***	I
23	−0.088	0.124	−0.710	I	***	I
24	0.090	0.125	0.717	I	***	I
				+ − − − − − − − − − − − − − − − +		

BOX-PIERCE TEST CHISQUARE STATISTIC WITH 20 DEGREES OF FREEDOM = 15.590

(Continued on page 199.)

d. What does the $t_{\hat{\phi}_1}$-statistic say about whether or not the parameter ϕ_1 should be included in the model? Justify your answer by using an appropriate rejection point.

e. Using $\bar{z}$ and $\hat{\phi}_1$, calculate $\hat{\delta}$.

f. Using the sample autocorrelations of the residuals, calculate the Box-Pierce Q statistic (with K such that we have $K - n_p = 20$ degrees of freedom).

g. Using the Box-Pierce Q statistic and the appropriate rejection point, can we reject the adequacy of the above model? Explain your answer.

h. Using the sample autocorrelations of the residuals, calculate the Ljung-Box Q^* statistic with $K = 6$.

i. Using the Ljung-Box Q^* statistic calculated in part (h) and an appropriate rejection point, can we reject the adequacy of the above model? Explain your answer.

FIGURE 1 *Continued*

(b) The RSPAC

	VALUE	S.E.	T VALUE	−1.0	0.0	1.0
1	0.093	0.106	0.879		***	
2	−0.221	0.106	−2.089		******	
3	0.027	0.106	0.252		*	
4	−0.020	0.106	−0.190		*	
5	0.055	0.106	0.520		**	
6	0.171	0.106	1.610		*****	
7	0.015	0.106	0.142		*	
8	0.061	0.106	0.579		**	
9	−0.060	0.106	−0.569		**	
10	−0.001	0.106	−0.014		*	
11	−0.053	0.106	−0.496		**	
12	0.031	0.106	0.293		*	
13	−0.005	0.106	−0.049		*	
14	−0.094	0.106	−0.887		***	
15	0.162	0.106	1.529		*****	
16	0.033	0.106	0.310		*	
17	−0.054	0.106	−0.513		**	
18	0.102	0.106	0.965		***	
19	0.153	0.106	1.447		****	
20	0.015	0.106	0.145		*	
21	0.030	0.106	0.284		*	
22	−0.154	0.106	−1.450		****	
23	−0.079	0.106	−0.741		**	
24	0.027	0.106	0.253		*	

j. Algebraically expand the above model and express y_t as a function of past time series observations and of the random shock a_t. Show that the $t_{\hat{\phi}_1}$-statistic measures the importance of the independent variable $(y_{t-1} - y_{t-2})$.

4.2 Write the SAS program needed to perform estimation, diagnostic checking, and forecasting (for the next 6 time periods) for the Ultra-Shine Toothpaste model given in Exercise 4.1.

4.3 (Optional) Write the TSERIES program needed to perform estimation, diagnostic checking, and forecasting (for the next 6 time periods) for the Ultra-Shine Toothpaste model given in Exercise 4.1.

4.4 Table 2 and Figure 2 present the TSERIES output of estimation and diagnostic checking for the model

$$z_t = (1 - \theta_1 B)(1 - \theta_{1,12} B^{12}) a_t$$

TABLE 2   TSERIES Output of Least Squares Estimation of θ_1 and $\theta_{1,12}$ in the Model $z_t = (1 - \theta_1 B)(1 - \theta_{1,12}B^{12})a_t$
where $z_t = y_t - y_{t-1} - y_{t-12} + y_{t-13}$

(a)   The Iterative Search

NO CONSTANT TERM IN THE MODEL
MAX BACKORDER = 13

BEGINING ESTIMATION

ITERATION	SUM OF SQUARES	PARAMETER VALUES	
0	0.3396408D+03	0.100000	0.100000
1	0.3117625D+03	−0.010000	0.199405
2	0.2959662D+03	−0.120000	0.308503
3	0.2915650D+03	−0.179856	0.398560
4	0.2908074D+03	−0.178018	0.433371
5	0.2906637D+03	−0.174793	0.449111
6	0.2906299D+03	−0.173245	0.456701
7	0.2906253D+03	−0.172877	0.458529
8	0.2906249D+03	−0.172843	0.458698
9	0.2906247D+03	−0.172826	0.458780
10	0.2906246D+03	−0.172822	0.458800

RELATIVE CHANGE IN EACH ESTIMATE LESS THAN 0.1000E−03
ESTIMATION COMPLETED

(b)   The Final Point Estimates and Their Associated t-Values

TERM #	TYPE	ORDER	ESTIMATE	STD. ERROR	T VALUE	95% LIMITS LOWER	UPPER
1	REG. MA	1	−0.1728	0.0772	−2.2401	−0.3241	−0.0216
2	SEA. MA	12	0.4588	0.0713	6.4351	0.3190	0.5986

STATISTICS COMPUTED FROM THE RESIDUALS

RESIDUAL MEAN   = 0.0015
RESIDUAL VARIANCE = 1.7558
RESIDUAL STD. DEV. = 1.3251

CORRELATION MATRIX OF THE ESTIMATED PARAMETERS

	1	2
1	0.100E+01	−0.839E−01
2	−0.839E−01	0.100E+01

FIGURE 2   TSERIES Output of the RSAC and RSPAC for the Model $z_t = (1 - \theta_1 B)(1 - \theta_{1,12} B^{12}) a_t$ where $z_t = y_t - y_{t-1} - y_{t-12} + y_{t-13}$

(a)   The RSAC

	VALUE	S.E.	T VALUE	−1.0	0.0	1.0
				+ − − − − − −	− − − − − − − −	− − − +
1	0.006	0.078	0.072	I	*	I
2	−0.020	0.078	−0.261	I	*	I
3	0.035	0.078	0.454	I	*	I
4	−0.100	0.078	−1.285	I	***	I
5	−0.007	0.079	−0.094	I	*	I
6	0.098	0.079	1.238	I	***	I
7	−0.063	0.079	−0.788	I	**	I
8	0.022	0.080	0.270	I	*	I
9	0.094	0.080	1.175	I	***	I
10	−0.016	0.080	−0.204	I	*	I
11	0.018	0.081	0.219	I	*	I
12	0.090	0.081	1.118	I	***	I
13	−0.085	0.081	−1.043	I	***	I
14	0.059	0.082	0.724	I	**	I
15	−0.037	0.082	−0.455	I	*	I
16	0.070	0.082	0.857	I	**	I
17	0.037	0.082	0.454	I	*	I
18	−0.061	0.083	−0.745	I	**	I
19	−0.151	0.083	−1.822	I	****	I
20	−0.050	0.084	−0.591	I	**	I
21	0.025	0.085	0.299	I	*	I
22	−0.020	0.085	−0.239	I	*	I
23	0.102	0.085	1.199	I	***	I
24	−0.060	0.085	−0.707	I	**	I
				+ − − − − − −	− − − − − − − −	− − − +

BOX-PIERCE TEST CHISQUARE STATISTIC WITH 20 DEGREES OF FREEDOM = 15.118

(Part b, the RSPAC, follows on page 202.)

where
$$z_t = y_t - y_{t-1} - y_{t-12} + y_{t-13}$$

which is the tentative model describing the trade time series described in Exercise 3.1 of Chapter 3.

a.   Find and identify the least squares point estimates of the parameters in the above model.

b.   Do the least squares point estimates of θ_1 and $\theta_{1,12}$ satisfy the stationarity and invertibility conditions on these parameters? What are these conditions?

c.   Using $\hat{\theta}_1$ and $s_{\hat{\theta}_1}$, calculate $t_{\hat{\theta}_1}$.

d.   What does the $t_{\hat{\theta}_1}$-statistic say about whether or not the parameter θ_1 should be included in the model? Justify your answer by using an appropriate rejection point.

FIGURE 2   Continued

(b)   The RSPAC

	VALUE	S.E.	T VALUE	−1.0	0.0	1.0
				+ — — — — — — — — — — — — — — — — +		
1	0.006	0.078	0.072	I	*	I
2	−0.020	0.078	−0.261	I	*	I
3	0.036	0.078	0.458	I	*	I
4	−0.101	0.078	−1.300	I	***	I
5	−0.004	0.078	−0.056	I	*	I
6	0.093	0.078	1.196	I	***	I
7	−0.059	0.078	−0.755	I	**	I
8	0.017	0.078	0.220	I	*	I
9	0.086	0.078	1.104	I	***	I
10	0.004	0.078	0.052	I	*	I
11	0.009	0.078	0.110	I	*	I
12	0.082	0.078	1.049	I	***	I
13	−0.060	0.078	−0.769	I	**	I
14	0.057	0.078	0.736	I	**	I
15	−0.059	0.078	−0.753	I	**	I
16	0.109	0.078	1.399	I	***	I
17	0.009	0.078	0.120	I	*	I
18	−0.072	0.078	−0.926	I	**	I
19	−0.145	0.078	−1.860	I	****	I
20	−0.058	0.078	−0.740	I	**	I
21	0.035	0.078	0.454	I	*	I
22	−0.041	0.078	−0.528	I	**	I
23	0.080	0.078	1.032	I	***	I
24	−0.071	0.078	−0.908	I	**	I
				+ — — — — — — — — — — — — — — — — +		

e. Using $\hat{\theta}_{1,12}$ and $s_{\hat{\theta}_{1,12}}$, calculate $t_{\hat{\theta}_{1,12}}$.

f. What does the $t_{\hat{\theta}_{1,12}}$-statistic say about whether or not the parameter $\theta_{1,12}$ should be included in the model? Justify your answer by using an appropriate rejection point.

g. What does the correlation matrix of the estimated parameters say about the adequacy of the above model?

h. Using the sample autocorrelations of the residuals, calculate the Box-Pierce Q statistic (with K such that we have $K - n_p = 20$ degrees of freedom).

i. Using the Box-Pierce Q statistic and the appropriate rejection point, can we reject the adequacy of the above model? Explain your answer.

j. Using the sample autocorrelations of the residuals, calculate the Ljung-Box Q^* statistic with $K = 12$.

k. Using the Ljung-Box Q^* statistic calculated in part (j) and an appropriate rejection point, can we reject the adequacy of the above model? Explain your answer.

l. Describe the behaviors of the RSAC and RSPAC. What do these functions say about the adequacy of the above model?

4.5 Consider the trade data of Exercises 3.1 and 4.4.

a. Write the SAS program needed to compute the SAC and SPAC for the time series values produced by each of the transformations

$$z_t = y_t, \qquad z_t = y_t - y_{t-1}, \qquad z_t = y_t - y_{t-12}$$

and

$$z_t = y_t - y_{t-1} - y_{t-12} + y_{t-13}$$

b. Write the SAS program needed to perform estimation, diagnostic checking, and forecasting (for the next 12 time periods) for the trade model given in Exercise 4.4.

4.6 (Optional) Consider the trade data given in Exercise 3.1.

a. Write the TSERIES program needed to compute the SAC and SPAC for the time series values produced by each of the transformations

$$z_t = y_t, \qquad z_t = y_t - y_{t-1}, \qquad z_t = y_t - y_{t-12}$$

and

$$z_t = y_t - y_{t-1} - y_{t-12} + y_{t-13}$$

b. Write the TSERIES program needed to perform estimation, diagnostic checking, and forecasting (for the next 12 time periods) for the trade model given in Exercise 4.4.

4.7 Suppose that the following Box-Jenkins model has been tentatively identified:

$$(1 - \phi_1 B)(1 - \phi_{1,12}B^{12})y_t = \delta + a_t$$

Assume that the sample mean of the 120 observed time series values—$y_1, y_2, \ldots, y_{120}$ —is $\bar{y} = 38$.

a. For the above model, find preliminary estimates (for use in a Box-Jenkins program) of the parameters ϕ_1, $\phi_{1,12}$, and δ.

b. The RSAC and RSPAC obtained from the above model are given below (for lags 1–6)

lag k	1	2	3	4	5	6
residual autocorrelation	.71	.09	.07	.10	.08	.04

lag k	1	2	3	4	5	6
residual partial autocorrelation	.07	.04	.02	.01	.02	.01

1. Calculate Q^*, the Ljung-Box statistic, by using the above $K = 6$ residual auto-correlations. Then, by using an appropriate rejection point, show that we can, by setting α equal to .05, reject

H_0: The model is adequate

in favor of

H_1: The model is inadequate

    2. Since the model is inadequate, we will write it as

$$(1 - \phi_1 B)(1 - \phi_{1,12}B^{12})y_t = \delta + \eta_t$$

    Using the above RSAC and RSPAC, identify a model describing η_t (which represents the residuals).

    3. Combine the model describing η_t (that you identified in part (3)) with the above model describing y_t to arrive at a new model describing y_t.

  c. Algebraically expand the combined model of part b (2) and express y_t as a function of past time series observations and of past and present random shocks.

4.8   Consider the data (given by Miller and Wichern [1977]) in Table 4B of Appendix B which consists of 120 monthly values of average weekly total investments per month of large New York City banks. In Example 4.9 we have found that stationary time series values were produced by using the stationarity transformation

$$z_t = y_t - y_{t-1}$$

and we have tentatively identified two possible models:

Model 1: $z_t = (1 - \theta_6 B^6)(1 - \theta_{1,12}B^{12})a_t$

Model 2: $z_t = (1 - \theta_6 B^6 - \theta_{12}B^{12})a_t$

  a. Write and run the SAS program needed to perform identification, estimation, diagnostic checking, and forecasting (for the next 12 months) for both of Models 1 and 2 above.

  b. (Optional) Write and run the TSERIES program needed to perform identification, estimation, diagnostic checking, and forecasting (for the next 12 months) for both of Models 1 and 2 above.

  c. Use the output obtained in part (a) or (b) above to write an essay that carefully compares the adequacy of Models 1 and 2 above. Verify that Model 2 provides the best combination of significant $t_{\hat{\theta}}$-statistics, a small standard error, a small Ljung-Box Q^* statistic, and short prediction interval forecasts.

TIME SERIES REGRESSION

5.1 INTRODUCTION

In this chapter we will discuss several ways to combine regression analysis with the Box-Jenkins methodology to forecast time series. Specifically, Section 5.2 describes using *deterministic functions of time* to model trend and seasonal effects. Included in our discussion here are the use of *seasonal dummy variables* and *trigonometric terms*. Section 5.4 covers *growth curve models*, while Section 5.6 explains an intuitive approach to time series analysis—*the multiplicative decomposition method*. Sections 5.3, 5.5, and 5.7 deal with using SAS to implement the techniques presented in this chapter.

5.2 MODELING TREND AND SEASONAL EFFECTS BY USING DETERMINISTIC FUNCTIONS OF TIME

It can sometimes be useful to describe a time series y_t by the model

$$y_t = TR_t + SN_t + \varepsilon_t$$

where TR_t denotes the trend in time period t, SN_t denotes the seasonal factor in time period t, and ε_t denotes the *error term* in time period t. The above model says that the time series y_t can be represented by an average level (denoted μ_t) that changes over time according to the equation

$$\mu_t \;=\; TR_t \,+\, SN_t$$

combined with random fluctuations (represented by the error term ε_t) which cause the observations of the time series to deviate from the average level. Some common trends are

1. *No trend*, which is modeled as

    $$TR_t \;=\; \beta_0$$

    and implies that there is *no long-run growth or decline* in the time series over time (see Figure 5.1a).

2. *Linear trend*, which is modeled as

    $$TR_t \;=\; \beta_0 \,+\, \beta_1 t$$

    and implies that there is a *straight line* long-run growth (if the slope β_1 is greater than zero) or *decline* (β_1 less than zero) over time (see Figure 5.1b and c).

3. *Quadratic trend*, which is modeled as

    $$TR_t \;=\; \beta_0 \,+\, \beta_1 t \,+\, \beta_2 t^2$$

    and implies that there is a *quadratic (or curvilinear)* long-run change over time. This quadratic change can either be *growth at an increasing or decreasing rate* or *decline at an increasing or decreasing rate* (see Figure 5.1d, e, f, and g).

Although these trends are probably most commonly used, other, more complicated, trends also exist.

Seasonal patterns can be modeled by using *dummy variables*. Supposing that there are L seasons (months, quarters, etc.) per year, we will assume that SN_t is given by the equation

$$SN_t \;=\; \beta_{s1} x_{s1,t} \,+\, \beta_{s2} x_{s2,t} \,+\, \cdots \,+\, \beta_{s(L-1)} x_{s(L-1),t}$$

where

$$x_{s1,t},\, x_{s2,t},\, \ldots ,\, x_{s(L-1),t}$$

are *dummy variables* which are defined as follows:

$$x_{s1,t} \;=\; \begin{cases} 1 & \text{if time period } t \text{ is season 1} \\ 0 & \text{otherwise} \end{cases}$$

$$x_{s2,t} \;=\; \begin{cases} 1 & \text{if time period } t \text{ is season 2} \\ 0 & \text{otherwise} \end{cases}$$

$$\vdots$$

$$x_{s(L-1),t} \;=\; \begin{cases} 1 & \text{if time period } t \text{ is season } L-1 \\ 0 & \text{otherwise} \end{cases}$$

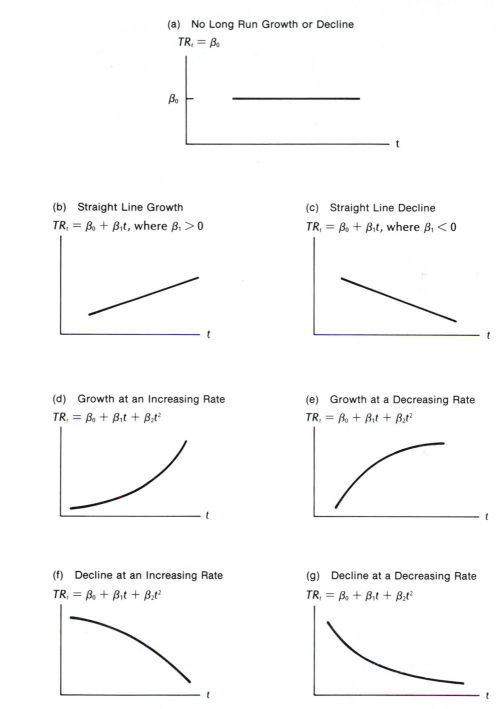

(a)   No Long Run Growth or Decline
$$TR_t = \beta_0$$

(b)   Straight Line Growth
$TR_t = \beta_0 + \beta_1 t$, where $\beta_1 > 0$

(c)   Straight Line Decline
$TR_t = \beta_0 + \beta_1 t$, where $\beta_1 < 0$

(d)   Growth at an Increasing Rate
$TR_t = \beta_0 + \beta_1 t + \beta_2 t^2$

(e)   Growth at a Decreasing Rate
$TR_t = \beta_0 + \beta_1 t + \beta_2 t^2$

(f)   Decline at an Increasing Rate
$TR_t = \beta_0 + \beta_1 t + \beta_2 t^2$

(g)   Decline at a Decreasing Rate
$TR_t = \beta_0 + \beta_1 t + \beta_2 t^2$

FIGURE 5.1   *Different Types of Trend*

For example, if $L = 12$ (we have monthly data) and period t is season 2 (February), we have

$$y_t = TR_t + SN_t + \varepsilon_t$$

$$= TR_t + \beta_{s1}x_{s1,t} + \beta_{s2}x_{s2,t} + \beta_{s3}x_{s3,t} + \cdots + \beta_{s11}x_{s11,t} + \varepsilon_t$$

$$= TR_t + \beta_{s1}(0) + \beta_{s2}(1) + \beta_{s3}(0) + \cdots + \beta_{s11}(0) + \varepsilon_t$$

$$= TR_t + \beta_{s2} + \varepsilon_t$$

So, use of the dummy variables insures that a seasonal parameter for season 2 is added to the trend in each appropriate time period. This seasonal parameter, β_{s2}, accounts for the seasonality of the time series in season 2. Therefore, in general, the purpose of the dummy variables is to insure that the appropriate seasonal parameter is included in the regression model in each time period.

We have, quite arbitrarily, set the seasonal parameter for season L equal to zero. Thus, the other seasonal parameters—$\beta_{s1}, \beta_{s2}, \ldots, \beta_{s(L-1)}$—are defined with respect to season L. Intuitively, β_{sj} is the difference, excluding trend, between the level of the time series in season j and the level of the time series in season L. A positive β_{sj} implies that, excluding trend, the value of the time series in season j can be expected to be greater than the value in season L. A negative β_{sj} implies that, excluding trend, the value of the time series in season j can be expected to be smaller than the value in season L. We do not have to set the seasonal parameter for season L equal to zero. We must, however, set the seasonal parameter for one of the seasons equal to zero (and thus define the other seasonal parameters with respect to that season).

Note that the dummy variable regression model

$$y_t = TR_t + SN_t + \varepsilon_t$$

where

$$SN_t = \beta_{s1}x_{s1,t} + \beta_{s2}x_{s2,t} + \cdots + \beta_{s(L-1)}x_{s(L-1),t}$$

assumes that we have *additive* (or *constant*) *seasonal variation*. We saw in Example 3.1, however, that a time series can exhibit increasing seasonal variability (and it can also exhibit decreasing seasonal variability). As illustrated in Example 3.1, we can sometimes transform a time series exhibiting changing seasonal variability into one exhibiting constant seasonal variability by taking natural logarithms.

Finally, note that each error term in the model is assumed to be a random variable that is *normally distributed* with mean zero and a *variance* that is the *same* for each and every time period t. Moreover, the error terms $\varepsilon_t, \varepsilon_{t-1}, \ldots$ are assumed to be *statistically independent*. If, as will often occur, the RSAC, RSPAC, and Q^* (and the related prob-value) indicate that the error terms are not statistically independent, we should use the guidelines in Tables 2.3 and 3.3 to identify a Box-Jenkins model describing the error terms:

$$\phi_p(B)\phi_P(B^L)\varepsilon_t = \delta + \theta_q(B)\theta_Q(B^L)a_t$$

Then, as will be illustrated in the following examples, we forecast future values of the time series by combining the model

$$y_t = TR_t + SN_t + \varepsilon_t$$

with the Box-Jenkins model describing ε_t.

EXAMPLE 5.1   Since the plot of the monthly hotel room averages in Figure 3.1 indicates a linear trend and seasonal variation with increasing variability, we might consider the regression model

$$
\begin{aligned}
y_t^* &= TR_t + SN_t + \varepsilon_t \\
&= \beta_0 + \beta_1 t + \beta_{s1} x_{s1,t} + \beta_{s2} x_{s2,t} + \ldots + \beta_{s11} x_{s11,t} + \varepsilon_t
\end{aligned}
$$

where $y_t^* = \ln y_t$ and $x_{s1,t}, x_{s2,t}, \ldots, x_{s11,t}$ are dummy variables. For example

$$
x_{s1,t} = \begin{cases} 1 & \text{if period } t \text{ is January} \\ 0 & \text{otherwise} \end{cases}
$$

Note that we have not defined a dummy variable for December. Assuming independent error terms $\varepsilon_1, \varepsilon_2, \ldots$, the SAS output of estimation and diagnostic checking for the model is given in Table 5.1.

Since the prob-values associated with the values of Q^* for $k = 6, 12, 18, 24$, and 30 are extremely small, we reject the adequacy of this model. To find a model

---

TABLE 5.1   *SAS Output of Estimation and Diagnostic Checking for the Logged Hotel Room Occupancy Dummy Variable Regression Model, Assuming Independent Error Terms $\varepsilon_1, \varepsilon_2, \ldots$.*

(a) Estimation

PARAMETER	ESTIMATE	STD ERROR	T RATIO	LAG	VARIABLE
MU	6.28756	0.00642719	978.27	0	LY
NUM1	0.00272528	0.00003379	80.65	0	T
NUM2	−0.0416063	0.00801623	−5.19	0	D1
NUM3	−0.112079	0.00801474	−13.98	0	D2
NUM4	−0.084459	0.00801338	−10.54	0	D3
NUM5	0.0398331	0.00801217	4.97	0	D4
NUM6	0.0203951	0.0080111	2.55	0	D5
NUM7	0.146909	0.00801018	18.34	0	D6
NUM8	0.289023	0.00800939	36.09	0	D7
NUM9	0.311195	0.00800875	38.86	0	D8
NUM10	0.0559872	0.00800825	6.99	0	D9
NUM11	0.0395438	0.0080079	4.94	0	D10
NUM12	−0.112215	0.00800768	−14.01	0	D11

Parts b and c follow on pages 210 and 211.

TABLE 5.1 *Continued*   (b) The RSAC

AUTOCORRELATION CHECK OF RESIDUALS

TO LAG	CHI SQUARE	DF	PROB	AUTOCORRELATIONS					
6	42.86	5	0.000	0.236	0.391	-0.074	-0.089	-0.156	-0.056
12	69.16	11	0.000	-0.041	-0.020	0.027	0.138	0.206	0.283
18	96.33	17	0.000	0.198	0.266	0.050	-0.034	-0.060	-0.169
24	102.99	23	0.000	-0.063	-0.019	0.038	-0.003	0.110	0.127
30	116.41	29	0.000	0.108	0.197	0.068	-0.055	-0.043	-0.081

AUTOCORRELATIONS

LAG	COVARIANCE	CORRELATION	STD
0	.00448859	1.00000	0
1	.000175359	0.39068	0.0771517
2	.000106121	0.23642	0.0881441
3	-3.316E-05	-0.07387	0.0918413
4	-3.980E-05	-0.08867	0.0921942
5	-7.009E-05	-0.15614	0.0927004
6	-2.509E-05	-0.05590	0.0942529
7	-8.786E-06	-0.01957	0.0944501
8	-.00001854	-0.04130	0.0944742
9	.000012306	0.02742	0.0945816
10	.000061748	0.13757	0.0946289
11	.000092555	0.20620	0.0958119
12	.000126953	0.28284	0.098418
13	0.0001195	0.26623	0.103143
14	.000088902	0.19806	0.107155
15	.000022357	0.04981	0.109313
16	-1.536E-05	-0.03421	0.109448
17	-2.713E-05	-0.06043	0.109511
18	-.00007598	-0.16927	0.10971
19	-8.483E-06	-0.01890	0.111253
20	-2.818E-05	-0.06278	0.111272
21	.000017221	0.03837	0.111483
22	-1.542E-06	-0.00344	0.111562
23	.000049197	0.10960	0.111562
24	0.00005679	0.12652	0.112201

AUTOCORRELATION PLOT OF RESIDUALS

'.' MARKS TWO STANDARD ERRORS

(c) The RSPAC

PARTIAL AUTOCORRELATIONS

LAG	CORRELATION
1	0.39068
2	0.09889
3	−0.23327
4	−0.01591
5	−0.06263
6	0.03008
7	0.01899
8	−0.09525
9	0.06886
10	0.16066
11	0.10012
12	0.16426
13	0.11244
14	0.07149
15	0.00392
16	−0.00912
17	0.04133
18	−0.14261
19	0.11167
20	−0.06850
21	−0.01991
22	−0.04747
23	0.00119
24	0.04178

describing ε_t (which represents the residuals), note that

1. At the *nonseasonal level*, the RSAC dies down quickly, and the RSPAC has spikes at lags 1 and 3 and cuts off after lag 3. By Guideline 2 in Table 2.3, we should use one of the following nonseasonal autoregressive operators

    $$\phi_p(B) = (1 - \phi_1 B - \phi_3 B^3)$$

    or

    $$\phi_p(B) = (1 - \phi_1 B - \phi_2 B^2 - \phi_3 B^3)$$

2. At the *seasonal level*, the RSAC dies down, and the RSPAC has a spike at lag 12 and cuts off after lag 12. By Guideline 7 in Table 3.3, we should use the seasonal autoregressive operator of order 1

    $$\phi_1(B^{12}) = (1 - \phi_{1,12} B^{12})$$

By inserting the second of the nonseasonal operators and the seasonal operator into the general model

$$\phi_p(B)\phi_P(B^L)\varepsilon_t = \delta + \theta_q(B)\theta_Q(B^L)a_t$$

and by noting that the mean of the residuals from the regression model is not statistically different from zero (implying that we should set δ equal to zero), we are led to tentatively conclude that the model

$$(1 - \phi_1 B - \phi_2 B^2 - \phi_3 B^3)(1 - \phi_{1,12} B^{12})\varepsilon_t = a_t$$

describes the error term ε_t in the regression model

$$y_t^* = \beta_0 + \beta_1 t + \beta_{s1} x_{s1,t} + \beta_{s2} x_{s2,t} + \cdots + \beta_{s11} x_{s11,t} + \varepsilon_t$$

The SAS output of estimation, diagnostic checking, and forecasting resulting from the *time series regression model* obtained by combining the above two models is given in Table 5.2, Figure 5.2, and Table 5.3. The prob-values associated with the values of Q^* do not indicate that we should reject the adequacy of the time series regression model. Furthermore, comparison of Table 5.3 and Table 4.7 indicates that the above time series regression model and Model 3 of Example 4.4 yield extremely similar point forecasts and prediction interval forecasts.

We will complete this example by discussing exactly how the model

$$y_t^* = \beta_0 + \beta_1 t + \beta_{s1} x_{s1,t} + \beta_{s2} x_{s2,t} + \cdots + \beta_{s11} x_{s11,t} + \varepsilon_t$$

where

$$(1 - \phi_1 B - \phi_2 B^2 - \phi_3 B^3)(1 - \phi_{1,12} B^{12})\varepsilon_t = a_t$$

or (equivalently)

$$\varepsilon_t = \phi_1 \varepsilon_{t-1} + \phi_2 \varepsilon_{t-2} + \phi_3 \varepsilon_{t-3} + \phi_{1,12} \varepsilon_{t-12}$$
$$- \phi_1 \phi_{1,12} \varepsilon_{t-13} - \phi_2 \phi_{1,12} \varepsilon_{t-14} - \phi_3 \phi_{1,12} \varepsilon_{t-15} + a_t$$

has produced the point forecasts in Table 5.3. Noting that we have oberved 168

TABLE 5.2 *SAS Output of the Least Squares Point Estimates of the Parameters in the Logged Hotel Room Occupancy Dummy Variable Regression Model, Assuming the Error Terms ε_1, ε_2, ... are Described by the Model $(1 - \phi_1 B - \phi_2 B^2 - \phi_3 B^3)(1 - \phi_{1,12} B^{12})\varepsilon_t = a_t$*

PARAMETER	ESTIMATE	STD ERROR	T RATIO	LAG	VARIABLE
MU	6.28438	0.00784847	800.71	0	LY
AR1,1	0.343109	0.0802359	4.28	1	LY
AR1,2	0.164314	0.0836629	1.96	2	LY
AR1,3	−0.238192	0.080685	−2.95	3	LY
AR2,1	0.191572	0.0857906	2.23	12	LY
NUM1	0.00274122	.000048039	57.06	0	T
NUM2	−0.0391343	0.00773845	−5.06	0	D1
NUM3	−0.109521	0.00848485	−12.91	0	D2
NUM4	−0.0825612	0.00986648	−8.37	0	D3
NUM5	0.0416737	0.00973649	4.28	0	D4
NUM6	0.0213048	0.00986569	2.16	0	D5
NUM7	0.148119	0.00964278	15.36	0	D6
NUM8	0.290676	0.0098997	29.36	0	D7
NUM9	0.312378	0.00976541	31.99	0	D8
NUM10	0.0575828	0.00989799	5.82	0	D9
NUM11	0.0399709	0.00837898	4.77	0	D10
NUM12	−0.112346	0.00761127	−14.76	0	D11

CONSTANT ESTIMATE   = 3.71264

VARIANCE ESTIMATE   = .000352022
STD ERROR ESTIMATE   = 0.0187622
NUMBER OF RESIDUALS = 168

time series observations y_1^*, y_2^*, ..., y_{168}^*, and noting that

$$y_{169}^* = \beta_0 + \beta_1(169) + \beta_{s1}(1) + \beta_{s2}(0) + \cdots + \beta_{s11}(0) + \varepsilon_{169}$$

where

$$\varepsilon_{169} = \phi_1 \varepsilon_{168} + \phi_2 \varepsilon_{167} + \phi_3 \varepsilon_{166} + \phi_{1,12} \varepsilon_{157}$$
$$- \phi_1 \phi_{1,12} \varepsilon_{156} - \phi_2 \phi_{1,12} \varepsilon_{155} - \phi_3 \phi_{1,12} \varepsilon_{154} + a_{169}$$

it follows that a point forecast of y_{169}^* is

$$\hat{y}_{169}^* = b_0 + b_1(169) + b_{s1} + \hat{\varepsilon}_{169}$$

where

$$\hat{\varepsilon}_{169} = \hat{\phi}_1 \varepsilon_{168} + \hat{\phi}_2 \hat{\varepsilon}_{167} + \hat{\phi}_e \hat{\varepsilon}_{166} + \hat{\phi}_{1,12} \hat{\varepsilon}_{157}$$
$$- \hat{\phi}_1 \hat{\phi}_{1,12} \hat{\varepsilon}_{156} - \hat{\phi}_2 \hat{\phi}_{1,12} \hat{\varepsilon}_{155} - \hat{\phi}_3 \hat{\phi}_{1,12} \hat{\varepsilon}_{154} + \hat{a}_{169}$$

Here, b_0, b_1, b_{s1}, $\hat{\phi}_1$, $\hat{\phi}_2$, $\hat{\phi}_3$, and $\hat{\phi}_{1,12}$ are the least squares point estimates given in Table 5.2, and the point prediction of a_{169} is $\hat{a}_{169} = 0$ (since a_{169} is a future random

FIGURE 5.2 SAS Output of the RSAC and RSPAC for the Logged Hotel Room Occupancy Dummy Variable Regression Model, Assuming the Error Terms $\varepsilon_1, \varepsilon_2, \ldots$ Are Described by the Model $(1 - \phi_1 B - \phi_2 B^2 - \phi_3 B^3)(1 - \phi_{1,12} B^{12})\varepsilon_t = a_t$

(a) The RSAC

AUTOCORRELATION CHECK OF RESIDUALS

TO LAG	CHI SQUARE	DF	PROB	AUTOCORRELATIONS					
6	3.07	1	0.080	-0.004	-0.018	0.028	0.013	-0.118	0.048
12	8.18	7	0.317	0.065	-0.074	0.004	0.112	0.078	-0.005
18	20.99	13	0.073	0.146	0.122	-0.020	0.024	0.049	-0.169
24	22.84	19	0.244	0.067	-0.011	0.028	-0.019	0.056	0.025
30	27.18	25	0.347	0.100	0.031	0.062	-0.068	-0.042	0.017

AUTOCORRELATION PLOT OF RESIDUALS

LAG	COVARIANCE	CORRELATION	-1 9 8 7 6 5 4 3 2 1 0 1 2 3 4 5 6 7 8 9 1	STD
0	.000352022	1.00000	|********************	0
1	-1.412E-06	-0.00401	. | .	0.0771517
2	-6.305E-06	-0.01791	. | .	0.0771529
3	9.935E-06	0.02822	. |* .	0.0771776
4	4.419E-06	0.01255	. | .	0.077239
5	-4.155E-05	-0.11804	. **| .	0.0772512
6	.000016784	0.04768	. |* .	0.0783175
7	.000022987	0.06530	. |* .	0.0784901
8	-.00002618	-0.07437	. *| .	0.0788128
9	1.384E-06	0.00393	. | .	0.0792294
10	.000039491	0.11218	. |** .	0.0792306
11	.000027287	0.07751	. |** .	0.0801705
12	-1.775E-06	-0.00504	. | .	0.0806154
13	.000051549	0.14644	. |*** .	0.0806172
14	.000042889	0.12184	. |** .	0.0821853
15	-7.205E-06	-0.02047	. | .	0.0832534
16	8.617E-06	0.02448	. | .	0.0832834
17	.000017286	0.04910	. |* .	0.0833262
18	-5.959E-05	-0.16929	. ***| .	0.0834983
19	.000023697	0.06732	. |* .	0.085517
20	-3.979E-06	-0.01130	. | .	0.0858318
21	9.853E-06	0.02799	. | .	0.0858407
22	-6.611E-06	-0.01878	. | .	0.085895
23	.000019549	0.05553	. |* .	0.0859194
24	8.921E-06	0.02534	. | .	0.0861328

'.' MARKS TWO STANDARD ERRORS

FIGURE 5.2 *Continued*

(b) The RSPAC

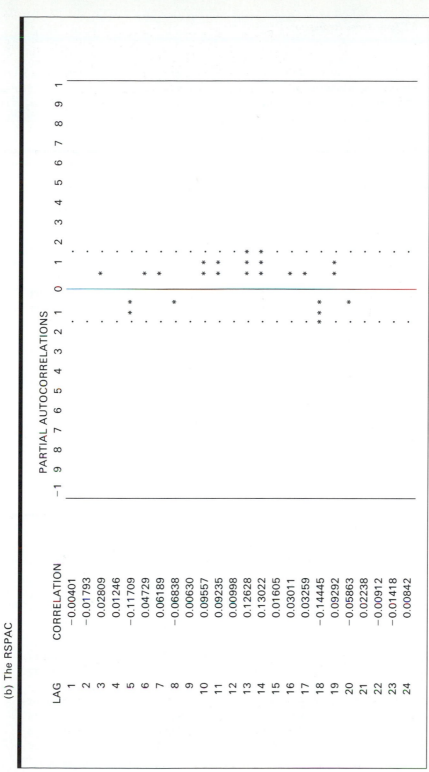

TABLE 5.3   SAS Output of Point Forecasts and Prediction Interval Forecasts of Logged and Actual Hotel Room Occupancies in Periods 169–180 Made Using the Logged Hotel Room Occupancy Dummy Variable Regression
Model, Assuming the Error Terms ε_1, ε_2, ... Are Described by the Model $(1 - \phi_1 B - \phi_2 B^2 - \phi_3 B^3)(1 - \phi_{1,12} B^{12})\varepsilon_t = a_t$

(a) Forecasts of Logged Hotel Room Occupancies

			SAS	
FORECASTS FOR VARIABLE LY				
OBS	FORECAST	STD ERROR	LOWER 95%	UPPER 95%
— — — —FORECAST BEGINS — — —				
169	6.7322	0.0188	6.6954	6.7689
170	6.6482	0.0198	6.6094	6.6871
171	6.6644	0.0205	6.6241	6.7046
172	6.7889	0.0206	6.7486	6.8293
173	6.7719	0.0206	6.7315	6.8124
174	6.9011	0.0207	6.8604	6.9417
175	7.0526	0.0207	7.0120	7.0932
176	7.0756	0.0207	7.0350	7.1162
177	6.8223	0.0207	6.7817	6.8629
178	6.8083	0.0207	6.7677	6.8489
179	6.6639	0.0207	6.6233	6.7046
180	6.7838	0.0207	6.7432	6.8244

(b) Forecasts of Actual Hotel Room Occupancies

		SAS		
OBS	Y	L95CI	FY	U95CI
169	.	808.67	838.96	870.39
170	.	742.02	771.43	802.01
171	.	753.06	783.98	816.16
172	.	852.84	887.97	924.54
173	.	838.40	872.99	909.01
174	.	953.80	993.32	1034.49
175	.	1109.85	1155.86	1203.77
176	.	1135.68	1182.76	1231.79
177	.	881.53	918.07	956.14
178	.	869.30	905.34	942.88
179	.	752.44	783.64	816.13
180	.	848.26	883.43	920.06

shock). Moreover, since the model

$$y_t^* = \beta_0 + \beta_1 t + \beta_{s1} x_{s1,t} + \beta_{s2} x_{s2,t} + \cdots + \beta_{s11} x_{s11,t} + \varepsilon_t$$

implies that

$$\varepsilon_t = y_t^* - (\beta_0 + \beta_1 t + \beta_{s1} x_{s1,t} + \beta_{s2} x_{s2,t} + \cdots + \beta_{s11} x_{s11,t})$$

and since we have observed y_{168}^*, y_{167}^*, y_{166}^*, y_{157}^*, y_{156}^*, y_{155}^*, and y_{154}^*, the point prediction of each of ε_{168}, ε_{167}, ε_{166}, ε_{157}, ε_{156}, ε_{155}, and ε_{154} is given by the equation

$$\hat{\varepsilon}_t = y_t^* - (b_0 + b_1 t + b_{s1} x_{s1,t} + b_{s2} x_{s2,t} + \cdots + b_{s11} x_{s11,t})$$

Next, since

$$y_{170}^* = \beta_0 + \beta_1(170) + \beta_{s1}(0) + \beta_{s2}(1) + \cdots + \beta_{s11}(0) + \varepsilon_{170}$$

where

$$\varepsilon_{170} = \phi_1 \varepsilon_{169} + \phi_2 \varepsilon_{168} + \phi_3 \varepsilon_{167} + \phi_{1,12} \varepsilon_{158}$$
$$- \phi_1 \phi_{1,12} \varepsilon_{157} - \phi_2 \phi_{1,12} \varepsilon_{156} - \phi_3 \phi_{1,12} \varepsilon_{155} + a_{170}$$

it follows that a point forecast of y_{170}^* is

$$\hat{y}_{170}^* = b_0 + b_1(170) + b_{s2} + \hat{\varepsilon}_{170}$$

where

$$\hat{\varepsilon}_{170} = \hat{\phi}_1 \hat{\varepsilon}_{169} + \hat{\phi}_2 \hat{\varepsilon}_{168} + \hat{\phi}_3 \hat{\varepsilon}_{167} + \hat{\phi}_{1,12} \hat{\varepsilon}_{158}$$
$$- \hat{\phi}_1 \hat{\phi}_{1,12} \hat{\varepsilon}_{157} - \hat{\phi}_2 \hat{\phi}_{1,12} \hat{\varepsilon}_{156} - \hat{\phi}_3 \hat{\phi}_{1,12} \hat{\varepsilon}_{155} + \hat{a}_{170}$$

Here, the point prediction of a_{170} is $\hat{a}_{170} = 0$ (since a_{170} is a future random shock). Furthermore, since we have observed y_{168}^*, y_{167}^*, y_{158}^*, y_{157}^*, y_{156}^*, and y_{155}^*, the point prediction of each of ε_{168}, ε_{167}, ε_{158}, ε_{157}, ε_{156}, and ε_{155} is given by the equation

$$\hat{\varepsilon}_t = y_t^* - (b_0 + b_1 t + b_{s1} x_{s1,t} + b_{s2} x_{s2,t} + \cdots + b_{s11} x_{s11,t})$$

and, since ε_{169} is a future error term (because there are 168 observations), the point prediction of ε_{169} is

$$\hat{\varepsilon}_{169} = \hat{y}_{169}^* - [b_0 + b_1(169) + b_{s1}(1)]$$

where $\hat{y}_{169}^*$ is the previously calculated point forecast of y_{169}^*. Point forecasts of other future values of y_t^* are made similarly.

EXAMPLE 5.2

Since the plot of the quarterly gas bills in Figure 4.8 indicates the existence of a quadratic trend and constant seasonal variability, we might consider the regression model

$$y_t = TR_t + SN_t + \varepsilon_t$$
$$= \beta_0 + \beta_1 t + \beta_2 t^2 + \beta_{s1} x_{s1,t} + \beta_{s2} x_{s2,t} + \beta_{s3} x_{s3,t} + \varepsilon_t$$

TABLE 5.4 *SAS Output of Estimation and Diagnostic Checking for the Gas Bill Dummy Variable Regression Model, Assuming Independent Error Terms* $\varepsilon_1, \varepsilon_2, \dots$.

(a) Estimation

ARIMA: LEAST SQUARES ESTIMATION

ITERATION	SSE	MU	NUM1	NUM2	NUM3	NUM4	NUM5	CONSTANT	LAMBDA
0	1208984	265.496	4.62838	0.124325	111.615	−21.1197	−133.985	265.496	1.0E−05
1	124400	276.353	−7.43036	0.300693	65.7896	−38.0584	−127.606	276.353	1.0E−06
2	124399	276.441	−7.44167	0.300956	65.7866	−38.06	−127.607	276.441	1.0E−07
3	124399	276.441	−7.44167	0.300956	65.7866	−38.0601	−127.607	276.441	1.0E−08

PARAMETER	ESTIMATE	STD ERROR	T RATIO	LAG	VARIABLE
MU	276.441	35.0574	7.89	0	Y
NUM1	−7.44167	3.39688	−2.19	0	T
NUM2	0.300956	0.0803246	3.75	0	TSQ
NUM3	65.7866	27.1661	2.42	0	D1
NUM4	−38.0601	27.1027	−1.40	0	D2
NUM5	−127.607	27.0644	−4.71	0	D3

CONSTANT ESTIMATE   = 276.441

VARIANCE ESTIMATE   = 3658.81
STD ERROR ESTIMATE   = 60.4881
NUMBER OF RESIDUALS   = 40

CORRELATIONS OF THE ESTIMATES

	MU	NUM1	NUM2	NUM3	NUM4	NUM5
MU	1.000	−0.763	0.655	−0.432	−0.413	−0.398
NUM1	−0.763	1.000	−0.970	0.023	0.009	0.002
NUM2	0.655	−0.970	1.000	0.000	0.006	0.006
NUM3	−0.432	0.023	0.000	1.000	0.503	0.500
NUM4	−0.413	0.009	0.006	0.503	1.000	0.501
NUM5	−0.398	0.002	0.006	0.500	0.501	1.000

(Continued)

where $x_{s1,t}, x_{s2,t}$, and $x_{s3,t}$ are dummy variables. For example,

$$x_{s1,t} = \begin{cases} 1 & \text{if period } t \text{ is Quarter 1} \\ 0 & \text{otherwise} \end{cases}$$

Assuming independent error terms $\varepsilon_1, \varepsilon_2, \dots$, the SAS output of estimation and diagnostic checking for the model is given in Table 5.4. Since the prob-values associated with the values of Q^* are extremely small, we reject the adequacy of the model. To find a model describing ε_t (which represents the residuals), note that

1. At the nonseasonal level (perhaps lags 1 and 2), the RSAC dies down quickly, and the RSPAC has a spike at lag 1 and cuts off after lag 1. By

TABLE 5.4   *Continued*   (b) The RSAC

AUTOCORRELATION CHECK OF RESIDUALS

TO LAG	CHI SQUARE	DF	PROB	AUTOCORRELATIONS					
6	24.99	5	0.000	0.553	0.208	0.061	−0.213	−0.362	−0.165
12	39.17	11	0.000	−0.035	−0.146	−0.091	−0.083	−0.259	−0.372
18	47.11	17	0.000	−0.200	−0.120	0.030	0.166	0.168	0.079
24	51.40	23	0.001	0.053	0.112	0.067	0.106	0.118	0.043

AUTOCORRELATION PLOT OF RESIDUALS

LAG	COVARIANCE	CORRELATION	−1 9 8 7 6 5 4 3 2 1 0 1 2 3 4 5 6 7 8 9 1	STD
0	3658.81	1.00000	| |* * * ** * * * * * * * * * * * * * * *	0
1	2023.42	0.55303	| . |* * * ** * * * * * *	0.158114
2	760.77	0.20793	| . |* * * * .	0.200728
3	222.599	0.06084	| . |* .	0.206043
4	−779.878	−0.21315	| . * * * * .	0.206491
5	−1322.83	−0.36155	| . * * * * * * * .	0.211921
6	−603.563	−0.16496	| . * * * .	0.226817
7	−128.499	−0.03512	| . * | .	0.229797
8	−535.688	−0.14641	| . * * * .	0.229931
9	−332.585	−0.09090	| . * * .	0.23225
10	−303.67	−0.08300	| . * * .	0.233138
11	−946.399	−0.25866	| . * * * * * .	0.233875
12	−1359.52	−0.37158	| . * * * * * * * .	0.240921
13	−730.631	−0.19969	| . * * * * .	0.254846
14	−439.639	−0.12016	| . * * .	0.258728
15	111.049	0.03035	| . |* .	0.26012
16	605.603	0.16552	| . |* * * .	0.260208
17	615.84	0.16832	| . |* * * .	0.262827
18	287.69	0.07863	| . |* * .	0.265508
19	192.645	0.05265	| . |* .	0.26609
20	410.823	0.11228	| . |* * .	0.26635
21	244.848	0.06692	| . |* .	0.267531
22	388.153	0.10609	| . |* * .	0.267949
23	430.374	0.11763	| . |* * .	0.268997
24	157.155	0.04295	| . |* .	0.27028

'·' MARKS TWO STANDARD ERRORS

(Part (c), the RSPAC, follows on page 220.)

TABLE 5.4 Continued

(c) The RSPAC

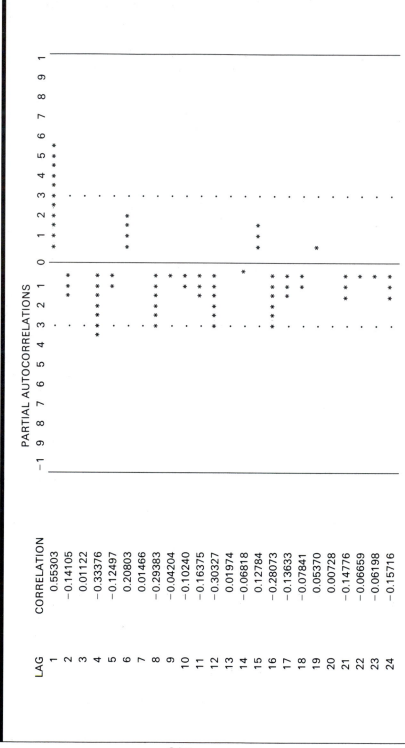

PARTIAL AUTOCORRELATIONS

LAG	CORRELATION
1	0.55303
2	−0.14105
3	0.01122
4	−0.33376
5	−0.12497
6	0.20803
7	0.01466
8	−0.29383
9	−0.04204
10	−0.10240
11	−0.16375
12	−0.30327
13	0.01974
14	−0.06818
15	0.12784
16	−0.28073
17	−0.13633
18	−0.07841
19	0.05370
20	0.00728
21	−0.14776
22	−0.06659
23	−0.06198
24	−0.15716

Guideline 2 in Table 2.3, we should use the nonseasonal autoregressive operator of order 1

$$\phi_1(B) = (1 - \phi_1 B)$$

2. At the seasonal level (lags 4, 8, 12, and 16), the RSAC dies down quickly (with the exception of a fairly large t-value at lag 5, which is equal to roughly $-.3616/.2119 = -1.7065$), and the RSPAC has spikes at lags 4, 8, 12, and 16.

Because of this confusing seasonal behavior, we will proceed by *arbitrarily* assuming that the error term ε_t in the regression model

$$y_t = \beta_0 + \beta_1 t + \beta_2 t^2 + \beta_{s1} x_{s1,t} + \beta_{s2} x_{s2,t} + \beta_{s3} x_{s3,t} + \varepsilon_t$$

might be described by one of the models

$$(1 - \phi_1 B)(1 - \phi_{1,4} B^4)\varepsilon_t = a_t$$

or

$$(1 - \phi_1 B)(1 - \phi_{1,5} B^5)\varepsilon_t = a_t$$

or

$$(1 - \phi_1 B)(1 - \phi_{1,4} B^4 - \phi_{1,5} B^5)\varepsilon_t = a_t$$

The SAS output of estimation, diagnostic checking, and forecasting resulting from the time series regression models obtained by combining the regression model with each of the proposed models describing ε_t is presented in Figures 5.3, 5.4, and 5.5. Examining this output, we see that the time series regression model

$$y_t = \beta_0 + \beta_1 t + \beta_2 t^2 + \beta_{s1} x_{s1,t} + \beta_{s2} x_{s2,t} + \beta_{s3} x_{s3,t} + \varepsilon_t$$

where

$$(1 - \phi_1 B)(1 - \phi_{1,4} B^4 - \phi_{1,5} B^5)\varepsilon_t = a_t$$

is probably the best model. Note that, although the absolute t-value associated with the estimate of $\phi_{1,4}$ in this model is fairly small (.91), this model has reasonably large prob-values associated with the values of Q^*, has the smallest standard error ($s = 46.3117$), and yields the shortest prediction interval forecasts. Also, note that this time series regression model has a smaller standard error and yields shorter prediction intervals than the best Box-Jenkins model discussed in Example 4.11, which is

$$(1 - \phi_1 B)z_t = (1 - \theta_{1,4} B^4 - \theta_{1,5} B^5)a_t$$

where

$$z_t = y_t - y_{t-4}.$$

To complete this section, we should make several comments. First, although using the time series regression model

$$y_t = TR_t + SN_t + \varepsilon_t$$

FIGURE 5.3   *SAS Output of Estimation, Diagnostic Checking, and Forecasting for the Gas Bill Dummy Variable Regression Model, Assuming the Error Terms $\varepsilon_1, \varepsilon_2, \ldots$ Are Described by the Model $(1 - \phi_1 B)(1 - \phi_{1,4} B^4)\varepsilon_t = a_t$*

(a) Estimation

PARAMETER	ESTIMATE	STD ERROR	T RATIO	LAG	VARIABLE
MU	276.027	43.8059	6.30	0	Y
AR1.1	0.579035	0.155643	3.72	1	Y
AR2.1	−0.183574	0.188695	−0.97	4	Y
NUM1	−8.21729	4.94004	−1.66	0	T
NUM2	0.32769	0.119437	2.74	0	TSQ
NUM3	71.3634	15.0887	4.73	0	D1
NUM4	−34.9722	16.8861	−2.07	0	D2
NUM5	−125.337	14.6933	−8.53	0	D3

CONSTANT ESTIMATE   = 137.529

VARIANCE ESTIMATE   = 2534.35
STD ERROR ESTIMATE   = 50.3423
NUMBER OF RESIDUALS = 40

CORRELATIONS OF THE ESTIMATES

	MU	AR1,1	AR2,1	NUM1	NUM2	NUM3	NUM4	NUM5
MU	1.000	0.028	0.049	−0.814	0.681	−0.269	−0.203	−0.146
AR1,1	0.028	1.000	−0.253	−0.103	0.137	0.057	0.050	0.041
AR2,1	0.049	−0.253	1.000	−0.065	0.079	−0.058	−0.032	−0.046
NUM1	−0.814	−0.103	−0.065	1.000	−0.966	0.057	−0.020	−0.040
NUM2	0.681	0.137	0.079	−0.966	1.000	−0.027	0.039	0.049
NUM3	−0.269	0.057	−0.058	0.057	−0.027	1.000	0.586	0.363
NUM4	−0.203	0.050	−0.032	−0.020	0.039	0.586	1.000	0.575
NUM5	−0.146	0.041	−0.046	−0.040	0.049	0.363	0.575	1.000

(*Continued*)

where TR_t and SN_t are deterministic functions of time, provided models in the previous examples that were as good or better than the Box-Jenkins models of Chapters 2, 3, and 4, this result cannot in general be expected to happen. Using a model in which TR_t and SN_t are deterministic functions of time implies that the trend and seasonal natures of the time series under consideration are largely deterministic (that is, unchanging) through time. Since the hotel room occupancy data exhibited a definite linear trend with well-defined seasonal variation, and since the gas bill data exhibited a definite quadratic trend (with perhaps less well defined seasonal variation), modeling the trend and seasonal effects of these time series by using deterministic functions of time provided good models. However, many time series exhibit trend and seasonal effects that are largely stochastic (that is, changing through time). The Box-Jenkins models of Chapters 2, 3, and 4 are particularly effective when used to model such time series because the autoregressive and differencing operators of these models are designed to describe stochastic trend and seasonal effects. For example, it is best to use the Box-Jenkins

(b) The RSAC

AUTOCORRELATION CHECK OF RESIDUALS

TO LAG	CHI SQUARE	DF	PROB	AUTOCORRELATIONS					
6	11.31	3	0.010	0.217	-0.143	0.063	-0.036	-0.409	-0.003
12	28.81	9	0.001	-0.050	-0.271	0.196	0.139	-0.148	-0.391
18	32.49	15	0.006	-0.044	-0.028	0.013	0.134	0.175	0.009
24	34.68	21	0.031	-0.018	0.125	-0.032	0.009	0.084	0.002

AUTOCORRELATIONS

AUTOCORRELATION PLOT OF RESIDUALS

-1 9 8 7 6 5 4 3 2 1 0 1 2 3 4 5 6 7 8 9 1

LAG	COVARIANCE	CORRELATION	STD
0	2534.35	1.00000	0
1	159.951	0.06311	0.158114
2	-363.585	-0.14346	0.158742
3	550.025	0.21703	0.161951
4	-90.9989	-0.03591	0.169066
5	-1036.69	-0.40906	0.169256
6	-6.57629	-0.00259	0.19239
7	496.97	0.19609	0.192391
8	-687.379	-0.27123	0.197325
9	-127.679	-0.05038	0.206435
10	351.977	0.13888	0.206742
11	-375.931	-0.14833	0.209061
12	-991.397	-0.39118	0.211676
13	32.5182	0.01283	0.229037
14	-71.144	-0.02807	0.229055
15	-111.842	-0.04413	0.229141
16	339.803	0.13408	0.229353
17	442.632	0.17465	0.231305
18	23.5167	0.00928	0.234578
19	-80.5964	-0.03180	0.234587
20	317.683	0.12535	0.234695
21	-46.056	-0.01817	0.236363
22	23.8725	0.00942	0.236398
23	213.643	0.08430	0.236407
24	55.4277	0.02187	0.237158

. MARKS TWO STANDARD ERRORS

(Part (c), the RSPAC, and Part (d), the Forecast, follow on page 224.)

FIGURE 5.3 *Continued*   (c) The RSPAC

PARTIAL AUTOCORRELATIONS

LAG	CORRELATION
1	0.06311
2	-0.14804
3	0.24304
4	-0.10750
5	-0.35438
6	0.00775
7	0.15976
8	-0.19465
9	-0.02808
10	-0.10647
11	-0.08338
12	-0.35352
13	-0.15916
14	-0.15906
15	0.15110
16	-0.09117
17	-0.12920
18	-0.12607
19	-0.07048
20	-0.04068
21	-0.06273
22	-0.03332
23	-0.00770
24	-0.21815

(d) Forecasts (of y_{41}, y_{42}, y_{43}, and y_{44})

FORECASTS FOR VARIABLE Y

OBS	FORECAST	STD ERROR	LOWER 95%	UPPER 95%
	FORECAST BEGINS			
41	604.3814	50.3423	505.7125	703.0503
42	493.2384	58.1727	379.2221	607.2546
43	413.7545	60.5719	295.0359	532.4731
44	542.6808	61.3554	422.4267	662.9348

FIGURE 5.4 SAS Output of Estimation, Diagnostic Checking, and Forecasting for the Gas Bill Dummy Variable Regression Model. Assuming the Error Terms $\varepsilon_1, \varepsilon_2, \ldots$ Are Described by the Model $(1 - \phi_1 B)(1 - \phi_{1,5}B^5)\varepsilon_t = a_t$

(a) Estimation

PARAMETER	ESTIMATE	STD ERROR	T RATIO	LAG	VARIABLE
MU	273.18	36.2526	7.54	0	Y
AR1,1	0.533541	0.145151	3.68	1	Y
AR1,2	-0.295875	0.147481	-2.01	5	Y
NUM1	-7.52568	3.94604	-1.91	0	T
NUM2	0.309858	0.0955924	3.24	0	TSQ
NUM3	69.4026	19.282	3.60	0	D1
NUM4	-35.5666	20.8421	-1.71	0	D2
NUM5	-128.097	18.9668	-6.75	0	D3

CONSTANT ESTIMATE   = 208.254

VARIANCE ESTIMATE   = 2323.94
STD ERROR ESTIMATE   = 48.2072
NUMBER OF RESIDUALS = 40

CORRELATIONS OF THE ESTIMATES

	MU	AR1,1	AR1,2	NUM1	NUM2	NUM3	NUM4	NUM5
MU	1.000	0.071	0.052	-0.825	0.716	-0.317	-0.298	-0.252
AR1,1	0.071	1.000	0.205	-0.139	0.170	0.059	0.036	0.027
AR1,2	0.052	0.205	1.000	-0.099	0.119	0.020	-0.005	0.017
NUM1	-0.825	-0.139	-0.099	1.000	-0.971	0.020	-0.013	-0.025
NUM2	0.716	0.170	0.119	-0.971	1.000	0.005	0.029	0.033
NUM3	-0.317	0.059	0.020	0.020	0.005	1.000	0.556	0.408
NUM4	-0.298	0.036	-0.005	-0.013	0.029	0.556	1.000	0.549
NUM5	-0.252	0.027	0.017	-0.025	0.033	0.408	0.549	1.000

(Parts (b), the RSAC, (c), the RSPAC, and (d), the Forecast follow on pages 226 and 227.)

FIGURE 5.4   Continued   (b) The RSAC

AUTOCORRELATION CHECK OF RESIDUALS

TO LAG	CHI SQUARE	DF	AUTOCORRELATIONS						PROB
6	5.70	3	0.026	-0.230	-0.211	0.079	-0.051	0.124	0.127
12	19.19	9	-0.392	-0.211	0.062	0.025	-0.126	0.108	0.024
18	25.90	15	0.025	0.166	0.187	0.088	-0.110	-0.120	0.039
24	27.74	21	0.007	0.074	0.041	-0.042	0.095	-0.053	0.148

AUTOCORRELATION PLOT OF RESIDUALS

LAG	COVARIANCE	CORRELATION	-1 9 8 7 6 5 4 3 2 1 0 1 2 3 4 5 6 7 8 9 1	STD
0	2323.94	1.00000		0
1	288.76	0.12425		0.158114
2	-118.293	-0.05090		0.160536
3	183.8	0.07909		0.160939
4	-489.743	-0.21074		0.161908
5	-533.663	-0.22964		0.168626
6	59.2915	0.02551		0.176271
7	250.397	0.10775		0.176363
8	-292.557	-0.12589		0.178001
9	58.636	0.02523		0.180213
10	143.622	0.06180		0.180302
11	-491.4	-0.21145		0.180831
12	-911.67	-0.39230		0.18691
13	-278.499	-0.11984		0.20647
14	-254.488	-0.10951		0.208202
15	205.036	0.08823		0.209637
16	435.15	0.18725		0.210563
17	385.523	0.16589		0.214686
18	58.488	0.02517		0.217867
19	-123.163	-0.05300		0.21794
20	219.976	0.09466		0.218262
21	-97.8276	-0.04210		0.219285
22	96.0221	0.04132		0.219487
23	172.36	0.07417		0.219682
24	16.6798	0.00718		0.220307

"." MARKS TWO STANDARD ERRORS

(c) The RSPAC

PARTIAL AUTOCORRELATIONS

LAG	CORRELATION
1	0.12425
2	−0.06738
3	0.09613
4	−0.24486
5	−0.16479
6	0.04220
7	0.12932
8	−0.18000
9	−0.01949
10	0.00176
11	−0.14680
12	−0.44261
13	−0.17198
14	−0.13428
15	0.15546
16	−0.16429
17	−0.04618
18	−0.09995
19	−0.02517
20	−0.02495
21	−0.07352
22	0.00627
23	−0.07432
24	−0.26119

(d) Forecasts (of y_{41}, y_{42}, y_{43}, and y_{44})

FORECASTS FOR VARIABLE Y

OBS	FORECAST	STD ERROR	LOWER 95%	UPPER 95%
41	612.1893	48.2072	517.7051	706.6736
42	514.5688	54.6396	407.4775	621.6602
43	419.0120	56.3365	308.5947	529.4293
44	552.7186	56.8103	441.3727	664.0645

227

FIGURE 5.5 SAS Output of Estimation, Diagnostic Checking, and Forecasting for the Gas Bill Dummy Variable Regression Model. Assuming the Error Terms $\varepsilon_1, \varepsilon_2, \ldots$ Are Described by the Model $(1 - \phi_1 B)(1 - \phi_{1,4}B^4 - \phi_{1,5}B^5)\varepsilon_t = a_t$

(a) Estimation

PARAMETER	ESTIMATE	STD ERROR	T RATIO	LAG	VARIABLE
MU	271.811	37.4773	7.25	0	Y
AR1,1	0.64091	0.146591	4.37	1	Y
AR2,1	-0.15533	0.17015	-0.91	4	Y
AR2,2	-0.448241	0.169139	-2.65	5	Y
NUM1	-7.46674	4.18937	-1.78	0	T
NUM2	0.308676	0.102115	3.02	0	TSQ
NUM3	69.3674	15.9591	4.35	0	D1
NUM4	-34.1481	14.5221	-2.35	0	D2
NUM5	-128.276	15.8072	-8.12	0	D3

CONSTANT ESTIMATE   = 156.516
VARIANCE ESTIMATE   = 2144.77
STD ERROR ESTIMATE = 46.3117
NUMBER OF RESIDUALS = 40

CORRELATIONS OF THE ESTIMATES

	MU	AR1,1	AR2,1	AR2,2	NUM1	NUM2	NUM3	NUM4	NUM5
MU	1.000	0.045	0.037	-0.004	-0.817	0.689	-0.284	-0.238	-0.224
AR1,1	0.045	1.000	-0.174	-0.124	-0.114	0.145	0.042	0.031	0.022
AR2,1	0.037	-0.174	1.000	-0.020	-0.045	0.062	-0.085	-0.019	-0.082
AR2,2	-0.004	-0.124	-0.020	1.000	-0.029	0.040	0.054	-0.007	0.062
NUM1	-0.817	-0.114	-0.045	-0.029	1.000	-0.967	0.039	0.014	-0.003
NUM2	0.689	0.145	0.062	0.040	-0.967	1.000	-0.013	0.006	0.012
NUM3	-0.284	0.042	-0.085	0.054	0.039	-0.013	1.000	0.468	0.583
NUM4	-0.238	0.031	-0.019	-0.007	0.014	0.006	0.468	1.000	0.462
NUM5	-0.224	0.022	-0.082	0.062	-0.003	0.012	0.583	0.462	1.000

FIGURE 5.5 Continued   (b) The RSAC

AUTOCORRELATION CHECK OF RESIDUALS

TO LAG	CHI SQUARE	DF	PROB						
6	0.83	2	0.660	0.028	-0.029	0.088	-0.064	-0.021	-0.063
12	12.74	8	0.121	-0.016	-0.240	-0.108	0.035	-0.111	-0.352
18	16.12	14	0.306	-0.084	-0.039	0.136	0.089	0.112	0.042
24	19.14	20	0.512	-0.070	0.152	0.001	0.063	0.057	-0.002

AUTOCORRELATIONS

AUTOCORRELATION PLOT OF RESIDUALS

LAG	COVARIANCE	CORRELATION	-1 9 8 7 6 5 4 3 2 1 0 1 2 3 4 5 6 7 8 9 1	STD
0	2144.77	1.00000	|****************|	0
1	59.0355	0.02753		0.158114
2	-61.4368	-0.02864		0.158234
3	188.895	0.08807		0.158363
4	-136.835	-0.06380		0.159583
5	-45.1743	-0.02106		0.160219
6	-135.594	-0.06322		0.160289
7	-33.3274	-0.01554		0.160911
8	-515.22	-0.24022		0.160948
9	-232.489	-0.10840		0.169675
10	74.1441	0.03457		0.171398
11	-238.297	-0.11111		0.171572
12	-755.966	-0.35247		0.173361
13	-180.193	-0.08402		0.190436
14	-84.354	-0.03933		0.19136
15	292.272	0.13627		0.191562
16	189.938	0.08856		0.193971
17	239.273	0.11156		0.194979
18	89.7833	0.04186		0.196568
19	-150.135	-0.07000		0.196791
20	326.07	0.15203		0.197413
21	2.82401	0.00132		0.200318
22	134.723	0.06281		0.200318
23	121.798	0.05679		0.20081
24	-5.00802	-0.00233		0.201211

"." MARKS TWO STANDARD ERRORS

(Parts (c), the RSPAC, and (d), the Forecast, follow on page 230.)

FIGURE 5.5 Continued   (c) The RSPAC

PARTIAL AUTOCORRELATIONS

LAG	CORRELATION
1	0.02753
2	-0.02942
3	0.08984
4	-0.07060
5	-0.01125
6	-0.07545
7	0.00044
8	-0.25120
9	-0.08802
10	0.00922
11	-0.08713
12	-0.41143
13	-0.16051
14	-0.15328
15	0.14368
16	-0.08297
17	0.03638
18	-0.10069
19	-0.13795
20	-0.14191
21	-0.14499
22	-0.00115
23	0.01854
24	-0.18397

(d) Forecasts (of y_{41}, y_{42}, y_{43}, and y_{44})

FORECASTS FOR VARIABLE Y

OBS	FORECAST	STD ERROR	LOWER 95%	UPPER 95%
	—FORECAST BEGINS	— —		
41	618.2118	46.3117	527.4428	708.9808
42	510.0080	55.0070	402.1965	617.8195
43	403.1972	58.2036	289.1206	517.2739
44	532.8463	59.4668	416.2937	649.3990

230

models of Chapters 2, 3, and 4 to forecast the Absorbent Paper Towel sales time series (see Example 2.1) and the Wisconsin "Trade" time series (see Exercise 3.1)—both of which exhibit stochastic trend and/or seasonal effects. Second, seasonal effects can be modeled in time series regression models by using methods other than dummy variables (although the authors personally feel that dummy variables are more effective than the other methods to be described). Two models involving *trigonometric terms* that are useful for modeling additive (or constant) seasonal variation are (here L is the number of seasons in a year):

1. $y_t = \beta_0 + \beta_1 t + \beta_2 \sin\left(\dfrac{2\pi t}{L}\right) + \beta_3 \cos\left(\dfrac{2\pi t}{L}\right) + \varepsilon_t$

and

2. $y_t = \beta_0 + \beta_1 t + \beta_2 \sin\left(\dfrac{2\pi t}{L}\right) + \beta_3 \cos\left(\dfrac{2\pi t}{L}\right) + \beta_4 \sin\left(\dfrac{4\pi t}{L}\right)$
$+ \beta_5 \cos\left(\dfrac{4\pi t}{L}\right) + \varepsilon_t$

These models assume a linear trend, but they can be altered to handle other trends. The first of these models is useful in modeling a very regular additive seasonal pattern, while the second model possesses terms that allow modeling of a more complicated additive seasonal pattern. Trigonometric models can also be used to measure *multiplicative seasonal variation*, which is a special type of changing seasonal variability. Specifically, if a time series displays *multiplicative seasonal variation*, the magnitude of the "seasonal swing" is proportional to the level of the trend. Thus, if the trend of the time series is increasing, so is the magnitude of the seasonal swing; if the trend of the time series is decreasing, the magnitude of the seasonal swing is also decreasing.

Two trigonometric regression models useful in modeling multiplicative seasonal variation are

3. $y_t = \beta_0 + \beta_1 t + \beta_2 \sin\left(\dfrac{2\pi t}{L}\right) + \beta_3 t \sin\left(\dfrac{2\pi t}{L}\right) + \beta_4 \cos\left(\dfrac{2\pi t}{L}\right)$
$+ \beta_5 t \cos\left(\dfrac{2\pi t}{L}\right) + \varepsilon_t$

and

4. $y_t = \beta_0 + \beta_1 t + \beta_2 \sin\left(\dfrac{2\pi t}{L}\right) + \beta_3 t \sin\left(\dfrac{2\pi t}{L}\right) + \beta_4 \cos\left(\dfrac{2\pi t}{L}\right)$
$+ \beta_5 t \cos\left(\dfrac{2\pi t}{L}\right) + \beta_6 \sin\left(\dfrac{4\pi t}{L}\right) + \beta_7 t \sin\left(\dfrac{4\pi t}{L}\right)$
$+ \beta_8 \cos\left(\dfrac{4\pi t}{L}\right) + \beta_9 t \cos\left(\dfrac{4\pi t}{L}\right) + \varepsilon_t$

Again, these models assume a linear trend but can be altered to handle other trends. The first of these models is useful for a very regular multiplicative seasonal pattern, while the second model possesses terms that allow modeling of complicated multiplicative seasonal pattern. Finally, it should be noted intuitive method—called the *multiplicative decomposition method*— to analyze time series exhibiting multiplicative seasonal variation. Th cussed in Section 5.4.

5.3   USING SAS

The following SAS program forecasts the hotel room occupancy data by using the model

$$y_t^* = \beta_0 + \beta_1 t + \beta_{s1} x_{s1,t} + \beta_{s2} x_{s2,t} + \cdots + \beta_{s11} x_{s11,t} + \varepsilon_t$$

where

$$y_t^* = \ln y_t$$

and

$$(1 - \phi_1 B - \phi_2 B^2 - \phi_3 B^3)(1 - \phi_{1,12} B^{12})\varepsilon_t = a_t$$

```
DATA DHOTEL;
INPUT  Y  T  D1  D2  D3  D4  D5  ⎫
            D6  D7  D8  D9  D10  D11; ⎬  Defines variables, including dummy variables D1, D2, . . . , D11.
LY=LOG(Y);  }   Transformation to obtain logged data.
CARDS;
501     1    1  0  0  0  0  0  0  0  0  0  0 ⎫
488     2    0  1  0  0  0  0  0  0  0  0  0 ⎪
  ⋮                                          ⎪
877    168   0  0  0  0  0  0  0  0  0  0  0 ⎬   Data, including input for dummy variables.
  ·    169   1  0  0  0  0  0  0  0  0  0  0 ⎪
  ·    170   0  1  0  0  0  0  0  0  0  0  0 ⎪
  ⋮                                          ⎪
  ·    180   0  0  0  0  0  0  0  0  0  0  0 ⎭
PROC ARIMA DATA=DHOTEL;   }   Declares ARIMA Procedure and Data File.

IDENTIFY VAR=LY NOPRINT CROSSCOR=(T D1 D2 D3 D4 D5 D6 D7 ⎫   Specifies
D8 D9 D10 D11);                                          ⎬   regression
ESTIMATE INPUT=(T D1 D2 D3 D4 D5 D6 D7 D8 D9 D10 D11)    ⎭   model to be fit.

   P=(1,2,3)(12) PRINTALL PLOT;  ⎫  P = (1, 2, 3)(12)
                                 ⎬    Specifies model to describe the residuals.
FORECAST LEAD=24 OUT=FCAST2;  }   Generates forecasts for y_t* (24 future time periods).

DATA FORE2;                  ⎫
  SET FCAST2;                ⎪
  Y = EXP(LY);               ⎬   Exponentiates point forecasts and prediction intervals.
  FY = EXP(FORECAST);        ⎪
  L95CI = EXP(L95);          ⎪
  U95CI = EXP(U95);          ⎭

PROC PRINT DATA = FORE2;  ⎫   Prints point forecasts and
  VAR Y L95CI FY U95CI;   ⎬      prediction intervals for y_t.
```

Note that, *before* obtaining the model describing ε_t

$$(1 - \phi_1 B - \phi_2 B^2 - \phi_3 B^3)(1 - \phi_{1,12} B^{12})\varepsilon_t = a_t$$

and thus *before* running the program, we would have to run a "first" SAS program assuming independent error terms. That program would be almost exactly the same as the given program, with P = (1, 2, 3)(12) removed from the ESTIMATE statement. This first SAS program would give the RSAC and RSPAC in Table 5.1, and it is from them that we would obtain the given model describing ε_t.

5.4   GROWTH CURVE MODELS

All of the models presented so far in this chapter describe deterministic trend and seasonal effects by using deterministic functions of time that are *linear in the parameters*. For example, the model

$$y_t = TR_t + SN_t + \varepsilon_t$$
$$= \beta_0 + \beta_1 t + \beta_2 t^2 + \beta_{s1} x_{s1,t} + \beta_{s2} x_{s2,t} + \beta_{s3} x_{s3,t} + \varepsilon_t$$

is linear in the parameters β_0, β_1, β_2, β_{s1}, β_{s2}, and β_{s3}. That is, each term in the model is the product of a model parameter and a numeric value determined by the observed data.

Sometimes, however, useful models are not linear in the parameters. For instance, the model

$$y_t = \beta_0 (\beta_1^t)\varepsilon_t$$

is not linear in the parameters since the independent variable t enters as an exponent and β_0 is multiplied by β_1^t. This model further departs from the usual linear model since the error term ε_t is multiplicative rather than additive [that is, multiplied by—rather than added to—$\beta_0 (\beta_1^t)$]. To apply the techniques of estimation and prediction we have presented in previous discussions, we must transform such a nonlinear model to one that is linear in the parameters. The model

$$y_t = \beta_0 (\beta_1^t)\varepsilon_t$$

can be transformed to a linear model by taking logarithms on both sides. Either base 10 logarithms (denoted log) or natural (base e) logarithms (denoted ln) can be used. Both base 10 and natural logarithms can easily be obtained on many modern pocket calculators or by using computer routines. Two important properties of logarithms are given by

$$\log (AB) = \log (A) + \log (B)$$

and

$$\log (A^r) = r \log (A)$$

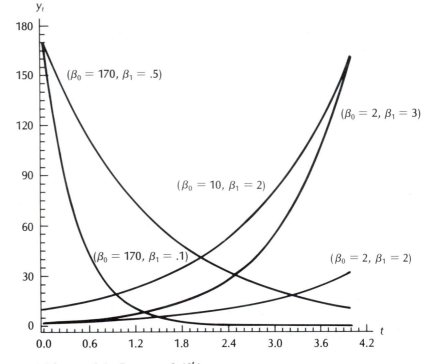

FIGURE 5.6   Exponential Curves of the Form $y_t = \beta_0(\beta_1^t)$

where A and B are positive numbers. These properties allow us to transform some nonlinear models to achieve models that are linear in the parameters.

If $\beta_0 > 0$ and $\beta_1 > 0$, applying a logarithmic transformation to the model $y_t = \beta_0(\beta_1^t)\varepsilon_t$ yields

$$\log(y_t) = \log(\beta_0) + \log(\beta_1)t + \log(\varepsilon_t)$$

If we let $\alpha_0 = \log(\beta_0)$, $\alpha_1 = \log(\beta_1)$, and $u_t = \log(\varepsilon_t)$, the transformed version of the model becomes

$$\log(y_t) = \alpha_0 + \alpha_1 t + u_t$$

Thus the model with the dependent variable $\log(y_t)$ is linear in the parameters α_0 and α_1.

Cases in which a model

$$y_t = \beta_0(\beta_1^t)\varepsilon_t$$

may be appropriate can be identified by data plots of y_t versus t. Plots of the expression $\beta_0(\beta_1^t)$ for several combinations of β_0 and β_1 are shown in Figure 5.6. Plots of observed data would have points "scattered about" such a function. The multiplicative error term would cause more variation around the high parts of the curve and less variation around the low parts, since the variation in y_t will be dependent upon the level of $\beta_0(\beta_1^t)$—that is, given the same error term ε_t, the larger $\beta_0(\beta_1^t)$ is, the larger will be the variation in y_t.

Figure 5.6 shows that the curves described by

$$y_t = \beta_0(\beta_1^t)$$

may be increasing ($\beta_1 > 1$) or decreasing ($0 < \beta_1 < 1$) functions of t. We can see that β_0 is the intercept and that β_1 determines the amount of curvature in the plot. The curvature gets more pronounced as β_1 moves away from 1 in either direction. We now present an example which illustrates why we call the model

$$y_t = \beta_0(\beta_1^t)\varepsilon_t$$

a *growth curve model*.

EXAMPLE 5.3

Western Steakhouses, a fast food chain, opened in 1973. Each year from 1973 to 1987 the number of steakhouses in operation, y_t, is recorded. For convenience, we will let $t = 0$ for the year 1973, $t = 1$ for the year 1974, etc. An analyst for the firm wishes to use this data (presented in Table 5.5) to predict the number of steakhouses that will be in operation in 1988.

The steakhouse data (y_t) is plotted against time (t) in Figure 5.7. This plot shows an exponential increase reminiscent of the plots in Figure 5.6 where β_1 is greater than 1. This suggests that the model

$$y_t = \beta_0(\beta_1^t)\varepsilon_t$$

may be appropriate. The natural logarithms of the steakhouse data ($\ln y_t$) are given in Table 5.5 and plotted in Figure 5.8. This plot suggests that the relationship between $\ln y_t$ and t is linear. Applying the logarithmic transformation to the model

$$y_t = \beta_0(\beta_1^t)\varepsilon_t$$

we obtain

$$\ln y_t = \ln(\beta_0) + t \ln(\beta_1) + \ln \varepsilon_t$$

TABLE 5.5 *Number of Western Steakhouses (y_t) in Operation for the Years 1973–1987*

Year	t	y_t	$\ln(y_t)$	Year	t	y_t	$\ln(y_t)$
1973	0	11	2.398	1981	8	82	4.407
1974	1	14	2.639	1982	9	99	4.595
1975	2	16	2.773	1983	10	119	4.779
1976	3	22	3.091	1984	11	156	5.050
1977	4	28	3.332	1985	12	257	5.549
1978	5	36	3.584	1986	13	284	5.649
1979	6	46	3.829	1987	14	403	5.999
1980	7	67	4.205				

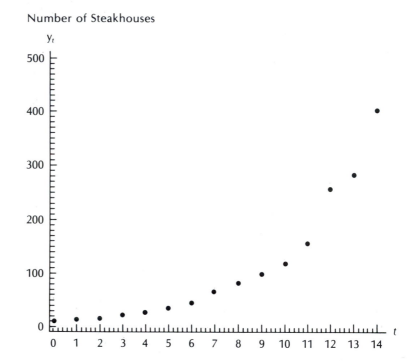

Number of Steakhouses

FIGURE 5.7   The Number of Western Steakhouses in Operation for the Years 1973–1987 Plotted versus Time

Defining $\alpha_0 = \ln(\beta_0)$, $\alpha_1 = \ln(\beta_1)$,   and $u_t = \ln\varepsilon_t$ we have

$$\ln y_t = \alpha_0 + \alpha_1 t + u_t$$

Using SAS (see Section 5.5), it can be shown that

1.   The least squares point estimates of α_0 and α_1 are

$$\hat{\alpha}_0 = 2.3270 \quad \text{and} \quad \hat{\alpha}_1 = .2569$$

Since $\alpha_1 = \ln(\beta_1)$, it follows that

$$\beta_1 = e^{\alpha_1}$$

and thus a point estimate of β_1 is

$$\hat{\beta}_1 = e^{\hat{\alpha}_1} = e^{.2569} = 1.293$$

Since the model

$$y_t = \beta_0(\beta_1^t)\varepsilon_t = [\beta_0(\beta_1^{t-1})]\beta_1\varepsilon_t \approx (y_{t-1})\beta_1\varepsilon_t$$

implies that we expect y_t to be approximately β_1 times y_{t-1}, and since

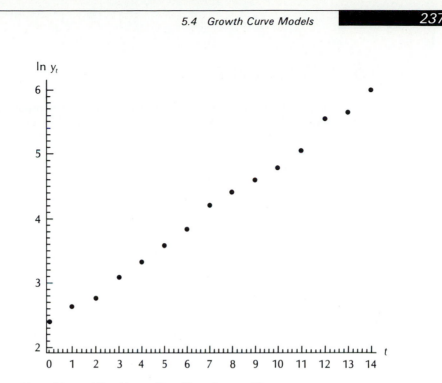

FIGURE 5.8   Natural Logarithms of Steakhouse Data Plotted versus Time

the point estimate of β_1 is 1.293, we estimate y_t to be approximately 1.293 times y_{t-1}, and thus we estimate y_t to be

$$100(\hat{\beta}_1 - 1)\% = 100(1.293 - 1)\% = 29.3\%$$

greater than y_{t-1}. Here, $100(\hat{\beta}_1 - 1)\% = 29.3\%$ is the point estimate of the *growth rate* $100(\beta_1 - 1)\%$ in the growth curve model

$$y_t = \beta_0(\beta_1^t)\varepsilon_t$$

2. The point prediction of $\ln y_{15}$, where y_{15} is the number of steakhouses that will be in operation in period 15 (1988), is

$$\widehat{\ln y}_{15} = \hat{\alpha}_0 + \hat{\alpha}_1 t = 2.3270 + .2569(15) = 6.1850$$

and thus a point prediction of y_{15} is

$$\hat{y}_{15} = e^{6.1850} = 485.41 \quad \text{(steakhouses)}$$

3. The 95% prediction interval (obtained by using SAS) for $\ln y_{15}$ is

$$[6.1850 \pm .1861] = [5.9989, 6.3711]$$

and thus a 95% prediction interval for y_{15} is

$$[e^{5.9989}, e^{6.3711}] = [402.99 \text{ steakhouses}, 584.70 \text{ steakhouses}]$$

5.5 USING SAS

The following SAS program calculates $\hat{\alpha}_0$, $\hat{\alpha}_1$, a point prediction of $\ln y_{15}$, and a 95% prediction interval for $\ln y_{15}$, where we assume that the model

$$\ln y_t = \alpha_0 + \alpha_1 t + u_t$$

describes the Western Steakhouse data.

DATA STEAK;   }   Assigns the name STEAK to the data file.

INPUT Y T;   }   Defines variable names Y and T.

LY=LOG(Y);   }   Transformation to obtain logged data.

CARDS;

```
 11    0
 14    1
  .    .
  .    .
  .    .
403   14
```
Historical Western Steakhouse Data.
One observation per data line.

```
  .   15
```
Generates prediction for period 15.
· denotes missing Y value.

PROC PRINT;   }   Prints the data file.

PROC REG;   }   Declares regression procedure.

MODEL LY=T / P CLI;

Specifies model: $\ln y_t = \alpha_0 + \alpha_1 t + u_t$
Intercept α_0 is assumed.
P = residuals desired.
CLI = prediction interval desired.

The above program employs PROC REG. Below we show how to use PROC ARIMA to carry out (essentially) the same analysis. The advantage of PROC REG is that it gives a more precise 95% prediction interval than does PROC ARIMA. However, PROC REG assumes that the error terms u_1, u_2, ... are statistically independent but does not calculate the RSAC and RSPAC to verify this assumption. PROC ARIMA, does calculate the RSAC and RSPAC, which indicates that the error terms u_1, u_2, ... are statistically independent.

```
DATA STEAK;
INPUT Y T;
LY=LOG(Y);
CARDS;
  11    0
  14    1
   .    .
   .    .
   .    .
 403   14
PROC PRINT;
PROC ARIMA;        }    Declares ARIMA procedure
IDENTIFY VAR=LY NOPRINT CROSSCOR = (T); ⎱   Specifies regression
ESTIMATE INPUT=(T) PRINTALL PLOT;       ⎰   model : lny_t = α_0 + α_1 t + u_t
FORECAST LEAD=1;   }    Generates point forecast and prediction interval for period 15.
```

The lines with the regression model read:

IDENTIFY VAR=LY NOPRINT CROSSCOR = (T); $\Bigr\}$ Specifies regression

ESTIMATE INPUT=(T) PRINTALL PLOT; model: $\ln y_t = \alpha_0 + \alpha_1 t + u_t$

FORECAST LEAD=1; } Generates point forecast and prediction interval for period 15.

5.6   AN INTUITIVE APPROACH : TIME SERIES DECOMPOSITION

If a time series exhibits trend effects and seasonal effects, it can be useful to "decompose" it in order to isolate these effects. One model that allows us to do this is the *multiplicative decomposition model*. Although this model has no theoretical basis (it is strictly an intuitive approach), it has sometimes been found to be useful when modeling time series that display *multiplicative* (increasing or decreasing) *seasonal variation*. We express the model as follows.

> The *multiplicative decomposition model* is
> $$y_t = TR_t \times SN_t \times CL_t \times IR_t$$

where

y_t = the observed value of the time series in time period t

TR_t = the trend component (or factor) in time period t

SN_t = the seasonal component (or factor) in time period t

CL_t = the cyclical component (or factor) in time period t

IR_t = the irregular component (or factor) in time period t

Recalling that we have previously discussed the nature of trend effects, seasonal variations, and irregular fluctuations, we note that CL_t (the cyclical component) refers to recurring up and down movements around trend levels as caused, for example, by the business cycle. These fluctuations can last anywhere from two to longer than ten years as measured from peak to peak or trough to trough. In business, a peak would mark the end of an expansion in business activity, and a trough would mark the end of a contraction. The following example illustrates obtaining point estimates—denoted tr_t, sn_t, cl_t, and ir_t—of TR_t, SN_t, CL_t, and IR_t.

EXAMPLE 5.4

The Discount Soda Shop, Inc., owns and operates ten drive-in soft drink stores in York City, which has a population of roughly one million. Discount Soda has been selling Tasty Cola, a soft drink which was introduced on the market just three years ago and has been gaining in popularity. Periodically, Discount Soda orders a supply of Tasty Cola from the regional distributor. Discount Soda uses an inventory policy that attempts to insure that its stores will have enough Tasty Cola to meet practically all of the demand for Tasty Cola, while at the same time insuring that the company does not needlessly tie up its money by ordering much more Tasty Cola than can be reasonably expected to be sold. Although "sales" and "demand" are not the same, since Discount Soda has a policy of having enough Tasty Cola in stock to insure that it very seldom loses a potential sale, we will in this example consider sales and demand to be the same. In order to implement its inventory policy, Discount Soda needs to forecast Tasty Cola sales. "Tasty Cola

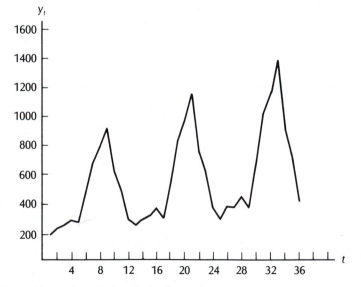

FIGURE 5.9   Monthly Sales of Tasty Cola (in hundreds of cases)

TABLE 5.6   *Monthly Sales of Tasty Cola (in hundreds of cases)*

Year	Month	t	Sales y_t	Year	Month	t	Sales y_t
1	1 (Jan.)	1	189	2	7	19	831
	2 (Feb.)	2	229		8	20	960
	3 (Mar.)	3	249		9	21	1152
	4 (Apr.)	4	289		10	22	759
	5 (May)	5	260		11	23	607
	6 (June)	6	431		12	24	371
	7 (July)	7	660	3	1	25	298
	8 (Aug.)	8	777		2	26	378
	9 (Sept.)	9	915		3	27	373
	10 (Oct.)	10	613		4	28	443
	11 (Nov.)	11	485		5	29	374
	12 (Dec.)	12	277		6	30	660
2	1	13	244		7	31	1004
	2	14	296		8	32	1153
	3	15	319		9	33	1388
	4	16	370		10	34	904
	5	17	313		11	35	715
	6	18	556		12	36	441

sales" for a particular period is defined to be the total sales in hundreds of cases of Tasty Cola to patrons shopping in Discount Soda's Ten York City stores during that period of time. At the end of each month, Discount Soda desires point forecasts and prediction interval forecasts of Tasty Cola sales in future months.

Discount Soda has recorded monthly Tasty Cola sales for the previous three years, which we will call year 1, year 2, and year 3. This time series is given in Table 5.6 and plotted in Figure 5.9. Notice that, in addition to having a linear trend, the Tasty Cola sales time series possesses seasonal variation, with sales of the soft drink being greatest in the summer and early fall months and lowest in the winter months. We will show later in this example that it is reasonable to conclude that y_t, the sales of Tasty Cola in period t, is adequately described by the model

$$y_t = TR_t \times SN_t \times CL_t \times IR_t$$

Therefore, we summarize in Table 5.7 the calculations needed to find estimates —denoted tr_t, sn_t, cl_t, and ir_t—of the components TR_t, SN_t, CL_t, and IR_t of this model.

To begin our consideration of the calculations in Table 5.7, we will explain the calculation of *moving averages* and *centered moving averages* (the centered moving averages are denoted CMA_t in Table 5.7). The purpose behind computing these averages is to eliminate seasonal variations and irregular fluctuations from

TABLE 5.7 Analysis of the Historical Tasty Cola Sales Time Series Using the Multiplicative Decomposition Method

(a) The Values of sn_t, d_t, and tr_t

t	y_t	12 Period Moving Average	$CMA_t = tr_t \times cl_t$	$sn_t \times ir_t = \dfrac{y_t}{tr_t \times cl_t}$	sn_t	$d_t = \dfrac{y_t}{sn_t}$	$tr_t = 380.163 + 9.489t$	$\hat{y}_t = tr_t \times sn_t$	$(y_t - \hat{y}_t)^2$
1	189				.493	383.37	389.652	192.10	$(-3.1)^2$
2	229				.596	384.23	399.141	237.89	$(-8.89)^2$
3	249				.595	418.49	408.630	243.13	$(5.87)^2$
4	289				.680	425	418.119	284.32	$(4.68)^2$
5	260				.564	460.99	427.608	241.17	$(18.83)^2$
6	431				.986	437.12	437.097	430.98	$(.02)^2$
7	660	447.833	450.1	1.466	1.467	449.9	446.586	655.14	$(4.86)^2$
8	777	452.417	455.2	1.707	1.693	458.95	456.075	772.13	$(4.87)^2$
9	915	458	460.9	1.985	1.990	459.79	465.564	926.47	$(-11.47)^2$
10	613	463.833	467.2	1.312	1.307	469.01	475.053	620.89	$(-7.89)^2$
11	485	470.583	472.8	1.026	1.029	471.33	489.542	498.59	$(-13.59)^2$
12	277	475	480.2	.577	.600	461.67	494.031	296.42	$(-19.42)^2$
13	244	485.417	492.5	.495	.493	494.97	503.520	248.24	$(-4.24)^2$
14	296	499.667	507.3	.583	.596	496.64	513.009	305.75	$(-9.75)^2$
15	319	514.917	524.8	.608	.595	536.13	522.498	310.89	$(8.11)^2$
16	370	534.667	540.7	.684	.680	544.12	531.987	361.75	$(8.25)^2$
17	313	546.833	551.9	.567	.564	554.97	541.476	305.39	$(7.61)^2$
18	556	557	560.9	.991	.986	563.89	550.965	543.25	$(12.65)^2$
19	831	564.833	567.1	1.465	1.467	566.46	560.454	822.19	$(8.81)^2$
20	960	569.333	572.7	1.676	1.693	567.04	569.943	964.91	$(-4.91)^2$
21	1152	576.167	578.4	1.992	1.990	578.89	579.432	1153.07	$(-1.07)^2$
22	759	580.667	583.7	1.300	1.307	580.72	588.921	769.72	$(-10.72)^2$
23	607	586.75	589.3	1.030	1.029	589.89	598.410	615.76	$(-8.76)^2$
24	371	591.833	596.2	.622	.600	618.33	607.899	364.74	$(6.26)^2$
25	298	600.5	607.7	.490	.493	604.46	617.388	304.37	$(-6.37)^2$
26	378	614.917	623.0	.607	.596	634.23	626.877	373.62	$(4.38)^2$
27	373	631	640.8	.582	.595	626.89	636.366	378.64	$(-5.64)^2$
28	443	650.667	656.7	.675	.680	651.47	645.855	439.18	$(3.82)^2$
29	374	662.75	667.3	.561	.564	663.12	655.344	369.61	$(4.39)^2$
30	660	671.75	674.7	.978	.986	669.37	664.833	655.53	$(4.47)^2$
31	1004	677.583			1.467	684.39	674.322	989.23	$(14.77)^2$
32	1153				1.693	681.04	683.811	1157.69	$(-4.69)^2$
33	1388				1.990	697.49	693.300	1379.67	$(8.33)^2$
34	904				1.307	691.66	702.789	918.55	$(-14.55)^2$
35	715				1.029	694.85	712.278	732.93	$(-17.93)^2$
36	441				.600	735	721.707	433.06	$(7.94)^2$

(b) The Values of cl_t and ir_t

t	y_t	$tr_t \times sn_t$	$cl_t \times ir_t = \dfrac{y_t}{tr_t \times sn_t}$	$cl_t = \dfrac{cl_{t-1}ir_{t-1} + cl_t ir_t + cl_{t+1}ir_{t+1}}{3}$	$ir_t = \dfrac{cl_t \times ir_t}{cl_t}$
1	189	192.10	.9839		
2	229	237.89	.9626	.9902	.9721
3	249	243.13	1.0241	1.0010	1.0231
4	289	284.32	1.0165	1.0396	.9778
5	260	241.17	1.0781	1.0315	1.0452
6	431	430.98	1.0000	1.0285	.9723
7	660	655.14	1.0074	1.0046	1.0028
8	777	772.13	1.0063	1.0004	1.0059
9	915	926.47	.9876	.9937	.9939
10	613	620.89	.9873	.9825	1.0063
11	485	498.59	.9727	.9648	1.0082
12	277	296.42	.9345	.9634	.9700
13	244	248.24	.9829	.9618	1.0219
14	296	305.75	.9681	.9924	.9755
15	319	310.89	1.0261	.9567	.9710
16	370	361.75	1.0228	1.0246	.9982
17	313	305.39	1.0249	1.0237	1.0012
18	556	543.25	1.0235	1.0197	1.0037
19	831	822.19	1.0107	1.0097	1.0010
20	960	964.91	.9949	1.0016	.9933
21	1152	1153.07	.9991	.9934	1.0057
22	759	769.72	.9861	.9903	.9958
23	607	615.76	.9858	.9964	.9894
24	371	364.74	1.0172	.9940	1.0233
25	298	304.37	.9791	1.0027	.9765
26	378	373.62	1.0117	.9920	1.0199
27	373	378.64	.9851	1.0018	.9833
28	443	439.18	1.0087	1.0030	1.0057
29	374	369.61	1.0119	1.0091	1.0028
30	660	655.53	1.0068	1.0112	.9956
31	1004	989.23	1.0149	1.0059	1.0089
32	1153	1157.69	.9959	1.0053	.9906
33	1388	1379.67	1.0060	.9954	1.0106
34	904	918.55	.9842	.9886	.9955
35	715	732.93	.9755	.9927	.9827
36	441	433.06	1.0183		

the data. The first moving average is the average of the first 12 Tasty Cola sales values

$$\frac{189 + 229 + 249 + 289 + 260 + 431 + 660 + 777 + 915 + 613 + 485 + 277}{12}$$

$$= 447.833$$

Here we use a "12-period moving average" because the Tasty Cola time series data is monthly ($L = 12$ time periods or "seasons" per year). If the data were quarterly, we would, of course, compute only a "4-period moving average." The second moving average is obtained by dropping the first sales value (y_1) from the average and by including the next sales value (y_{13}) in the average. Thus, we obtain

$$\frac{229 + 249 + 289 + 260 + 431 + 660 + 777 + 915 + 613 + 485 + 277 + 244}{12}$$

$$= 452.417$$

The third moving average is obtained by dropping y_2 from the average and by including y_{14} in the average. We obtain

$$\frac{249 + 289 + 260 + 431 + 660 + 777 + 915 + 613 + 485 + 277 + 244 + 296}{12}$$

$$= 458$$

Successive moving averages are computed similarly until we include y_{36} in the last moving average. Note that we use the term "moving average" here because, as we calculate these averages, we move along by dropping the most remote observation in the previous average and by including the "next" observation in the new average.

The first moving average corresponds to a time that is midway between periods 6 and 7, the second moving average corresponds to a time that is midway between periods 7 and 8, and so forth. In order to obtain averages corresponding to time periods in the original Tasty Cola time series, we calculate *centered moving averages*. The centered moving averages are two-period moving averages of the previously computed 12-period moving averages. Thus the first centered moving average is

$$\frac{447.833 + 452.417}{2} = 450.1$$

The second centered moving average is

$$\frac{452.417 + 458}{2} = 455.2$$

Successive centered moving averages are calculated in a similar fashion. The 12-period moving averages and centered moving averages for the Tasty Cola sales time series are given in Table 5.7a.

If the original moving averages had been based on an odd number of time series values, the centering procedure would not have been necessary. For example, if we had three seasons per year, we would compute 3-period moving averages. Then, the first moving average would correspond to period 2, the second moving average would correspond to period 3, and so on. However, most seasonal time series are quarterly, monthly, or weekly, and the centering procedure is necessary.

The centered moving average in time period t, CMA_t, is considered to be equal to $tr_t \times cl_t$, the estimate of $TR_t \times CL_t$, because the averaging procedure is assumed to have removed seasonal variations (note that each moving average is computed using exactly one observation from each season) and (short term) irregular fluctuations. The (longer term) trend effects and cyclical effects, that is, $tr_t \times cl_t$, remain.

Since the model

$$y_t = TR_t \times SN_t \times CL_t \times IR_t$$

implies that

$$SN_t \times IR_t = \frac{y_t}{TR_t \times CL_t}$$

and since we have computed the centered moving average CMA_t, the estimate $tr_t \times cl_t$ of $TR_t \times CL_t$, it follows that the estimate $sn_t \times ir_t$ of $SN_t \times IR_t$ is

$$sn_t \times ir_t = \frac{y_t}{tr_t \times cl_t} = \frac{y_t}{CMA_t}$$

Noting that the values of $sn_t \times ir_t$ are calculated in Table 5.7a, we can find $\overline{sn}_t$ by grouping the values of $sn_t \times ir_t$ by months and calculating an average, $\overline{sn}_t$, for each month. These seasonal factors are then normalized so that they add to $L = 12$, the number of periods in a year. This normalization is accomplished by multiplying each value of $\overline{sn}_t$ by the quantity

$$\frac{L}{\sum_{t=1}^{L} \overline{sn}_t} = \frac{12}{11.9895} = 1.0008758$$

This normalization process results in the estimate $sn_t = 1.0008758(\overline{sn}_t)$, which is the estimate of SN_t. These calculations are summarized in Table 5.8.

Having calculated the values of sn_t and placed them in Table 5.7a, we next define the deseasonalized observation in time period t to be

$$d_t = \frac{y_t}{sn_t}$$

TABLE 5.8   Estimates of the Seasonal Factors of the Tasty Cola Sales Time Series

$$sn_t \times ir_t = y_t/(tr_t \times cl_t)$$

		Year 1	Year 2	$\overline{sn_t}$	$sn_t = 1.0008758(\overline{sn_t})$
1	Jan.	.495	.490	.4925	.493
2	Feb.	.583	.607	.595	.596
3	Mar.	.608	.582	.595	.595
4	Apr.	.684	.675	.6795	.680
5	May	.567	.561	.564	.564
6	June	.991	.978	.9845	.986
7	July	1.466	1.465	1.4655	1.467
8	Aug.	1.707	1.676	1.6915	1.693
9	Sept.	1.985	1.992	1.9885	1.990
10	Oct.	1.312	1.300	1.306	1.307
11	Nov.	1.026	1.030	1.028	1.029
12	Dec.	.577	.622	.5995	.600

Deseasonalized observations are computed in order to better estimate the trend component TR_t. Dividing y_t by the estimated seasonal factor removes the seasonality from the data and allows us to better understand the nature of the trend factor. The deseasonalized observations are calculated in Table 5.7a and are plotted in Figure 5.10. Here the deseasonalized observations indicate that the trend in Tasty Cola sales is linear. That is, since the deseasonalized observations plotted in Figure 5.10 have a straight line appearance, it seems reasonable to assume that

$$TR_t = \beta_0 + \beta_1 t$$

So we will estimate TR_t by fitting a straight line to the deseasonalized data using d_t as the dependent variable and t as the independent variable. We obtain tr_t, the estimate of TR_t, by computing (see Section 8.2 for the formulas that allow us to compute estimates b_0 and b_1 of β_0 and β_1)

$$b_1 = \frac{36 \sum_{t=1}^{36} td_t - \left(\sum_{t=1}^{36} t\right)\left(\sum_{t=1}^{36} d_t\right)}{36 \sum_{t=1}^{36} t^2 - \left(\sum_{t=1}^{36} t\right)^2} = 9.489$$

and

$$b_0 = \frac{\sum_{t=1}^{36} d_t}{36} - b_1 \left(\frac{\sum_{t=1}^{36} t}{36}\right) = 380.163$$

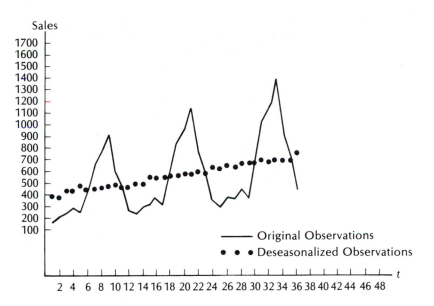

FIGURE 5.10   Deseasonalized Tasty Cola Sales

which gives us

$$tr_t = b_0 + b_1 t = 380.163 + 9.489t$$

The values of tr_t are calculated in Table 5.7a. Note that, for example, although $y_{22} = 759$, Tasty Cola sales in period 22 (October of year 2) are larger than $tr_{22} = 588.921$ (the estimated trend in period 22), $d_{22} = 580.72$ is smaller than $tr_{22} = 588.921$. This implies that, on a deseasonalized basis, Tasty Cola sales were slightly down in October of year 2. This might have been caused by a slightly colder October than usual.

Thus far, we have found estimates sn_t and tr_t of SN_t and TR_t. Since the model

$$y_t = TR_t \times SN_t \times CL_t \times IR_t$$

implies that

$$CL_t \times IR_t = \frac{y_t}{TR_t \times SN_t}$$

it follows that the estimate of $CL_t \times IR_t$ is

$$cl_t \times ir_t = \frac{y_t}{tr_t \times sn_t}$$

Moreover, experience has shown that, when considering either monthly or quarterly data, we can average out the irregular influence and thus calculate the estimate cl_t of CL_t by using

$$cl_t = \frac{cl_{t-1}ir_{t-1} + cl_t ir_t + cl_{t+1}ir_{t+1}}{3}$$

That is, cl_t is a three-period moving average of the $cl_t \times ir_t$ values.

Finally, we calculate the estimate ir_t of IR_t by using the equation

$$ir_t = \frac{cl_t \times ir_t}{cl_t}$$

The calculations of the values of cl_t and ir_t for the Tasty Cola data are summarized in Table 5.7b. Since there are only three years of data, and since most of the values of cl_t are near 1, we cannot discern a well-defined cycle. Furthermore, examining the values of ir_t, we cannot detect a pattern in the estimates of the irregular factors.

Traditionally, the estimates tr_t, sn_t, cl_t, and ir_t obtained by using the multiplicative decomposition method are used to describe the time series. However, we can also use these estimates to forecast future values of the time series. If there is no pattern in the irregular component in the model

$$y_t = TR_t \times SN_t \times CL_t \times IR_t$$

we predict this component to equal one. Therefore, the point forecast of y_t is

$$\hat{y}_t = tr_t \times sn_t \times cl_t$$

if a well-defined cycle exists and can be predicted, or is

$$\hat{y}_t = tr_t \times sn_t$$

if a well-defined cycle does not exist or cannot be predicted. For our Tasty Cola example, where

$$tr_t = b_0 + b_1 t = 380.163 + 9.489t$$

the point forecasts of the $n = 36$ historical Tasty Cola sales are given in Table 5.7a, and the point forecasts of future Tasty Cola sales in the twelve months of year 4 ("next year") are as given in Table 5.9. For example, the point forecast of Tasty Cola sales in August of next year (period 44) is

$$\hat{y}_{44} = tr_{44} \times sn_{44}$$

$$= (380.163 + 9.489(44))(1.693)$$

$$= 797.699 (1.693)$$

$$= 1350.50$$

Although there is no theoretically correct prediction interval for y_t, the authors have found that a fairly accurate (approximate) $100(1 - \alpha)\%$ prediction interval for y_t is

$$\{\hat{y}_t - B_t[100(1 - \alpha)], \hat{y}_t + B_t[100(1 - \alpha)]\}$$

TABLE 5.9 *Forecasts of Future Values of Tasty Cola Sales Calculated Using the Multiplicative Decomposition Method*

t	sn_t	$tr_t = 380.163 + 9.489t$	$\hat{y}_t = tr_t \times sn_t$	$B_t(95)$	$[\hat{y}_t - B_t(95), \hat{y}_t + B_t(95)]$	y_t	$(y_t - \hat{y}_t)^2$
37	.493	731.273	360.52	26.80	[333.72, 387.32]	352	$(-8.52)^2$
38	.596	740.762	441.48	26.92	[414.56, 468.40]	445	$(3.51)^2$
39	.595	750.252	446.40	27.04	[419.36, 473.44]	453	$(6.6)^2$
40	.680	759.741	516.62	27.17	[489.45, 543.79]	541	$(24.38)^2$
41	.564	769.231	433.85	27.30	[406.55, 461.15]	457	$(23.15)^2$
42	.986	778.720	767.82	27.44	[740.38, 795.26]	762	$(-5.82)^2$
43	1.467	788.209	1156.30	27.59	[1128.71, 1183.89]	1194	$(37.7)^2$
44	1.693	797.699	1350.50	27.74	[1322.76, 1378.24]	1361	$(10.5)^2$
45	1.990	807.188	1606.30	27.89	[1578.41, 1634.19]	1615	$(8.7)^2$
46	1.307	816.678	1067.40	28.05	[1039.35, 1095.45]	1059	$(-8.4)^2$
47	1.029	826.167	850.12	28.22	[821.90, 878.34]	824	$(-26.12)^2$
48	.600	835.657	501.39	28.39	[473, 529.78]	495	$(-6.39)^2$

where $B_t[100(1 - \alpha)]$ is the error bound in a $100(1 - \alpha)\%$ prediction interval

$$\{tr_t - B_t[100(1 - \alpha)], tr_t + B_t[100(1 - \alpha)]\}$$

for the deseasonalized observation

$$d_t = TR_t + \varepsilon_t$$
$$= \beta_0 + \beta_1 t + \varepsilon_t$$

For example, using SAS to predict d_t on the basis of t by using the above trend line model, we find that a 95% prediction interval for d_{44} is

$$[769.959, 825.439] = [tr_{44} - B_{44}[95], tr_{44} + B_{44}[95]]$$

which implies that

$$B_{44}[95] = \frac{825.439 - 769.959}{2}$$
$$= 27.74$$

which implies that an approximate 95% prediction interval for y_{44} is

$$[\hat{y}_{44} - B_{44}[95], \hat{y}_{44} + B_{44}[95]]$$
$$= [1350.50 - 27.74, 1350.50 + 27.74]$$
$$= [1322.76, 1378.24]$$

In Table 5.9 we present 95% prediction intervals (calculated by the above method) for Tasty Cola sales in the twelve months of year 4.

Suppose that we actually observe the Tasty Cola sales in year 4 and that these sales are as given in Table 5.9. Note that eleven of twelve prediction intervals calculated on the basis of the sales observed in years 1, 2, and 3 contain the actual

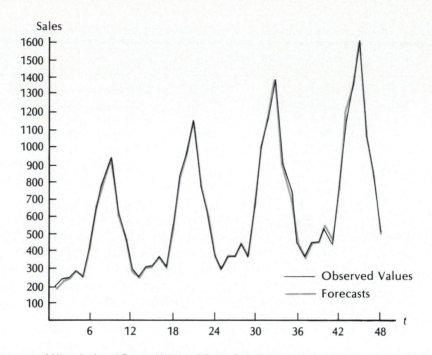

FIGURE 5.11   *Forecasts of Historical and Future Values of Tasty Cola Sales Calculated Using the Multiplicative Decomposition Method*

future sales in year 4. In Figure 5.11 we plot the observed and forecasted sales for all 48 sales periods. In practice, the comparison of the observed and forecasted sales in years 1 through 3 would be used by the analyst to determine whether the forecasting equation adequately fits the historical data. An adequate fit (as indicated by Figure 5.11, for example) might prompt an analyst to use this equation to calculate forecasts for future time periods. One reason that the Tasty Cola forecasting equation

$$\hat{y}_t = tr_t \times sn_t$$

$$= (380.163 + 9.489t)sn_t$$

provides reasonable forecasts is that, since this equation *multiplies sn_t by tr_t*, it assumes that as the average level of the time series (determined by the trend) increases, the seasonal swing of the time series increases, which is consistent with the data plots in Figures 5.9 and 5.11. For example, note from Table 5.8 that the estimated seasonal factor for August is 1.693. The forecasting equation yields a prediction of Tasty Cola sales in August of year 1

$$\hat{y}_8 = [380.163 + 9.489(8)]\ 1.693$$

$$= (456.075)(1.693)$$

$$= 772.13$$

which implies a seasonal swing of $772.13 - 456.075 = 316.055$ (hundreds of cases) above 456.075, the estimated trend, while this forecasting equation yields a prediction of Tasty Cola sales in August of year 2

$$\hat{y}_{20} = [380.163 + 9.489(20)]\, 1.693$$

$$= (569.943)\,(1.693)$$

$$= 964.91$$

which implies an *increased* seasonal swing of $964.91 - 569.943 = 394.967$ (hundreds of cases) above 569.943, the estimated trend. In general, then, the forecasting equation is appropriate for forecasting a time series for which the seasonal swing of the time series is proportional to the average level of the time series as determined by the trend—that is, a time series exhibiting multiplicative seasonal variation.

5.7   USING SAS

Below we present the SAS program that computes the point estimates b_0 and b_1 of the parameters β_0 and β_1 in the trend line model

$$d_t = \beta_0 + \beta_1 t + \varepsilon_t$$

describing the deseasonalized Tasty Cola sales in Table 5.7a. The program also computes point predictions and 95% prediction intervals for $d_{37}, d_{38}, \ldots, d_{48}$, the deseasonalized Tasty Cola sales in months 37 through 48.

```
DATA DESEA;    }   Assigns the name DESEA to the data file.
INPUT D T;     }   Defines variable names D and T.
CARDS;

383.37    1  ⎫
384.23    2  ⎬    Historical Tasty Cola sales data.
418.49    3  ⎭        One observation per data line.
   ⋮
735      36


  ·      37  ⎫
  ·      38  ⎬    Generates predictions for periods 37 through 48.
   ⋮     ⎭        · denotes missing D values.
  ·      48

PROC REG;   }   Declares regression procedure.

                     ⎫  Specifies model: dₜ = β₀ + β₁t + εₜ
MODEL D = T / CLI;   ⎬  Intercept β₀ is assumed.
                     ⎭  CLI = Prediction intervals desired.
```

The above program will fit only a straight line to the deseasonalized data. However, the SAS package does include a procedure that will carry out the entire analysis involved in a version of the multiplicative decomposition method. The procedure employed by SAS differs somewhat from the procedure presented in Section 5.6. We refer the reader to Appendix B of Bowerman and O'Connell [1984] for a listing of a program that carries out the decomposition method described in this book. This program, which is called DECOMP, was written by Professor Michael Broida of Miami University, Oxford, Ohio.

EXERCISES

5.1   Consider the time series regression model of Example 5.2

$$y_t = \beta_0 + \beta_1 t + \beta_2 t^2 + \beta_{s1} x_{s1,t} + \beta_{s2} x_{s2,t} + \beta_{s3} x_{s3,t} + \varepsilon_t$$

where

$$(1 - \phi_1 B)(1 - \phi_{1,4} B^4 - \phi_{1,5} B^5)\varepsilon_t = a_t$$

a.   Using the gas bill data in Table 4.10 and the least squares point estimates of the parameters in the above model in Figure 5.5, hand calculate the point forecasts of y_{41} and y_{42}. Note that, because SAS calculations are carried out with more decimal place accuracy than your hand calculations, your point forecasts will differ slightly from the results in Figure 5.5d.

b.   Write the SAS program to obtain the results in Figure 5.5.

5.2   Bargain Department Stores, Inc., is a chain of department stores in the Midwest. Sales ($= y_t$) of the "Bargain 8000 B.T.U. Air Conditioner" over the past three years are as follows (by quarters)

Year	Quarter	Period $= t$	$y_t =$ Sales
1	1	1	2915
	2	2	8032
	3	3	10411
	4	4	2427
2	1	5	4381
	2	6	9138
	3	7	11386
	4	8	3382
3	1	9	5105
	2	10	9894
	3	11	12300
	4	12	4013

a.   Plot sales versus time.

b.   From your data plot, what kind of trend appears to exist?

c.   The following SAS program analyzes the above data by using the regression model

$$y_t = \beta_0 + \beta_1 t + \beta_2 t^2 + \beta_3 x_{s2,t} + \beta_4 x_{s3,t} + \beta_5 x_{s4,t} + \varepsilon_t$$

where $x_{s2,t}$, $x_{s3,t}$, and $x_{s4,t}$ are appropriately defined dummy variables for quarters

2, 3, and 4, respectively. Note that the SAS program contains data lines allowing us to predict y_{13}, y_{14}, y_{15}, and y_{16}.

```
DATA BARGAIN;
INPUT Y T TSQ D2 D3 D4;
CARDS;
 2915   1    1  0  0  0
 8032   2    4  1  0  0
10411   3    9  0  1  0
 2427   4   16  0  0  1
 4381   5   25  0  0  0
 9138   6   36  1  0  0
11386   7   49  0  1  0
 3382   8   64  0  0  1
 5105   9   81  0  0  0
 9894  10  100  1  0  0
12300  11  121  0  1  0
 4013  12  144  0  0  1
   .   13  169  0  0  0
   .   14  196  1  0  0
   .   15  225  0  1  0
   .   16  256  0  0  1

PROC PRINT;
PROC REG;
 MODEL Y=T TSQ D2 D3 D4 / P CLM CLI;
```

The above SAS program gives the following output.

DEP VARIABLE: Y

SOURCE	DF	SUM OF SQUARES	MEAN SQUARE	F VALUE	PROB > F
MODEL	5	142952179	28590436	3346.937	0.0001
ERROR	6	51253.616	8542.269		
C TOTAL	11	143003433			

			R-SQUARE	0.9996	
ROOT MSE		92.424398			
DEP MEAN		6948.667	ADJ R-SQ	0.9993	
C.V.		1.330103			

VARIABLE	DF	PARAMETER ESTIMATE	STANDARD ERROR	T FOR HO: PARAMETER = 0	PROB > ITI
INTERCEP	1	2624.527	100.364	26.150	0.0001
T	1	382.819	34.032350	11.249	0.0001

TSQ	1	−11.353831	2.541332	−4.468	0.0042
D2	1	4629.740	76.075069	60.858	0.0001
D3	1	6738.855	77.379749	87.088	0.0001
D4	1	−1565.323	79.344026	−19.728	0.0001

OBS	ACTUAL	PREDICT VALUE	STD ERR PREDICT	LOWER95% MEAN	UPPER95% MEAN	LOWER95% PREDICT	UPPER95% PREDICT	RESIDUAL
1	2915	2996	76.522	2809	3183	2702	3290	−80.992
2	8032	7974	66.916	7811	8138	7695	8254	57.511
3	10411	10410	62.663	10256	10563	10136	10683	1.347
4	2427	2409	64.825	2250	2567	2133	2685	18.183
5	4381	4255	59.852	4108	4401	3985	4524	126.226
6	9138	9142	59.852	8996	9289	8873	9412	−4.441
7	11386	11487	59.852	11340	11633	11217	11756	−100.774
8	3382	3395	59.852	3249	3542	3126	3665	−13.108
9	5105	5150	64.825	4992	5309	4874	5426	−45.234
10	9894	9947	62.663	9794	10100	9674	10220	−53.070
11	12300	12201	66.916	12037	12364	11921	12480	99.427
12	4013	4018	76.522	3831	4205	3724	4312	−5.075
13	·	5682	112.586	5407	5958	5326	6039	·
14	·	10388	142.798	10039	10738	9972	10805	·
15	·	12551	177.239	12117	12985	12062	13040	·
16	·	4278	213.875	3754	4801	3708	4848	·

1. Do all of the independent variables in the model seen important? Justify your answer.
2. Find and identify $\hat{y}_{13}$, $\hat{y}_{14}$, $\hat{y}_{15}$, and $\hat{y}_{16}$, the point predictions of y_{13}, y_{14}, y_{15}, and y_{16}.
3. Calculate $\hat{y}_{15}$ by using the appropriate least squares point estimates.
4. Calculate $\hat{y}_{16}$ by using the appropriate least squares point estimates.
5. Find and identify the 95% prediction intervals for y_{13}, y_{14}, y_{15}, and y_{16}. Interpret these intervals.

5.3   The data in the table below gives quarterly sales of the popular game Oligopoly at the J-Mart variety store.

Year	Quarter	Oligopoly Sales
1	1	20
	2	25
	3	35
	4	44
2	1	28
	2	29
	3	43
	4	48
3	1	24
	2	37
	3	39
	4	56

Consider using the multiplicative decomposition method to forecast Oligopoly sales for year 4.

a. Compute appropriate 4 period moving averages for this data.

b. Compute centered moving averages for this data.

c. Calculate $sn_t \times ir_t$ values for this data.

d. Calculate estimates of the seasonal factors for quarterly Oligopoly sales (that is, compute sn_t values for this data).

e. Compute the deseasonalized observations for this data.

f. Plot the deseasonalized observations versus time. From your data plot, what kind of trend appears to exist?

g. Assuming that a linear trend

$$TR_t = \beta_0 + \beta_1 t$$

describes the deasonalized observations, compute least squares point estimates of β_0 and β_1.

h. Write and run a SAS program that will

1. Compute least squares point estimates of B_0 and B_1, the parameters of the trend line

$$TR_t = \beta_0 + \beta_1 t$$

describing the deseasonalized observations.

2. Compute point forecasts of deseasonalized Oligopoly sales for the four quarters of year 4.

3. Compute 95% prediction interval forecasts of deseasonalized Oligopoly sales for the four quarters of year 4.

i. Compute $cl_t \times ir_t$ values for the Oligopoly data.

j. Compute estimates of the cyclical factors for the Oligopoly data (that is, compute cl_t values for this data).

k. Compute estimates of the irregular factors for the Oligopoly data (that is, compute ir_t values for this data).

l. Do the cl_t values determine any well defined cycle? Explain your answer.

m. Using estimated trend and seasonal factors, compute point forecasts of Oligopoly sales for each quarter of year 4.

n. Using estimated trend and seasonal factors, compute approximate 95% prediction interval forecasts of Oligopoly sales for each quarter of year 4.

5.4 The data in the table below concerns four years of demand for the Deep-Freeze Freezer.

Year	Quarter	Period	Demand
1	1	1	10
	2	2	31
	3	3	43
	4	4	16
2	1	5	11
	2	6	33
	3	7	45
	4	8	17

3	1	9	13
	2	10	34
	3	11	48
	4	12	19
4	1	13	15
	2	14	37
	3	15	51
	4	16	21

a. Plot the Deep-Freeze demand data versus time.

b. From your data plot, what kind of trend appears to exist?

c. From your data plot, does seasonal variation appear to exist? If seasonal variation exists, is it constant, increasing, or decreasing as time advances?

d. Define an appropriate dummy variable regression model that might be used to forecast future Deep-Freeze Freezer demand.

e. Write and run a SAS program that analyzes the above data using the regression model you defined in part d. Use the SAS program to compute point and interval predictions for each quarter in year 5.

f. Do all of the independent variables in the model seem important? Justify your answer.

g. Find and identify $\hat{y}_{17}$, $\hat{y}_{18}$, $\hat{y}_{19}$, and $\hat{y}_{20}$, the point predictions of y_{17}, y_{18}, y_{19}, and y_{20}.

h. Calculate $\hat{y}_{17}$ by using the appropriate least squares point estimates.

i. Calculate $\hat{y}_{18}$ by using the appropriate least squares point estimates.

j. Find and identify the 95% prediction intervals for y_{17}, y_{18}, y_{19}, and y_{20}. Interpret these intervals.

5.5 International Machinery, Inc. produces a tractor and wishes to use tractor sales data observed in 1981 through 1984 to predict tractor sales in 1985. The DECOMP output of the multiplicative decomposition analysis of this data is given in Tables 1, 2, and 3 on pages 257–259.

a. Find and identify the four seasonal factors for Quarters 1, 2, 3, and 4.

b. What is the equation of the estimated trend that has been calculated using the deseasonalized data?

c. Do the cyclical factors determine a well-defined cycle for this data? Explain your answer.

d. Find and identify point forecasts of tractor sales (based on trend and seasonal factors) for each of the quarters in 1985.

e. Find a point forecast of total tractor sales for 1985.

f. Write and run a SAS program that will compute 95% prediction interval forecasts of deseasonalized tractor demand for each of the four quarters in 1985.

g. Compute (using estimated trend and seasonal factors) $\hat{y}_{17}$, $\hat{y}_{18}$, $\hat{y}_{19}$, and $\hat{y}_{20}$. Verify that the point forecasts for 1985 given in Table 3 are correct.

h. Use your results from parts d. and f. to compute approximate 95% prediction interval forecasts of tractor sales for each of the four quarters in 1985.

i. For the four years of historical tractor sales data, compute the forecast error

$$[y_t - (tr_t \times sn_t)]$$

for each time period. By examining these forecast errors, does the multiplicative decomposition method provide accurate predictions, or is there a systematic bias in these predictions? Can you suggest a reason for this behavior?

TABLE 1

		TIME PERIOD	ORIGINAL DATA	CENTERED MOVING AVG	C.M.A. (%) = Y/M.A.	SEASONAL INDEX	DESEASONALIZED DATA
1	QUARTER1	1981	293.0			1.191	245.9
2	QUARTER2	1981	392.0			1.521	257.7
3	QUARTER3	1981	221.0	275.1	0.803	0.804	275.0
4	QUARTER4	1981	147.0	302.0	0.487	0.484	303.9
5	QUARTER1	1982	388.0	325.3	1.193	1.191	325.7
6	QUARTER2	1982	512.0	338.1	1.514	1.521	336.6
7	QUARTER3	1982	287.0	354.1	0.810	0.804	357.1
8	QUARTER4	1982	184.0	381.5	0.482	0.484	380.4
9	QUARTER1	1983	479.0	405.0	1.183	1.191	402.0
10	QUARTER2	1983	640.0	417.4	1.533	1.521	420.7
11	QUARTER3	1983	347.0	435.0	0.798	0.804	431.8
12	QUARTER4	1983	223.0	462.1	0.483	0.484	461.0
13	QUARTER1	1984	581.0	484.4	1.199	1.191	487.7
14	QUARTER2	1984	755.0	497.6	1.517	1.521	496.3
15	QUARTER3	1984	410.0			0.804	510.2
16	QUARTER4	1984	266.0			0.484	549.9

TREND USING DESEASONALIZED DATA

TREND = 220.54 + 19.9499 * TIME

5.6. The data in the table below gives the number of reported cases y_t of a newly discovered disease over the last 11 months.

Month (t)	Number of Reported Cases (y_t)
1	1
2	1
3	2
4	3
5	4
6	6
7	8
8	13
9	21
10	27
11	45

a. Plot y_t versus time. Does the use of a growth curve model for forecasting future y_t values seem appropriate? Explain your answer.

TABLE 2

TIME PERIOD	NAME	ORIG DATA Y	TREND T	SEASONAL S	T*S	C*I = Y/T*S
1	QUARTER1	293.0	240.5	1.191	286.5	1.023
2	QUARTER2	392.0	260.4	1.521	396.2	0.989
3	QUARTER3	221.0	280.4	0.804	225.3	0.981
4	QUARTER4	147.0	300.3	0.484	145.3	1.012
5	QUARTER1	388.0	320.3	1.191	381.6	1.017
6	QUARTER2	512.0	340.2	1.521	517.6	0.989
7	QUARTER3	287.0	360.2	0.804	289.4	0.992
8	QUARTER4	184.0	380.1	0.484	183.9	1.001
9	QUARTER1	479.0	400.1	1.191	476.7	1.005
10	QUARTER2	640.0	420.0	1.521	639.0	1.002
11	QUARTER3	347.0	440.0	0.804	353.6	0.981
12	QUARTER4	223.0	459.9	0.484	222.5	1.002
13	QUARTER1	581.0	479.9	1.191	571.7	1.016
14	QUARTER2	755.0	499.8	1.521	760.4	0.993
15	QUARTER3	410.0	519.8	0.804	417.7	0.982
16	QUARTER4	266.0	539.7	0.484	261.1	1.019

CYCLE C	IRREGULAR I = C*I/C	SIMULATED DATA=T*S*C	ERROR = ORIG−SIM
0.998	0.992	450.6	−58.648
0.994	0.987	247.8	−26.750
1.003	1.009	156.9	−9.918
1.006	1.011	403.4	−15.377
0.999	0.990	531.7	−19.670
0.994	0.998	289.9	−2.893
0.999	1.002	181.8	2.190
1.002	1.003	465.1	13.948
0.996	1.006	610.0	29.982
0.995	0.986	332.5	14.470
1.000	1.002	207.5	15.457
1.004	1.012	528.9	52.086
0.997	0.996	690.7	64.277
0.998	0.984	375.8	34.211

b. Using natural logarithms, define a transformed growth curve model that will be linear in its parameters.

c. Plot the natural logarithms of the y_t values versus time. Has the logarithmic transformation linearized the data?

d. Write and run a SAS program that will calculate least squares point estimates of the parameters in the transformed model. Also, use SAS to compute a point

TRACTOR SALES FORECASTS

TABLE 3

PERIOD	YEAR	FORECAST=T·S
QUARTER1	1985	666.8
QUARTER2	1985	881.8
QUARTER3	1985	481.8
QUARTER4	1985	299.7
QUARTER1	1986	761.9
QUARTER2	1986	1003.2
QUARTER3	1986	546.0
QUARTER4	1986	338.3
QUARTER1	1987	857.0
QUARTER2	1987	1124.5
QUARTER3	1987	610.1
QUARTER4	1987	376.9
QUARTER1	1988	952.0
QUARTER2	1988	1245.9
QUARTER3	1988	674.2
QUARTER4	1988	415.5

prediction and 95% prediction interval for $\ln y_{12}$ (the natural logarithm of next month's number of reported cases of the disease).

e. Estimate the growth rate for this disease.

f. Calculate a forecast for y_{12}, the number of reported cases of the disease in month 12.

g. Calculate a 95% prediction interval for y_{12}. Interpret this interval.

EXPONENTIAL SMOOTHING

6.1 INTRODUCTION

Exponential smoothing is a forecasting technique that attempts to "track" changes in a time series by using newly observed time series values to "update" the estimates of the parameters describing the time series. This chapter presents several exponential smoothing techniques—*simple exponential smoothing* (in Section 6.2), *Winter's Method*, which is an exponential smoothing approach to handling seasonal data (in Section 6.4), and *one- and two-parameter double exponential smoothing* (also in Section 6.4). Here we note that, when using exponential smoothing, it is often useful to employ *adaptive control procedures* to check the accuracy of the forecasting system. Such procedures are explained in Section 6.3. In Section 6.5 we will find that the exponential smoothing techniques of Sections 6.2 and 6.4 are essentially equivalent to some special Box-Jenkins models. Sections 6.5 and 6.6 discuss the practical implications of the relationships between exponential smoothing, the Box-Jenkins methodology, and time series regression, and also show how to implement the techniques of this chapter by using SAS.

6.2 SIMPLE EXPONENTIAL SMOOTHING

EXAMPLE 6.1

The Bay City Seafood Company, which owns a fleet of fishing trawlers and operates a fish processing plant, wishes to make monthly predictions of its catch of cod (measured in tons). In order to forecast its minimum and maximum possible revenues from cod sales and in order to plan the operations of its fish processing plant, the Bay City Seafood Company desires to make both point forecasts and prediction interval forecasts of its monthly cod catch. The company has recorded monthly cod catch in tons for the previous two years, which we will call year 1 and year 2. The cod catch history is given in Table 6.1. When this data is plotted, it appears to randomly fluctuate around a constant average level (see Figure 6.1). Since the company subjectively believes that this data pattern will continue in the future, it seems reasonable to use the model

$$y_t = \beta_0 + \varepsilon_t$$

to forecast cod catch in future months.

The least squares point estimate of the average level β_0 in the model

$$y_t = \beta_0 + \varepsilon_t$$

can be shown to equal

$$\bar{y} = \frac{\sum_{t=1}^{n} y_t}{n}$$

$$= \frac{\sum_{t=1}^{24} y_t}{24} = \frac{y_1 + y_2 + \cdots + y_{24}}{24} = \frac{362 + 381 + \cdots + 365}{24}$$

$$= 351.29$$

TABLE 6.1   *Cod Catch (in tons)*

	Year 1	Year 2
Jan.	362	276
Feb.	381	334
Mar.	317	394
Apr.	297	334
May	399	384
June	402	314
July	375	344
Aug.	349	337
Sept.	386	345
Oct.	328	362
Nov.	389	314
Dec.	343	365

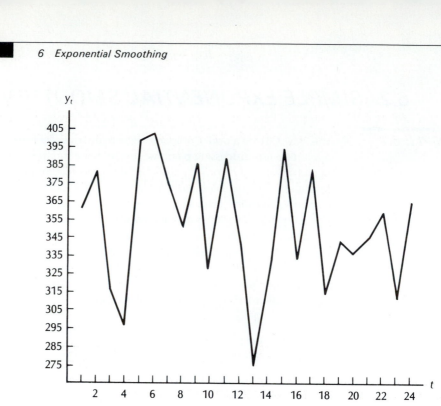

FIGURE 6.1   *Cod Catch (in tons)*

which is the mean of the $n = 24$ observed cod catches. It follows that

$$\hat{y}_t = \bar{y} = 351.29$$

is the point prediction of the cod catch (y_t) in any future month. Moreover, a $100(1 - \alpha)\%$ prediction interval for y_t is

$$[\bar{y} - t_{[\alpha/2]}^{(n-1)} s \sqrt{1 + (1/n)}, \bar{y} + t_{[\alpha/2]}^{(n-1)} s \sqrt{1 + (1/n)}]$$

where

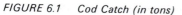

$$s = \sqrt{\frac{\sum_{t=1}^{n} (y_t - \bar{y})^2}{n - 1}}$$

$$= \sqrt{\frac{\sum_{t=1}^{30} (y_t - \bar{y})^2}{n - 1}}$$

$$= \sqrt{\frac{(362 - 351.29)^2 + (381 - 351.29)^2 + \cdots + (365 - 351.29)^2}{24 - 1}}$$

$$= 33.82$$

and where $t_{[\alpha/2]}^{(n-1)}$ is the point on the scale of the t-distribution having $(n - 1)$

degrees of freedom so that the area under the curve of this t-distribution to the right of $t_{[\alpha/2]}^{(n-1)}$ is $\alpha/2$. If we wish to calculate a 95% prediction interval for y_t, which implies that $\alpha = .05$, then Table A2 (in Appendix A) tells us to use the point

$$t_{[\alpha/2]}^{(n-1)} = t_{[.05/2]}^{(24-1)} = t_{[.025]}^{(23)} = 2.069$$

Therefore, a 95% prediction interval for the cod catch (y_t) in any future time period is

$$[\bar{y} \pm t_{[.025]}^{(23)} s\sqrt{1 + (1/n)}] = [351.29 \pm 2.069(33.82)\sqrt{1 + (1/24)}]$$

$$= [351.29 \pm 71.37]$$

$$= [279.29, \quad 422.66]$$

Using the sample mean

$$\bar{y} = \frac{\sum\limits_{t=1}^{n} y_t}{n}$$

as the point estimate of β_0 in the model

$$y_t = \beta_0 + \varepsilon_t$$

implies that we are calculating the point estimate of β_0 by *equally weighting* each of the previously observed time series values $y_1, y_2, \ldots, y_n$. In ongoing forecasting systems, forecasts of a time series are made each period for succeeding periods. Hence, the forecasting equation and the estimates of the model parameters need to be updated at the end of each period to account for the most recent observations. This updating must take into account the fact that the model parameters may be changing over time. Therefore, it may not be reasonable to continually give equal weight to each of the previously observed values of the time series. *Exponential smoothing* is a forecasting method that *weights* each of the observed time series values *unequally*, with more recent observations being weighted more heavily than more remote observations. This unequal weighting is achieved through one or more *smoothing constants*, which determine how much weight is given to each observation. This procedure allows the forecaster to update the estimates of the model parameters so that changes in the values of these parameters can be detected and incorporated into the forecasting system.

In the following example we will illustrate *simple exponential smoothing*, which should be used when the model

$$y_t = \beta_0 + \varepsilon_t$$

describes the time series under consideration and when β_0 may be changing over time.[1]

---

[1] Since β_0 may be changing over time, it might be appropriate to denote this fact by rewriting β_0 as $\beta_{0,t}$ and thus by rewriting the model $y_t = \beta_0 + \varepsilon_t$ as $y_t = \beta_{0,t} + \varepsilon_t$. Indeed, since all of the exponential smoothing procedures to be discussed in this chapter take into account the fact that the parameter(s) of the underlying models may be changing over time, it might be appropriate to add a "t" subscript to the parameters of the models. However, for notational simplicity, we will omit this subscript.

EXAMPLE 6.2

To forecast the cod catch data in Table 6.1 by using simple exponential smoothing, we begin by calculating an initial estimate (denoted $a_0(0)$) of the average level β_0 of the time series by averaging the first six time series values:

$$a_0(0) \; = \; \frac{\sum\limits_{t=1}^{6} y_t}{6} \; = \; \frac{362 + 381 + \cdots + 365}{6} \; = \; 359.67$$

Note that, since simple exponential smoothing attempts to track changes over time in the average level β_0 by using newly observed values to update the estimates of β_0, we use only six of the $n = 24$ time series observations to calculate the initial estimate of β_0. If we do this, then 18 observations remain to tell us how β_0 may be changing over time. Experience has shown that, in general, it is reasonable to calculate initial estimates in exponential smoothing procedures by using half of the historical data. However, it can be shown that in simple exponential smoothing using six observations is reasonable (it would not, however, be reasonable to use a very small number of observations because doing so might make the initial estimate so different from the true value of β_0 that the exponential smoothing procedure would be adversely affected).

Exactly how does simple exponential smoothing work? Assume that at the end of time period $T - 1$ we have an estimate $a_0(T - 1)$ of the average level of the time series. Then, assuming that in time period T we obtain a new observation y_T, we can update $a_0(T - 1)$ to $a_0(T)$, which is an estimate made in time period T of the average level of the time series, by using the *smoothing equation*

$$a_0(T) \; = \; \alpha y_T + (1 - \alpha)a_0(T - 1)$$

Here, α is a smoothing constant between 0 and 1 (α will be discussed in more detail later). The updating equation says that $a_0(T)$, the estimate made in time period T of the average level of the time series, equals a fraction α (for example, .1) of the newly observed time series observation y_T plus a fraction $(1 - \alpha)$ (for example, .9) of $a_0(T - 1)$, the estimate made in time period $T - 1$ of the average level of the process. The more the average level of the process is changing, the more a newly observed time series value should influence our estimate, and thus the larger the smoothing constant α should be set. We will soon see how to use historical data to determine an appropriate value of α.

We will now begin with the initial estimate $a_0(0) = 359.67$ and update this initial estimate by applying the smoothing equation to the 24 observed cod catches. To do this, we will arbitrarily set α equal to .02, and to judge the appropriateness of this choice of α we will calculate "one-period-ahead" forecasts of the historical cod catches as we carry out the smoothing procedure. Since the initial estimate of β_0 is $a_0(0) = 359.67$, it follows that 360 is the rounded forecast made at time 0 for y_1, the value of the time series in period 1 (January of year 1). Since we see from Table 6.1 that $y_1 = 362$, we have a forecast error of $362 - 360 = 2$. Using $y_1 = 362$, we can update $a_0(0)$ to $a_0(1)$, an estimate made in period 1 of the average level of the time series, by using the equation

$$a_0(1) = \alpha y_1 + (1 - \alpha)a_0(0)$$

$$= .02(362) + .98(359.67)$$

$$= 359.72$$

Since this implies that 360 is the rounded forecast made in period 1 for y_2, the value of the time series in period 2 (February of year 1), and since we see from Table 6.1 that $y_2 = 381$, we have a forecast error of $360 - 381 = -21$. Using $y_2 = 381$, we can update $a_0(1)$ to $a_0(2)$, an estimate made in period 2 of the average level of the time series, by using the equation

$$a_0(2) = \alpha y_2 + (1 - \alpha)a_0(1)$$

$$= .02(381) + .98(359.72)$$

$$= 360.14$$

Since this implies that 360 is the rounded forecast made in period 2 for y_3, the value of the time series in period 3 (March of year 1), and since we see from Table 6.1 that $y_3 = 317$, we have a forecast error of $317 - 360 = -43$. This procedure is continued through the entire two years (24 periods) of historical data. The results are summarized in Table 6.2. Using the results in this table, we find that, for $\alpha = .02$, the sum of squared forecast errors is 27,744. To find a "good" value of α, we evaluate the sum of squared forecast errors for values of α ranging from .02 to .30 in increments of .02 (in most exponential smoothing applications, the value of the smoothing constant used is between .01 and .30). Noting that the results are given in Table 6.3, we find that $\alpha = .02$ minimizes the sum of squared forecast errors. This small value of α indicates that the average level of the time series is not changing much over time.

In general, if $a_0(T)$ is the estimate made in time period T of the average level β_0 of the process

$$y_t = \beta_0 + \varepsilon_t$$

then

A *point forecast* made in time period T for $y_{T+\tau}$ is

$$\hat{y}_{T+\tau}(T) = a_0(T)$$

Furthermore, defining $z_{[\alpha/2]}$ to be the point on the scale of the standard normal curve such that the area under this curve to the right of $z_{[\alpha/2]}$ is $\alpha/2$, it can be shown that

A $100(1 - \alpha)\%$ *prediction interval* made in time period T for $y_{T+\tau}$ is

$$[\hat{y}_{T+\tau}(T) - B_{T+\tau}^{[100(1-\alpha)]}(T), \hat{y}_{T+\tau}(T) + B_{T+\tau}^{[100(1-\alpha)]}(T)]$$

where

$$B_{T+\tau}^{[100(1-\alpha)]}(T) = z_{[\alpha/2]}1.25\Delta(T)$$

and

$$\Delta(T) = \frac{\sum_{t=1}^{T} |y_t - a_0(t - 1)|}{T}$$

Moreover, if we observe the time series value y_{T+1} in period $T + 1$,

We can *update* $a_0(T)$ to $a_0(T + 1)$ by using the equation

$$a_0(T + 1) = \alpha y_{T+1} + (1 - \alpha)a_0(T)$$

and we can *update* $\Delta(T)$ to $\Delta(T + 1)$ by using the equation

$$\Delta(T + 1) = \frac{T\Delta(T) + |y_{T+1} - a_0(T)|}{T + 1}$$

Therefore,

A *point forecast* made in time period $T + 1$ for $y_{T+1+\tau}$ is

$$\hat{y}_{T+1+\tau}(T + 1) = a_0(T + 1)$$

and a $100(1 - \alpha)\%$ *prediction interval* made in time period $T + 1$ for $y_{T+1+\tau}$ is

$$[\hat{y}_{T+1+\tau}(T + 1) - B_{T+1+\tau}^{[100(1-\alpha)]}(T + 1), \hat{y}_{T+1+\tau}(T + 1) + B_{T+1+\tau}^{[100(1-\alpha)]}(T + 1)]$$

where

$$B_{T+1+\tau}^{[100(1-\alpha)]}(T + 1) = z_{[\alpha/2]}1.25\Delta(T + 1)$$

EXAMPLE 6.3

Since we saw in Example 6.2 that a "good" value of α is .02, we will use this value to forecast future monthly cod catches. Since Table 6.2 indicates that $a_0(24)$, the estimate made in month 24 of the average level of the time series, is 356.13, it follows that the point forecast made in month 24 of the cod catch in month 25

TABLE 6.2 One-Period-Ahead Forecasting of the Historical Cod Catch Time Series Using Simple Exponential Smoothing with $\alpha = .02$

Year	Month	Actual Cod Catch y_T	Smoothed Estimate $a_0(T)$	Forecast Made Last Period	Forecast Error	Squared Forecast Error
			$(a_0(0) = 359.67)$			
1	Jan.	362	359.72	360	2	4
	Feb.	381	360.14	360	21	441
	Mar.	317	359.28	360	-43	1849
	Apr.	297	358.03	359	-62	3844
	May	399	358.85	358	41	1681
	June	402	359.71	359	43	1849
	July	375	360.02	360	15	225
	Aug.	349	359.80	360	-11	121
	Sept.	386	360.32	360	26	676
	Oct.	328	359.68	360	-32	1024
	Nov.	389	360.26	360	29	841
	Dec.	343	359.92	360	-17	289
2	Jan.	276	358.24	360	-84	7056
	Feb.	334	357.75	358	-24	576
	Mar.	394	358.48	358	36	1296
	Apr.	334	357.99	358	-24	576
	May	384	358.51	358	26	676
	June	314	357.62	359	-45	2025
	July	344	357.35	358	-14	196
	Aug.	337	356.94	357	-20	400
	Sept.	345	356.70	357	-12	144
	Oct.	362	356.81	357	5	25
	Nov.	314	355.95	357	-43	1849
	Dec.	365	356.13	356	9	81

Sum of squared errors = 27,744

(January of year 3) and of any other future monthly cod catch is

$$\hat{y}_{24+\tau}(24) = a_0(24) = 356.13$$

Moreover, Table A1 (in Appendix A) tells us that $z_{[.025]} = 1.96$. Therefore, since

$$\Delta(24) = \frac{\sum_{t=1}^{24} |y_t - a_0(t-1)|}{24}$$

$$= \frac{|y_1 - a_0(0)| + |y_2 - a_0(1)| + \cdots + |y_{24} - a_0(23)|}{24}$$

TABLE 6.3   *The Sums of Squared Forecast Errors for Different Values of α*

Smoothing Constant α	Sum of Squared Errors
.02	27744
.04	28021
.06	28036
.08	28459
.10	28792
.12	28987
.14	29496
.16	29756
.18	30274
.20	30774
.22	31269
.24	31725
.26	32203
.28	32596
.30	33316

$$= \frac{|362 - 359.67| + |381 - 359.72| + \cdots + |365 - 355.95|}{24}$$

$$= 28.62$$

it follows that

$$B^{[95]}_{24+\tau}(24) = z_{[.025]} 1.25 \Delta(24)$$

$$= 1.96(1.25)(28.62)$$

$$= 70.12$$

and thus that a 95% prediction interval made in month 24 for the cod catch in month 25 and for any other future monthly cod catch is

$$[\hat{y}_{24+\tau}(24) \pm B^{[95]}_{24+\tau}(24)] = [356.13 \pm 70.12]$$

$$= [286.01, 426.25]$$

Now, assuming that we observe a cod catch in January of year 3 of $y_{25} = 384$, we can update $a_0(24)$ to $a_0(25)$, an estimate made in month 25 of the average level of the time series, by using the equation

$$a_0(25) = \alpha y_{25} + (1 - \alpha)a_0(24)$$

$$= .02(384) + .98(356.13)$$

$$= 356.69$$

This implies that the point forecast made in month 25 of the cod catch in month 26 (February of year 3) and of any other future monthly cod catch is

$$\hat{y}_{25+\tau}(25) \;=\; a_0(25) \;=\; 356.69$$

Furthermore, since we can update $\Delta(24)$ to $\Delta(25)$ by using the equation

$$\Delta(25) \;=\; \frac{24\Delta(24) + |y_{25} - a_0(24)|}{25}$$

$$=\; \frac{24(28.62) + |384 - 356.13|}{25}$$

$$=\; 28.59$$

which implies that

$$B^{[95]}_{25+\tau}(25) \;=\; z_{[.025]}1.25\Delta(25)$$

$$=\; 1.96(1.25)(28.59)$$

$$=\; 70.05$$

it follows that a 95% prediction interval made in month 25 for the cod catch in month 26 and for any other future monthly cod catch is

$$[\,\hat{y}_{25+\tau}(25) \;\pm\; B^{[95]}_{25+\tau}(25)] \;=\; [356.69 \;\pm\; 70.05]$$

$$=\; [286.64,\; 426.74]$$

In general, note that the smoothing equation

$$a_0(T) \;=\; \alpha y_T + (1 - \alpha)a_0(T - 1)$$

implies that

$$a_0(T - 1) \;=\; \alpha y_{T-1} + (1 - \alpha)a_0(T - 2)$$

Substitution therefore gives us

$$a_0(T) \;=\; \alpha y_T + (1 - \alpha)[\alpha y_{T-1} + (1 - \alpha)a_0(T - 2)]$$

$$=\; \alpha y_T + \alpha(1 - \alpha)y_{T-1} + (1 - \alpha)^2 a_0(T - 2)$$

Substituting recursively for $a_0(T - 2)$, $a_0(T - 3)$, . . . , $a_0(1)$, and $a_0(0)$, we obtain

$$a_0(T) \;=\; \alpha y_T + \alpha(1 - \alpha)y_{T-1} + \alpha(1 - \alpha)^2 y_{T-2} + \cdots + \alpha(1 - \alpha)^{T-1}y_1$$

$$+\; (1 - \alpha)^T a_0(0)$$

Thus we see that $a_0(T)$, the estimate made in time period T of the average level β_0 of the time series, can be expressed in terms of the observations $y_T, y_{T-1}, y_{T-2}, \ldots, y_1$, and the initial estimate $a_0(0)$. Since the coefficients measuring the contributions of the observations $y_T, y_{T-1}, y_{T-2}, \ldots, y_1$—that is, $\alpha, \alpha(1 - \alpha), \alpha(1 - \alpha)^2, \ldots, \alpha(1 - \alpha)^{T-1}$ —decrease *exponentially* with age (for example, if $\alpha = 0.1$, these coefficients equal .1, .09, .081, and so on), we refer to the forecasting procedure using the above smoothing equation as *simple exponential smoothing*.

6.3 ADAPTIVE CONTROL PROCEDURES

A forecasting system will never produce perfect forecasts of a time series. Therefore, we will now discuss some methods that can be used to determine when something is wrong with the forecasting system. That is, we wish to determine if the forecast errors are larger than an "accurate" forecasting system can reasonably be expected to produce. To do this, assume we have a history of T single-period-ahead forecast errors, $e_1(\alpha)$, $e_2(\alpha)$, . . . , $e_T(\alpha)$. Here, (α) denotes the fact that the single-period-ahead forecast errors are obtained from an exponential smoothing forecasting system employing a smoothing constant of α. We next define the following sum (Y) of the single-period-ahead forecast errors

$$Y(\alpha, T) = \sum_{t=1}^{T} e_t(\alpha)$$

It is obvious that

$$Y(\alpha, T) = Y(\alpha, T - 1) + e_T(\alpha)$$

and we define the following mean absolute deviation (D)

$$D(\alpha, T) = \frac{\sum_{t=1}^{T} |e_t(\alpha)|}{T}$$

Then:

The *tracking signal* $TS(\alpha, T)$ is defined as

$$TS(\alpha, T) = \left| \frac{Y(\alpha, T)}{D(\alpha, T)} \right|$$

If $TS(\alpha, T)$ is "large," this means that $Y(\alpha, T)$ is large relative to the mean absolute deviation $D(\alpha, T)$, which in turn says that, since

$$Y(\alpha, T) = \sum_{t=1}^{T} e_T(\alpha)$$

the forecasting system is producing errors that are either consistently positive or consistently negative. That is, a large value of $TS(\alpha, T)$ implies that the forecasting system is producing forecasts that are either consistently smaller or consistently larger than the time series values that are being forecasted. Since an "accurate" forecasting system should be producing roughly one half positive errors and one half negative errors, a large value of $TS(\alpha, T)$ indicates that the forecasting system is not performing accurately. In practice, if $TS(\alpha, T)$ exceeds a control limit, denoted by K_1, for two or more consecutive periods, this is taken as a strong indication that the forecast errors have been larger than an accurate forecasting system can reasonably be expected to produce. The control limit K_1 is generally taken to be between 4 and 6 for most exponential smoothing models.

More precise methods for determining the control limits for particular exponential smoothing models can be found by the interested reader in Johnson and Montgomery [1976], which also discusses tracking signals slightly different from the one discussed above. If the tracking signal exceeds a large control limit (say near 6), it is a very strong indication that the forecasting system is not performing accurately. If the tracking signal indicates that corrective action is needed, several possibilities exist. One possibility is that the model needs to be changed. To do this, variables may be added or deleted to obtain a better representation of the time series. Another possibility is that the model being used does not need to be changed, but the estimates of the parameter(s) of the model do need to be changed. This can be accomplished by changing the smoothing constant, a subject which is further discussed in the next paragraphs. Before continuing, we discuss the initial values to be used for $Y(\alpha, T)$ and $D(\alpha, T)$ when starting the forecasting procedure. Since it is reasonable to assume that the original model is correct, $Y(\alpha, 0) = 0$ is the starting value for $Y(\alpha, T)$. Since there is some random variation in the process, however, it is not reasonable to use $D(\alpha, 0) = 0$ as the starting value for $D(\alpha, T)$. One possibility is to calculate an initial value $D(\alpha, 0)$ by employing the one-step-ahead forecast errors used in the forecasting of historical data to determine an appropriate smoothing constant.

It is common procedure to use different values of the smoothing constant at different times in the analysis of a time series. For example, a large value of α may be appropriate at the start of an exponential smoothing procedure when the initial values used to initialize the procedure are based on only a few observations. This allows new observations to weigh heavily in early forecasts. Later, a smaller value of α may be used since more information concerning the behavior of the time series has built up. It is also common to use a small value of α when a time series is stable and to use a larger value of α when the parameters in the time series are suspected to be changing. Forecasters often use procedures that automatically change the value of the smoothing constant. These procedures are known as *adaptive-control procedures* since they allow the smoothing constant to adapt itself to changes in the parameters of the time series.

Several techniques have been developed to automatically control the values of smoothing constants. We discuss a method developed by W. M. Chow [1965]. This method involves the use of three values for the smoothing constant. Suppose that a value for the smoothing constant in our smoothing procedure has been determined using simulation. We call this value α. Chow's method employs two other values of the smoothing constant: an upper value α_U and a lower value α_L such that

$$\alpha_U = \alpha + d$$
$$\alpha_L = \alpha - d$$

where d is a positive constant. Chow suggests that a value of $d = .05$ be used. The smoothing procedure is now performed using each of these values for the smoothing constant. That is, three different forecasts are generated in each period, one for each of the smoothing constants α_L, α, and α_U. The forecast generated using the smoothing constant α is the forecast to be used in practice. The other two forecasts are used to monitor the time series in an attempt to detect substantial changes in it. This is done in the following manner. We define $D(\alpha_L, T)$, $D(\alpha, T)$, and $D(\alpha_U, T)$ to be the mean absolute deviations of forecast errors calculated in period T using the streams of forecasts generated using, respectively, the smoothing constants α_L, α, and α_U. Whenever

$$D(\alpha, T) < \begin{cases} D(\alpha_L, T) \\ D(\alpha_U, T) \end{cases}$$

the forecast errors generated using α have been smaller than the forecast errors generated using α_L and α_U. Therefore, it is reasonable to continue using the smoothing constant α to generate forecasts. If, however,

$$D(\alpha_L, T) < \begin{cases} D(\alpha, T) \\ D(\alpha_U, T) \end{cases}$$

the smoothing constant to be used in forecasting should be reduced by setting α at α_L and adding and subtracting d to fix new limits, or, if

$$D(\alpha_U, T) < \begin{cases} D(\alpha, T) \\ D(\alpha_L, T) \end{cases}$$

α should be increased to α_U and new limits fixed by adding and subtracting d. Each time the smoothing constant is changed, the mean absolute deviations $D(\alpha_L, T)$, $D(\alpha, T)$, and $D(\alpha_U, T)$ are set equal to zero and the monitoring procedure is started over.

Chow reported good results with his method. However, a disadvantage of this method is that three forecasts must be computed each period, and hence a large amount of data must be stored in the information system maintained by the forecaster.

Chow's procedure can be extended to deal with exponential smoothing procedures that employ several smoothing constants. The interested reader is again referred to Johnson and Montgomery [1976], which also discusses other adaptive control procedures.

6.4 WINTERS' METHOD AND RELATED PROCEDURES

6.4.1 Winters' Method

We next present an exponential smoothing method called Winters' Method, and then we will discuss several other exponential smoothing techniques that are special cases of Winters' Method.

Winters' Method is an exponential smoothing approach to handling seasonal data. It assumes that the time series to be predicted can be adequately described using the following model:

Winters' (Multiplicative) Model is

$$y_t = (\beta_0 + \beta_1 t) \times SN_t + \varepsilon_t$$

Note that this model assumes a linear trend and multiplicative seasonal variation.

We will illustrate the use of Winters' Method by forecasting the Tasty Cola time series (see Table 5.6). We begin Winters' Method by calculating

1. An initial estimate $b_1(0)$ of β_1,
2. An initial estimate $a_0(0)$ of β_0, and
3. An initial estimate $sn_t(0)$ of SN_t

Suppose that historical data for the last m years is available. We define $\bar{y}_i$ to be the average of the observations in the ith year, where i takes on values 1, 2, . . . , m.

The *initial estimate of the trend component*, β_1, is

$$b_1(0) = \frac{\bar{y}_m - \bar{y}_1}{(m - 1)L}$$

Here $\bar{y}_m$, the average of the observations in year m, measures the average level[2] of the time series in the middle of year m. Similarly, $\bar{y}_1$ measures the average level of the time series in the middle of year 1. Thus $\bar{y}_m - \bar{y}_1$ measures the difference in these average levels. The total number of seasons elapsed between the middle of year 1 and the middle of year m is $(m - 1)L$. Thus the initial estimate $b_1(0)$ is simply the change in average level per season from the middle of year 1 to the middle of year m. For example, since the historical Tasty Cola data consists of 3 years of monthly data, we have $m = 3$ and $L = 12$. The average sales for year 1 is $\bar{y}_1 = 447.82$, while the average sales for year 3 is $\bar{y}_3 = 677.58$. Thus, the initial estimate of the trend component is

$$b_1(0) = \frac{\bar{y}_m - \bar{y}_1}{(m - 1)L} = \frac{677.58 - 447.82}{(3 - 1)12} = 9.57$$

---

[2] For Winters' Method, the "average level" of a time series refers to the average *deseasonalized* level of the time series.

The *initial estimate of the permanent component* β_0 (which represents the average level of the time series at time 0) is

$$a_0(0) = \bar{y}_1 - \frac{L}{2} b_1(0)$$

Again, $\bar{y}_1$, the average of the observations in year 1, measures the average level of the time series in the middle of year 1. The number of seasons that have elapsed from the start of year 1 to the middle of year 1 is $L/2$. The initial estimate $a_0(0)$ is, therefore, the average level of the time series at the middle of year 1 less the amount this average level has changed from the start of year 1 to the middle of year 1. For the Tasty Cola data, the initial estimate of the permanent component is

$$a_0(0) = \bar{y}_1 - \frac{L}{2} [b_1(0)] = 447.82 - \frac{12}{2} (9.57) = 390.40$$

Obtaining initial estimates for the L seasonal factors is done as follows. The expression

$$S_t = \frac{y_t}{\bar{y}_i - [(L + 1)/2 - j]b_1(0)}$$

must be computed for each season (month, quarter, etc.) t occurring in years 1 through m. Here $\bar{y}_i$ is the average of the observations for the year in which season t occurs. So, if $1 \leqslant t \leqslant L$, then $i = 1$; if $L + 1 \leqslant t \leqslant 2L$, then $i = 2$; etc. Thus $\bar{y}_i$ measures the average level of the time series in the middle of the year in which season t occurs. The letter j denotes the position of season t within the year. If the time series consists of monthly data, then, for January $j = 1$, for February $j = 2$, for March $j = 3$, and so on. Thus

$$- [(L + 1)/2 - j]$$

measures the number of seasons that season t is from the middle of the year. A positive value for this expression indicates that season t occurs after the middle of the year, while a negative value indicates that season t occurs before the middle of the year. Hence, the expression

$$\bar{y}_i - [(L + 1)/2 - j]b_1(0)$$

measures the average level of the time series in season t. If season t occurs before the middle of the year, we subtract the appropriate trend from the average level at midyear in order to obtain the average level in season t. If season t occurs after the middle of the year, we add the appropriate trend to the average level at midyear to obtain the average level in season t. To illustrate the calculations, we use the Tasty Cola data to calculate

S_t values for each April in the three-year sales history. For April of year 1 we have

$$S_4 = \frac{y_4}{\bar{y}_1 - [(L + 1)/2 - j]b_1(0)}$$

$$= \frac{289}{447.82 - [(12 + 1)/2 - 4](9.57)}$$

$$= .6818$$

For April of year 2 we have

$$S_{16} = \frac{y_{16}}{\bar{y}_2 - [(L + 1)/2 - j]b_1(0)}$$

$$= \frac{370}{564.83 - [(12 + 1)/2 - 4](9.57)}$$

$$= .6840$$

For April of year 3 we have

$$S_{28} = \frac{y_{28}}{\bar{y}_3 - [(L + 1)/2 - j]b_1(0)}$$

$$= \frac{443}{677.58 - [(12 + 1)/2 - 4](9.57)}$$

$$= .6777$$

The equation for S_t yields m estimates of each distinct seasonal factor, one for each year. If the time series is monthly, we obtain m estimates of the seasonal factor for each month. These m estimates are then averaged to yield one estimate for each different season. Thus we obtain

$$\overline{sn}_t = \frac{1}{m} \sum_{K=0}^{m-1} S_{t+KL} \qquad \text{for} \qquad t = 1, 2, \ldots, L$$

which is the average seasonal index for each different season. Finally

The *initial estimate* $sn_t(0)$ of SN_t is

$$sn_t(0) = \overline{sn}_t \left[\frac{L}{\sum_{t=1}^{L} \overline{sn}_t} \right] \qquad \text{for} \qquad t = 1, \ldots, L$$

This guarantees that the initial estimates $sn_1(0)$, $sn_2(0)$, . . . , $sn_L(0)$ add to L. For example, recalling that we have calculated S_t values for April in the Tasty Cola data, it follows that

$$\overline{sn}_4 = \frac{S_4 + S_{16} + S_{28}}{3} = \frac{.6818 + .6840 + .6777}{3} = .6812$$

Similarly, values of $\overline{sn}_t$ may be obtained for the other eleven months. If this is done, we find that

$$\sum_{t=1}^{12} \overline{sn}_t = 11.8285$$

Thus, the initial estimate $sn_t(0)$ is given by

$$sn_t(0) = \overline{sn}_t \left[\frac{L}{\displaystyle\sum_{t=1}^{L} \overline{sn}_t}\right] = \overline{sn}_t \left(\frac{12}{11.8285}\right) = 1.0145\overline{sn}_t$$

For example,

$$sn_4(0) = 1.0145\overline{sn}_4 = 1.0145(.6812) = .6911$$

The initial estimates obtained using the above method are

$a_0(0)$	= 390.40	$b_1(0)$	= 9.57
$sn_1(0)$	= .4841	$sn_7(0)$	= 1.4842
$sn_2(0)$	= .5847	$sn_8(0)$	= 1.6927
$sn_3(0)$	= .6022	$sn_9(0)$	= 1.9869
$sn_4(0)$	= .6911	$sn_{10}(0)$	= 1.2897
$sn_5(0)$	= .5859	$sn_{11}(0)$	= 1.0074
$sn_6(0)$	= .9965	$sn_{12}(0)$	= .5946

Notice that these initial estimates are quite close to the estimates obtained using the multiplicative decomposition method in Section 5.6.

Note that the initial estimates $a_0(0)$, $b_1(0)$, and $sn_t(0)$, for $t = 1, 2, . . . , L$ are based on a time origin that immediately precedes the first period accounted for in the m years of historical data used in obtaining them. For forecasting purposes, however, a time origin at the end of the m years of historical data is desired. Estimates at this point in time can be generated by repeatedly smoothing the initial estimates using updating equations. To discuss these updating equations, assume that at the end of time period $T - 1$ we have an estimate $a_0(T - 1)$ of the permanent component of the time series (that is, $a_0(T - 1)$ is the estimate of the average level of the time series at time $T - 1$ —see Figure 6.2). Assume also that we have an estimate $b_1(T - 1)$ of the trend component. Then, assuming that in time period T we obtain a new observation y_T, we wish to use y_T to update the estimates $a_0(T - 1)$ and $b_1(T - 1)$ to new estimates $a_0(T)$

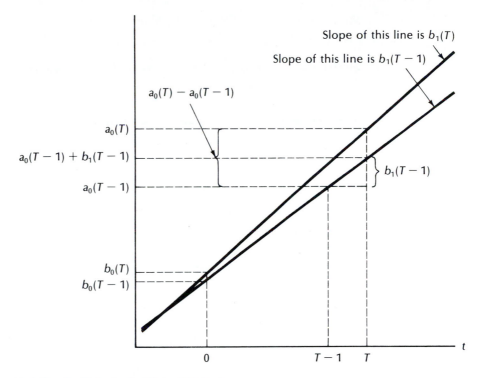

FIGURE 6.2   Updating the Estimates in Winters' Method

and $b_1(T)$ and to update $sn_T(T - L)$, the estimate of the seasonal factor obtained L periods ago, to a new estimate $sn_T(T)$.

The *updated estimate* $a_0(T)$ *of the permanent component*, which is the estimate of the average level of the process at time period T, is calculated by using the equation

$$a_0(T) = \alpha \frac{y_T}{sn_T(T - L)} + (1 - \alpha)[a_0(T - 1) + b_1(T - 1)]$$

where α is a smoothing constant between 0 and 1.

This updating equation says that $a_0(T)$, the estimate of the average level of the process at time period T, equals a fraction α (for example, .1) of the deseasonalized, newly observed time series value

$$\frac{y_T}{sn_T(T - L)}$$

plus a fraction $(1 - \alpha)$ (for example, .9) of

$$[a_0(T - 1) + b_1(T - 1)]$$

which is the estimate of the average level of the time series at time T, as calculated using the estimates computed in the last period (see Figure 6.2).

The updated estimate $b_1(T)$ of the trend component is

$$b_1(T) = \beta[a_0(T) - a_0(T - 1)] + (1 - \beta)b_1(T - 1)$$

where β is a smoothing constant between 0 and 1.

Here $[a_0(T) - a_0(T - 1)]$ is simply the difference between the estimate of the permanent component made in the current period and the estimate of the permanent component made in the last period. This difference smoothed with the estimate of the trend made last period yields the estimate of the trend in the current period. Again, see Figure 6.2.

The updated estimate of the seasonal factor is

$$sn_T(T) = \gamma \frac{y_T}{a_0(T)} + (1 - \gamma)sn_T(T - L)$$

where γ is a smoothing constant between 0 and 1.

Here we obtain the updated estimate by smoothing the estimate of the current season's factor computed L periods ago with the current observed seasonal variation. The estimate made L periods ago is used since this was the last time this particular season was observed. The current observed seasonal variation is obtained by dividing the observation y_T by the current estimate of the average level of the time series, $a_0(T)$.

EXAMPLE 6.4

To smooth the previously obtained initial estimates up to a time origin at the end of the three years of historical Tasty Cola data, we will use these updating equations, and we will arbitrarily set α equal to .2, β equal to .15, and γ equal to .05. Moreover, in order to judge the appropriateness of the choice of values for α, β, and γ, we will calculate one-period-ahead forecasts of the historical Tasty Cola sales as we carry out the smoothing procedure. To begin, note that, by using the previously calculated initial estimates, that is,

$$a_0(0) = 390.40, \quad b_1(0) = 9.57, \quad \text{and} \quad sn_1(0) = .4841$$

a forecast made at time 0 for the value of the time series in period 1 (January of year 1) is

$$\hat{y}_1(0) = [a_0(0) + b_1(0) \cdot 1]sn_1(0)$$

$$= [390.40 + 9.57][.4841]$$

$$= 193.63$$

for which, since the observation in January of year 1 is $y_1 = 189$, we have a forecast error of $y_1 - \hat{y}_1(0) = 189 - 193.63 = -4.63$ (see Table 6.4).

We can now obtain the updated estimates, $a_0(1)$, $b_1(1)$, and $sn_1(1)$, as shown below.

$$a_0(1) = \alpha \frac{y_1}{sn_1(0)} + (1 - \alpha)[a_0(0) + b_1(0)]$$

$$= .2 \left[\frac{189}{.4841} \right] + .8[390.40 + 9.57]$$

$$= 398.06$$

$$b_1(1) = \beta[a_0(1) - a_0(0)] + (1 - \beta)b_1(0)$$

$$= .15[398.06 - 390.40] + .85(9.57)$$

$$= 9.28$$

$$sn_1(1) = \gamma \frac{y_1}{a_0(1)} + (1 - \gamma)sn_1(0)$$

$$= .05 \left[\frac{189}{398.06} \right] + .95(.4841)$$

$$= .4836$$

Using these updated initial estimates, a forecast made in period 1, January of year 1, for period 2, February of year 1, is

$$\hat{y}_2(1) = [a_0(1) + b_1(1) \cdot 1]sn_2(0)$$

$$= [398.06 + 9.28][.5847]$$

$$= 238.17$$

for which, since the observation in February of year 1 is $y_2 = 229$, we have a forecast error of $y_2 - \hat{y}_2(1) = 229 - 238.17 = -9.17$ (see Table 6.4).

Using the estimates just obtained, we can go on to compute the updated estimates, $a_0(2)$, $b_1(2)$, and $sn_2(2)$, as shown below.

$$a_0(2) = \alpha \frac{y_2}{sn_2(0)} + (1 - \alpha)[a_0(1) + b_1(1)]$$

$$= .2 \left[\frac{229}{.5847} \right] + .8[398.06 + 9.28]$$

$$= 404.20$$

TABLE 6.4   One-Period-Ahead Forecasting of the Historical Tasty Cola Sales Time Series Using Winters' Method with
$\alpha = .2,$ $\beta = .15,$ and $\gamma = .05.$

T	y_T	$a_0(T)$	$b_1(T)$	$sn_T(T - 12)$	$sn_T(T)$	$\hat{y}_T(T - 1)$	$y_T - \hat{y}_T(T - 1)$
		$a_0(0) = 390.40$	$b_1(0) = 9.57$				
1	189	398.06	9.28	.4841	.4836	193.63	-4.63
2	229	404.20	8.81	.5847	.5838	238.17	-9.17
3	249	413.11	8.83	.6022	.6022	248.72	.28
4	289	421.18	8.71	.6911	.6909	291.60	-2.60
5	260	432.67	9.13	.5859	.5867	251.88	8.12
6	431	439.94	8.85	.9965	.9957	440.25	-9.25
7	660	447.97	8.73	1.4842	1.4837	666.10	-6.10
8	777	457.16	8.80	1.6927	1.6930	773.05	3.95
9	915	464.87	8.63	1.9869	1.9860	925.82	-10.82
10	613	473.87	8.69	1.2897	1.2899	610.68	2.32
11	485	482.33	8.65	1.0074	1.0073	486.12	-1.12
12	277	485.96	7.90	.5946	.5934	291.94	-14.94
13	244	495.99	8.22	.4836	.4841	238.85	5.15
14	296	504.77	8.30	.5838	.5839	294.35	1.65
15	319	516.40	8.80	.6022	.6030	308.99	10.01
16	370	527.28	9.11	.6909	.6914	362.84	7.16
17	313	535.82	9.03	.5867	.5865	314.68	-1.68
18	556	547.56	9.44	.9957	.9966	542.48	13.52
19	831	557.62	9.53	1.4837	1.4840	826.39	4.61
20	960	567.12	9.52	1.6930	1.6930	960.21	$-.21$
21	1152	577.33	9.63	1.9860	1.9864	1145.21	6.79
22	759	587.25	9.67	1.2899	1.2900	757.12	1.88
23	607	598.06	9.84	1.0073	1.0077	601.28	5.72
24	371	611.37	10.36	.5934	.5940	360.71	10.29
25	298	620.51	10.18	.4841	.4839	300.95	-2.95
26	378	634.02	10.68	.5839	.5845	368.27	9.73
27	373	639.47	9.89	.6030	.6020	388.75	-15.75
28	443	647.64	9.64	.6914	.6910	448.97	-5.97
29	374	653.35	9.05	.5865	.5858	385.51	-11.51
30	660	662.36	9.04	.9966	.9966	660.18	$-.18$
31	1004	672.43	9.20	1.4840	1.4844	996.35	7.65
32	1153	681.51	9.18	1.6930	1.6930	1154.02	-1.02
33	1388	692.30	9.42	1.9864	1.9874	1372.01	15.99
34	904	701.53	9.39	1.2900	1.2900	905.23	-1.23
35	715	710.64	9.35	1.0077	1.0076	716.38	-1.38
36	441	724.47	10.02	.5940	.5948	427.71	13.29

$$\Sigma_{T=1}^{36}(y_T - \hat{y}_T(T - 1)^2) = 2254.3228$$

$$b_1(2) = \beta[a_0(2) - a_0(1)] + (1 - \beta)b_1(1)$$

$$= .15[404.20 - 398.06] + .85(9.28)$$

$$= 8.81$$

$$sn_2(2) = \gamma \frac{y_2}{a_0(2)} + (1 - \gamma)sn_2(0)$$

$$= .05\left[\frac{229}{404.20}\right] + (.95)(.5847)$$

$$= .5838$$

This procedure is continued through the entire three years (36 periods) of historical data. The results are summarized in Table 6.4. In this table, $sn_T(T - L)$ is considered to be obtained before y_T is observed and is used to find the forecast $\hat{y}_T(T - 1)$ of y_T. For $T = 1, 2, \ldots, 12$, we have

$$sn_T(T - L) = sn_T(0)$$

When y_T is observed, $sn_T(T - L)$ is updated to $sn_T(T)$.

Using the results in Table 6.4, we find that, for $\alpha = .2$, $\beta = .15$, and $\gamma = .05$, the sum of the squared forecast errors is

$$\sum_{T=1}^{36} (y_T - \hat{y}_T(T - 1))^2 = 2254.3228$$

To find a good combination of α, β, and γ, we evaluate the sum of the squared forecast errors for each combination in a set of 125 combinations formed by arbitrarily ranging α, β, and γ from .05 to .25 in increments of .05. The combination of α, β, and γ that minimizes the sum of the squared forecast errors is $\alpha = .2$, $\beta = .15$, and $\gamma = .05$. This is the combination for which we have demonstrated the relevant calculations and which we will use to forecast future values of the time series.

When updated estimates $a_0(T)$, $b_1(T)$, and $sn_T(T)$ are obtained, forecasts of future time series values are generated using the following equation:

A *point forecast* made at time T for $y_{T+\tau}$ is

$$\hat{y}_{T+\tau}(T) = [a_0(T) + b_1(T)\tau]sn_{T+\tau}(T + \tau - L)$$

Note that in order to forecast for period $T + \tau$, we use the estimate of the seasonal factor computed in period $T + \tau - L$, which is denoted $sn_{T+\tau}(T + \tau - L)$. However, if a forecast is to be made for a period more than L periods in the future, the index $T + \tau - L$ refers to a seasonal factor that has not yet been computed. In such a case the latest estimate of the appropriate seasonal factor should be used in calculating the forecast. Furthermore, although there is not a theoretical, statistical basis behind Winters' Method, it can be shown that

An *approximate* $100(1 - \alpha)\%$ *prediction interval* made at time T for $y_{T+\tau}$ is

$$[\hat{y}_{T+\tau}(T) - B_{T+\tau}^{[100(1-\alpha)]}(T), \; \hat{y}_{T+\tau}(T) + B_{T+\tau}^{[100(1-\alpha)]}(T)]$$

where

$$\hat{y}_{T+\tau}(T) = [a_0(T) + b_1(T)\tau]sn_{T+\tau}(T + \tau - L) \quad \text{and}$$

$$B_{T+\tau}^{[100(1-\alpha)]}(T) = z_{\alpha/2}d_\tau\Delta_W(T)$$

Here $z_{\alpha/2}$ is as previously defined; d_τ is a constant (depending only on τ and not upon T) which is given by the formula

$$d_\tau = 1.25 \left[\frac{1 + \dfrac{\theta}{(1+v)^3}[(1 + 4v + 5v^2) + 2\theta(1 + 3v)\tau + 2\theta^2\tau^2]}{1 + \dfrac{\theta}{(1+v)^3}[(1 + 4v + 5v^2) + 2\theta(1 + 3v) + 2\theta^2]} \right]^{1/2}$$

where θ equals the maximum of α, β, and γ, the smoothing constants employed by Winters' Method, and where $v = 1 - \theta$; and

$$\Delta_W(T) = \frac{\displaystyle\sum_{t=1}^{T} \left| \dfrac{y_t}{sn_t(t - L)} - (a_0(t - 1) + b_1(t - 1)) \right|}{T}$$

where $sn_t(t - L)$ is the latest estimate of the appropriate seasonal factor for period t, an estimate obtained before y_t is observed. Since $\Delta_W(T)$ can be easily updated to $\Delta_W(T + 1)$ using the formula

$$\Delta_W(T + 1) = \frac{T\Delta_W(T) + \left| y_{T+1}/(sn_{T+1}(T + 1 - L)) - (a_0(T) + b_1(T)) \right|}{T + 1}$$

$B_{T+\tau}^{[100(1-\alpha)]}(T)$ can be easily updated to $B_{T+1+\tau}^{[100(1-\alpha)]}(T+1)$ using the formula

$$B_{T+1+\tau}^{[100(1-\alpha)]}(T+1) = z_{\alpha/2}d_\tau\Delta_w(T+1)$$

Hence, we can update the previous approximate $100(1-\alpha)\%$ prediction interval, which is made at time T for $y_{T+\tau}$, to the following approximate $100(1-\alpha)\%$ prediction interval, which is made at time $T+1$ for $y_{T+1+\tau}$.

$$[\hat{y}_{T+1+\tau}(T+1) - B_{T+1+\tau}^{[100(1-\alpha)]}(T+1), \quad \hat{y}_{T+1+\tau}(T+1) + B_{T+1+\tau}^{[100(1-\alpha)]}(T+1)]$$

Here

$$\hat{y}_{T+1+\tau}(T+1) = [a_0(T+1) + b_1(T+1)\tau]sn_{T+1+\tau}(T+1+\tau-L)$$

The intuitive reasoning behind this method will be discussed later.

EXAMPLE 6.5

Consider calculating approximate 95% one-period-ahead and three-period-ahead prediction intervals for Tasty Cola sales in months 37 through 48. Table A1 (in Appendix A) tells us that $z_{[\alpha/2]} = z_{[.025]} = 1.96$. Since $\theta = .2$ is the maximum of the smoothing constants used in the final forecasting of Tasty Cola sales, we have $v = 1 - \theta = .8$. Hence, for generating one-period-ahead forecasts ($\tau = 1$) we have

$$d_1 = 1.25$$

and

$$B_{T+1}^{[95]}(T) = z_{[.025]}d_1\Delta_w(T) = 1.96(1.25)\Delta_w(T) = 2.45\Delta_w(T)$$

For generating three-period-ahead forecasts ($\tau = 3$) we have

$$d_3 = 1.25\left[\frac{1 + \dfrac{.2}{(1.8)^3}\{[1 + 4(.8) + 5(.8)^2] + 2(.2)[1 + 3(.8)](3) + 2(.2)^2(3)^2\}}{1 + \dfrac{.2}{(1.8)^3}\{[1 + 4(.8) + 5(.8)^2] + 2(.2)[1 + 3(.8)] + 2(.2)^2\}}\right]^{1/2}$$

$$= 1.3041$$

and

$$B_{T+3}^{[95]}(T) = z_{[.025]}d_3\Delta_w(T) = 1.96(1.3041)\Delta_w(T) = 2.556036\Delta_w(T)$$

Since we will begin the generation of prediction intervals from a time origin at

$T = 36$, we now refer to Table 6.4 and compute

$$\Delta_W(36) = \frac{\sum_{t=1}^{36} \left| \dfrac{y_t}{sn_t(t-12)} - (a_0(t-1) + b_1(t-1)) \right|}{36}$$

$$= \frac{1}{36}\left[\left| \frac{189}{.4841} - (390.40 + 9.57) \right| \right.$$

$$+ \left| \frac{229}{.5847} - (398.06 + 9.28) \right| + \cdots$$

$$+ \left. \left| \frac{441}{.5940} - (710.64 + 9.35) \right| \right]$$

$$= 8.2354$$

Hence,

$$B_{37}^{[95]}(36) = 2.45\Delta_W(36) = 20.1767$$

and

$$B_{39}^{[95]}(36) = 2.556036\Delta_W(36) = 21.05$$

Therefore, noting from Table 6.4 that

$$a_0(36) = 724.47, \qquad b_1(36) = 10.02, \qquad \text{and} \qquad sn_{37}(25) = .4839$$

it follows, since

$$\hat{y}_{37}(36) = [a_0(36) + b_1(36) \cdot 1]sn_{37}(25)$$

$$= [724.47 + 10.02(1)][.4839]$$

$$= 355.39$$

that an approximate 95% prediction interval made at time 36 for y_{37} is

$$[355.39 - 20.1767, \ 355.39 + 20.1767] \qquad \text{or} \qquad [335.21, \ 375.57]$$

And since

$$\hat{y}_{39}(36) = [a_0(36) + b_1(36) \cdot 3]sn_{39}(27)$$

$$= [724.47 + (10.02)3](.6020)$$

$$= 454.2271$$

an approximate 95% prediction interval made at time 36 for y_{39} is

$$[454.2271 - 21.05, \ 454.2271 + 21.05] \qquad \text{or} \qquad [433.18, \ 475.28]$$

Assuming we now observe $y_{37} = 352$, we can obtain updated values $a_0(37)$, $b_1(37)$, $sn_{37}(37)$, and $\Delta_W(37)$ as follows.

$$a_0(37) = \alpha \frac{y_{37}}{sn_{37}(25)} + (1 - \alpha)[a_0(36) + b_1(36)]$$

$$= .2\left[\frac{352}{.4839}\right] + .8[724.47 + 10.02]$$

$$= 733.09$$

$$b_1(37) = \beta[a_0(37) - a_0(36)] + (1 - \beta)b_1(36)$$

$$= .15[733.09 - 724.47] + (.85)(10.02)$$

$$= 9.81$$

$$sn_{37}(37) = \gamma \frac{y_{37}}{a_0(37)} + (1 - \gamma)sn_{37}(25)$$

$$= .05\left[\frac{352}{733.09}\right] + (.95)(.4839)$$

$$= .4837$$

$$\Delta_W(37) = \frac{36\Delta_W(36) + |352/.4839 - (724.47 + 10.02)|}{37} = 8.2038$$

Hence,

$$B_{38}^{[95]}(37) = 2.45\Delta_W(37) = 20.0993$$

and

$$B_{40}^{[95]}(37) = 2.556036\Delta_W(37) = 20.9692$$

Therefore, since

$$\hat{y}_{38}(37) = [a_0(37) + b_1(37) \cdot 1]sn_{38}(26)$$

$$= [733.09 + 9.81][.5845]$$

$$= 434.25$$

an approximate 95% prediction interval made at time 37 for y_{38} is

$$[434.25 - 20.0993, 434.25 + 20.0993] \quad \text{or} \quad [414.15, 454.35]$$

And since

$$\hat{y}_{40}(37) = [a_0(37) + b_1(37) \cdot 3]sn_{40}(28)$$

$$= [733.09 + 9.81(3)][.6910]$$

$$= 526.9013$$

an approximate 95% prediction interval made at time 37 for y_{40} is

$$[526.9013 - 20.9692, 526.9013 + 20.9692] \quad \text{or} \quad [505.93, 547.87]$$

TABLE 6.5 One-Period-Ahead and Three-Period-Ahead Forecasts of Future Values of Tasty Cola Sales Calculated Using Winters' Method with $\alpha = .2$ $\beta = .15$ and $\gamma = .05$

T	y_T	$a_0(T)$	$b_1(T)$	$sn_T(T-12)$	$sn_T(T)$	$\Delta_W(T)$	$B_T^{[95]}(T-1)$	$\hat{y}_T(T-1)$	$[\hat{y}_T(T-1) - B_T^{[95]}(T-1),$ $\hat{y}_T(T-1) + B_T^{[95]}(T-1)]$	$B_T^{[95]}(T-3)$	$\hat{y}_T(T-3)$	$[\hat{y}_T(T-3) - B_T^{[95]}(T-3),$ $\hat{y}_T(T-3) + B_T^{[95]}(T-3)]$
36	762	724.47	10.02			8.2354						
37	352	733.09	9.81	.4839	.4837	8.2038	20.1767	355.39	[335.21, 375.57]			
38	445	746.57	10.36	.5845	.5851	8.4730	20.0993	434.25	[414.15, 454.35]			
39	453	756.04	10.23	.6020	.6019	8.3696	20.7589	455.67	[434.91, 476.43]	21.05	454.2271	[433.18, 475.28]
40	541	769.60	10.73	.6910	.6916	8.5767	20.5055	529.51	[509.00, 550.02]	20.9692	526.9013	[505.93, 547.87]
41	457	780.28	10.72	.5858	.5858	8.3724	21.0130	457.13	[436.12, 478.14]	21.6573	455.5474	[433.89, 477.20]
42	762	785.71	9.93	.9966	.9953	8.8016	20.5124	788.34	[767.83, 808.85]	21.3930	784.0551	[762.66, 805.45]
43	1194	797.38	10.19	1.4844	1.4851	8.7998	21.5639	1181.08	[1159.52, 1202.64]	21.9224	1190.1770	[1168.25, 1212.10]
44	1361	806.84	10.08	1.6930	1.6927	8.6833	21.5596	1367.19	[1345.63, 1388.75]	21.4002	1375.4609	[1354.06, 1396.86]
45	1615	816.06	9.95	1.9874	1.9869	8.5859	21.2740	1623.51	[1602.24, 1644.78]	22.4972	1620.7247	[1598.23, 1643.22]
46	1059	825.00	9.80	1.2900	1.2896	8.5097	21.0355	1065.51	[1044.47, 1086.54]	22.4926	1068.0555	[1045.56, 1090.55]
47	824	831.39	9.29	1.0076	1.0068	8.6906	20.8487	841.15	[820.30, 862.00]	22.1948	843.4418	[821.25, 865.64]
48	495	838.99	9.03	.5948	.5945	8.6860	21.2921	500.02	[478.73, 521.31]	21.9459	503.1473	[481.20, 525.09]

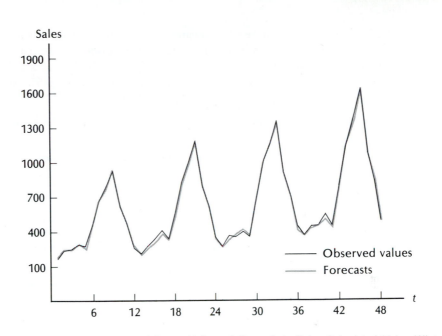

FIGURE 6.3   *One-Period-Ahead Forecasts of Future Values of Tasty Cola Sales Calculated Using Winters' Method with α = .2, β = .15, and γ = .05*

This procedure may be continued for as many periods as desired. Table 6.5 summarizes the one-period-ahead and three-period-ahead point forecasts and prediction interval forecasts of Tasty Cola sales in months 37 through 48. It should be noted that 11 out of 12 of the one-period-ahead prediction intervals and 10 out of 11 of the three-period-ahead prediction intervals contain the actual observed values of the time series. Moreover, the sum of squared one-period-ahead forecast errors is

$$\sum_{T=37}^{48} [y_T - \hat{y}_T(T - 1)]^2 = 1599.4889$$

In Figure 6.3 the observed values of the time series and the one-period-ahead forecasts are plotted.

Recall that we calculated the initial estimates of the permanent, trend, and seasonal components by using all three years of historical Tasty Cola data. Generally, since we start with the initial estimates when making one-period-ahead forecasts to determine a good combination of α, β, and γ, it is best to use only a portion of the historical data to calculate the initial estimates so that part of the historical data remains to tell us how the time series model parameters are changing. As previously discussed, it is reasonable to calculate the initial estimates by using half of the historical data. In the Tasty Cola situation, however, we used all of the historical data to calculate the initial estimates because it was deemed that half of the data (which is one and one-half years of seasonal data) would not be enough to obtain reasonable initial estimates of the seasonal factors.

To complete this subsection, we present two modifications of Winters' Method.

1. *No Trend Winters' Method*
    Winters' Method can also be used to forecast a time series having no trend.

No Trend Winters' Model is

$$y_t = (\beta_0) \times SN_t + \varepsilon_t$$

Using this model, updated estimates are obtained as follows.

The *updated estimate of the permanent component* is

$$a_0(T) = \alpha \frac{y_t}{sn_T(T - L)} + (1 - \alpha)a_0(T - 1)$$

where α is a smoothing constant between 0 and 1.

The *updated estimate of the seasonal factor, $sn_T(T)$, is*

$$sn_T(T) = \gamma \frac{y_T}{a_0(T)} + (1 - \gamma)sn_T(T - L)$$

where γ is a smoothing constant between 0 and 1.

Here the initial estimate of the permanent component, $a_0(0)$, is the average of the values of the observations in $m/2$ years of historical data. The initial estimate of SN_t, $sn_t(0)$, is determined using the method previously discussed for the situation in which there is a linear trend, with the exception that we compute S_t using the formula

---

$$S_t = \frac{y_t}{a_0(0)}$$

Finally,

A *point forecast* made at time T for $y_{T+\tau}$ is

$$\hat{y}_{T+\tau}(T) \;=\; [a_0(T)]sn_{T+\tau}(T + \tau - L)$$

2. *Additive Winters' Method*

Winters' Method—which as described above assumes multiplicative seasonal variation—can be modified to forecast time series exhibiting *additive* (constant) seasonal variation. The appropriate modification consists of obtaining the updating and forecasting equations for additive Winters' Method from the same equations for the multiplicative version by replacing division operations with subtraction operations and multiplication operations with addition operations. Thus

Winters' (Additive) Model is

$$y_t \;=\; (\beta_0 + \beta_1 t) + SN_t + \varepsilon_t$$

Using this model, updated estimates of the model parameters are obtained as follows:

The *updated estimate of the permanent component* is

$$a_0(T + 1) \;=\; \alpha[\,y_{T+1} - sn_{T+1}(T + 1 - L)] + (1 - \alpha)[a_0(T) + b_1(T)]$$

where α is a smoothing constant between 0 and 1.

The *updated estimate of the trend component* is

$$b_1(T + 1) \;=\; \beta[a_0(T + 1) - a_0(T)] + (1 - \beta)b_1(T)$$

where β is a smoothing constant between 0 and 1.

> The *updated estimate of the seasonal factor, $sn_T(T)$, is*
>
> $$sn_{T+1}(T + 1) = \gamma[y_{T+1} - a_0(T + 1)]$$
> $$+ (1 - \gamma)sn_{T+1}(T + 1 - L)$$
>
> where γ is a smoothing constant between 0 and 1.

Moreover, we can obtain initial estimates $a_0(0)$, $b_1(0)$, and $sn_t(0)$ (for $t = 1$, 2, . . . , L) by finding the least squares point estimates of the parameters in the dummy variable regression model (see Section 5.2)

$$y_t = \beta_0 + \beta_1 t + \beta_{s1} x_{s1,t} + \beta_{s2} x_{s2,t} + \cdots + \beta_{s(L-1)} x_{s(L-1),t} + \varepsilon_t$$

where we assume independent error terms.

Finally,

> A point forecast made at time T for $y_{T+\tau}$ is
>
> $$\hat{y}_{T+\tau}(T) = a_0(T) + b_1(T)\tau + sn_{T+\tau}(T + \tau - L)$$

6.4.2 Related Procedures: One- and Two-Parameter Double Exponential Smoothing

Double exponential smoothing is used to forecast a time series described by a linear trend and irregular fluctuations (but having no seasonal variations):

> $$y_t = \beta_0 + \beta_1 t + \varepsilon_t$$

Initial estimates $a_0(0)$ of β_0 and $b_1(0)$ of β_1 are found by fitting a trend line to half of the historical data.

> *Updating the Estimates in One- and Two-Parameter Double Exponential Smoothing*:
>
> 1. In *two-parameter double exponential smoothing*, we update the permanent and trend components by using the equations
>
> $$a_0(T) = \alpha y_T + (1 - \alpha)[a_0(T - 1) + b_1(T - 1)]$$

and

$$b_1(T) = \beta[a_0(T) - a_0(T - 1)] + (1 - \beta)b_1(T - 1)$$

These equations are the Winters' Method updating equations with the seasonal factors eliminated.

2. In *one-parameter double exponential smoothing*, we use these updating equations but we let

$$\alpha = 1 - w^2$$

and

$$\beta = 1 - \frac{2w}{(1 + w)}$$

where $w = 1 - \delta$, and δ is a smoothing constant that is chosen to be between 0 and 1.

The formulas for point forecasts and prediction interval forecasts of future observations are the same formulas given for Winters' Method, with the seasonal factors eliminated. Therefore,

A *point forecast* and a $100(1 - \alpha)\%$ *prediction interval forecast* made at time T for $y_{T+\tau}$ are, respectively,

$$\hat{y}_{T+\tau}(T) = a_0(T) + b_1(T)\tau$$

and

$$[\hat{y}_{T+\tau}(T) - B_{T+\tau}^{[100(1-\alpha)]}(T), \hat{y}_{T+\tau}(T) + B_{T+\tau}^{[100(1-\alpha)]}(T)]$$

Here

$$B_{T+\tau}^{[100(1-\alpha)]} = z_{[\alpha/2]}d_\tau\Delta(T)$$

$$d_\tau = 1.25\left[\frac{1 + \dfrac{\theta}{(1 + v)^3}[(1 + 4v + 5v^2) + 2\theta(1 + 3v)\tau + 2\theta^2\tau^2]}{1 + \dfrac{\theta}{(1 + v)^3}[(1 + 4v + 5v^2) + 2\theta(1 + 3v) + 2\theta^2]}\right]^{1/2}$$

$$\Delta(T) = \frac{\sum\limits_{t=1}^{T}|y_t - [a_0(t - 1) + b_1(t - 1)]|}{T}$$

Moreover,

$\Delta(T)$ may be updated to $\Delta(T + 1)$ by using

$$\Delta(T + 1) = \frac{T\Delta(T) + |y_{T+1} - [a_0(T) + b_1(T)]|}{T + 1}$$

In two-parameter double exponential smoothing, θ equals the maximum of α and β, and $\nu = 1 - \theta$. In one-parameter double exponential smoothing, θ is the smoothing constant δ defined for one-parameter double exponential smoothing, and $\nu = 1 - \theta$. It can be proven that the resulting prediction interval formula for one-parameter double exponential smoothing is theoretically correct. Therefore, the prediction interval formula for two-parameter double exponential smoothing may be regarded as a slight, "intuitive" modification of the theoretically correct formula, and the prediction interval formula for Winters' Method may be regarded as a fairly substantial, intuitive modification of the theoretically correct formula. We obtain the Winters' Method prediction interval formula by modifying the double exponential smoothing prediction interval formula because Winters' Method updates $a_0(T)$ and $b_1(T)$ by applying double exponential smoothing to the deseasonalized data.

In the next section we will show how to implement simple and double exponential smoothing and a special version of Winters' Method by using the Box-Jenkins methodology.

6.5 *RELATIONSHIPS BETWEEN EXPONENTIAL SMOOTHING, THE BOX-JENKINS METHODOLOGY, AND TIME SERIES REGRESSION*

We will begin this section by discussing the practical aspects of several relationships between exponential smoothing and the Box-Jenkins methodology. The relationships are based on a paper by McKenzie [1984], who summarizes and gives excellent references concerning the research on these relationships.

We begin with an example illustrating the following fact.

It can be proven that *forecasting with simple exponential smoothing is equivalent to forecasting with the Box-Jenkins model*

$$z_t = (1 - \theta_1 B)a_t$$

where

$$z_t = (1 - B)y_t = y_t - y_{t-1}$$

It follows that we can implement simple exponential smoothing by using SAS to implement this Box-Jenkins model.

EXAMPLE 6.6

Figure 6.4 presents the SAS output of the SAC and SPAC of the cod catch values (see Table 6.1) obtained by using the transformation

$$z_t \ = \ (1 \ - \ B)y_t \ = \ y_t \ - \ y_{t-1}$$

At the nonseasonal level, the SAC has a spike at lag 1 and cuts off after lag 1, and the SPAC dies down. It follows, by Guideline 1 in Table 2.3, that we should use the nonseasonal moving average operator of order 1

$$\theta_1(B) \ = \ (1 \ - \ \theta_1 B)$$

By inserting this operator into the general model

$$\phi_p(B)\phi_P(B^L)z_t \ = \ \delta \ + \ \theta_q(B)\theta_Q(B^L)a_t$$

we are led to consider the model

$$z_t \ = \ (1 \ - \ \theta_1 B)a_t$$

where

$$z_t \ = \ y_t \ - \ y_{t-1}$$

This model is the Box-Jenkins model equivalent to simple exponential smoothing. Figure 6.5 presents the SAS output of estimation, diagnostic checking, and forecasting for the model. Note that whereas the point and 95% prediction interval forecasts of y_{25} given by the Box-Jenkins model are (see Figure 6.5d)

$$\hat{y}_{25} \ = \ 348.9471 \quad \text{and} \quad [264.0193, \ 433.8748]$$

the point and 95% prediction interval forecasts of y_{25} given by the simple exponential smoothing procedure carried out in Examples 6.2 and 6.3 are

$$\hat{y}_{25} \ = \ 356.13 \quad \text{and} \quad [286.01, \ 426.25]$$

In the next example we will illustrate the following fact.

It can be proven that *forecasting with either one- or two-parameter double exponential smoothing is essentially equivalent to forecasting with the Box-Jenkins model*

$$z_t \ = \ (1 \ - \ \theta_1 B)(1 \ - \ \theta_{(1)} B)a_t$$

where

$$z_t \ = \ (1 \ - \ B)^2 y_t \ = \ y_t \ - \ 2y_{t-1} \ + \ y_{t-2}$$

FIGURE 6.4 *SAS Output of the SAC and SPAC of the Cod Catch Values Obtained by Using the Transformation* $z_t = y_t - y_{t-1}$

(a) The SAC

NAME OF VARIABLE   = Y
PERIODS OF DIFFERENCING = 1.

MEAN OF WORKING SERIES = 0.130435
STANDARD DEVIATION   = 57.5985
NUMBER OF OBSERVATIONS = 23

AUTOCORRELATIONS

LAG	COVARIANCE	CORRELATION	STD
0	3317.59	1.00000	0
1	−1430.88	−0.43130	0.208514
2	2.91724	0.00088	0.244242
3	−815.448	−0.24580	0.244242
4	876.61	0.26423	0.25477
5	−502.37	−0.15143	0.266418
6	675.428	0.20359	0.270134
7	−313.725	−0.09456	0.276725
8	−770.754	−0.23232	0.278127
9	254.88	0.07683	0.28644
10	1007.83	0.30378	0.287334
11	−327.454	−0.09870	0.300975
12	−502.128	−0.15135	0.302379
13	231.068	0.06965	0.305655
14	−229.766	−0.06926	0.306344
15	296.042	0.08923	0.307024
16	−98.6394	−0.02973	0.30815
17	46.2067	0.01393	0.308274
18	−249.386	−0.07517	0.308302
19	226.852	0.06838	0.309098
20	103.518	0.03120	0.309755
21	−181.326	−0.05466	0.309891
22	41.7342	0.01258	0.31031

'.' MARKS TWO STANDARD ERRORS

FIGURE 6.4 Continued

(b) The SPAC

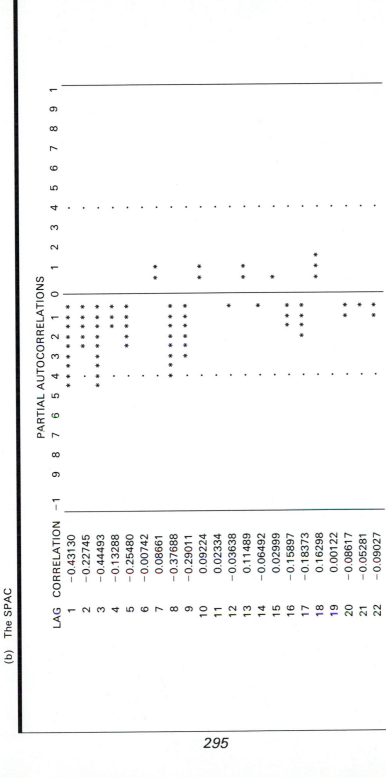

FIGURE 6.5 SAS Output of Estimation, Diagnostic Checking, and Forecasting for the Model $z_t = (1 - \theta_1 B)a_t$ where $z_t = y_t - y_{t-1}$

(a) Estimation

ARIMA: PRELIMINARY ESTIMATION

MOVING AVERAGE ESTIMATES

1   0.43130

WHITE NOISE VARIANCE EST = 2797.25

ARIMA: LEAST SQUARES ESTIMATION

ITERATION	SSE	MA1,1	LAMBDA
0	56377.5	0.431301	1.0E−05
1	45979.1	0.771636	1.0E−06
2	42000.4	0.979529	0.01
3	41335.8	0.953298	1.0E−03
4	41311.7	0.961245	0.1
5	41307.8	0.958443	0.1
6	41307.3	0.959361	0.01

PARAMETER	ESTIMATE	STD ERROR	T RATIO	LAG
MA1,1	0.959361	0.0709816	13.52	1

VARIANCE ESTIMATE   = 1877.61
STD ERROR ESTIMATE   = 43.3314
NUMBER OF RESIDUALS = 23

CORRELATIONS OF THE ESTIMATES

	MA1,1
MA1,1	1.000

(b) The RSAC

AUTOCORRELATION CHECK OF RESIDUALS

TO LAG	CHI SQUARE	DF	PROB	AUTOCORRELATIONS					
6	2.25	5	0.814	-0.202	-0.077	0.030	0.136	-0.032	0.094
12	11.58	11	0.396	0.116	-0.259	-0.156	0.309	-0.011	-0.146
18	12.34	17	0.779	0.071	-0.054	-0.010	-0.009	-0.005	-0.043

AUTOCORRELATION PLOT OF RESIDUALS

LAG	COVARIANCE	CORRELATION	-1 ⋯ 0 ⋯ 1	STD
0	1875.4	1.00000	|********************|	0
1	56.1789	0.02996	*	0.208514
2	-144.207	-0.07689	**	0.208701
3	-378.479	-0.20181	****	0.20993
4	255.581	0.13628	***	0.218202
5	-59.6795	-0.03182	*	0.221872
6	175.984	0.09384	**	0.22207
7	-292.676	-0.15606	***	0.223787
8	-485.13	-0.25868	*****	0.22847
9	218.199	0.11635	**	0.240868
10	578.799	0.30863	******	0.243299
11	-20.6764	-0.01103	*	0.259763
12	-273.688	-0.14594	***	0.259784
13	-18.076	-0.00964	*	0.263324
14	-100.879	-0.05379	*	0.263339
15	132.307	0.07055	*	0.263817
16	-17.1559	-0.00915	*	0.264636
17	-8.66357	-0.00462	*	0.264649
18	-80.1975	-0.04276	*	0.264653

'.' MARKS TWO STANDARD ERRORS

(Parts (c), the RSPAC, and (d), the Forecasts, follow on pages 298 and 299.)

297

FIGURE 6.5 Continued

(c) The RSPAC

PARTIAL AUTOCORRELATIONS

LAG	CORRELATION
1	0.02996
2	−0.07786
3	−0.19838
4	0.14689
5	−0.07525
6	0.08445
7	−0.12615
8	−0.29649
9	0.20716
10	0.20904
11	−0.10613
12	−0.01337
13	0.03388
14	−0.13111
15	0.04445
16	−0.09344
17	0.10862
18	0.17001

(d) Forecasts of y_{25} through y_{36}

FORECASTS FOR VARIABLE Y

OBS	FORECAST	STD ERROR	LOWER 95%	UPPER 95%
- - - - - - -	- - - - - - FORECAST BEGINS - - - - -		- -	
25	348.9471	43.3314	264.0193	433.8748
26	348.9471	43.3671	263.9492	433.9449
27	348.9471	43.4029	263.8792	434.0149
28	348.9471	43.4386	263.8092	434.0849
29	348.9471	43.4742	263.7393	434.1548
30	348.9471	43.5099	263.6694	434.2247
31	348.9471	43.5455	263.5996	434.2945
32	348.9471	43.5811	263.5298	434.3643
33	348.9471	43.6167	263.4601	434.4340
34	348.9471	43.6522	263.3905	434.5036
35	348.9471	43.6877	263.3209	434.5732
36	348.9471	43.7232	263.2514	434.6427

299

It follows that we can implement double exponential smoothing by using SAS to implement this Box-Jenkins model.

EXAMPLE 6.7

We will analyze a time series of $n = 52$ weekly thermostat sales observations originally analyzed using one-parameter double exponential smoothing by R. G. Brown (1962), who initially developed exponential smoothing. The sales observations are presented in Table 6.6 and are plotted in Figure 6.6. Figure 6.7 presents the SAC and SPAC of the sales values obtained by using the transformation

$$z_t = (1 - B)^2 y_t = y_t - 2y_{t-1} + y_{t-2}$$

At the nonseasonal level, the SAC has a spike at lag 1 and cuts off after lag 1, and the SPAC dies down. It follows, by Guideline 1 in Table 2.3, that we should use the nonseasonal moving average operator of order 1

$$\theta_1(B) = (1 - \theta_1 B)$$

By inserting this operator into the general model

$$\phi_p(B)\phi_P(B^L)z_t = \delta + \theta_q(B)\theta_Q(B^L)a_t$$

we are led to consider the model

$$z_t = (1 - \theta_1 B)a_t$$
where
$$z_t = y_t - 2y_{t-1} + y_{t-2}$$

TABLE 6.6   *Weekly Thermostat Sales[a]*

206	189	172	255
245	244	210	303
185	209	205	282
169	207	244	291
162	211	218	280
177	210	182	255
207	173	206	312
216	194	211	296
193	234	273	307
230	156	248	281
212	206	262	308
192	188	258	280
162	162	233	345

[a] Read downwards, left to right.

Source: Reprinted from R. G. Brown (1962), *Smoothing, Forecasting and Prediction of Discrete Time Series*, p. 431 by permission of Prentice-Hall, Inc. Suggested by an example in Abraham and Ledolter (1983).

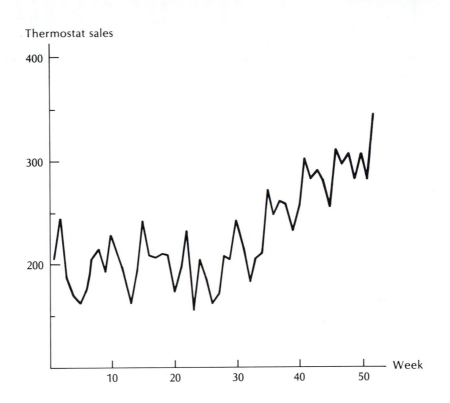

FIGURE 6.6   *Plot of Weekly Thermostat Sales*

This model is not the Box-Jenkins model equivalent to double exponential smoothing (although it is similar). However, the plot of the weekly sales values in Figure 6.6 indicates that this time series is described by the linear trend model

$$y_t = \beta_0 + \beta_1 t + \varepsilon_t$$

where the trend $\beta_0 + \beta_1 t$ is changing over time (for the first half of the data, the trend is slightly downward; for the last half, the trend is upward). Since double exponential smoothing is the appropriate forecasting technique for such a changing linear trend model, we will (arbitrarily) forecast the thermostat sales time series by using the Box-Jenkins model that is essentially equivalent to double exponential smoothing:

$$z_t = (1 - \theta_1 B)(1 - \theta_{(1)} B) a_t$$

where

$$z_t = y_t - 2y_{t-1} + y_{t-2}$$

Figure 6.8 presents the SAS output of estimation, diagnostic checking, and forecasting for the model.

FIGURE 6.7 SAS Output of the SAC and SPAC of the Thermostat Sales Values Obtained by Using the Transformation $z_t = y_t - 2y_{t-1} + y_{t-2}$

(a) The SAC

```
                          NAME OF VARIABLE           = Y
                          PERIODS OF DIFFERENCING    = 1, 1.

                          MEAN OF WORKING SERIES     = 0.52
                          STANDARD DEVIATION         = 52.733
                          NUMBER OF OBSERVATIONS     = 50

                                        AUTOCORRELATIONS
LAG COVARIANCE CORRELATION -1 9 8 7 6 5 4 3 2 1 0 1 2 3 4 5 6 7 8 9 1       STD
  0    2870.77     1.00000   |                    |********************|         0
  1   -1578.38    -0.56760   |         ***********|     .              |  0.141421
  2    195.342     0.07025   |             .      |*     .             |  0.181348
  3    149.681     0.05383   |             .      |*     .             |  0.181891
  4    -179.92    -0.06470   |             .     *|      .             |  0.182209
  5   -162.149    -0.05831   |             .     *|      .             |  0.182668
  6    309.628     0.11135   |             .      |**    .             |   0.18304
  7    71.5029     0.02571   |             .      |*     .             |   0.18439
  8   -229.564    -0.08255   |             .    **|      .             |  0.184462
  9    307.851     0.11071   |             .      |**    .             |  0.185199
 10   -465.144    -0.16727   |             .   ***|      .             |  0.186518
 11    438.989     0.15787   |            .       |***    .            |  0.189494
 12   -604.733    -0.21747   |            .   ****|       .            |  0.192107
 13    733.582     0.26381   |            .       |*****  .            |  0.196969
 14   -332.698    -0.11964   |            .     **|       .            |  0.203913
 15   -60.2475    -0.02167   |            .       |       .            |  0.205312
 16    114.895     0.04132   |            .       |*      .            |  0.205358
 17   -40.6671    -0.01462   |            .       |       .            |  0.205524
 18   -142.434    -0.05122   |            .      *|       .            |  0.205545
 19    -10.414    -0.00374   |            .       |       .            |    0.2058
 20    610.942     0.21970   |            .       |****   .            |  0.205801
 21   -831.874    -0.29915   |            . ******|       .            |   0.21044
 22    559.195     0.20109   |           .        |****    .           |   0.21878
 23    -232.42    -0.08358   |           .      **|        .           |  0.222446
 24    34.7046     0.01248   |           .        |        .           |  0.223073
                              '.' MARKS TWO STANDARD ERRORS
```

(b) The SPAC

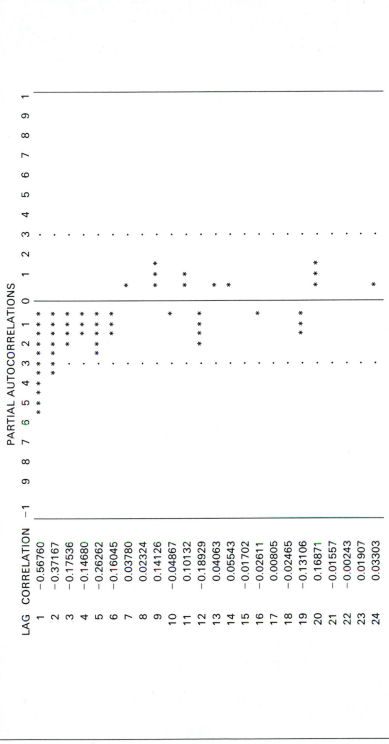

303

FIGURE 6.8 SAS Output of Estimation, Diagnostic Checking, and Forecasting for the Model $z_t = (1 - \theta_1 B)(1 - \theta_{(1)} B)a_t$ where $z_t = y_t - 2y_{t-1} + y_{t-2}$

(a) Estimation

ARIMA: LEAST SQUARES ESTIMATION

ITERATION	SSE	MA1,1	MA2,1	LAMBDA
0	112569	0.1	0.1	1.0E−05
1	72457.4	0.386663	0.386663	1.0E−06
2	62824.9	0.550406	0.550365	1.0E−07
3	61540.2	0.648956	0.56754	1.0E−05
4	61455.6	0.689868	0.540022	0.01
5	61378.4	0.724487	0.498745	0.01
6	61305.6	0.780497	0.421355	1.0E−03
7	61277.8	0.766267	0.420762	1.0E−04
8	61274	0.777439	0.406613	1.0E−05
9	61273.4	0.773714	0.408315	1.0E−06
10	61273.1	0.776011	0.405809	1.0E−07
11	61273.1	0.775391	0.406048	0.01

PARAMETER	ESTIMATE	STD ERROR	T RATIO	LAG
MA1,1	0.775391	0.169851	4.57	1
MA2,1	0.406048	0.24888	1.63	1

VARIANCE ESTIMATE = 1276.52
STD ERROR ESTIMATE = 35.7285
NUMBER OF RESIDUALS = 50

CORRELATIONS OF THE ESTIMATES

	MA1,1	MA2,1
MA1,1	1.000	−0.842
MA2,1	−0.842	1.000

(b) The RSAC

AUTOCORRELATION CHECK OF RESIDUALS

AUTOCORRELATIONS

TO LAG	CHI SQUARE	DF	PROB						
6	4.42	4	0.352	-0.080	-0.040	0.033	-0.107	0.002	0.236
12	7.14	10	0.712	0.022	0.025	0.134	-0.132	0.027	-0.074
18	9.84	16	0.875	-0.031	0.010	0.180	-0.005	-0.013	-0.063
24	14.85	22	0.869	-0.124	0.134	0.074	0.085	-0.069	-0.069

AUTOCORRELATION PLOT OF RESIDUALS

LAG	COVARIANCE	CORRELATION	STD
0	1276.52	1.00000	0
1	42.2676	0.03311	0.141421
2	-50.5089	-0.03957	0.141576
3	-101.538	-0.07954	0.141797
4	-136.974	-0.10730	0.142637
5	2.87471	0.00225	0.144292
6	300.789	0.23563	0.144292
7	170.787	0.13379	0.151793
8	32.0919	0.02514	0.154134
9	27.9243	0.02188	0.154216
10	-169.066	-0.13244	0.154278
11	34.8575	0.02731	0.156535
12	-94.5119	-0.07404	0.15663
13	229.947	0.18014	0.157329
14	12.7298	0.00997	0.161401
15	-39.4836	-0.03093	0.161413
16	-5.88704	-0.00461	0.161532
17	-16.9268	-0.01326	0.161534
18	-80.5891	-0.06313	0.161556
19	94.4829	0.07402	0.162049
20	170.861	0.13385	0.162724
21	-158.387	-0.12408	0.164911
22	108.951	0.08535	0.166768
23	-88.2007	-0.06909	0.167639
24	-88.6482	-0.06945	0.168207

CORRELATION -1   0   1

'.' MARKS TWO STANDARD ERRORS

(Parts (c), the RSPAC, and (d), the Forecasts, follow on pages 306 and 307.)

FIGURE 6.8 Continued

(c) The RSPAC

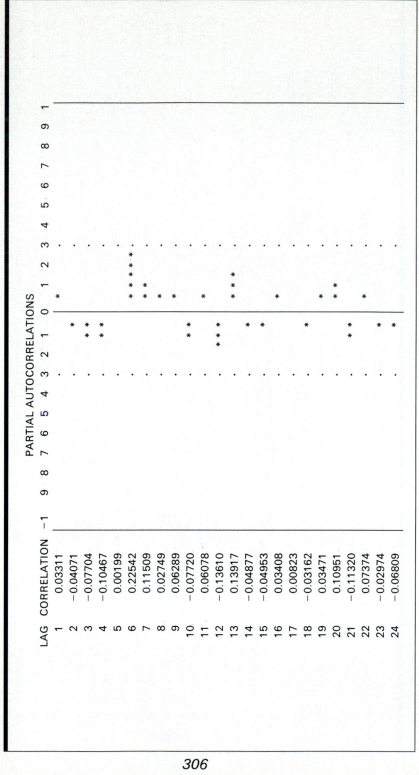

(d) Forecasts of y_{53} through y_{64}

FORECASTS FOR VARIABLE Y

OBS	FORECAST	STD ERROR	LOWER 95%	UPPER 95%
- - - - -	- - - - -	- - - FORECAST BEGINS - - -	- - - - -	
53	334.6917	35.7285	264.6653	404.7180
54	342.2359	46.1719	251.7408	432.7311
55	349.7802	57.3471	237.3823	462.1782
56	357.3245	69.2277	221.6410	493.0081
57	364.8688	81.7842	204.5751	525.1626
58	372.4131	94.9877	186.2410	558.5853
59	379.9574	108.8115	166.6913	593.2236
60	387.5017	123.2312	145.9736	629.0299
61	395.0460	138.2246	124.1314	665.9607
62	402.5903	153.7717	101.2040	703.9767
63	410.1346	169.8542	77.2272	743.0420
64	417.6789	186.4554	52.2338	783.1240

307

Next, we note that (it can be proven that) *forecasting with a special version of the additive Winters' Model* (basically, this version uses only one smoothing constant) *is essentially equivalent to forecasting with the Box-Jenkins model*

$$z_t = (1 - \theta_1 B)(1 - \theta_{1,L} B^L) a_t$$
where
$$z_t = (1 - B)(1 - B^L) y_t$$
$$= y_t - y_{t-1} - y_{t-L} + y_{t-L-1}$$

This model (as we have previously stated) describes many seasonal time series (see Example 3.5 and Exercise 3.1). It can also be proven that *forecasting with a special version of the multiplicative Winters' Model* (basically, this version again uses only one smoothing constant) *is essentially equivalent to forecasting with the Box-Jenkins model*

$$z_t = (1 - \theta_{1,L} B^L)(1 - \theta_{(1,L)} B^L) a_t$$
where
$$z_t = (1 - B^L)^2 y_t = y_t - 2y_{t-L} + y_{t-2L}$$

Although when carrying out traditional Box-Jenkins modeling we very seldom employ second seasonal differencing

$$z_t = (1 - B^L)^2 y_t = y_t - 2y_{t-L} + y_{t-2L}$$

such differencing can be shown to model increasing seasonal variation quite well. Therefore, if a time series exhibits increasing seasonal variation, instead of using the traditional method of initially taking natural logarithms to approximately equalize the seasonal variation, we can directly model the increasing seasonal variation by analyzing the second seasonal differences. Specifically,

1. If the time series seems to exhibit increasing seasonal variation and an essentially *deterministic* trend, we should analyze the values obtained by using the transformation

   $$z_t = (1 - B^L)^2 y_t$$

2. If the time series seems to exhibit increasing seasonal variation and an essentially *stochastic* trend, we should analyze the values obtained by using the transformation

   $$z_t = (1 - B)(1 - B^L)^2 y_t$$

EXAMPLE 6.8

The original hotel room occupancy values plotted in Figure 3.1 exhibit increasing seasonal variation and an essentially deterministic trend. Figure 6.9 presents the SAS output of the SAC and SPAC of the hotel room occupancy values obtained by using the transformation

$$z_t = (1 - B^{12})^2 y_t$$

At the seasonal level, the SPAC has spikes at lags 1, 3, 4, and 5 and cuts off after lag 5, and the SAC dies down. It follows, by using Guideline 2 in Table 2.3, that we will use the following nonseasonal autoregressive operator

$$\phi_5(B) = (1 - \phi_1 B - \phi_3 B^3 - \phi_5 B^5)$$

where we have arbitrarily ignored the spike at lag 4. At the seasonal level, the SAC has a spike at lag 12 and cuts off after lag 12, and the SPAC dies down. This implies, by using Guideline 6 in Table 3.3, that we should use the seasonal moving average operator of order 1

$$\theta_1(B^{12}) = (1 - \theta_{1,12} B^{12})$$

By inserting the appropriate operators into the general model

$$\phi_p(B)\phi_P(B^L)z_t = \delta + \theta_q(B)\theta_Q(B^L)a_t$$

we are led to consider the model

$$(1 - \phi_1 B - \phi_3 B^3 - \phi_5 B^5)z_t = (1 - \theta_{1,12} B^{12})a_t$$

where

$$z_t = (1 - B^{12})^2 y_t$$

However, since the Box-Jenkins model that is essentially equivalent to the (special version of) multiplicative Winters' Model

$$z_t = (1 - \theta_{1,12} B^{12})(1 - \theta_{(1,12)} B^{12})a_t$$

where

$$z_t = (1 - B^{12})^2 y_t$$

uses two seasonal moving average operators, we will add another seasonal moving average operator to the above model to form the model

Model 1: $(1 - \phi_1 B - \phi_3 B^3 - \phi_5 B^5)z_t = (1 - \theta_{1,12} B^{12})(1 - \theta_{(1,12)} B^{12})a_t$

where

$$z_t = (1 - B^{12})^2 y_t$$

Figure 6.10 presents the SAS output of estimation, diagnostic checking, and forecasting for this model. The small *t*-statistics associated with $\hat{\theta}_{1,12}$ and $\hat{\theta}_{(1,12)}$ indicate that it might be wise to remove one of the two seasonal moving average operators from the model. However, since it can be verified that doing this produces a very inadequate model, we will forecast using Model 1. Recalling that the actual future hotel room averages for periods 169 through 180 are given in

FIGURE 6.9 SAS Output of the SAC and SPAC of the Hotel Room Occupancy Values Obtained by Using the Transformation $z_t = (1 - B^{12})^2 y_t$
(a) The SAC

NAME OF VARIABLE   = Y
PERIODS OF DIFFERENCING   = 12, 12.

MEAN OF WORKING SERIES   = 0.131944
STANDARD DEVIATION   = 27.8979
NUMBER OF OBSERVATIONS   = 144

AUTOCORRELATIONS

LAG	COVARIANCE	CORRELATION	STD
0	778.295	1.00000	0
1	144.674	0.18589	0.0833333
2	-17.9098	-0.02301	0.0861647
3	-216.054	-0.27760	0.0862074
4	-194.669	-0.25012	0.0922063
5	-204.828	-0.26318	0.0968034
6	68.5964	0.08814	0.101651
7	90.3886	0.11614	0.10218
8	149.293	0.19182	0.103093
9	58.8262	0.07558	0.105542
10	74.8508	0.09617	0.105917
11	-88.9447	-0.11428	0.106522
12	-401.227	-0.51552	0.10737
13	-139.47	-0.17920	0.123367
14	7.45105	0.00957	0.125162
15	80.8061	0.10382	0.125167
16	144.249	0.18534	0.125763
17	148.662	0.19101	0.127646
18	-93.8348	-0.12056	0.129616
19	22.8011	0.02930	0.130392
20	-25.2305	-0.03242	0.130438
21	14.1959	0.01824	0.130494
22	-100.862	-0.12959	0.130512
23	-27.3484	-0.03514	0.131402
24	4.93918	0.00635	0.131467

'.' MARKS TWO STANDARD ERRORS

(b) The SPAC

PARTIAL AUTOCORRELATIONS

LAG	CORRELATION
1	0.18589
2	−0.05963
3	−0.27233
4	−0.16781
5	−0.23766
6	0.08627
7	−0.02648
8	0.03476
9	0.00366
10	0.10853
11	−0.03390
12	−0.53264
13	0.03081
14	−0.00460
15	−0.16844
16	−0.09052
17	−0.08115
18	−0.12605
19	0.15428
20	0.04153
21	−0.03735
22	−0.04212
23	−0.09827
24	−0.29963

FIGURE 6.10 SAS Output of Estimation, Diagnostic Checking, and Forecasting for the Model $(1 - \phi_1 B - \phi_3 B^3 - \phi_5 B^5)z_t = (1 - \theta_{1,12}B^{12})(1 - \theta_{(1,12)}B^{12})a_t$ where $z_t = (1 - B^{12})^2 y_t$

(a) Estimation

ARIMA: LEAST SQUARES ESTIMATION

ITERATION	SSE	MA1,1	MA2,1	AR1,1	AR1,2	AR1,3	LAMBDA
0	101301	0.1	0.1	0.1	0.1	0.1	1.0E−05
1	49667.6	0.357607	0.357607	.0718197	−.216278	−.190365	1.0E−06
2	32197.1	0.653842	0.653926	.0607174	−.306324	−.247219	1.0E−07
3	30617.7	0.736187	0.73625	0.09241	−.294223	−.253669	0.1
4	30332.8	0.719012	0.718952	0.124795	−.270244	−.241221	0.01
5	30327.6	0.721069	0.722978	0.131542	−.269553	−.244109	1.0E−03
6	30327.3	0.722632	0.718764	0.131741	−.268379	−.243996	0.01
7	30327	0.722351	0.719698	0.131808	−0.26846	−.244151	0.1

PARAMETER	ESTIMATE	STD ERROR	T RATIO	LAG
MA1,1	0.722351	3.90648	0.18	12
MA2,1	0.719698	3.97294	0.18	12
AR1,1	0.131808	0.0789276	1.67	1
AR1,2	−0.26846	0.077137	−3.48	3
AR1,3	−0.244151	0.0783992	−3.11	5

VARIANCE ESTIMATE = 218.18
STD ERROR ESTIMATE = 14.7709
NUMBER OF RESIDUALS = 144

CORRELATIONS OF THE ESTIMATES

	MA1,1	MA2,1	AR1,1	AR1,2	AR1,3
MA1,1	1.000	−1.000	−0.056	−0.000	0.038
MA2,1	−1.000	1.000	0.055	0.001	−0.038
AR1,1	−0.056	0.055	1.000	−0.096	0.200
AR1,2	−0.000	0.001	−0.096	1.000	−0.092
AR1,3	0.038	−0.038	0.200	−0.092	1.000

(b) The RSAC

AUTOCORRELATION CHECK OF RESIDUALS

AUTOCORRELATIONS

TO LAG	CHI SQUARE	DF	PROB						
6	3.02	1	0.082	0.013	-0.030	-0.000	-0.127	-0.043	-0.031
12	8.80	7	0.267	-0.037	-0.134	-0.074	0.038	0.066	-0.081
18	16.90	13	0.204	0.068	0.010	0.052	0.043	0.021	-0.198
24	22.38	19	0.266	0.022	0.077	0.013	-0.047	-0.008	-0.150

AUTOCORRELATION PLOT OF RESIDUALS

LAG	COVARIANCE	CORRELATION	-1 0 1	STD
0	208.757	1.00000		0
1	-0.0790936	-0.00038		0.0833333
2	2.75703	0.01321		0.0833333
3	-6.26964	-0.03003		0.0833479
4	-26.611	-0.12747		0.083423
5	-8.8826	-0.04255		0.0847648
6	-6.47368	-0.03101		0.084913
7	-15.4394	-0.07396		0.0849917
8	-7.7574	-0.03716		0.0854374
9	-27.9328	-0.13381		0.0855496
10	7.84019	0.03756		0.0869908
11	13.6834	0.06555		0.0871033
12	-16.8574	-0.08075		0.0874452
13	10.8667	0.05205		0.0879615
14	14.1549	0.06781		0.0881751
15	2.18697	0.01048		0.0885365
16	9.0473	0.04334		0.0885451
17	4.28647	0.02053		0.0886923
18	-41.272	-0.19770		0.0887253
19	2.66113	0.01275		0.0917336
20	4.49624	0.02154		0.0917459
21	16.0788	0.07702		0.091781
22	-9.85187	-0.04719		0.0922287
23	-1.70196	-0.00815		0.0923963
24	-31.3944	-0.15039		0.0924013

'.' MARKS TWO STANDARD ERRORS

(Parts (c), the RSPAC, and (d), the Forecasts, follow on pages 314 and 315.)

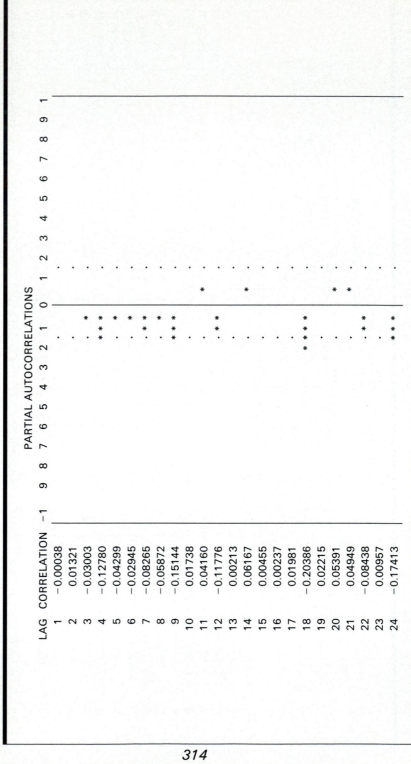

FIGURE 6.10 Continued

(c) The RSPAC

(d) Forecasts of y_{169} through y_{180}

FORECASTS FOR VARIABLE Y

OBS	FORECAST	STD ERROR	LOWER 95%	UPPER 95%
	- - - - - - - FORECAST BEGINS - - - - - - -			
169	834.2862	15.1981	804.4986	864.0738
170	758.9667	15.3997	728.7840	789.1494
171	760.8775	15.5011	730.4959	791.2591
172	853.9142	16.0627	822.4319	885.3965
173	844.5216	16.1153	812.9362	876.1070
174	969.4843	16.1557	937.8199	1001.1488
175	1144.6892	16.1869	1112.9635	1176.4149
176	1170.7745	16.1936	1139.0357	1202.5133
177	890.0234	16.2020	858.2681	921.7786
178	893.5568	16.2030	861.7995	925.3141
179	773.3939	16.2035	741.6356	805.1522
180	884.0261	16.2048	852.2654	915.7868

315

Figure 3.7, it follows that when we compare the forecasts given by Model 1 (see Figure 6.10d) with the forecasts given by

$$\text{Model 2:} \quad (1 - \phi_1 B - \phi_2 B^2 - \phi_3 B^3)z_t \;=\; \delta + (1 - \theta_{1,12} B^{12})a_t$$

where

$$z_t \;=\; (1 - B^{12})y_t^* \qquad \text{and} \qquad y_t^* \;=\; \ln y_t$$

which was the best Box-Jenkins model obtained in Chapter 3, we find that

1. The mean absolute point forecast error (over periods 169 through 180) is 12.4 for Model 1 and 14.7 for Model 2.
2. The 95% prediction intervals given by Model 1 are roughly 10% shorter than the 95% prediction intervals given by Model 2.

Thus Model 1, which uses second seasonal differencing and two seasonal moving average operators (motivated by Winters' Model), is a slight improvement over Model 2, the "traditional" Box-Jenkins model (which employs natural logarithms).

EXAMPLE 6.9

Reconsider Example 3.5. Note that a plot of the monthly passenger totals (see Box and Jenkins [1976]) exhibits increasing seasonal variation and an essentially stochastic trend. Figure 6.11 presents the SAS output of the SAC and SPAC of the passenger totals obtained by using the transformation

$$z_t \;=\; (1 - B)(1 - B^{12})^2 y_t$$

At the nonseasonal level, the SAC has a spike at lag 1 and (with the exception of a spike at lag 3) cuts off after lag 1, and the SPAC dies down. This implies, by using Guideline 1 in Table 2.3, that we should use the nonseasonal moving average operator of order 1

$$\theta_1(B) \;=\; (1 - \theta_1 B)$$

At the seasonal level, the SAC has a spike at lag 12 and cuts off after lag 12, and the SPAC dies down. This implies, by using Guideline 6 in Table 3.3, that we should use the seasonal moving average operator of order 1

$$\theta_1(B^{12}) \;=\; (1 - \theta_{1,12} B^{12})$$

By inserting the appropriate operators into the general model

$$\phi_p(B)\phi_P(B^L)z_t \;=\; \delta + \theta_q(B)\theta_Q(B^L)a_t$$

we are led to consider the model

$$z_t \;=\; (1 - \theta_1 B)(1 - \theta_{1,12} B^{12})a_t$$

where

$$z_t \;=\; (1 - B)(1 - B^{12})^2 y_t$$

However, since the Box-Jenkins model that is essentially equivalent to (a special version of) the multiplicative Winters' Model

$$z_t = (1 - \theta_{1,12}B^{12})(1 - \theta_{(1,12)}B^{12})a_t$$

where

$$z_t = (1 - B^{12})^2 y_t$$

uses two seasonal moving average operators, we will add another seasonal moving average operator to our previous model to form the model

$$\text{Model 1:} \quad z_t = (1 - \theta_1 B)(1 - \theta_{1,12}B^{12})(1 - \theta_{(1,12)}B^{12})a_t$$

where

$$z_t = (1 - B)(1 - B^{12})^2 y_t$$

Figure 6.12 presents the SAS output of estimation, diagnostic checking, and forecasting for this model. The small t-statistics associated with $\hat{\theta}_{1,12}$ and $\hat{\theta}_{(1,12)}$ indicate that it might be wise to remove one of the two seasonal moving average operators from the model. However, since it can be verified that doing this produces a very inadequate model, we will forecast using Model 1. Noting that the actual future monthly passenger totals for periods 133 through 144 are given in Appendix B (see Table B1), it follows that when we compare the forecasts given by Model 1 (see Figure 6.12d) with the forecasts given by

$$\text{Model 2:} \quad z_t = (1 - \theta_1 B)(1 - \theta_{1,12}B^{12})a_t$$

where

$$z_t = (1 - B)(1 - B^{12})y_t^* \quad \text{and} \quad y_t^* = \ln y_t$$

which is the model obtained by Box and Jenkins [1976] (see Example 3.5 and note that we can obtain the forecasts of the actual future values by exponentiating the forecasts in Table 3.5 of the logged future values), we find that

1. The mean absolute point forecast error (over periods 133 through 144) is 14.7 for Model 1 and 12.4 for Model 2.
2. The 95% prediction intervals given by Model 1 are roughly 40% shorter than those given by Model 2!

Thus Model 1, which uses second seasonal differencing and two seasonal moving average operators (motivated by Winters' Model), yields only slightly less accurate point forecasts and substantially shorter 95% prediction interval forecasts than Model 2, the "traditional" Box-Jenkins model (which employs natural logarithms).

From these discussions and examples we see that we can often implement a particular exponential smoothing procedure by using SAS to implement the equivalent (or somewhat modified) Box-Jenkins model. For simple and double exponential smoothing, this implementation can be done for moderately sized data sets. For Winters' Method (that is, when dealing with seasonal data), larger data sets are usually needed to implement the equivalent (or somewhat modified) Box-Jenkins model. If a data set is not large enough for SAS to obtain least squares point estimates of model parameters,

FIGURE 6.11 SAS Output of the SAC and SPAC of the Monthly Passenger Totals Obtained by Using the Transformation $z_t = (1-B)(1-B^{12})^2 y_t$

(a) The SAC

NAME OF VARIABLE = Y
PERIODS OF DIFFERENCING = 1, 12, 12.

MEAN OF WORKING SERIES = 0.373832
STANDARD DEVIATION = 16.2294
NUMBER OF OBSERVATIONS = 107

AUTOCORRELATIONS

LAG	COVARIANCE	CORRELATION	STD
0	263.393	1.00000	0
1	−79.6423	−0.30237	0.0966736
2	43.904	0.16669	0.105141
3	−55.1519	−0.20939	0.107583
4	8.8835	0.03373	0.111326
5	11.3581	0.04312	0.111422
6	11.9982	0.04555	0.111578
7	−28.6886	−0.10892	0.111751
8	20.0227	0.07602	0.112739
9	32.6369	0.12391	0.113217
10	−17.7083	−0.06723	0.114478
11	20.9423	0.07951	0.114846
12	−132.9	−0.50457	0.115359
13	52.0492	0.19761	0.134412
14	−13.4518	−0.05107	0.1371
15	41.9017	0.15908	0.137278
16	−23.3525	−0.08866	0.13899
17	−0.0611587	−0.00023	0.139517
18	−10.7167	−0.04069	0.139517
19	30.0693	0.11416	0.139628
20	−44.2157	−0.16787	0.140498
21	22.8168	0.08663	0.14236
22	−30.0065	−0.11392	0.142852
23	48.9605	0.18588	0.143698
24	13.7806	0.05232	0.145928

'.' MARKS TWO STANDARD ERRORS

(b) The SPAC

PARTIAL AUTOCORRELATIONS

LAG	CORRELATION
1	-0.30237
2	0.08283
3	-0.15307
4	-0.08477
5	0.07462
6	0.05878
7	-0.12081
8	0.03955
9	0.22864
10	-0.04632
11	0.02534
12	-0.46295
13	-0.09043
14	0.08072
15	0.03167
16	-0.04459
17	-0.00144
18	0.03244
19	0.02615
20	-0.11275
21	0.17750
22	-0.07304
23	0.10767
24	-0.13889

FIGURE 6.12 SAS Output of Estimation, Diagnostic Checking, and Forecasting for the Model $z_t = (1 - \theta_1 B)(1 - \theta_{1,12} B^{12})(1 - \theta_{(1,12)} B^{12})a_t$ where $z_t = (1 - B)(1 - B^{12})^2 a_t$

(a) Estimation

ARIMA: LEAST SQUARES ESTIMATION

ITERATION	SSE	MA1,1	MA2,1	MA3,1	LAMBDA
0	22147.1	0.1	0.1	0.1	1.0E−0.5
1	13457.4	0.297858	0.389035	0.389035	1.0E−0.6
2	10284.5	0.342761	0.623974	0.623922	1.0E−07
3	10240.7	0.344921	0.64057	0.739771	1.0E−05
4	10103.8	0.32989	0.704884	0.649965	0.01
5	10098	0.330523	0.639704	0.704571	0.01
6	10088.9	0.330321	0.644513	0.694037	0.1
7	10085	0.330967	0.650605	0.687622	0.1
8	10082.8	0.331446	0.65551	0.683057	0.1
9	10081.6	0.331725	0.659179	0.679655	0.1
10	10081	0.331881	0.66189	0.677103	0.1
11	10080.6	0.331965	0.663891	0.675192	0.1
12	10080.4	0.332012	0.665369	0.673763	0.1
13	10080.3	0.332037	0.666463	0.672697	0.1
14	10080.2	0.332051	0.667272	0.671902	0.1

PARAMETER	ESTIMATE	STD ERROR	T RATIO	LAG
MA1,1	0.332051	0.0940647	3.53	1
MA2,1	0.667272	3.76368	0.18	12
MA3,1	0.671902	3.6957	0.18	12

VARIANCE ESTIMATE = 96.9252
STD ERROR ESTIMATE = 9.84506
NUMBER OF RESIDUALS = 107

CORRELATIONS OF THE ESTIMATES

	MA1,1	MA2,1	MA3,1
MA1,1	1.000	0.083	−0.084
MA2,1	0.083	1.000	−1.000
MA3,1	−0.084	−1.000	1.000

(b)   The RSAC

AUTOCORRELATION CHECK OF RESIDUALS

TO LAG	CHI SQUARE	DF	PROB	AUTOCORRELATIONS					
6	3.72	3	0.294	−0.024	0.094	−0.060	−0.113	0.022	0.081
12	10.52	9	0.310	0.013	−0.023	0.175	−0.160	−0.008	0.006
18	13.57	15	0.558	−0.042	0.097	−0.018	−0.111	0.000	−0.016
24	22.28	21	0.384	0.005	−0.033	0.023	−0.146	0.176	−0.095

AUTOCORRELATION PLOT OF RESIDUALS

−1   0   1   2   3   4   5   6   7   8   9   1

LAG	COVARIANCE	CORRELATION	STD
0	93.8727	1.00000	0
1	−2.29691	−0.02447	0.0966736
2	8.83921	0.09416	0.0967315
3	−5.67452	−0.06045	0.0975843
4	−10.6141	−0.11307	0.0979337
5	2.09621	0.02233	0.0991462
6	7.5788	0.08073	0.0991932
7	1.26328	0.01346	0.0998054
8	−2.11396	−0.02252	0.0998224
9	16.393	0.17463	0.0998699
10	−15.0357	−0.16017	0.102684
11	−0.785065	−0.00836	0.104993
12	0.52232	0.00556	0.104999
13	−3.92993	−0.04186	0.105002
14	9.09393	0.09688	0.105158
15	−1.69869	−0.01810	0.105989
16	−10.4136	−0.11093	0.106018
17	0.00406446	0.00004	0.107097
18	−1.46829	−0.01564	0.107097
19	0.457554	0.00487	0.107118
20	−3.08536	−0.03287	0.10712
21	2.19982	0.02343	0.107214
22	−13.7147	−0.14610	0.107262
23	16.4982	0.17575	0.109106
24	−8.90769	−0.09489	0.111721

'.' MARKS TWO STANDARD ERRORS

(Parts (c), the RSPAC, and (d), the Forecasts follow on pages 322 and 323.)

FIGURE 6.12 *Continued*

(c) The RSPAC

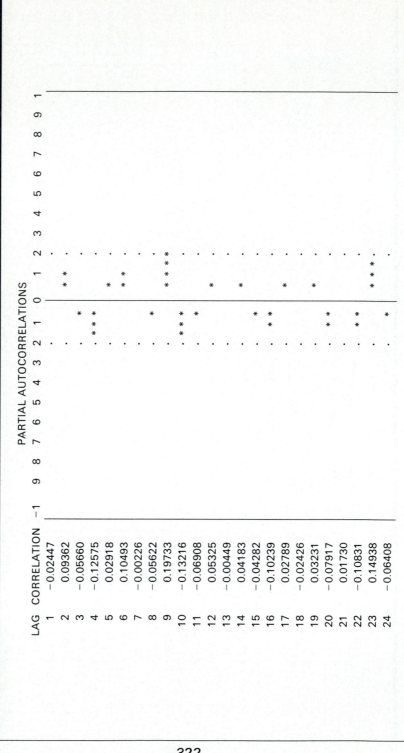

(d) Forecasts of y_{133} through y_{144}

FORECASTS FOR VARIABLE Y

OBS	FORECAST	STD ERROR	LOWER 95%	UPPER 95%
- - - - - - - - FORECAST BEGINS - - - - -
133	421.9914	9.8451	402.6955	441.2873
134	400.5568	11.8393	377.3522	423.7613
135	463.5381	13.5430	436.9944	490.0818
136	451.5399	15.0551	422.0325	481.0473
137	473.8333	16.4286	441.5338	506.0327
138	543.4905	17.6959	508.8073	578.1737
139	618.8521	18.8782	581.8515	655.8526
140	630.2559	19.9908	591.0747	669.4370
141	525.2321	21.0446	483.9856	566.4787
142	464.7052	22.0481	421.4918	507.9186
143	413.5453	23.0079	368.4508	458.6398
144	456.0152	23.9292	409.1149	502.9154

323

then we must use the exponential smoothing procedures discussed in Sections 6.2 through 6.4. For computer programs to do this, see Johnson and Montgomery [1976].

Whereas the Box-Jenkins methodology relies heavily on the SAC and SPAC to tentatively identify a time series model, exponential smoothing relies heavily on the examination of data plots. For example:

1. The data plot in Figure 6.1 indicates that the model

    $$y_t = \beta_0 + \varepsilon_t$$

    where β_0 might be changing over time, describes the cod catch time series in Table 6.1. Thus we might forecast this time series by using simple exponential smoothing, or the equivalent Box-Jenkins model

    $$(1 - B)y_t = (1 - \theta_1 B)a_t$$

2. The data plot in Figure 6.6 indicates that the model

    $$y_t = \beta_0 + \beta_1 t + \varepsilon_t$$

    where β_0 and β_1 are changing over time, describes the thermostat sales time series in Table 6.6. Thus we might forecast this time series by using double exponential smoothing, or the essentially equivalent Box-Jenkins model

    $$(1 - B)^2 y_t = (1 - \theta_1 B)(1 - \theta_{(1)} B)a_t$$

3. The data plot in Figure 3.1 indicates that the model

    $$y_t = (\beta_0 + \beta_1 t)SN_t + \varepsilon_t$$

    where β_0, β_1, and SN_t might be changing over time, describes the hotel room occupancy time series in Table 3.1. Thus we might forecast this time series by using (multiplicative) Winters' Method, or the essentially equivalent (and somewhat modified) Box-Jenkins model (see Example 6.8)

    $$(1 - \phi_1 B - \phi_3 B^3 - \phi_5 B^5)(1 - B^{12})^2 y_t = (1 - \theta_{1,12} B^{12})(1 - \theta_{(1,12)} B^{12})a_t$$

Although the SAC and SPAC sometimes lead us to tentatively identify the simple exponential smoothing model

$$(1 - B)y_t = (1 - \theta_1 B)a_t$$

(for example, the SAC and SPAC led us to tentatively identify this model for the Absorbent Paper Towels sales time series in Chapter 2), the SAC and SPAC rarely lead us to

1. Use second regular differencing to achieve stationarity, and thus rarely lead us to tentatively identify the double exponential smoothing model

    $$(1 - B)^2 y_t = (1 - \theta_1 B)(1 - \theta_{(1)} B)a_t$$

2. Use second seasonal differencing to achieve stationarity, and thus rarely lead

us to tentatively identify a Winters' Method model such as

$$(1 - \phi_1 B - \phi_3 B^3 - \phi_5 B^5)(1 - B^{12})^2 y_t = (1 - \theta_{1,12} B^{12})(1 - \theta_{(1,12)} B^{12}) a_t$$

Because of this fact, Box-Jenkins advocates often conclude that we should rarely (if ever) use double exponential smoothing or Winters' Method models. However, we have seen that data plots (along with the SAC and SPAC—see Example 6.8) can lead us to tentatively identify double exponential smoothing and Winters' Method models. Furthermore, many forecasting competitions that have been carried out indicate that, while the "traditional" Box-Jenkins models tentatively identified by using the SAC and SPAC provide the most accurate forecasts in some situations, double exponential smoothing and Winters' Method models (tentatively identified by data plots and the SAC and SPAC) provide the most accurate forecasts in other situations. For example, we found in Example 6.8 that the Winters' Method model

$$(1 - \phi_1 B - \phi_3 B^3 - \phi_5 B^5)(1 - B^{12})^2 y_t = (1 - \theta_{1,12} B^{12})(1 - \theta_{(1,12)} B^{12}) a_t$$

provides somewhat more accurate point forecasts and somewhat shorter prediction interval forecasts of the hotel room occupancy time series than does the "traditional" Box-Jenkins model

$$(1 - \phi_1 B - \phi_2 B^2 - \phi_3 B^3)(1 - B^{12})(\ln y_t) = \delta + (1 - \theta_{1,12} B^{12}) a_t$$

Therefore, we suggest using all of the tools at our disposal—data plots, the SAC and SPAC, and intuition—to tentatively identify all of the reasonable (Box-Jenkins, exponential smoothing, and other) models that might be used to forecast a time series. Then, we would use the comparison techniques presented in this book to choose a final model that will be used to compute "final" forecasts. Alternatively (as has been suggested by some and found to be surprisingly effective) we can simply average the forecasts obtained by the most reasonable models.

To conclude this section, we discuss how the time series regression models of Chapter 5 fit into our discussion of model selection. Whereas the exponential smoothing methods of this chapter are most appropriately used when the parameters of the underlying models

$$y_t = \beta_0 + \varepsilon_t \qquad \text{(simple exponential smoothing)}$$

$$y_t = \beta_0 + \beta_1 t + \varepsilon_t \qquad \text{(double exponential smoothing)}$$

$$y_t = \beta_0 + \beta_1 t + SN_t + \varepsilon_t \qquad \text{(additive Winters' Model)}$$

$$y_t = (\beta_0 + \beta_1 t) SN_t + \varepsilon_t \qquad \text{(multiplicative Winters' Model)}$$

are changing over time, time series regression models are most appropriately used when the parameters of the underlying models

$$y_t = \beta_0 + \varepsilon_t \qquad \text{(no trend regression models)}$$

$$y_t = \beta_0 + \beta_1 t + \varepsilon_t \qquad \text{(linear trend regression models)}$$

$$y_t = \beta_0 + \beta_1 t + \beta_2 t^2 + \varepsilon_t \quad \text{(quadratic trend regression models[3])}$$

$$y_t = \beta_0 + \beta_1 t + SN_t + \varepsilon_t \quad \text{(dummy variable regression models)}$$

$$y_t = (\beta_0 + \beta_1 t)SN_t + \varepsilon_t \quad \text{(logged, dummy variable regression models, or multiplicative decomposition methods)}$$

are essentially remaining constant over time. However, exponential smoothing models can (and have) been used in situations where the model parameters are essentially unchanging over time. Moreover, because time series regression models allow us (as seen in Chapter 5) to model dependent error terms by a Box-Jenkins model, time series regression models can be used in situations where the model parameters are changing (slightly) over time.

6.6 USING SAS

Below we present the SAS instructions needed to forecast the monthly passenger totals discussed in Example 6.9 using the model

$$z_t = (1 - \theta_1 B)(1 - \theta_{1,12} B^{12})(1 - \theta_{(1,12)} B^{12})a_t$$

where

$$z_t = (1 - B)(1 - B^{12})^2 y_t$$

```
PROC ARIMA DATA = PASS;   }   Declares ARIMA procedure and Data File.
IDENTIFY VAR=Y(1,12,12);  }   Specifies transformation z_t = (1 − B)(1 − B^12)^2 y_t
ESTIMATE Q=(1)(12)(12) NOCONSTANT PRINTALL ⎫   Specifies Model z_t =
              PLOT MAXIT=30;               ⎬   (1 − θ_1 B)(1 − θ_{1,12} B^12)(1 − θ_{(1,12)} B^12)a_t
                                           ⎭
FORECAST LEAD=12;   }   Generates forecasts for 12 future time periods.
```

EXERCISES

6.1   Consider the Bay City Seafood Company cod catch data which was analyzed in Examples 6.2 and 6.3.

    a. Verify that $a_0(3)$, an estimate made in period 3 (March of Year 1) of the average level of the cod catch time series, is 359.28 as shown in Table 6.2.

    b. Verify that the one-period-ahead forecast error for period 4 (April of Year 1) is -62 as shown in Table 6.2.

---

[3] Note that there is an exponential smoothing "version" of the quadratic regression model, called *triple exponential smoothing*. This model is essentially equivalent to the Box-Jenkins model $(1 - B)^3 y_t = (1 - \theta_1 B)(1 - \theta_{(1)} B)(1 - \theta_{(1,1)} B)a_t$. We do not discuss triple exponential smoothing here because it has not been found to be very useful in practice. The interested reader is referred to Bowerman and O'Connell (1979).

c.   Verify that $a_0(4)$, an estimate made in period 4 (April of Year 1) of the average level of the cod catch time series, is 358.03 as shown in Table 6.2.

d.   Verify that the one-period-ahead forecast error for period 5 (May of Year 1) is 41 as shown in Table 6.2.

6.2   Consider the Bay City Seafood Company cod catch data presented in Table 6.1. Simulate one-period-ahead forecasts for the historical cod catch data using a smoothing constant of $\alpha = .10$. Verify that the sum of squared forecast errors obtained using this smoothing constant is 28,792 as shown in Table 6.3.

6.3   Consider the Bay City Seafood Company cod catch data which was analyzed in Examples 6.2 and 6.3.

a.   If we observe a cod catch in February of year 3 of $y_{26} = 375$, update $a_0(25)$ to $a_0(26)$. Recall that we have set $\alpha = .02$.

b.   If $y_{26} = 375$, update $\Delta(25)$ to $\Delta(26)$.

c.   Find a point forecast made in period 26 of any future monthly cod catch.

d.   Find a 95% prediction interval forecast calculated in period 26 for any future monthly cod catch.

e.   If we observe a cod catch in March of year 3 of $y_{27} = 350$, update $a_0(26)$ to $a_0(27)$.

f.   If $y_{27} = 350$, update $\Delta(26)$ to $\Delta(27)$.

g.   Find a point forecast made in period 27 of any future monthly cod catch.

h.   Find a 95% prediction interval forecast calculated in period 27 for any future monthly cod catch.

6.4   Consider Section 6.4 in which we forecasted the Tasty Cola sales time series using Winters' Method. Verify that the initial estimate of the seasonal factor for the month of June is $sn_6(0) = .9965$.

6.5   Consider the Tasty Cola sales data which was analyzed in Section 6.4.

a.   Verify that $a_0(3)$ equals 413.11 as shown in Table 6.4.

b.   Verify that $b_1(3)$ equals 8.83 as shown in Table 6.4.

c.   Verify that $sn_3(3)$ equals .6022 as shown in Table 6.4.

d.   Verify that $\hat{y}_4(3)$ equals 291.60 as shown in Table 6.4.

e.   Verify that $a_0(4)$ equals 421.18 as shown in Table 6.4.

f.   Verify that $b_1(4)$ equals 8.71 as shown in Table 6.4.

g.   Verify that $sn_4(4)$ equals .6909 as shown in Table 6.4.

h.   Verify that $\hat{y}_5(4)$ equals 251.88 as shown in Table 6.4.

6.6   Consider the Tasty Cola sales data and Table 6.5.

a.   Given the observation of Tasty Cola sales in February of year 4, $y_{38} = 445$, verify that the updated estimate of the permanent component is $a_0(38) = 746.57$.

b.   If $y_{38} = 445$, verify that the updated estimate of the trend component is $b_1(38) = 10.36$.

c.   If $y_{38} = 445$, verify that the updated seasonal factor for February is $sn_{38}(38) = .5851$.

d.   Verify that the forecast of Tasty Cola sales which is made in period 38 for y_{39} is $\hat{y}_{39}(38) = 455.67$.

e.   If $y_{39} = 453$, verify that $a_0(39) = 756.04$.

f.   If $y_{39} = 453$, verify that $b_1(39) = 10.23$.

g.   If $y_{39} = 453$, verify that $sn_{39}(39) = .6019$.

h.   Verify that $\hat{y}_{40}(39) = 529.51$.

6.7   Consider Table 6.5, in which prediction intervals for future values of the Tasty Cola sales time series were calculated.

    a.   Suppose that we observe a sales figure for period 38 of $y_{38} = 445$. Verify that when $\Delta_W(37)$ is updated, we obtain $\Delta_W(38) = 8.4730$ as presented in Table 6.5.

    b.   Verify that the updated error in an approximate 95% prediction interval for y_{39} is $B_{39}^{(95)}(38) = 20.7589$.

    c.   Verify that the updated error in an approximate 95% prediction interval for y_{41} is $B_{41}^{(95)}(38) = 21.6573$.

    d.   Verify that an updated approximate 95% prediction interval for y_{39} is [434.91, 476.43].

    e.   Verify that an updated approximate 95% prediction interval for y_{41} is [433.89, 477.20].

6.8   In this problem we consider annual U.S. lumber production from 1947 to 1976. The data were obtained from the U.S. Department of Commerce *Survey of Current Business* and are listed in Table 1. (This exercise suggested by an example in Abraham and Ledolter (1983).)

TABLE 1   *Annual Total U.S. Lumber Production (Millions of Board Feet), 1947–1976[a]*

35,404	36,762	32,901	38,902	37,515
37,462	36,742	36,356	37,858	38,629
32,901	33,385	37,166	32,926	32,019
33,178	34,171	35,733	35,697	35,710
34,449	36,124	35,791	34,548	36,693
38,044	38,658	34,592	32,087	37,153

[a]Table reads from left to right.

    a.   The SAS output of the SAC and SPAC of the lumber production values obtained by using the transformation $z_t = y_t - y_{t-1}$ is given in Figure 1.

        1.   Describe the behavior of the SAC.

        2.   Should the lumber production values produced by the transformation $z_t = y_t - y_{t-1}$ be considered stationary? Why or why not?

        3.   Describe the behavior of the SPAC.

        4.   Discuss why the SAC and SPAC indicate that the model

$$z_t = (1 - \theta_1 B)a_t$$

where

$$z_t = y_t - y_{t-1}$$

might be appropriate for forecasting lumber production.

        5.   What guideline(s) (in Tables 2.3 and 3.3) did you use in answering part 4?

    b.   Plot the lumber production data and explain why simple exponential smoothing might be a reasonable technique to use to forecast this data.

    c.   Figure 2 presents the SAS output of estimation, diagnostic checking, and forecasting for the model

$$z_t = (1 - \theta_1 B)a_t$$

where

$$z_t = y_t - y_{t-1}$$

Note that this Box-Jenkins model is equivalent to simple exponential smoothing.

        1.   Find and identify the least squares point estimate $\hat{\theta}_1$.

FIGURE 1 SAS Output of the SAC and SPAC of the Lumber Production Values Obtained by Using the Transformation $z_t = y_t - y_{t-1}$

(a) The SAC

NAME OF VARIABLE = Y
PERIODS OF DIFFERENCING = 1.

MEAN OF WORKING SERIES = 60.3103
STANDARD DEVIATION = 2561.92
NUMBER OF OBSERVATIONS = 29

AUTOCORRELATIONS

LAG	COVARIANCE	CORRELATION	STD
0	6563454	1.00000	0
1	-1988050	-0.30290	0.185695
2	-1964588	-0.29932	0.202015
3	525848	0.08012	0.216769
4	508364	0.07745	0.217788
5	490164	0.07468	0.218736
6	-435831	-0.06640	0.219613
7	-1186850	-0.18083	0.220305
8	932359	0.14205	0.225365
9	930734	0.14181	0.228431
10	-1161264	-0.17693	0.231447
11	-690943	-0.10527	0.236065
12	225663	0.03438	0.237678
13	592410	0.09026	0.237849
14	755759	0.11515	0.239028
15	-790121	-0.12038	0.240933
16	-946134	-0.14415	0.242998
17	1161434	0.17695	0.245929
18	967648	0.14743	0.250281
19	-1611502	-0.24553	0.253258
20	-206885	-0.03152	0.261337
21	918462	0.13994	0.261468
22	-95330.4	-0.01452	0.264038
23	113452	0.01729	0.264066
24	-753924	-0.11487	0.264105

"." MARKS TWO STANDARD ERRORS

(Part (b), the SPAC, follows on page 330.)

FIGURE 1 Continued

(b) The SPAC

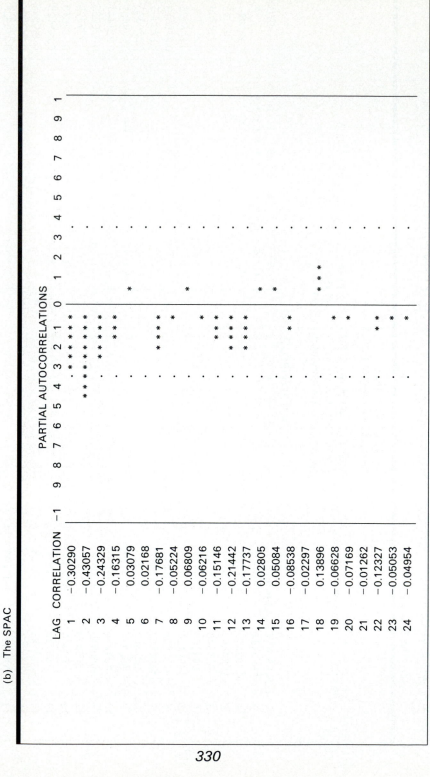

FIGURE 2   SAS Output of Estimation, Diagnostic Checking, and Forecasting for the Model

$$z_t = (1 - \theta_1 B) a_t$$

where

$$z_t = y_t - y_{t-1}$$

(a) Estimation

ARIMA: LEAST SQUARES ESTIMATION

ITERATION	SSE	MA1,1	LAMBDA
0	159168666	0.302897	1.0E−05
1	135781355	0.632687	1.0E−06
2	131095869	0.76149	1.0E−07
3	129801028	0.813531	1.0E−08
4	128964112	0.849347	1.0E−09
5	128144824	0.880505	1.0E−10
6	127087233	0.912333	1.0E−11
7	126052874	0.946153	1.0E−12
8	126051588	0.947117	1.0E−13

PARAMETER	ESTIMATE	STD ERROR	T RATIO	LAG
MA1,1	0.947117	0.0576355	16.43	1

VARIANCE ESTIMATE   = 4501842
STD ERROR ESTIMATE   = 2121.75
NUMBER OF RESIDUALS = 29

CORRELATIONS OF THE ESTIMATES

	MA1,1
MA1,1	1.000

(Parts (b), the RSAC, (c), the RSPAC, and (d), the Forecast follow on pages 332–334.)

FIGURE 2 Continued (b) The RSAC

AUTOCORRELATION CHECK OF RESIDUALS

TO LAG	CHI SQUARE	DF	PROB	AUTOCORRELATIONS					
6	3.81	5	0.577	0.131	-0.075	0.198	0.151	0.055	-0.148
12	12.98	11	0.295	-0.026	-0.009	-0.219	-0.277	-0.260	-0.058
18	18.80	17	0.340	-0.110	0.062	0.067	-0.071	0.196	0.127
24	24.84	23	0.359	0.126	0.005	-0.130	0.030	-0.032	-0.117

AUTOCORRELATION PLOT OF RESIDUALS

LAG	COVARIANCE	CORRELATION	-1 9 8 7 6 5 4 3 2 1 0 1 2 3 4 5 6 7 8 9 1	STD
0	4486509	1.00000	|****************|	0
1	886260	0.19754	| ****|	0.185695
2	-336581	-0.07502	| **|	0.192805
3	585497	0.13050	| ***|	0.193809
4	676717	0.15083	| ***|	0.196816
5	245038	0.05462	| *|	0.200762
6	-662883	-0.14775	| ***|	0.201274
7	-982205	-0.21892	| ****|	0.20498
8	-39936.1	-0.00890	| |	0.21289
9	-114919	-0.02561	| *|	0.212903
10	-1241025	-0.27661	| *****|	0.213009
11	-1167188	-0.26016	| *****|	0.225055
12	-258788	-0.05768	| *|	0.235196
13	302474	0.06742	| *|	0.235684
14	277490	0.06185	| *|	0.236348
15	-495559	-0.11046	| **|	0.236905
16	-318021	-0.07088	| **|	0.238675
17	880164	0.19618	| ****|	0.239399
18	569563	0.12695	| ***|	0.24488
19	-584983	-0.13039	| ***|	0.247139
20	22076.2	0.00492	| |	0.2495
21	565088	0.12595	| ***|	0.249503
22	133481	0.02975	| *|	0.251686
23	-143670	-0.03202	| *|	0.251808
24	-524643	-0.11694	| **|	0.251948

'.' MARKS TWO STANDARD ERRORS

(c) The RSPAC

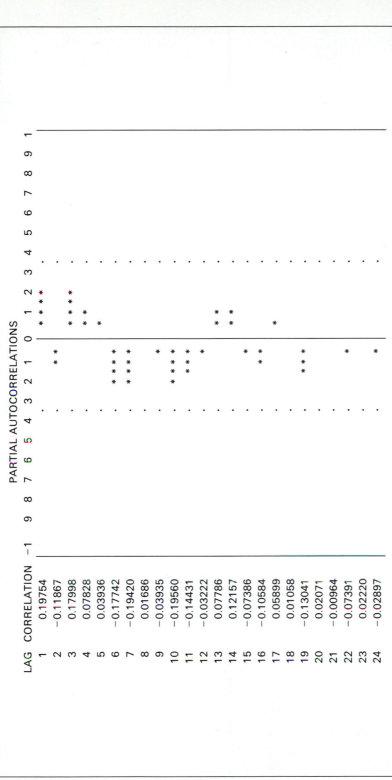

PARTIAL AUTOCORRELATIONS

LAG	CORRELATION
1	0.19754
2	-0.11867
3	0.17998
4	0.07828
5	0.03936
6	-0.17742
7	-0.19420
8	0.01686
9	-0.03935
10	-0.19560
11	-0.14431
12	-0.03222
13	0.07786
14	0.12157
15	-0.07386
16	-0.10584
17	0.05899
18	0.01058
19	-0.13041
20	0.02071
21	-0.00964
22	-0.07391
23	0.02220
24	-0.02897

(Part (d), the Forecast, follows on page 334.)

FIGURE 2 Continued

(d) Forecasts of y_{31} through y_{42}

FORECASTS FOR VARIABLE Y

OBS	FORECAST	STD ERROR	LOWER 95%	UPPER 95%
- - - - - - FORECAST BEGINS - - - - - - -				
31	35683.9824	2121.7546	31525.4284	39842.5365
32	35683.9824	2124.7194	31519.6174	39848.3474
33	35683.9824	2127.6801	31513.8146	39854.1503
34	35683.9824	2130.6366	31508.0198	39859.9450
35	35683.9824	2133.5891	31502.2331	39865.7318
36	35683.9824	2136.5375	31496.4543	39871.5105
37	35683.9824	2139.4819	31490.6836	39877.2813
38	35683.9824	2142.4221	31484.9207	29883.0441
39	35683.9824	2145.3584	31479.1658	39888.7991
40	35683.9824	2148.2906	31473.4187	39894.5462
41	35683.9824	2151.2189	31467.6794	39900.2854
42	35683.9824	2154.1432	31461.9480	39906.0169

2. What do the RSAC and RSPAC say about the adequacy of the model? Explain your answer.
3. What do the values of the Q^* statistics for $K = 6, 12, 18,$ and 24 (and the associated prob-values) say about the adequacy of the model? Explain your answer.
4. Find and identify point forecasts of lumber production for 1977, 1978, 1979, and 1980.
5. Find and identify 95% prediction interval forecasts for lumber production for 1981 and 1982. Interpret these intervals.

6.9   Smith's Department Stores, Inc. owns and operates three stores in Central City. This small chain of stores has sold personal electronic calculators since the early 1970s. For the last two years Smith's has carried a relatively new type of electronic calculator called the Bismark X-12. Sales of the Bismark X-12 have been generally increasing over these two years. Periodically, Smith's orders a supply of the Bismark X-12 from Bismark Electronics Company and then distributes the supply to its three stores. Smith's uses an inventory policy that attempts to insure that Smith's stores will have enough Bismark X-12 calculators to meet practically all of the demand for the Bismark X-12, while at the same time insuring that Smith's does not needlessly tie up its money by ordering many more calculators than can be reasonably expected to be sold. In order to implement its inventory policy, Smith's must have both a point forecast and a prediction interval forecast of total monthly demand for the Bismark X-12. Smith's has recorded monthly sales of the Bismark X-12 for the previous two years, which we will call year 1 and year 2. The calculator sales history is given in Table 2.

TABLE 2   *Calculator Sales*

	Year 1	Year 2
Jan.	197	296
Feb.	211	276
Mar.	203	305
Apr.	247	308
May	239	356
June	269	393
July	308	363
Aug.	262	386
Sept.	258	443
Oct.	256	308
Nov.	261	358
Dec.	288	384

a.   The SAS output of the SAC and SPAC of the calculator sales values obtained by using the transformation

$$z_t = (1 - B)^2 y_t = y_t - 2y_{t-1} + y_{t-2}$$

are given in Figure 3.

1.   Describe the behavior of the SAC.

FIGURE 3 SAS Output of the SAC and SPAC of the Calculator Sales Values Obtained by Using the Transformation $z_t = y_t - 2y_{t-1} + y_{t-2}$

(a) The SAC

NAME OF VARIABLE   = Y
PERIODS OF DIFFERENCING = 1, 1.

MEAN OF WORKING SERIES = 0.545455
STANDARD DEVIATION   = 68.5643
NUMBER OF OBSERVATIONS = 22

AUTOCORRELATIONS

LAG	COVARIANCE	CORRELATION	STD
0	4701.07	1.00000	0
1	-2786.09	-0.59265	0.213201
2	-461.362	-0.03432	0.278181
3	1413.8	0.30074	0.278374
4	-1129.38	-0.24024	0.29277
5	23.9159	0.00509	0.301598
6	860.406	0.18302	0.301601
7	-1172.88	-0.24949	0.306608
8	933.813	0.19864	0.315702
9	-200.626	-0.04268	0.321332
10	-579.47	-0.12326	0.32159
11	786.124	0.16722	0.32373
12	-387.009	-0.08232	0.327633
13	-423.99	-0.09019	0.328572
14	1285.57	0.27346	0.329695
15	-1051.96	-0.22377	0.339849
16	-49.4767	-0.01052	0.346482
17	788.968	0.16783	0.346496
18	-966.962	-0.20569	0.350172
19	687.355	0.14621	0.355621
20	-246.436	-0.05242	0.358343
21	25.154	0.00535	0.358692

'.' MARKS TWO STANDARD ERRORS

FIGURE 3 Continued

(b) The SPAC

PARTIAL AUTOCORRELATIONS

LAG	CORRELATION
1	-0.59265
2	-0.59430
3	-0.19994
4	-0.16315
5	-0.23927
6	-0.08130
7	-0.25506
8	-0.09243
9	-0.06233
10	-0.14118
11	-0.10973
12	-0.12364
13	-0.26952
14	0.00466
15	0.11519
16	0.05084
17	0.05430
18	-0.08299
19	0.07169
20	-0.05996
21	0.10895

FIGURE 4 SAS Output of Estimation, Diagnostic Checking, and Forecasting for the Model

$$z_t = (1 - \theta_1 B)(1 - \theta_{(1)} B)a_t$$

where

$$z_t = y_t - 2y_{t-1} + y_{t-2}$$

(a) Estimation

ARIMA: LEAST SQUARES ESTIMATION

ITERATION	SSE	MA1,1	MA2,1	LAMBDA
0	82403.2	0.1	0.1	1.0E−0.5
1	46270.8	0.403754	0.403754	1.0E−06
2	32654.6	0.674013	0.674005	1.0E−07
3	29434.9	0.926269	0.603189	1.0E−06
4	28940	0.899531	0.704481	1.0E−07
5	28557.3	0.891559	0.780766	1.0E−08
6	28356.3	0.857198	0.859035	0.01
7	28329.3	0.852409	0.849304	0.01
8	28327.8	0.850157	0.856588	0.01
9	28326.8	0.850257	0.85481	0.1
10	28326.6	0.850946	0.854126	0.1

PARAMETER	ESTIMATE	STD ERROR	T RATIO	LAG
MA1,1	0.850946	3.73563	0.23	1
MA2,1	0.854126	3.61027	0.24	1

VARIANCE ESTIMATE   = 1416.33
STD ERROR ESTIMATE = 37.6341
NUMBER OF RESIDUALS = 22

CORRELATIONS OF THE ESTIMATES

	MA1,1	MA2,1
MA1,1	1.000	−1.000
MA2,1	−1.000	1.000

(b) The RSAC

AUTOCORRELATION CHECK OF RESIDUALS

AUTOCORRELATIONS

TO LAG	CHI SQUARE	DF	PROB						
6	4.28	4	0.369	0.077	-0.262	-0.036	0.126	-0.215	-0.083
12	8.41	10	0.589	-0.080	-0.076	-0.010	-0.215	0.153	0.128
18	18.89	16	0.274	-0.104	-0.105	0.273	0.180	0.011	-0.119

AUTOCORRELATION PLOT OF RESIDUALS

LAG	COVARIANCE	CORRELATION	-1 ... 0 ... 1	STD
0	1415.41	1.00000		0
1	178.677	0.12624		0.213201
2	-50.5949	-0.03575		0.216572
3	109.508	0.07737		0.21684
4	-370.869	-0.26202		0.218091
5	-304.569	-0.21518		0.231959
6	-117.703	-0.08316		0.240862
7	-303.806	-0.21464		0.242163
8	-14.0762	-0.00994		0.250662
9	-113.396	-0.08012		0.25068
10	-108.238	-0.07647		0.251841
11	216.865	0.15322		0.252894
12	180.599	0.12759		0.257079
13	254.155	0.17956		0.259942
14	386.272	0.27290		0.26552
15	-147.031	-0.10388		0.277977
16	-149.054	-0.10531		0.279736
17	15.0029	0.01060		0.281533
18	-168.659	-0.11916		0.281551

'.' MARKS TWO STANDARD ERRORS

(Parts (c), the RSPAC, and (d), the Forecasts, follow on pages 340 and 341.)

FIGURE 4 Continued

(c) The RSPAC

PARTIAL AUTOCORRELATIONS

LAG	CORRELATION
1	0.12624
2	−0.05252
3	0.09043
4	−0.29424
5	−0.13932
6	−0.08613
7	−0.18816
8	−0.01891
9	−0.22728
10	−0.11878
11	−0.00030
12	0.01868
13	0.10904
14	0.14169
15	−0.15461
16	−0.06361
17	0.06446
18	0.07480

(d) Forecasts of y_{25} through y_{36}

FORECASTS FOR VARIABLE Y

OBS	FORECAST	STD ERROR	LOWER 95%	UPPER 95%
- - - - - -	FORECAST BEGINS - - - - - -			
25	397.6314	37.6341	323.8700	471.3928
26	405.6138	39.2368	328.7113	482.5164
27	413.5962	41.0068	333.2246	493.9679
28	421.5786	42.9390	337.4198	505.7374
29	429.5610	45.0275	341.3089	517.8132
30	437.5434	47.2657	344.9045	530.1824
31	445.5258	49.6469	348.2200	542.8317
32	453.5082	52.1642	351.2685	555.7480
33	461.4906	54.8112	354.0630	568.9183
34	469.4730	57.5815	356.6155	582.3306
35	477.4554	60.4694	358.9378	595.9731
36	485.4378	63.4693	361.0405	609.8352

341

2. Should the calculator sales values produced by the transformation $z_t = y_t - 2y_{t-1} + y_{t-2}$ be considered stationary? Why or why not?
3. Describe the behavior of the SPAC.
4. Discuss why the SAC and SPAC indicate that the model

$$z_t = (1 - \theta_1 B)a_t$$

where

$$z_t = (1 - B)^2 y_t = y_t - 2y_{t-1} + y_{t-2}$$

might be appropriate for forecasting calculator sales.
5. What guideline(s) (in Tables 2.3 and 3.3) did you use in answering part 4?

b. Plot the calculator sales data and explain why double exponential smoothing might be a reasonable technique to use to forecast this data.

c. Figure 4 presents the SAS output of estimation, diagnostic checking, and forecasting for the model

$$z_t = (1 - \theta_1 B)(1 - \theta_{(1)} B)a_t$$

where

$$z_t = (1 - B)^2 y_t = y_t - 2y_{t-1} + y_{t-2}$$

Note that this Box-Jenkins model is essentially equivalent to double exponential smoothing.
1. Find and identify the least squares point estimates $\hat{\theta}_1$ and $\hat{\theta}_{(1)}$.
2. What do the RSAC and RSPAC say about the adequacy of the model? Explain your answer.
3. What do the values of the Q^* statistics for $K = 6, 12, 18$, and 24 (and the associated prob-values) say about the adequacy of the model? Explain your answer.
4. Find and identify point forecasts of calculator sales for the first 3 months of year 3.
5. Find and identify 95% prediction interval forecasts of calculator sales for the first 3 months of year 3. Interpret these intervals.

6.10 Use additive Winters' Method to forecast the Deep-Freeze Freezer data of Problem 5.4. Forecast demand for each of the four quarters in year 5.

6.11 Use (multiplicative) Winters' Method to forecast the Oligopoly sales data of Problem 5.3. Forecast Oligopoly sales for each of the four quarters in year 4.

6.12 Use traditional simple exponential smoothing to forecast the lumber production sales of Problem 6.8. Forecast lumber production for 1977 and 1987. Also compute 95% prediction intervals for lumber production in 1977 and 1987.

6.13 Use traditional two-parameter double exponential smoothing to forecast the calculator sales data of Problem 6.9. Forecast calculator sales for time periods 25, 26, and 27. Also compute 95% prediction intervals for calculator sales in time periods 25, 26, and 27.

7 7 7 7 7 7 7

TRANSFER FUNCTION MODELS

7.1 INTRODUCTION

When employing the Box-Jenkins methodology, we use the term *transfer function model* to refer to a model that predicts future values of a time series on the basis of past values of the time series and on the basis of values of one or more other time series related to the time series to be predicted. In this chapter we will study using transfer function models to forecast future values of a time series (called the *output series*) on the basis of past values of the time series and on the basis of *one related time series* (called the *input series*). Section 7.2 presents a *three-step procedure for building a transfer function model*. We will see that this involves (1) identifying a model to describe the input series, (2) identifying a preliminary transfer function model describing the output series, and (3) using the residuals for the preliminary model to identify a model describing the error structure of the preliminary model and to form a final transfer function model. Section 7.3 shows how to use SAS to implement transfer function modeling.

7.2   A THREE-STEP PROCEDURE FOR BUILDING A TRANSFER FUNCTION MODEL

In this chapter we will denote the tth values of the output series and the input series by the symbols y_t and x_t, and we will assume that the same stationarity transformation can be used to transform both the output series and the input series into stationary time series values. To illustrate the use of transfer functions, we will analyze data presented by Makridakis, Wheelwright, and McGee [1983]. This data consists of $n = 100$ observations of

$$y_t = \text{total sales (in thousands of cases) in month } t$$

and

$$x_t = \text{advertising expenditure (in thousands of dollars) in month } t$$

and is presented in Appendix B (see Table B5). The data provides a good example of transfer function modeling. In addition, we can obtain a model that may be an alternative to the model obtained by Makridakis, Wheelwright, and McGee [1983].

The development of an appropriate transfer function model consists of three basic steps.

Step 1   Identification of a Model Describing x_t and "Prewhitening" of x_t and y_t

Figure 7.1 presents the SAC of the original advertising expenditure values and indicates that these values are nonstationary. Figure 7.2 presents the SAC and SPAC of the advertising expenditure values produced by the transformation

$$z_t^{(x)} = x_t - x_{t-1}$$

where the use of the superscript "(x)" refers to the fact that the transformation is applied to x_t.

1.   At the *nonseasonal level*, the SAC dies down quickly, and the SPAC dies down quickly. By Guideline 5 in Table 2.3, we could use both the nonseasonal autoregressive operator of order 1

$$\phi_1^{(x)}(B) = (1 - \phi_1^{(x)}B)$$

and the nonseasonal moving average operator of order 1

$$\theta_1^{(x)}(B) = (1 - \theta_1^{(x)}B)$$

2.   At the *seasonal level*, there are no spikes in either the SAC or SPAC, indicating (by Guideline 9 in Table 3.3) that we should use no seasonal operator.

By inserting the nonseasonal operators into the general model

$$\phi_p^{(x)}(B)\phi_P^{(x)}(B^L)z_t^{(x)} = \delta^{(x)} + \theta_q^{(x)}(B)\theta_Q^{(x)}(B^L)a_t^{(x)}$$

FIGURE 7.1 *SAS Output of the SAC of the Original Advertising Expenditure Values*

```
                                                 AUTOCORRELATIONS
LAG  COVARIANCE  CORRELATION  -1 9 8 7 6 5 4 3 2 1 0 1 2 3 4 5 6 7 8 9 1        STD
  0     131.441      1.00000   |                    |********************|         0
  1     117.706      0.89550   |                .   |******************  |       0.1
  2     95.6414      0.72764   |              .     |***************     |  0.161365
  3     79.4591      0.60452   |            .       |************        |  0.191384
  4     69.6083      0.52958   |            .       |***********         |  0.209611
  5     64.3172      0.48933   |           .        |**********          |  0.222589
  6     54.9882      0.41835   |           .        |********.           |  0.233098
  7     43.8861      0.33389   |          .         |*******  .          |  0.240489
  8     37.4047      0.28458   |          .         |******   .          |  0.245081
  9     33.2731      0.25314   |          .         |*****    .          |  0.248363
 10     29.6722      0.22575   |          .         |*****    .          |   0.25093
 11     26.1298      0.19880   |          .         |****     .          |  0.252953
 12     22.0423      0.16770   |          .         |***      .          |   0.25451
 13     19.5899      0.14904   |          .         |***      .          |  0.255613
 14     15.9991      0.12172   |          .         |**       .          |   0.25648
 15     10.5106      0.07996   |          .         |**       .          |  0.257057
 16     6.25595      0.04760   |          .         |*        .          |  0.257306
 17     3.58061      0.02724   |          .         |*        .          |  0.257394
 18     4.54457      0.03458   |          .         |*        .          |  0.257423
 19     4.62295      0.03517   |          .         |*        .          |  0.257469
 20   -0.148355     -0.00113   |          .         |         .          |  0.257517
 21    -5.19441     -0.03952   |          .        *|         .          |  0.257517
 22     -8.1535     -0.06203   |          .        *|         .          |  0.257578
 23    -12.3665     -0.09408   |          .       **|         .          |  0.257727
 24    -17.5958     -0.13387   |          .      ***|         .          |  0.258071
                               '.' MARKS TWO STANDARD ERRORS
```

FIGURE 7.2 *SAS Output of the SAC and SPAC of the Advertising Expenditure Values Obtained by Using the Transformation* $z_t^{(x)} = x_t - x_{t-1}$

(a) The SAC

```
                                               AUTOCORRELATIONS
LAG  COVARIANCE  CORRELATION  -1 9 8 7 6 5 4 3 2 1 0 1 2 3 4 5 6 7 8 9 1         STD
  0      27.528      1.00000   |                    |********************|          0
  1     8.35594      0.30354   |                 .  |******              |   0.100504
  2    -5.73655     -0.20839   |                ****|   .                |   0.109373
  3    -6.47002     -0.23503   |               *****|    .               |   0.113312
  4    -4.46801     -0.16231   |               . ***|    .               |   0.118134
  5     4.01102      0.14571   |               .    |***  .              |   0.120365
  6     1.77471      0.06447   |               .    |*    .              |   0.122134
  7    -4.41149     -0.16025   |               . ***|    .               |   0.122477
  8    -2.97005     -0.10789   |               .  **|    .               |   0.124577
  9    -0.82607     -0.03001   |               .   *|    .               |   0.125518
 10   0.0662574      0.00241   |               .    |    .               |    0.12559
 11    0.876662      0.03185   |               .    |*   .               |   0.125591
 12    -1.30032     -0.04724   |               .   *|    .               |   0.125672
 13    0.799269      0.02903   |               .    |*   .               |   0.125851
 14     1.55823      0.05661   |               .    |*   .               |   0.125919
 15    -1.25632     -0.04564   |               .   *|    .               |   0.126176
 16    -1.48435     -0.05392   |               .   *|    .               |   0.126342
 17    -3.66605     -0.13318   |               . ***|    .               |   0.126575
 18    0.639727      0.02324   |               .    |    .               |   0.127982
 19     4.80398      0.17451   |               .    |***  .              |   0.128025
 20    0.407492      0.01480   |               .    |    .               |   0.130405
 21    -1.83885     -0.06680   |               .   *|    .               |   0.130422
 22     1.58369      0.05753   |               .    |*   .               |   0.130768
 23     1.36361      0.04954   |               .    |*   .               |   0.131023
 24     -1.8969     -0.06891   |               .   *|    .               |   0.131212
                              '.' MARKS TWO STANDARD ERRORS
```

(b) The SPAC

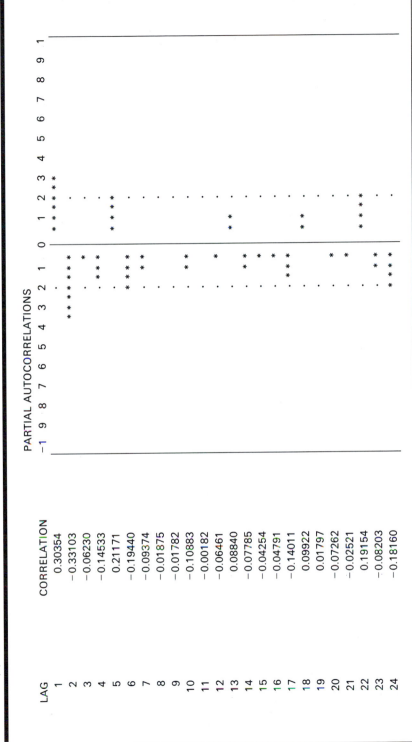

LAG	CORRELATION
1	0.30354
2	-0.33103
3	-0.06230
4	-0.14533
5	0.21171
6	-0.19440
7	-0.09374
8	-0.01875
9	-0.01782
10	-0.10883
11	-0.00182
12	-0.06461
13	0.08840
14	-0.07785
15	-0.04254
16	-0.04791
17	-0.14011
18	0.09922
19	0.01797
20	-0.07262
21	-0.02521
22	0.19154
23	-0.08203
24	-0.18160

PARTIAL AUTOCORRELATIONS

TABLE 7.1   *SAS Output of Estimation and Diagnostic Checking for the Model*

$$(1 - \phi_1^{(x)} B) z_t^{(x)} = (1 - \theta_1^{(x)} B) a_t^{(x)}$$

where

$$z_t^{(x)} = x_t - x_{t-1}$$

(a) Estimation

ARIMA: LEAST SQUARES ESTIMATION

PARAMETER	ESTIMATE	STD ERROR	T RATIO	LAG
MA1,1	−0.787259	0.0995646	−7.91	1
AR1,1	−0.300413	0.154771	−1.94	1

VARIANCE ESTIMATE   = 22.3698
STD ERROR ESTIMATE   = 4.72968
NUMBER OF RESIDUALS = 99

CORRELATIONS OF THE ESTIMATES

	MA1,1	AR1,1
MA1,1	1.000	0.775
AR1,1	0.775	1.000

(b) The RSAC

AUTOCORRELATION CHECK OF RESIDUALS

TO LAG	CHI SQUARE	DF	PROB	AUTOCORRELATIONS					
6	8.68	4	0.070	−0.008	−0.066	−0.184	−0.143	0.141	0.062
12	12.21	10	0.271	−0.157	−0.036	−0.023	−0.009	0.044	−0.061
18	17.86	16	0.332	0.016	0.094	−0.095	0.038	−0.160	0.040
24	21.87	22	0.468	0.134	−0.008	−0.066	0.058	0.039	−0.065

---

and by noting that the mean of the values produced by the transformation $z_t^{(x)} = x_t - x_{t-1}$ is not statistically different from zero (which implies that we should set $\delta^{(x)}$ equal to zero), we are led to tentatively consider the model

$$(1 - \phi_1^{(x)} B) z_t^{(x)} = (1 - \theta_1^{(x)} B) a_t^{(x)}$$

where

$$z_t^{(x)} = x_t - x_{t-1}$$

The SAS output in Table 7.1 of estimation and diagnostic checking indicates that this model is adequate.

Box and Jenkins [1976] suggest that, once we have obtained an appropriate model describing x_t, we should use this model to *"prewhiten"* the values of x_t and y_t (this prewhitening is a preliminary step to determining the relationship between x_t and y_t). Noting that the model

$$(1 - \phi_1^{(x)} B)z_t^{(x)} = (1 - \theta_1^{(x)} B)a_t^{(x)}$$

where

$$z_t^{(x)} = x_t - x_{t-1}$$

implies that

$$a_t^{(x)} = \frac{(1 - \phi_1^{(x)} B)}{(1 - \theta_1^{(x)} B)} z_t^{(x)}$$

and noting that Table 7.1 tells us that the least squares point estimates of $\phi_1^{(x)}$ and $\theta_1^{(x)}$ are $\hat{\phi}_1^{(x)} = -.3004$ and $\hat{\theta}_1^{(x)} = -.7873$, we calculate the prewhitened x_t values by the equation

$$\alpha_t = \frac{(1 - \hat{\phi}_1^{(x)} B)}{(1 - \hat{\theta}_1^{(x)} B)} z_t^{(x)}$$

$$= \frac{(1 + .3004B)}{(1 + .7873B)} z_t^{(x)}$$

which is equivalent to

$$\alpha_t(1 + .7873B) = (1 + .3004B)z_t^{(x)}$$

or

$$\alpha_t = -.7873\alpha_{t-1} + z_t^{(x)} + .3004z_{t-1}^{(x)}$$

$$= -.7873\alpha_{t-1} + (x_t - x_{t-1}) + .3004(x_{t-1} - x_{t-2})$$

Moreover, noting that Figure 7.3 indicates that the sales values obtained by using the transformation

$$z_t = y_t - y_{t-1}$$

are stationary, and noting that this is the same stationarity transformation used to obtain stationary advertising expenditure values, we calculate the prewhitened y_t values by the equation

$$\beta_t = \frac{(1 + .3004B)}{(1 + .7873B)} z_t$$

which is equivalent to

$$\beta_t(1 + .7873B) = (1 + .3004B)z_t$$

or

$$\beta_t = -.7873\beta_{t-1} + z_t + .3004z_{t-1}$$

$$= -.7873\beta_{t-1} + (y_t - y_{t-1}) + .3004(y_{t-1} - y_{t-2})$$

FIGURE 7.3 *SAS Output of the SAC and SPAC of the Sales Values Obtained by Using the Transformation* $z_t = y_t - y_{t-1}$

(a) The SAC

```
                                                 AUTOCORRELATIONS
LAG  COVARIANCE  CORRELATION  -1 9 8 7 6 5 4 3 2 1 0 1 2 3 4 5 6 7 8 9 1        STD
  0     475.947      1.00000   |                    |********************|          0
  1     302.819      0.63625   |                .   |*************       |   0.100504
  2    -28.8335     -0.06058   |               .   *|    .               |     0.1352
  3    -244.864     -0.51448   |          **********|    .               |   0.135474
  4    -245.007     -0.51478   |          **********|     .              |   0.153949
  5    -105.591     -0.22185   |             .  ****|      .             |   0.170452
  6     15.7543      0.03310   |             .      |*     .             |   0.173344
  7     45.5328      0.09567   |             .      |**    .             |   0.173408
  8     25.1876      0.05292   |             .      |*     .             |    0.17394
  9     4.30568      0.00905   |             .      |      .             |   0.174103
 10     8.18623      0.01720   |             .      |      .             |   0.174107
                               '.' MARKS TWO STANDARD ERRORS
```

(b) The SPAC

```
                                      PARTIAL AUTOCORRELATIONS
LAG            CORRELATION       -1 9 8 7 6 5 4 3 2 1 0 1 2 3 4 5 6 7 8 9 1
  1                0.63625        |                .   |*************       |
  2               -0.78191        |    ****************|   .                |
  3                0.22350        |                .   |****                |
  4               -0.21165        |                ****|   .                |
  5               -0.02055        |                .   |   .                |
  6               -0.29259        |              ******|   .                |
  7                0.03156        |                .   |*  .                |
  8               -0.06148        |                .  *|   .                |
  9               -0.12817        |                .***|   .                |
 10                0.04132        |                .   |*  .                |
```

Step 2 Calculation of the Sample Cross-Correlation Function Between the α_t Values and the β_t Values, and Identification of a Preliminary Transfer Function Model

To identify a preliminary transfer function model describing the relationship between y_t and x_t, we calculate the *sample cross-correlation function between the α_t values and the β_t values* (denoted SCC). The SCC is a listing, for $k = -10, -9, \ldots, -1, 0, 1, \ldots, 10$, of the values of

$$r_k(\alpha_t, \beta_t) = \frac{\displaystyle\sum_{t=b}^{n-k} (\alpha_t - \bar{\alpha})(\beta_{t+k} - \bar{\beta})}{\left[\displaystyle\sum_{t=b}^{n} (\alpha_t - \bar{\alpha})^2\right]^{1/2} \left[\displaystyle\sum_{t=b}^{n} (\beta_t - \bar{\beta})^2\right]^{1/2}}$$

where

$$\bar{\alpha} = \frac{\displaystyle\sum_{t=b}^{n} \alpha_t}{n - b + 1} \quad \text{and} \quad \bar{\beta} = \frac{\displaystyle\sum_{t=b}^{n} \beta_t}{n - b + 1}$$

Here, $r_k(\alpha_t, \beta_t)$ is called the *sample cross-correlation* between the α_t values and β_t values at lag k and is a measure of the linear relationship between the values of α_t and the values of β_{t+k}.

In Figure 7.4 we present the SCC between the prewhitened values of advertising expenditure and the prewhitened values of sales. The first thing that we note is that there are no spikes in the SCC at lags less than zero, a prerequisite for using transfer function models. To understand why, suppose that, for example, there were a spike in the SCC at lag -1. This spike would say that $r_{-1}(\alpha_t, \beta_t)$ is statistically different from zero, which would say that values of α_t are related to values of β_{t-1}. Therefore, values of x_t would be related to the values of y_{t-1}, which would say that advertising expenditure in the present month is related to sales one month ago. Since this conclusion would say that present values of x_t are related to past values of y_t, we could not use the transfer function models of this chapter, because these models assume that present values of x_t are not related (at any lag) to past values of y_t. Said another way, these models assume that y_t is not a leading indicator of x_t (the lack of spikes at negative lags—as in Figure 7.4—would indicate that this is true). Rather, the transfer function models of this chapter assume that present values of y_t are related to present and/or past values of x_t. Said another way, these models assume that x_t might be a leading indicator of y_t. Figure 7.4 shows that there is no spike in the SCC at lag 0 and therefore implies that $r_0(\alpha_t, \beta_t)$ is not statistically different from zero, which says that present values of y_t are not related to present values of x_t and therefore sales in one month are not related to advertising expenditure in the same month. Similarly, the fact that there is no spike in the SCC at lag 1 implies that $r_1(\alpha_t, \beta_t)$ is not statistically different from zero, which says that sales in the present month are not related to advertising expenditure one month ago. However, since there is a

FIGURE 7.4 SAS Output of the SCC Between the Prewhitened Values of Advertising Expenditure and the Prewhitened Values of Sales

```
                                           CROSSCORRELATIONS
 LAG  COVARIANCE  CORRELATION  -1 9 8 7 6 5 4 3 2 1 0 1 2 3 4 5 6 7 8 9 1        STD
 -10     4.76104      0.06169  |                .   |*  .                |  0.100504
  -9     1.90313      0.02466  |                .   |   .                |  0.100504
  -8      1.3668      0.01771  |                .   |   .                |  0.100504
  -7    -4.74778     -0.06152  |                .  *|   .                |  0.100504
  -6    -4.22464     -0.05474  |                .  *|   .                |  0.100504
  -5    -7.90245     -0.10239  |                . **|   .                |  0.100504
  -4    -6.04532     -0.07833  |                . **|   .                |  0.100504
  -3     4.66658      0.06046  |                .   |*  .                |  0.100504
  -2     6.97545      0.09038  |                .   |** .                |  0.100504
  -1     -3.3581     -0.04351  |                .  *|   .                |  0.100504
   0    -9.15712     -0.11865  |                . **|   .                |  0.100504
   1    -7.39002     -0.09575  |                . **|   .                |  0.100504
   2     26.2736      0.34042  |                .   |*******             |  0.100504
   3     49.0419      0.63543  |                .   |*************       |  0.100504
   4     22.2382      0.28814  |                .   |******              |  0.100504
   5    -14.6822     -0.19023  |                ****|   .                |  0.100504
   6     -35.869     -0.46475  |           *********|   .                |  0.100504
   7    -31.1241     -0.40327  |            ********|   .                |  0.100504
   8    -2.80163     -0.03630  |                .  *|   .                |  0.100504
   9      10.287      0.13329  |                .   |***.                |  0.100504
  10     8.42225      0.10913  |                .   |** .                |  0.100504
                               '.' MARKS TWO STANDARD ERRORS
```

spike in the SCC at lag 2, it follows that $r_2(\alpha_t, \beta_t)$ is statistically different from zero, which says that sales in the present month are related to advertising expenditure two months ago. To summarize, since the first spike in the SCC occurs at lag 2, we conclude that it takes two months for advertising to affect sales.

We are now prepared to discuss the identification of a preliminary transfer function model.

The *general preliminary transfer function model* is of the form

$$z_t = \mu + \frac{Cw(B)}{\delta(B)} B^b z_t^{(x)} + \eta_t$$

Here:

1. z_t represents the stationary y_t values, and $z_t^{(x)}$ represents the stationary x_t values.
2. C is an unknown *scale parameter.*
3. η_t represents an error component.
4. b is the number of periods before the input series (x_t) begins to affect the output series (y_t) and equals the lag where we encounter the first spike in the SCC. Examining Figure 7.4, we see that $b = 2$, which implies that it takes two months before advertising affects sales—as previously discussed.

5. $w(B) = (1 - w_1 B - w_2 B^2 - \cdots - w_s B^s)$

   is called the $z_t^{(x)}$ *operator of order s*, where s represents the number of past $z_t^{(x)}$ values influencing z_t. Practice has shown that the first spike in the SCC will be followed (possibly after one or more lags) by a "clear dying down pattern" that may be exponential or sinusoidal. The value s is set equal to the number of lags that reside between the first spike in the SCC and the beginning of the clear dying down pattern. In our example, if we examine Figure 7.4, the value of s is not obvious. After the first spike in the SCC (at lag 2), a sinusoidal dying down pattern appears to begin. Whether the spikes at lags 3 and 4 are part of the clear pattern of the die-down is questionable. If both of these spikes are part of the clear dying down pattern, then $s = 0$, while if the spike at lag 4 is part of this clear pattern but the spike at lag 3 is not, then $s = 1$. If neither of these spikes is part of the clear dying down pattern, then $s = 2$. Here we will (somewhat arbitrarily) set $s = 2$. This conclusion implies that we should use the operator

   $$w(B) = (1 - w_1 B - w_2 B^2)$$

6. $\delta(B) = (1 - \delta_1 B - \delta_2 B^2 - \cdots - \delta_r B^r)$

   is called the z_t *operator of order r*, where r intuitively represents the number of its own past values to which z_t is related. Practice has shown that it is reasonable to determine r by examining the manner in which the sample cross-correlations die down at lags after lag $(b + s)$. Specifically, if the sample cross-correlations die down in a damped exponential fashion, it is reasonable to set $r = 1$. If the sample cross-correlations die down in a damped sine wave fashion, it is reasonable to set $r = 2$. Since in Figure 7.4 the sample cross-correlations die down in a damped sine wave fashion at lags after lag $(b + s) = (2 + 2) = 4$, it is reasonable to set $r = 2$. This implies that we should use the operator $\delta(B) = (1 - \delta_1 B - \delta_2 B^2)$.

7. μ is a constant term which should be included if, when it is included in the model, the absolute t-value associated with the least squares point estimate of μ is greater than 2.

Inserting the appropriate operators into the general model

$$z_t = \mu + \frac{Cw(B)}{\delta(B)} B^b z_t^{(x)} + \eta_t$$

and arbitrarily excluding μ, we are led to tentatively consider the model

$$z_t = \frac{C(1 - w_1 B - w_2 B^2)}{(1 - \delta_1 B - \delta_2 B^2)} B^2 z_t^{(x)} + \eta_t$$

where

$$z_t = y_t - y_{t-1} \quad \text{and} \quad z_t^{(x)} = x_t - x_{t-1}$$

Step 3   Identification of a Model Describing η_t and of a Final Transfer Function Model

The SAS output of estimation and diagnostic checking for the tentative model is given in Table 7.2. In addition to the RSAC and RSPAC, the SAS output in Table 7.2 gives the *sample cross-correlation function of the residuals with the values of the input x_t* (denoted RSCC). Since Table 7.2d indicates that the prob-values associated with the values of Q^* are large, we conclude that the prewhitened input series (α_t) is statistically independent of the error component η_t, which is a necessary condition for the validity of transfer function modeling. However, since the prob-values associated with the values of Q^* pertaining to the RSAC are extremely small, we conclude that the error terms $\eta_1, \eta_2, \ldots$ are statistically dependent. To find a model describing η_t (which represents the residuals), note that at the nonseasonal level the RSAC dies down quickly and the RSPAC, which has spikes at lags 1, 2, and 4, also seems to die down quickly. Using Guideline 5 in Table 2.3, we might employ both the nonseasonal autoregressive operator of order 2

$$\phi_2(B) = (1 - \phi_1 B - \phi_2 B^2)$$

and the nonseasonal moving average operator of order 1

$$\theta_1(B) = (1 - \theta_1 B)$$

TABLE 7.2   *SAS Output of Estimation and Diagnostic Checking for the Model*

$$z_t = \frac{C(1 - w_1 B - w_2 B^2)}{(1 - \delta_1 B - \delta_2 B^2)} B^2 z_t^{(x)} + \eta_t$$

where

$$z_t = y_t - y_{t-1} \quad \text{and} \quad z_t^{(x)} = x_t - x_{t-1}$$

(a) Estimation

ARIMA: LEAST SQUARES ESTIMATION

ITERATION	SSE	SCALE1	NUM1,1	NUM1,2	DEN1,1	DEN1,2	LAMBDA
0	36243.3	1.19875	0.1	0.1	0.1	0.1	1.0E–05
1	30075.5	1.24179	–.079898	.0096462	0.279825	0.190172	1
2	16152.2	1.47812	–0.71402	–0.12161	0.348688	–.219199	0.1
3	8592.14	1.30951	–.770864	–.281053	0.80424	–.433006	0.01
4	5578.61	1.27104	–.368583	0.556347	1.28014	–.846108	1.0E–03
5	676.587	1.32245	–0.29166	0.38521	1.21837	–.753262	1.0E–04
6	339.347	1.29567	–.459881	0.60412	1.20408	–.704252	1.0E–05
7	336.195	1.29587	–.464345	0.621472	1.20633	–.705093	1.0E–06
8	336.194	1.29581	–0.46438	0.621691	1.20637	–.705072	1.0E–07

PARAMETER	ESTIMATE	STD ERROR	T RATIO	LAG	VARIABLE
SCALE1	1.29581	0.0412667	31.40	0	X
NUM1,1	–0.46438	0.063409	–7.32	1	X
NUM1,2	0.621691	0.0606716	10.25	2	X
DEN1,1	1.20637	0.00701467	171.98	1	X
DEN1,2	–0.705072	0.00620486	–113.63	2	X

VARIANCE ESTIMATE   = 3.73549
STD ERROR ESTIMATE   = 1.93274
NUMBER OF RESIDUALS = 95

CORRELATIONS OF THE ESTIMATES

	SCALE1	NUM1,1	NUM1,2	DEN1,1	DEN1,2
SCALE1	1.000	0.926	–0.890	0.021	–0.311
NUM1,1	0.926	1.000	–0.909	0.235	–0.518
NUM1,2	–0.890	–0.909	1.000	0.067	0.499
DEN1,1	0.021	0.235	0.067	1.000	–0.549
DEN1,2	–0.311	–0.518	0.499	–0.549	1.000

(Parts (b), the RSAC, (c), the RSPAC, and (d), the RSCC, follow on pages 356 and 357.)

TABLE 7.2 Continued

(b) The RSAC

AUTOCORRELATION CHECK OF RESIDUALS

TO LAG	CHI SQUARE	DF	PROB		AUTOCORRELATIONS				
6	28.79	6	0.000	-0.154	-0.186	-0.230	-0.208	-0.051	
12	33.45	12	0.001	-0.039	-0.021	0.113	0.126	0.108	
18	46.44	18	0.000	-0.063	-0.235	-0.077	0.179	0.119	
24	50.28	24	0.001	-0.080	-0.137	-0.044	-0.032	-0.037	

AUTOCORRELATION PLOT OF RESIDUALS

LAG	COVARIANCE	CORRELATION	-1 9 8 7 6 5 4 3 2 1 0 1 2 3 4 5 6 7 8 9 1	STD
0	3.73549	1.00000	|******************	0
1	1.35591	0.36298	|******	0.102598
2	-0.573512	-0.15353	***|	0.115326
3	-0.694351	-0.18588	****|	0.117458
4	-0.859806	-0.23017	*****|	0.120514
5	-0.77804	-0.20828	****|	0.125056
6	-0.189472	-0.05072	*|	0.128656
7	0.081672	0.02186	|*	0.128866
8	-0.146369	-0.03918	*|	0.128905
9	-0.0772195	-0.02067	|	0.129031
10	0.420956	0.11269	|**	0.129065

'.' MARKS TWO STANDARD ERRORS

(c) The RSPAC

PARTIAL AUTOCORRELATIONS

LAG	CORRELATION
1	0.36298
2	-0.32858
3	0.00959
4	-0.25126
5	-0.08559
6	-0.04200
7	-0.07650
8	-0.13306
9	-0.03936
10	0.06842

(d) The RSCC

CROSSCORRELATION CHECK OF RESIDUALS WITH INPUT X

CROSSCORRELATIONS

TO LAG	CHI SQUARE	DF	PROB						
5	1.62	1	0.203	0.025	-0.054	0.009	0.047	-0.016	-0.105
11	1.96	7	0.962	-0.037	-0.007	-0.021	-0.036	-0.010	0.019
17	3.53	13	0.995	-0.050	0.087	0.032	-0.007	0.058	-0.044
23	5.30	19	0.999	-0.004	0.019	0.017	0.024	-0.018	-0.131

TABLE 7.3 SAS Output of Estimation, Diagnostic Checking, and Forecasting for the Model

$$z_t = \frac{C(1 - w_1 B - w_2 B^2)}{(1 - \delta_1 B - \delta_2 B^2)} B^2 z_t^{(x)} + \frac{1}{(1 - \phi_1 B - \phi_2 B^2)} a_t$$

where

$$z_t = y_t - y_{t-1} \quad \text{and} \quad z_t^{(x)} = x_t - x_{t-1}$$

(a) Estimation

ARIMA: LEAST SQUARES ESTIMATION

ITERATION	SSE	AR1,1	AR1,2	SCALE1	NUM1,1	NUM1,2	DEN1,1	DEN1,2	LAMBDA
0	12418.7	1.13373	−.781914	1.19875	0.1	0.1	0.1	0.1	1.0E-05
1	7315.67	1.11472	−.819399	0.94042	−.336353	.0148257	0.536745	0.184966	0.1
2	3160.15	1.21701	−.901852	0.873685	−1.11324	0.239843	0.447815	−.111459	0.01
3	1429.21	1.29915	−.972445	1.11811	−.975512	−.028686	0.716	−.511402	0.01
4	694.259	1.37778	−.973245	1.25899	−.536352	0.572273	1.11237	−.615995	1.0E-03
5	489.86	1.00783	−.675922	1.27331	−.345754	0.735528	1.29603	−.740985	1.0E-04
6	291.245	0.63711	−.507992	1.33262	−.465578	0.530646	1.19006	−.707151	1.0E-05
7	259.794	0.494984	−.331763	1.31797	−.431112	0.600018	1.2083	−.706185	1.0E-06
8	259.298	0.485738	−.341966	1.31559	−.440461	0.5989	1.20697	−.706661	1.0E-07
9	259.296	0.48628	−.339913	1.31551	−0.4407	0.599483	1.20698	−.706569	1.0E-08
10	259.296	0.486159	−.339935	1.31549	−.440734	0.599452	1.20697	−.706574	1.0E-09

PARAMETER	ESTIMATE	STD ERROR	T RATIO	LAG	VARIABLE
AR1,1	0.486159	0.100744	4.83	1	Y
AR1,2	-0.339935	0.101914	-3.34	2	Y
SCALE1	1.31549	0.0380814	34.54	0	X
NUM1,1	-0.440734	0.0538531	-8.18	1	X
NUM1,2	0.599452	0.0509336	11.77	2	X
DEN1,1	1.20697	0.00882814	136.72	1	X
DEN1,2	-0.706574	0.0074709	-94.58	2	X

VARIANCE ESTIMATE = 2.94654
STD ERROR ESTIMATE = 1.71655
NUMBER OF RESIDUALS = 95

CORRELATIONS OF THE ESTIMATES

	AR1,1	AR1,2	SCALE1	NUM1,1	NUM1,2	DEN1,1	DEN1,2
AR1,1	1.000	-0.356	-0.040	-0.062	0.037	-0.001	-0.012
AR1,2	-0.356	1.000	-0.124	-0.045	0.049	0.066	-0.043
SCALE1	-0.040	-0.124	1.000	0.845	-0.771	-0.058	-0.061
NUM1,1	-0.062	-0.045	0.845	1.000	-0.808	0.276	-0.356
NUM1,2	0.037	0.049	-0.771	-0.808	1.000	0.127	0.383
DEN1,1	-0.001	0.066	-0.058	0.276	0.127	1.000	-0.569
DEN1,2	-0.012	-0.043	-0.061	-0.356	0.383	-0.569	1.000

(Parts (b), the RSAC, (c), the RSPAC, (d), the RSCC and (e), the Forecasts, follow on pages 360, 361, and 362.)

TABLE 7.3 Continued

(b) The RSAC

AUTOCORRELATION CHECK OF RESIDUALS

TO LAG	CHI SQUARE	DF	PROB	AUTOCORRELATIONS					
6	7.92	4	0.095	0.003	−0.088	0.061	−0.162	−0.195	−0.047
12	11.87	10	0.294	0.022	−0.047	−0.025	0.143	0.034	0.107
18	24.16	16	0.086	−0.086	0.088	−0.213	−0.082	0.192	−0.042
24	26.67	22	0.224	0.049	−0.016	−0.122	0.005	−0.050	−0.013

AUTOCORRELATION PLOT OF RESIDUALS

LAG	COVARIANCE	CORRELATION	−1 ... 1	STD
0	2.94654	1.00000		0
1	0.00766257	0.00260		0.102598
2	−0.260304	−0.08834		0.102598
3	0.178358	0.06053		0.103396
4	−0.475939	−0.16152		0.103768
5	−0.575863	−0.19544		0.106382
6	−0.138145	−0.04688		0.110097
7	0.0654922	0.02223		0.110307
8	−0.138622	−0.04705		0.110354
9	−0.0722487	−0.02452		0.110565
10	0.420107	0.14258		0.110622

'.' MARKS TWO STANDARD ERRORS

360

(c) The RSPAC

PARTIAL AUTOCORRELATIONS

LAG	CORRELATION
1	0.00260
2	−0.08835
3	0.06149
4	−0.17199
5	−0.18777
6	−0.08795
7	0.00197
8	−0.07291
9	−0.09068
10	0.07266

(d) The RSCC

CROSSCORRELATION CHECK OF RESIDUALS WITH INPUT X

TO LAG	CHI SQUARE	DF	PROB	CROSSCORRELATIONS					
5	1.04	1	0.308	−0.010	−0.044	0.015	0.041	−0.029	−0.079
11	1.41	7	0.985	0.010	−0.019	−0.039	−0.042	0.005	0.011
17	4.99	13	0.975	−0.064	0.140	0.015	−0.032	0.079	−0.081
23	6.90	19	0.995	0.047	−0.006	0.006	0.027	−0.034	−0.126

TABLE 7.3 *Continued*

(e) Forecasts of y_{101} through y_{110}, where Future Values of x_t are Forecasted by the Model

$$(1 - \phi_1^{(x)} B) z_t^{(x)} = (1 - \theta_1^{(x)} B) a_t^{(x)}$$

where

$$z_t^{(x)} = x_t - x_{t-1}$$

FORECASTS FOR VARIABLE Y

OBS	FORECAST	STD ERROR	LOWER 95%	UPPER 95%
— — —	FORECAST BEGINS	— — —		
101	209.1096	1.7165	205.7452	212.4739
102	218.9221	3.0748	212.8956	224.9486
103	252.4891	7.3347	238.1133	266.8648
104	282.0121	20.9324	240.9854	323.0387
105	291.4256	34.8755	223.0711	359.7802
106	282.7257	43.8789	196.7249	368.7266
107	265.3520	48.1478	170.9843	359.7197
108	250.5891	49.7079	153.1636	348.0147
109	245.0191	50.3855	146.2656	343.7726
110	248.7334	51.0903	148.5985	348.8682

362

This choice implies that the model

$$(1 - \phi_1 B - \phi_2 B^2)\eta_t = (1 - \theta_1 B)a_t$$

describes η_t, which implies that

$$\eta_t = \frac{(1 - \theta_1 B)a_t}{(1 - \phi_1 B - \phi_2 B^2)}$$

and that an appropriate final transfer function model would be

$$\text{Model 1:} \quad z_t = \frac{C(1 - w_1 B - w_2 B^2)}{(1 - \delta_1 B - \delta_2 B^2)} B^2 z_t^{(x)} + \frac{(1 - \theta_1 B)}{(1 - \phi_1 B - \phi_2 B^2)} a_t$$

where

$$z_t = y_t - y_{t-1} \quad \text{and} \quad z_t^{(x)} = x_t - x_{t-1}$$

which is the model obtained by Makridakis, Wheelwright, and McGee [1983].

Table 7.4 presents the SAS output of estimation, diagnostic checking, and forecasting for Model 1. This table indicates that Model 1 is adequate and that the standard error for this model is $s = 1.7214$. Table 7.5 presents the SAS output of estimation for Model 1 with μ included. This table indicates that including μ is not warranted, because the absolute t-value corresponding to the point estimate of μ is only .29.

Although Model 1 is adequate, further examination of Table 7.4 shows that the absolute t-values associated with $\hat{\theta}_1$, $\hat{\phi}_1$, and $\hat{\phi}_2$ are small. These small t-values might indicate that at this stage of the modeling process the model describing η_t may be overly complex. With this in mind, we might reconsider the RSPAC of Table 7.2. If we ignore the spike at lag 4, we might conclude that the RSPAC cuts off after lag 2. Recalling that the RSAC dies down quickly, and using Guideline 2 in Table 2.3, we might decide to use (only) the nonseasonal autoregressive operator of order 2

$$\phi_2(B) = (1 - \phi_1 B - \phi_2 B^2)$$

This implies that the model

$$(1 - \phi_1 B - \phi_2 B^2)\eta_t = a_t$$

describes η_t, which implies that

$$\eta_t = \frac{a_t}{(1 - \phi_1 B - \phi_2 B^2)}$$

and implies that an appropriate final transfer function model would be

$$\text{Model 2:} \quad z_t = \frac{C(1 - w_1 B - w_2 B^2)}{(1 - \delta_1 B - \delta_2 B^2)} B^2 z_t^{(x)} + \frac{a_t}{(1 - \phi_1 B - \phi_2 B^2)}$$

where

$$z_t = y_t - y_{t-1} \quad \text{and} \quad z_t^{(x)} = x_t - x_{t-1}$$

Table 7.5 presents the SAS output of estimation, diagnostic checking, and forecasting for Model 2. This table indicates that Model 2 is also adequate, with a standard

TABLE 7.4 SAS Output of Estimation, Diagnostic Checking, and Forecasting for the Model

$$z_t = \frac{C(1 - w_1 B - w_2 B^2)}{(1 - \delta_1 B - \delta_2 B^2)} B^2 z_t^{(x)} + \frac{(1 - \theta_1 B)}{(1 - \phi_1 B - \phi_2 B^2)} a_t$$

where

$$z_t = y_t - y_{t-1} \quad \text{and} \quad z_t^{(x)} = x_t - x_{t-1}$$

(a) Estimation

ARIMA: LEAST SQUARES ESTIMATION

ITERATION	SSE	MA1,1	AR1,1	AR1,2	SCALE1	NUM1,1	NUM1,2	DEN1,1	DEN1,2	LAMBDA
0	27463.3	-.738125	-.173786	-.825162	1.19875	0.1	0.1	0.1	0.1	1.0E-05
1	5220.92	-.895163	0.529799	-.651453	1.32229	-.532801	0.339825	0.655591	-.320526	1.0E-04
2	1219.99	-.82124	0.763197	-.53826	1.44188	-.551496	0.329506	1.11988	-.764609	0.01
3	492.312	-.700304	0.476153	-.173721	1.41487	-.436529	0.55765	1.16636	-.652313	1.0E-03
4	375.892	-.603658	-.006203	-.146862	1.31471	-.333177	0.634796	1.2579	-.73461	1.0E-03
5	259.017	-.458859	.0893664	-.193801	1.31933	-.427205	0.603464	1.21128	-.707498	1.0E-04
6	257.812	-.424515	0.10383	-.187147	1.31668	-.435769	0.596059	1.20713	-.706598	1.0E-05
7	257.8	-.417457	0.110887	-.185832	1.31566	-.437322	0.596913	1.20697	-.70654	1.0E-06
8	257.8	-.415223	0.112789	-.18659	1.3156	-.437415	0.596918	1.20696	-.706546	1.0E-07
9	257.8	-.414479	0.113482	-.186901	1.31559	-.437434	0.596941	1.20696	-.706545	1.0E-08

364

PARAMETER	ESTIMATE	STD ERROR	T RATIO	LAG	VARIABLE
MA1,1	-0.414479	0.261666	-1.58	1	Y
AR1,1	0.113482	0.265654	0.43	1	Y
AR1,2	-0.186901	0.16076	-1.16	2	Y
SCALE1	1.31559	0.038362	34.29	0	X
NUM1,1	-0.437434	0.0543116	-8.05	1	X
NUM1,2	0.596941	0.0521449	11.45	2	X
DEN1,1	1.20696	0.00871376	138.51	1	X
DEN1,2	-0.706545	0.007198	-98.16	2	X

VARIANCE ESTIMATE = 2.96321
STD ERROR ESTIMATE = 1.7214
NUMBER OF RESIDUALS = 95

CORRELATIONS OF THE ESTIMATES

	MA1,1	AR1,1	AR1,2	SCALE1	NUM1,1	NUM1,2	DEN1,1	DEN1,2
MA1,1	1.000	0.918	-0.747	0.017	-0.037	0.028	-0.043	0.050
AR1,1	0.918	1.000	-0.710	0.008	-0.047	0.037	-0.038	0.046
AR1,2	-0.747	-0.710	1.000	-0.113	-0.045	0.041	0.055	-0.051
SCALE1	0.017	0.008	-0.113	1.000	0.856	-0.788	-0.046	-0.084
NUM1,1	-0.037	-0.047	-0.045	0.856	1.000	-0.815	0.269	-0.383
NUM1,2	0.028	0.037	0.041	-0.788	-0.815	1.000	0.155	0.368
DEN1,1	-0.043	-0.038	0.055	-0.046	0.269	0.155	1.000	-0.566
DEN1,2	0.050	0.046	-0.051	-0.084	-0.383	0.368	-0.566	1.000

(Parts (b), the RSAC, (c), the RSPAC, (d), the RSCC, and (e), the Forecasts follow on pages 366, 367 and 368.)

TABLE 7.4 *Continued*

(b) The RSAC

AUTOCORRELATION CHECK OF RESIDUALS

TO LAG	CHI SQUARE	DF	PROB	AUTOCORRELATIONS					
6	6.15	3	0.104	-0.010	-0.029	-0.062	-0.175	-0.155	-0.033
12	11.57	9	0.239	0.017	-0.027	-0.038	0.148	0.010	0.157
18	24.32	15	0.060	-0.116	0.076	-0.221	-0.074	0.191	-0.015
24	27.45	21	0.156	0.065	-0.034	-0.124	0.013	-0.062	0.018

AUTOCORRELATION PLOT OF RESIDUALS

```
LAG  COVARIANCE  CORRELATION  -1 9 8 7 6 5 4 3 2 1 0 1 2 3 4 5 6 7 8 9 1   STD
 0   2.96321     1.0000                         |********************       0
 1  -0.0294971  -0.00995                        *|                          0.102598
 2  -0.0870379  -0.02937                        *|                          0.102608
 3  -0.182828   -0.06170                       *|                           0.102696
 4  -0.519272   -0.17524                    ****|                           0.103086
 5  -0.458689   -0.15479                     ***|                           0.106175
 6  -0.0985828  -0.03327                       *|                           0.108525
 7   0.0508185   0.01715                        |*                          0.108632
 8  -0.0812624  -0.02742                       *|                           0.108661
 9  -0.112742   -0.03805                       *|                           0.108733
10   0.439589    0.14835                        |***                        0.108874
```

'.' MARKS TWO STANDARD ERRORS

366

(c) The RSPAC

PARTIAL AUTOCORRELATIONS

-1 9 8 7 6 5 4 3 2 1 0 1 2 3 4 5 6 7 8 9 1

LAG	CORRELATION
1	-0.00995
2	-0.02947
3	-0.06235
4	-0.17836
5	-0.17096
6	-0.06537
7	-0.02624
8	-0.09315
9	-0.12160
10	0.09211

(d) The RSCC

CROSSCORRELATION CHECK OF RESIDUALS WITH INPUT X

TO LAG	CHI SQUARE	DF	PROB	CROSSCORRELATIONS					
5	1.27	1	0.259	-0.009	-0.052	0.043	0.013	-0.023	-0.090
11	1.53	7	0.981	0.006	-0.017	-0.028	-0.037	0.006	0.012
17	4.98	13	0.976	-0.058	0.142	-0.045	0.030	0.058	-0.081
23	7.01	19	0.994	0.052	-0.020	0.023	0.024	-0.035	-0.126

(Continued)

TABLE 7.4 Continued

(e) Forecasts of y_{101} through y_{110}, where Future Values of x_t are Forecasted by the Model

$$(1 - \phi_1^{(x)} B)z_t^{(x)} = (1 - \theta_1^{(x)} B)a_t^{(x)}$$

where

$$z_t^{(x)} = x_t - x_{t-1}$$

FORECASTS FOR VARIABLE Y

OBS	FORECAST	STD ERROR	LOWER 95%	UPPER 95%
— —	— FORECAST BEGINS —	— — —		
101	209.1327	1.7214	205.7588	212.5066
102	218.9414	3.1435	212.7803	225.1024
103	252.3685	7.3766	237.9106	266.8264
104	281.7826	20.9509	240.7198	322.8455
105	291.1965	34.8773	222.8385	359.5546
106	282.5415	43.8744	196.5495	368.5335
107	265.2077	48.1455	170.8444	359.5710
108	250.4677	49.7138	153.0305	347.9048
109	244.9049	50.4020	146.1190	343.6908
110	248.6123	51.1177	148.4237	348.8008

TABLE 7.5 SAS Output of Estimation for the Model

$$z_t = \mu + \frac{C(1 - w_1 B - w_2 B^2)}{(1 - \delta_1 B - \delta_2 B^2)} B^2 z_t^{(x)} + \frac{(1 - \theta_1 B)}{(1 - \phi_1 B - \phi_2 B^2)} a_t$$

where

$$z_t = y_t - y_{t-1} \quad \text{and} \quad z_t^{(x)} = x_t - x_{t-1}$$

PARAMETER	ESTIMATE	STD ERROR	T RATIO	LAG	VARIABLE
MU	0.069189	0.236745	0.29	0	Y
MA1,1	−0.409965	0.264188	−1.55	1	Y
AR1,1	0.114987	0.267697	0.43	1	Y
AR1,2	−0.190233	0.161357	−1.18	2	Y
SCALE1	1.31591	0.0385855	34.10	0	X
NUM1,1	−0.437363	0.0546365	−8.00	1	X
NUM1,2	0.596538	0.0524304	11.38	2	X
DEN1,1	1.20684	0.00875178	137.90	1	X
DEN1,2	−0.70652	0.00723075	−97.71	2	X

CONSTANT ESTIMATE = 0.0743953

VARIANCE ESTIMATE = 2.99472

STD ERROR ESTIMATE = 1.73053

NUMBER OF RESIDUALS = 95

CORRELATIONS OF THE ESTIMATES

	MU	MA1,1	AR1,1	AR1,2	SCALE1	NUM1,1	NUM1,2	DEN1,1	DEN1,2
MU	1.000	0.021	0.009	−0.028	0.027	0.006	−0.032	−0.046	0.003
MA1,1	0.021	1.000	0.918	−0.746	0.019	−0.037	0.027	−0.046	0.051
AR1,1	0.009	0.918	1.000	−0.709	0.008	−0.048	0.036	−0.040	0.047
AR1,2	−0.028	−0.746	−0.709	1.000	−0.115	−0.045	0.043	0.059	−0.053
SCALE1	0.027	0.019	0.008	−0.115	1.000	0.856	−0.789	−0.047	−0.084
NUM1,1	0.006	−0.037	−0.048	−0.045	0.856	1.000	−0.815	0.269	−0.383
NUM1,2	−0.032	0.027	0.036	0.043	−0.789	−0.815	1.000	0.154	0.370
DEN1,1	−0.046	−0.046	−0.040	0.059	−0.047	0.269	0.154	1.000	−0.565
DEN1,2	0.003	0.051	0.047	−0.053	−0.084	−0.383	0.370	−0.565	1.000

error of $s = 1.7166$ (which is slightly smaller than the standard error for Model 1). Furthermore, the t-values associated with $\hat{\phi}_1$ and $\hat{\phi}_2$ in Model 2 are significant. Therefore, we will use the forecasts generated by

$$\text{Model 2:} \qquad z_t = \frac{C(1 - w_1 B - w_2 B^2)}{(1 - \delta_1 B - \delta_2 B^2)} B^2 z_t^{(x)} + \frac{1}{(1 - \phi_1 B - \phi_2 B^2)} a_t$$

where

$$z_t = y_t - y_{t-1} \qquad \text{and} \qquad z_t^{(x)} = x_t - x_{t-1}$$

which is equivalent to

$$z_t(1 - \delta_1 B - \delta_2 B^2)(1 - \phi_1 B - \phi_2 B^2)$$
$$= C(1 - w_1 B - w_2 B^2)(1 - \phi_1 B - \phi_2 B^2)B^2 z_t^{(x)} + (1 - \delta_1 B - \delta_2 B^2)a_t$$

Note that, in order to forecast future values of y_t by using Model 2 (or Model 1), we must use past and future values of x_t. If we know future values of x_t, we may insert them into the appropriate prediction equation. Alternatively, we can predict future values of x_t by using the model describing x_t obtained in Step 1

$$(1 - \phi_1^{(x)} B) z_t^{(x)} = (1 - \theta_1^{(x)} B) a_t^{(x)}$$

where

$$z_t^{(x)} = x_t - x_{t-1}$$

We have obtained the forecasts in Tables 7.3 and 7.4 by using this model to predict future values of x_t.

To conclude this section, we consider the following example, which has been adapted from Abraham and Ledolter [1983].

EXAMPLE 7.1

Hillmer and Tiao [1979] analyzed U.S. monthly housing starts y_t and houses sold x_t for the years 1965 to 1974.

Step 1: *Identification of a Model Describing x_t and Prewhitening of x_t and y_t*

The model

$$z_t^{(x)} = (1 - \theta_1^{(x)} B)(1 - \theta_{1,12}^{(x)} B^{12}) a_t^{(x)}$$

where

$$z_t^{(x)} = x_t - x_{t-1} - x_{t-12} + x_{t-13}$$

adequately describes the x_t values, and we can achieve stationary y_t values by using the transformation

$$z_t = y_t - y_{t-1} - y_{t-12} + y_{t-13}$$

Furthermore, since

$$a_t^{(x)} = \frac{1}{(1 - \theta_1^{(x)} B)(1 - \theta_{1,12}^{(x)} B^{12})} z_t^{(x)}$$

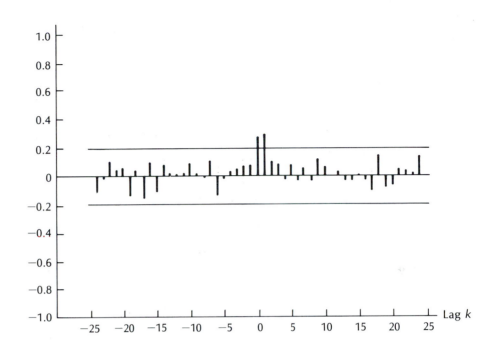

FIGURE 7.5 The SCC

and since the least squares point estimates of $\theta_1^{(x)}$ and $\theta_{1,12}^{(x)}$ are $\hat{\theta}_1^{(x)} = .20$ and $\hat{\theta}_{1,12}^{(x)} = .83$, we calculate prewhitened x_t values and prewhitened y_t values by the equations

$$\alpha_t = \frac{1}{(1 - .20B)(1 - .83B^{12})} z_t^{(x)}$$

and

$$\beta_t = \frac{1}{(1 - .20B)(1 - .83B^{12})} z_t$$

Step 2: *Calculation of the SCC and Identification of a Preliminary Transfer Function Model*

The SCC is presented in Figure 7.5. We note that

1. $b = 0$ is the lag at which we encounter the first spike in the SCC (which implies that the number of houses sold in a month is related to the number of housing starts in the same month).
2. The SCC appears to be dying down in a damped exponential fashion after lag 1. Since the spike at lag 1 does not appear to be part of this clear dying down pattern, it is reasonable to set $s = 1$. Thus we should use the operator $w(B) = (1 - w_1 B)$.

3. Since the sample cross-correlations die down in a damped exponential fashion at lags after lag $(b + s) = (0 + 1) = 1$, it is reasonable to set $r = 1$. This conclusion implies that we should use the operator $\delta(B) = (1 - \delta_1 B)$.

Inserting the appropriate operators into the general model

$$z_t = \mu + \frac{Cw(B)}{\delta(B)} B^b z_t^{(x)} + \eta_t$$

and arbitrarily excluding μ, we are led to tentatively consider the model

$$z_t = \frac{C(1 - w_1 B)}{(1 - \delta_1 B)} B^0 z_t^{(x)} + \eta_t$$

where

$$z_t = y_t - y_{t-1} - y_{t-12} + y_{t-13}$$

and

$$z_t^{(x)} = x_t - x_{t-1} - x_{t-12} + x_{t-13}$$

Step 3: *Identification of a Model Describing η_t and of a Final Transfer Function Model*

The RSAC for the above model is given in Figure 7.6. The prob-values associated with the values of Q^* pertaining to the RSAC can be calculated to be very small, indicating that the error terms $\eta_1, \eta_2, \ldots$ are statistically dependent. To find a model describing η_t (which represents the residuals), note that

1. At the *nonseasonal level*, the RSAC has a spike at lag 1 and cuts off after lag 1. This implies, by Guideline 1 or 3 in Table 2.3, that we should use the nonseasonal moving average operator of order 1

$$\theta_1(B) = (1 - \theta_1 B)$$

2. At the *seasonal level*, the RSAC has spikes at lags 12 and 14 and either cuts off after lag 14 or (possibly) dies down quickly. Although the behavior of the RSAC is not obvious here, it might be reasonable to conclude that the RSAC cuts off after lag 14. If we make this assumption, by Guideline 6 or 8 in Table 3.3 we should use the seasonal moving average operator of order 1

$$\theta_1(B^{12}) = (1 - \theta_{1,12} B^{12})$$

The conclusions imply that the model

$$\eta_t = (1 - \theta_1 B)(1 - \theta_{1,12} B^{12}) a_t$$

describes η_t, which implies that an appropriate final transfer function model would be

$$z_t = \frac{C(1 - w_1 B)}{(1 - \delta_1 B)} B^0 z_t^{(x)} + (1 - \theta_1 B)(1 - \theta_{1,12} B^{12}) a_t$$

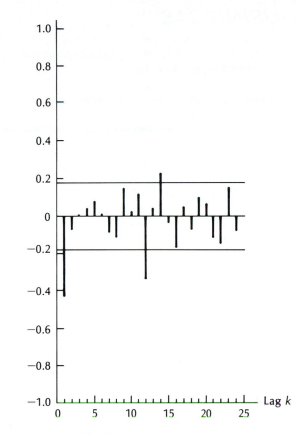

FIGURE 7.6   The RSAC Resulting from the Model

$$z_t = \frac{C(1 - w_1 B)}{(1 - \delta_1 B)} B^0 z_t^{(x)} + \eta_t$$

where

$$z_t = y_t - y_{t-1} - y_{t-12} + y_{t-13} \quad \text{and} \quad z_t^{(x)} = x_t - x_{t-1} - x_{t-12} + x_{t-13}$$

where

$$z_t = y_t - y_{t-1} - y_{t-12} + y_{t-13}$$

and

$$z_t^{(x)} = x_t - x_{t-1} - x_{t-12} + x_{t-13}$$

and where future values of x_t are forecasted by the model describing x_t obtained in Step 1

$$z_t^{(x)} = (1 - \theta_1^{(x)} B)(1 - \theta_{1,12}^{(x)} B^{12}) a_t^{(x)}$$

Examination of the RSAC, RSPAC, and RSCC indicates that this model is adequate.

7.3   USING SAS

In this section we present five SAS programs that are needed to analyze the sales and advertising expenditure data.

PROGRAM 1   *This SAS Program Computes the SAC and SPAC for Each of the Transformations $z_t^{(x)} = x_t$, $z_t^{(x)} = x_t - x_{t-1}$, $z_t = y_t$ and $z_t = y_t - y_{t-1}$.*

```
DATA SALES;   } Assigns name SALES to data file.
INPUT X Y;   } Defines variable names. X = advertising expenditure, Y = sales
CARDS;
116.44    202.66
119.58    232.91
    .            Data lines. Sales and advertising expenditure observations placed
    .                here (see Table B5 in Appendix B). One observation per line.
    .
117.09    230.56
PROC ARIMA;   } Declares ARIMA procedure.
IDENTIFY VAR=X;   } z_t^(x) = x_t
IDENTIFY VAR=X(1);   } z_t^(x) = x_t − x_{t−1}
IDENTIFY VAR=Y;   } z_t = y_t
IDENTIFY VAR=Y(1);   } z_t = y_t − y_{t−1}
```

After examining the output generated by Program 1, we tentatively identify the following model describing x_t

$$(1 - \phi_1^{(x)} B) z_t^{(x)} = (1 - \theta_1^{(x)} B) a_t^{(x)}$$

where

$$z_t^{(x)} = x_t - x_{t-1}$$

PROGRAM 2   *This Program Fits the Above Model Describing x_t*

```
DATA SALES;
INPUT X Y;
CARDS;
116.44    202.66
119.58    232.91

    .
    .

117.09    230.56
PROC ARIMA;
IDENTIFY VAR=X(1);   } z_t^(x) = x_t − x_{t−1}
ESTIMATE P=1 Q=1 NOCONSTANT PRINTALL PLOT;

            (1 − φ_1^(x) B) z_t^(x) = (1 − θ_1^(x) B) a_t^(x)
```

After examining the output from Program 2, we conclude that the above model describing x_t is adequate.

PROGRAM 3 This SAS Program Computes the Prewhitened α_t and β_t Values and also Computes the SCC

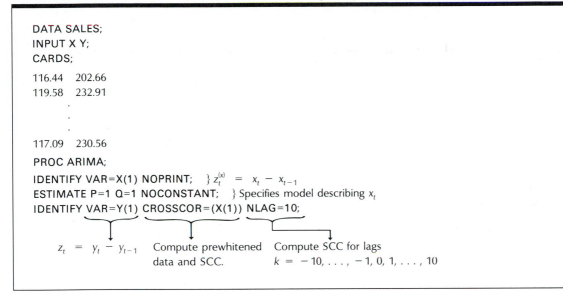

```
DATA SALES;
INPUT X Y;
CARDS;

116.44   202.66
119.58   232.91
         .
         .
         .
117.09   230.56

PROC ARIMA;
IDENTIFY VAR=X(1) NOPRINT;      } z_t^(x) = x_t − x_{t−1}
ESTIMATE P=1 Q=1 NOCONSTANT;    } Specifies model describing x_t
IDENTIFY VAR=Y(1) CROSSCOR=(X(1)) NLAG=10;
```

$$z_t = y_t - y_{t-1}$$ Compute prewhitened data and SCC.   Compute SCC for lags $k = -10, \ldots, -1, 0, 1, \ldots, 10$

After examining the SCC, we identify the preliminary transfer function model

$$z_t = \frac{C(1 - w_1 B - w_2 B^2)}{(1 - \delta_1 B - \delta_2 B^2)} B^2 z_t^{(x)} + \eta_t$$

where

$$z_t = y_t - y_{t-1} \quad \text{and} \quad z_t^{(x)} = x_t - x_{t-1}$$

PROGRAM 4 This SAS Program Fits The Above Preliminary Transfer Function Model

```
DATA SALES;
INPUT X Y;
CARDS;

116.44   202.66
119.58   232.91
         .
         .
         .
117.09   230.56

PROC ARIMA;
IDENTIFY VAR=X(1) NOPRINT;
ESTIMATE P=1 Q=1 NOCONSTANT;
IDENTIFY VAR=Y(1) CROSSCOR=(X(1)) NLAG=10;
```

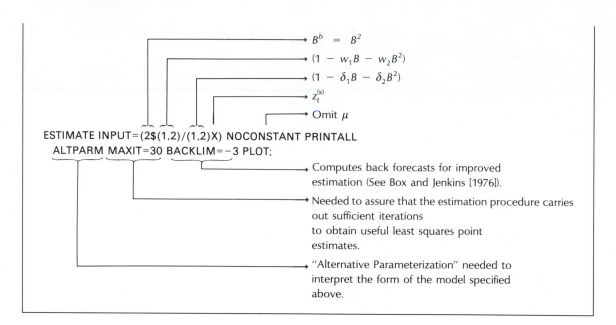

After examining the RSAC and RSPAC resulting from fitting the above preliminary transfer function model, we consider the following models:

$$z_t = \frac{C(1 - w_1B - w_2B^2)}{(1 - \delta_1B - \delta_2B^2)} B^2 z_t^{(x)} + \frac{1}{(1 - \phi_1B - \phi_2B^2)} a_t$$

$$z_t = \frac{C(1 - w_1B - w_2B^2)}{(1 - \delta_1B - \delta_2B^2)} B^2 z_t^{(x)} + \frac{(1 - \theta_1B)}{(1 - \phi_1B - \phi_2B^2)} a_t$$

$$z_t = \mu + \frac{C(1 - w_1B - w_2B^2)}{(1 - \delta_1B - \delta_2B^2)} B^2 z_t^{(x)} + \frac{(1 - \theta_1B)}{(1 - \phi_1B - \phi_2B^2)} a_t$$

where

$$z_t = y_t - y_{t-1} \quad \text{and} \quad z_t^{(x)} = x_t - x_{t-1}$$

PROGRAM 5   *This SAS Program Fits and Forecasts Using Each of the Three Tentative Models.*

```
DATA SALES;
INPUT X Y;
CARDS;
116.44    202.66
119.58    232.91
   .
   .
   .
117.09    230.56
```

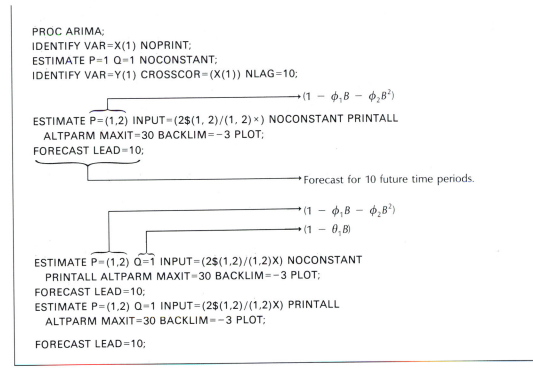

```
PROC ARIMA;
IDENTIFY VAR=X(1) NOPRINT;
ESTIMATE P=1 Q=1 NOCONSTANT;
IDENTIFY VAR=Y(1) CROSSCOR=(X(1)) NLAG=10;
```

$(1 - \phi_1 B - \phi_2 B^2)$

```
ESTIMATE P=(1,2) INPUT=(2$(1, 2)/(1, 2)×) NOCONSTANT PRINTALL
  ALTPARM MAXIT=30 BACKLIM=-3 PLOT;
FORECAST LEAD=10;
```

Forecast for 10 future time periods.

$(1 - \phi_1 B - \phi_2 B^2)$

$(1 - \theta_1 B)$

```
ESTIMATE P=(1,2) Q=1 INPUT=(2$(1,2)/(1,2)X) NOCONSTANT
  PRINTALL ALTPARM MAXIT=30 BACKLIM=-3 PLOT;
FORECAST LEAD=10;
ESTIMATE P=(1,2) Q=1 INPUT=(2$(1,2)/(1,2)X) PRINTALL
  ALTPARM MAXIT=30 BACKLIM=-3 PLOT;

FORECAST LEAD=10;
```

Note that in this program forecasting of future y_t values is done by forecasting future x_t values using the model

$$(1 - \phi_1^{(x)} B)z_t^{(x)} = (1 - \theta_1^{(x)} B)a_t^{(x)}$$

If we wished to specify future x_t values (say, $x_{101} = 119$ and $x_{102} = 121$) we could do so at the end of the data set. For example,

```
CARDS;
116.44   202.66
119.58   232.91
         .
         .
         .
117.09   230.56
  119      .
  121      .
```

EXERCISES

The exercises for this chapter deal with data analyzed by Box and Jenkins [1976]. They forecast sales (y_t) by using a leading economic indicator (x_t). The historical data consists of 150 observations and is given in Table B6 of Appendix B.

7.1   In Figure 1 we present the SAS output of the SAC of the leading indicator (x_t) values. In Figure 2 we present the SAS output of the SAC and SPAC of the leading indicator values produced by the transformation

$$z_t^{(x)} = x_t - x_{t-1}$$

a.   Describe the behavior of the SAC of the original leading indicator values.

b.   Should the original leading indicator values be considered stationary? Why or why not?

c.   Describe the behavior of the SAC of the leading indicator values produced by the transformation $z_t^{(x)} = x_t - x_{t-1}$.

d.   Should the leading indicator values produced by the transformation $z_t^{(x)} = x_t - x_{t-1}$ be considered stationary? Why or why not?

e.   Describe the behavior of the SPAC of the leading indicator values produced by the transformation $z_t^{(x)} = x_t - x_{t-1}$.

f.   Use the SAS output to rationalize the following tentative model describing x_t.

$$z_t^{(x)} = (1 - \theta_1^{(x)} B) a_t^{(x)}$$

where

$$z_t^{(x)} = x_t - x_{t-1}$$

g.   What guideline(s) (in Tables 2.3 and 3.3) did you use in identifying the tentative model of part f?

7.2   The SAS output of estimation and diagnostic checking for the tentative model given in Exercise 7.1f is presented in Table 1.

a.   Find and identify $\hat{\theta}_1^{(x)}$, the least squares point estimate of the model parameter $\theta_1^{(x)}$.

b.   Does the parameter $\theta_1^{(x)}$ seem to be significantly different from zero? Justify your answer.

c.   What do the values of the Q^* statistic for $K = 6, 12, 18,$ and 24 (and the associated prob-values) say about the adequacy of the tentative model describing x_t?

7.3   Figure 3 presents the SAS output of the SAC and SPAC of the sales values produced by the transformation

$$z_t = y_t - y_{t-1}$$

a.   Describe the behavior of the SAC of the sales values produced by the transformation $z_t = y_t - y_{t-1}$.

b.   Should the sales values produced by the transformation $z_t = y_t - y_{t-1}$ be considered stationary? Why or why not?

c.   Write equations that show exactly how the prewhitened x_t values and the prewhitened y_t values are calculated if the model given in Exercise 7.1f is used to describe x_t.

7.4   The SAS output of the SCC between the prewhitened values of the leading indicator and the prewhitened values of sales is given in Figure 4. Use this output to rationalize the following preliminary transfer function model (where we arbitrarily include μ)

$$z_t = \mu + \frac{C}{(1 - \delta_1 B)} B^3 z_t^{(x)} + \eta_t$$

FIGURE 1 SAS Output of the SAC of the Original Leading Indicator Values

```
                                                      AUTOCORRELATIONS
LAG  COVARIANCE  CORRELATION  -1 9 8 7 6 5 4 3 2 1 0 1 2 3 4 5 6 7 8 9 1          STD
  0     1.46844      1.00000  |                    |********************|          0
  1     1.39957      0.95310  |                 .  |******************* |  0.0816496
  2     1.37113      0.93373  |               .    |******************* |   0.137035
  3     1.33954      0.91222  |             .      |******************  |   0.174365
  4     1.30658      0.88977  |            .       |******************  |   0.203712
  5     1.26827      0.86368  |           .        |*****************   |   0.228154
  6     1.23858      0.84346  |          .         |*****************   |   0.248998
  7     1.20191      0.81849  |         .          |****************    |   0.267369
  8     1.17071      0.79725  |         .          |****************    |   0.283581
  9     1.12124      0.76355  |        .           |***************     |   0.298149
 10     1.07973      0.73529  |        .           |***************     |   0.310912
 11     1.05456      0.71815  |       .            |**************      |   0.322297
 12     1.01483      0.69109  |       .            |**************      |   0.332794
 13    0.987921      0.67277  |      .             |*************.      |   0.342228
 14    0.958522      0.65275  |      .             |*************.      |   0.350934
 15    0.929546      0.63301  |      .             |*************.      |   0.358937
 16    0.897181      0.61097  |     .              |************  .     |   0.366304
 17    0.865526      0.58942  |     .              |************  .     |   0.373036
 18    0.830381      0.56548  |     .              |***********   .     |   0.379194
 19    0.812529      0.55333  |     .              |***********   .     |   0.384775
 20    0.786461      0.53557  |    .               |***********    .    |   0.390043
 21    0.753314      0.51300  |    .               |**********     .    |   0.394916
 22    0.728836      0.49633  |    .               |**********     .    |   0.399333
 23    0.700193      0.47683  |    .               |**********     .    |   0.403425
 24    0.671081      0.45700  |    .               |*********      .    |   0.407165
                              '.' MARKS TWO STANDARD ERRORS
```

FIGURE 2 *SAS Output of the SAC and SPAC of the Leading Indicator Values Obtained by Using the Transformation* $z_t^{(x)} = x_t - x_{t-1}$

(a) The SAC

```
                                           AUTOCORRELATIONS
LAG  COVARIANCE  CORRELATION  -1 9 8 7 6 5 4 3 2 1 0 1 2 3 4 5 6 7 8 9 1        STD
  0   0.0993273      1.00000  |                    |********************|          0
  1   -0.044402     -0.44703  |           *********|  .                 |  0.0819232
  2  0.00848313      0.08541  |                .   |** .                |  0.0969213
  3  -0.0069778     -0.07025  |                .  *|   .                |  0.0974251
  4   0.0128692      0.12956  |                .   |***.                |  0.0977644
  5  -.00903002     -0.09091  |                . **|   .                |  0.0989101
  6  0.00771008      0.07762  |                .   |** .                |  0.0994693
  7  -.00776883     -0.07821  |                . **|   .                |   0.099875
  8   0.0119132      0.11994  |                .   |** .                |   0.100285
  9  -.00518369     -0.05219  |                .  *|   .                |   0.101243
 10  -0.0123953     -0.12479  |                . **|   .                |   0.101424
 11   0.0185692      0.18695  |                .   |****                |   0.102449
 12  -.00856292     -0.08621  |                . **|   .                |   0.104714
 13  0.00403515      0.04062  |                .   |*  .                |   0.105189
 14  -.00378746     -0.03813  |                .  *|   .                |   0.105294
 15  -.00178056     -0.01793  |                .   |   .                |   0.105387
 16  -.00017731     -0.00179  |                .   |   .                |   0.105407
 17  0.00508489      0.05119  |                .   |*  .                |   0.105407
 18  -.00938217     -0.09446  |                . **|   .                |   0.105574
 19  0.00194002      0.01953  |                .   |   .                |    0.10614
 20   0.0116358      0.11715  |                .   |** .                |   0.106164
 21  -.00763475     -0.07686  |                . **|   .                |   0.107028
 22  0.00219281      0.02208  |                .   |   .                |   0.107398
 23  0.00121266      0.01221  |                .   |   .                |   0.107428
 24  -.00409504     -0.04123  |                .  *|   .                |   0.107438
                               '.' MARKS TWO STANDARD ERRORS
```

(b) The SPAC

PARTIAL AUTOCORRELATIONS

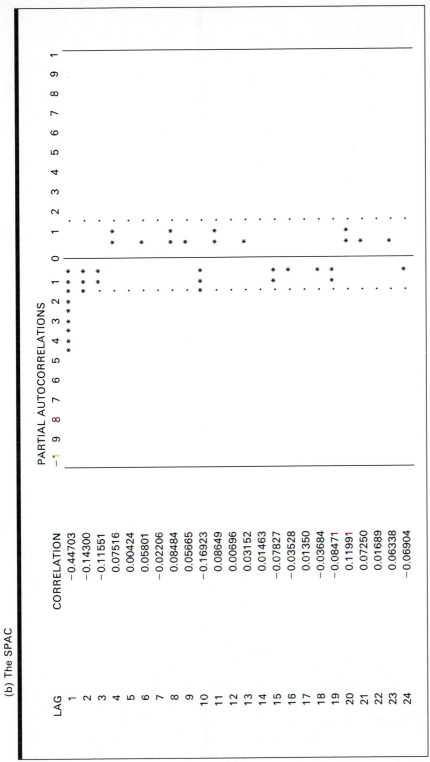

LAG	CORRELATION
1	-0.44703
2	-0.14300
3	-0.11551
4	0.07516
5	0.00424
6	0.05801
7	-0.02206
8	0.08484
9	0.05665
10	-0.16923
11	0.08649
12	0.00696
13	0.03152
14	0.01463
15	-0.07827
16	-0.03528
17	0.01350
18	-0.03684
19	-0.08471
20	0.11991
21	0.07250
22	0.01689
23	0.06338
24	-0.06904

FIGURE 3 SAS Output of the SAC and SPAC of the Sales Values Obtained by Using the Transformation $z_t = y_t - y_{t-1}$

(a) The SAC

AUTOCORRELATIONS

LAG	COVARIANCE	CORRELATION	STD
0	2.07114	1.00000	0
1	0.645779	0.31180	0.0819232
2	0.576179	0.27819	0.0895341
3	0.468885	0.22639	0.0951587
4	0.522143	0.25210	0.0987073
5	0.309832	0.14960	0.102938
6	0.276728	0.13361	0.104387
7	0.130231	0.06288	0.105528
8	0.274203	0.13239	0.10578
9	−0.039048	−0.01885	0.106886
10	−.00773587	−0.00374	0.106908

'.' MARKS TWO STANDARD ERRORS

(b) The SPAC

PARTIAL AUTOCORRELATIONS

LAG	CORRELATION
1	0.31180
2	0.20046
3	0.10910
4	0.13829
5	−0.00292
6	0.01034
7	−0.04773
8	0.07146
9	−0.11219
10	−0.03439

TABLE 1 *SAS Output of Estimation and Diagnostic Checking for the Model*

$$z_t^{(x)} = (1 - \theta_1^{(x)} B) a_t^{(x)}$$

where

$$z_t^{(x)} = x_t - x_{t-1}$$

(a) Estimation

ARIMA: LEAST SQUARES ESTIMATION

PARAMETER	ESTIMATE	STD ERROR	T RATIO	LAG
MA1,1	0.448888	0.0733836	6.12	1

VARIANCE ESTIMATE = 0.080382
STD ERROR ESTIMATE = 0.283517
NUMBER OF RESIDUALS = 149

CORRELATIONS OF THE ESTIMATES

	MA1,1
MA1,1	1.000

(b) The RSAC

AUTOCORRELATION CHECK OF RESIDUALS

TO LAG	CHI SQUARE	DF	PROB	AUTOCORRELATIONS					
6	5.93	5	0.313	−0.056	0.082	0.018	0.146	−0.001	0.083
12	13.82	11	0.243	−0.002	0.116	−0.029	−0.072	0.171	−0.009
18	15.07	17	0.590	0.029	−0.031	−0.023	0.010	0.038	−0.059
24	19.45	23	0.675	0.054	0.141	−0.017	0.028	0.015	−0.033

If we consider the general preliminary transfer function model

$$z_t = \mu + \frac{Cw(B)}{\delta(B)} B^b z_t^{(x)} + \eta_t$$

carefully explain how you arrived at:
a. A tentative value of b.
b. A tentative form of $w(B) = (1 - w_1 B - w_2 B^2 - \cdots - w_s B^s)$
c. A tentative form of $\delta(B) = (1 - \delta_1 B - \delta_2 B^2 - \cdots - \delta_r B^r)$

7.5   The SAS output of estimation and diagnostic checking for the preliminary transfer function model given in Exercise 7.4 is given in Table 2.
a. Find and identify the least squares point estimates of the model parameters.

TABLE 2 *SAS Output of Estimation and Diagnostic Checking for the Model*

$$z_t = \mu + \frac{C}{(1 - \delta_1 B)} B^3 z_t^{(x)} + \eta_t$$

where

$$z_t = y_t - y_{t-1} \quad \text{and} \quad z_t^{(x)} = x_t - x_{t-1}$$

(a) Estimation

ARIMA: LEAST SQUARES ESTIMATION

ITERATION	SSE	MU	SCALE1	DEN1,1	CONSTANT	LAMBDA
0	137.392	0.420134	4.70685	0.1	0.420134	1.0E−05
1	18.6386	0.243171	4.62261	0.636796	0.243171	1.0E−06
2	9.83283	0.041549	4.69373	0.741634	0.041549	1.0E−07
3	9.17109	0.030196	4.69141	0.725678	0.030196	1.0E−08
4	9.1698	.0295215	4.69215	0.724897	0.0295215	1.0E−09

PARAMETER	ESTIMATE	STD ERROR	T RATIO	LAG	VARIABLE
MU	0.0295215	0.0238907	1.24	0	Y
SCALE1	4.69215	0.0763434	61.46	0	X
DEN1,1	0.724897	0.00735106	98.61	1	X

CONSTANT ESTIMATE   = 0.0295215

VARIANCE ESTIMATE   = 0.0645761
STD ERROR ESTIMATE = 0.254118
NUMBER OF RESIDUALS = 145

CORRELATIONS OF THE ESTIMATES

	MU	SCALE1	DEN1,1
MU	1.000	−0.085	−0.370
SCALE1	−0.085	1.000	−0.405
DEN1,1	−0.370	−0.405	1.000

(*Continued on pages 385 and 386.*)

b. Do all of the model parameters seem to be significantly different from zero? Explain your answer.

c. Consider the RSCC. What do the prob-values associated with the Q^* statistic say about whether or not the prewhitened input series is statistically independent of the error component? Explain your answer.

d. Consider the RSAC. What do the values of the Q^* statistic (and the associated prob-values) say about the adequacy of the model?

e. Should we consider the error terms $\eta_1, \eta_2, \ldots$ to be statistically independent or statistically dependent? Explain your answer.

f. Describe the behavior of the RSAC.

TABLE 2 Continued

(b) The RSAC

AUTOCORRELATION CHECK OF RESIDUALS

TO LAG	CHI SQUARE	DF	PROB	AUTOCORRELATIONS					
6	28.69	5	0.000	-0.400	0.018	-0.042	-0.094	0.142	0.042
12	36.47	11	0.000	-0.093	0.076	-0.135	0.124	-0.013	-0.039
18	47.27	17	0.000	-0.088	0.055	-0.054	0.141	-0.161	0.075
24	54.63	23	0.000	0.043	-0.051	0.032	-0.065	-0.073	0.164

AUTOCORRELATION PLOT OF RESIDUALS

LAG	COVARIANCE	CORRELATION	-1 ... 0 ... 1	STD
0	0.0645761	1.0000	| *	0
1	-0.0258357	-0.40008	* * * * * * * * |	0.0830454
2	0.00116455	0.01803	. |	0.0954167
3	-.00269706	-0.04177	. |	0.0954402
4	-.00606356	-0.09390	. * * |	0.0955661
5	0.00919362	0.14237	| * * *	0.0962003
6	0.00271007	0.04197	| * .	0.0976426
7	-0.0059807	-0.09261	. * * |	0.0977669
8	0.00488184	0.07560	| * .	0.0983701
9	-.00874358	-0.13540	. * * * |	0.0987699
10	0.00800949	0.12403	| * * .	0.100042

'.' MARKS TWO STANDARD ERRORS

(Parts (c), the RSPAC, and (d), the RSCC, follow on page 386.)

385

TABLE 2 Continued

(c) The RSPAC

PARTIAL AUTOCORRELATIONS

LAG	CORRELATION
1	-0.40008
2	-0.16910
3	-0.12377
4	-0.19881
5	0.00940
6	0.11347
7	-0.02181
8	0.05657
9	-0.07268
10	0.04533

(d) The RSCC

CROSSCORRELATION CHECK OF RESIDUALS WITH INPUT X

TO LAG	CHI SQUARE	DF	PROB	CROSSCORRELATIONS					
5	10.59	4	0.032	-0.002	0.084	-0.170	0.079	-0.060	0.165
11	12.55	10	0.250	-0.054	-0.022	0.034	0.035	-0.033	0.082
17	14.60	16	0.554	-0.035	-0.032	-0.007	-0.064	0.088	0.001
23	21.92	22	0.465	-.0149	0.110	0.001	0.022	-0.067	0.105

386

FIGURE 4 SAS Output of the SCC Between the Prewhitened Values of the Leading Indicator and the Prewhitened Values of Sales

CROSSCORRELATIONS

LAG	COVARIANCE	CORRELATION	-1 9 8 7 6 5 4 3 2 1 0 1 2 3 4 5 6 7 8 9 1	STD
-10	-0.0282946	-0.05201	. | . . | *	0.0819232
-9	-0.0260465	-0.04788	. | . . | *	0.0819232
-8	0.0269448	0.04953	* . | . . |	0.0819232
-7	-.00130137	-0.00239	. | . . |	0.0819232
-6	-0.0346921	-0.06377	. | . . | *	0.0819232
-5	0.0130202	0.02393	. | . . |	0.0819232
-4	0.00124596	0.00229	. | . . |	0.0819232
-3	0.0220304	0.04050	* . | . . |	0.0819232
-2	0.00537559	0.00988	** | . . |	0.0819232
-1	0.0514406	0.09456	*** | . . |	0.0819232
0	0.0341657	0.06281	*** | . . |	0.0819232
1	0.0429644	0.07898	*** | . . |	0.0819232
2	0.00987332	0.01815	* | . . |	0.0819232
3	0.367312	0.67522	|*************	0.0819232
4	0.245902	0.45204	|*********	0.0819232
5	0.185243	0.34053	|*******	0.0819232
6	0.139976	0.25732	|*****	0.0819232
7	0.145713	0.26786	|*****	0.0819232
8	0.107673	0.19793	|****	0.0819232
9	0.0941315	0.17304	|***	0.0819232
10	0.0530217	0.09747	|**	0.0819232

'.' MARKS TWO STANDARD ERRORS

TABLE 3　SAS Output of Estimation, Diagnostic Checking, and Forecasting for the Model

$$z_t = \mu + \frac{C}{(1 - \delta_1 B)} B^3 z_t^{(x)} + (1 - \theta_1 B) a_t$$

where

$$z_t = y_t - y_{t-1} \quad \text{and} \quad z_t^{(x)} = x_t - x_{t-1}$$

(a) Estimation

ARIMA: LEAST SQUARES ESTIMATION

ITERATION	SSE	MU	MA1,1	SCALE1	DEN1,1	CONSTANT	LAMBDA
0	128.675	0.420134	−.311799	4.70685	0.1	0.420134	1.0E−05
1	25.1339	0.258565	−.617448	4.59937	0.676355	0.258565	1.0E−06
2	13.1004	−.009834	−0.31406	4.60491	0.748583	−.00983379	1.0E−07
3	8.58968	.0407448	.0969186	4.64781	0.724129	0.0407448	1.0E−08
4	7.33303	.0248704	0.451049	4.72075	0.725112	0.0248704	1.0E−09
5	7.27073	.0309749	0.523948	4.71104	0.723831	0.0309749	1.0E−10
6	7.27011	.0303955	0.524371	4.70829	0.724274	0.0303955	1.0E−11
7	7.2701	.0303963	0.524737	4.70826	0.724278	0.0303963	1.0E−12

PARAMETER	ESTIMATE	STD ERROR	T RATIO	LAG	VARIABLE
MU	0.0303963	0.0103186	2.95	0	Y
MA1,1	0.524737	0.0717734	7.31	1	Y
SCALE1	4.70826	0.0572984	82.17	0	X
DEN1,1	0.724278	0.00425259	170.31	1	X

CONSTANT ESTIMATE　　= 0.0303963

VARIANCE ESTIMATE　　= 0.051561
STD ERROR ESTIMATE　= 0.227071
NUMBER OF RESIDUALS = 145

CORRELATIONS OF THE ESTIMATES

	MU	MA1,1	SCALE1	DEN1,1
MU	1.000	−0.003	−0.042	−0.308
MA1,1	−0.003	1.000	−0.042	0.017
SCALE1	−0.042	−0.042	1.000	−0.690
DEN1,1	−0.308	0.017	−0.690	1.000

(Continued on pages 389, 390 and 391.)

TABLE 3 Continued

(b) The RSAC

AUTOCORRELATION CHECK OF RESIDUALS

AUTOCORRELATIONS

TO LAG	CHI SQUARE	DF	PROB						
6	5.61	4	0.230	0.013	-0.000	-0.064	-0.056	0.151	0.081
12	10.09	10	0.433	-0.052	0.012	-0.099	0.070	-0.025	-0.101
18	16.23	16	0.437	-0.128	-0.004	-0.016	0.089	-0.100	0.054
24	22.14	22	0.451	0.044	-0.050	-0.037	-0.112	-0.084	0.092

AUTOCORRELATION PLOT OF RESIDUALS

LAG	COVARIANCE	CORRELATION	-1 ... 0 ... 1	STD
0	0.051561	1.00000	******************************	0
1	.000655289	0.01271	*	0.0830454
2	-5.195E-07	-0.00001	*	0.0830588
3	-.00332317	-0.06445	**	0.0830588
4	-.00290591	-0.05636	**	0.083403
5	0.00778419	0.15097	***	0.0836653
6	0.00418502	0.08117	**	0.0855234
7	-.00268012	-0.05198	*	0.086053
8	.000619478	0.01201	*	0.0862693
9	-.00511501	-0.09920	**	0.0862808
10	0.00360648	0.06995	*	0.0870639

'.' MARKS TWO STANDARD ERRORS

(Parts (c), the RSPAC, (d), the RSCC, and (e), the Forecasts, follow on pages 390 and 391.)

TABLE 3   *Continued*

(c) The RSPAC

PARTIAL AUTOCORRELATIONS

LAG	CORRELATION
1	0.01271
2	-0.00017
3	-0.06446
4	-0.05496
5	0.15335
6	0.07535
7	-0.06501
8	0.02846
9	-0.07362
10	0.05239

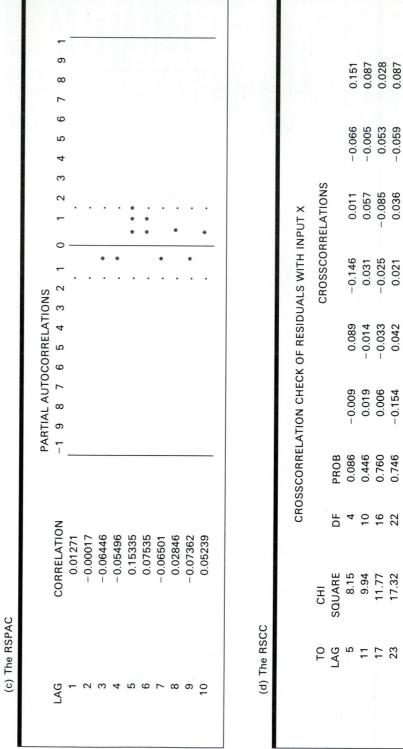

(d) The RSCC

CROSSCORRELATION CHECK OF RESIDUALS WITH INPUT X

TO LAG	CHI SQUARE	DF	PROB	CROSSCORRELATIONS						
5	8.15	4	0.086	-0.009	0.089	-0.154	-0.146	0.011	-0.066	0.151
11	9.94	10	0.446	0.019	-0.014	0.031	0.057	-0.005	0.087	
17	11.77	16	0.760	0.006	-0.033	-0.025	-0.085	0.053	0.028	
23	17.32	22	0.746	-0.154	0.042	0.021	0.036	-0.059	0.087	

(e) Forecasts of y_{151} through y_{160}, where Future Values of x_t are Forecasted by the Model

$$z_t^{(x)} = (1 - \theta_1^{(x)} B) a_t^{(x)}$$

where

$$z_t^{(x)} = x_t - x_{t-1}$$

FORECASTS FOR VARIABLE Y

OBS	FORECAST	STD ERROR	LOWER 95%	UPPER 95%
	FORECAST BEGINS			
151	262.8970	0.2271	262.4520	263.3421
152	264.2397	0.2514	263.7469	264.7325
153	263.4785	0.2736	262.9423	264.0147
154	263.4736	1.3669	260.7945	266.1526
155	263.4784	2.1860	259.1940	267.7628
156	263.4903	2.9438	257.7206	269.2601
157	263.5073	3.6538	256.3460	270.6686
158	263.5280	4.3194	255.0621	271.9939
159	263.5513	4.9437	253.8618	273.2409
160	263.5766	5.5300	252.7380	274.4153

g.   Describe the behavior of the RSPAC.

h.   Use the behavior of the RSAC and RSPAC to rationalize that the model

$$\eta_t = (1 - \theta_1 B)a_t$$

is a reasonable tentative model describing η_t. What guideline(s) (in Tables 2.3 and 3.3) did you use to identify this model?

i.   Use the model of part h describing η_t to specify an appropriate final transfer function model.

7.6   The SAS output of estimation, diagnostic checking, and forecasting for the final model is given in Table 3.

a.   Find and identify the least squares point estimates of the model parameters.

b.   Do all of the model parameters seem to be significantly different from zero? Explain your answer.

c.   What do the correlations of the least squares point estimates say about the adequacy of the model?

d.   What do the behaviors of the RSAC and the RSPAC say about the adequacy of the model? Explain your answer.

e.   What do the values of the Q^* statistics for $K = 6, 12, 18$, and 24 (and the associated prob-values) say about the adequacy of the model? Explain your answer.

f.   Find a point forecast and a 95% prediction interval forecast for y_{151}. Interpret the 95% prediction interval.

7.7   Write the five SAS programs needed to carry out the transfer function analysis of the sales and leading economic indicator data. That is, write SAS programs to:

a.   Compute the SAC and SPAC for each of the transformations

$$z_t^{(x)} = x_t \quad \text{and} \quad z_t^{(x)} = x_t - x_{t-1}$$

b.   Fit the tentative model describing x_t

$$z_t^{(x)} = (1 - \theta_1^{(x)} B)a_t^{(x)} \quad \text{where} \quad z_t^{(x)} = x_t - x_{t-1}$$

c.   Compute the prewhitened x_t and y_t values and compute the SCC.

d.   Fit the preliminary transfer function model

$$z_t = \mu + \frac{C}{(1 - \delta_1 B)} B^3 z_t^{(x)} + \eta_t$$

e.   Fit (and forecast with) the final transfer function model

$$z_t = \mu + \frac{C}{(1 - \delta_1 B)} B^3 z_t^{(x)} + (1 - \theta_1 B)a_t$$

CLASSICAL REGRESSION ANALYSIS

8.1 INTRODUCTION

In this chapter we discuss *classical regression analysis*, which is a very useful statistical technique that can be used to *describe and predict* a *dependent variable* on the basis of one or more *independent variables*. We begin in Section 8.2, which illustrates the use of regression models in estimation and prediction. Here we present the *simple linear regression model*, the *quadratic regression model*, and more complicated models. Section 8.3 addresses the topic of model building— how to use historical data to build an appropriate regression model. Included are discussions of R^2, *the multiple coefficient of determination*, testing the *importance of an independent variable*, comparing regression models, an *F-test* for the adequacy of the overall regression model, and a *partial F-test* for the importance of a portion of a regression model. In Section 8.4 we present the *assumptions* behind regression analysis, and we describe how to use *residual analysis* to check the validity of these assumptions. We discuss validation of the *normality*, *constant variance*, and *independence assumptions*, as well as detection of *outlying* and *influential observations*. Section 8.5 demonstrates how to use *dummy variables* in regression analysis to compare population means. Finally, we conclude this chapter with Section 8.6, which shows how to use SAS to implement regression analysis.

8.2 REGRESSION MODELS AND THEIR USE IN ESTIMATION AND PREDICTION

8.2.1 The Simple Linear and Quadratic Regression Models

EXAMPLE 8.1

Comp-U-Systems, a new computer manufacturer, sells and services the Comp-U-Systems Microcomputer. As part of its standard purchase contract, Comp-U-Systems agrees to perform regular service on its microcomputer. Because of its lack of experience in servicing computers, Comp-U-Systems wishes to obtain information concerning the time it takes to perform the required service and therefore has collected data for its first eleven service calls. This data is presented in Table 8.1, which lists the number of microcomputers serviced and the time (in minutes) required to perform the needed service for each of these service calls. Here, for $t = 1, 2, \ldots, 11$,

x_t = the number of microcomputers serviced on the tth service call, and
y_t = the number of minutes required to perform service on the tth service call.

Note that the data is listed in Table 8.1a in the time order in which it was observed, and thus the subscript "t" in this table refers to time. However, since we can perform any regression analysis by listing the observed data in any order, we (arbitrarily) list the data in Table 8.1b in order of increasing numbers of micro-computers serviced. Thus, the subscript "t" in Table 8.1b refers to the order, in terms of the number of microcomputers serviced, of an observation. Henceforth, we will analyze the Comp-U-Systems data as it is listed in Table 8.1b.

In Figure 8.1a the observed values of y_t are plotted against the increasing values of x_t. Inspecting this figure, we see that as the values of x_t increase, the observed values of y_t tend to increase in a straight line fashion. This implies that we can relate y_t to x_t by the *simple linear regression model*

$$y_t = \mu_t + \varepsilon_t$$
$$= \beta_0 + \beta_1 x_t + \varepsilon_t$$

This model says that the values of y_t can be represented by an *average level* (denoted μ_t) that changes as x_t increases according to the *straight line* defined by the equation

$$\mu_t = \beta_0 + \beta_1 x_t$$

combined with random fluctuations (represented by the *error term* ε_t) which cause the values of y_t to deviate from the average level. The slope of the straight line

$$\mu_t = \beta_0 + \beta_1 x_t$$

TABLE 8.1   *Comp-U-Systems Service Call Data*

(a)   Data Arranged in Time Order

Service Call, t	Number of Microcomputers Serviced, x_t	Number of Minutes Required on Call, y_t
1	4	109
2	2	58
3	5	138
4	7	189
5	1	37
6	3	82
7	4	103
8	5	134
9	2	68
10	4	112
11	6	154

(b)   Data Arranged in Order of Increasing Values of Number of Microcomputers Serviced

Service Call, t	Number of Microcomputers Serviced, x_t	Number of Minutes Required on Call, y_t
1	1	37
2	2	58
3	2	68
4	3	82
5	4	103
6	4	109
7	4	112
8	5	134
9	5	138
10	6	154
11	7	189

(Part (c), Calculation of the least squares point estimate, follows on page 396.)

is β_1, while the intercept at $x_t = 0$ is β_0. Since (as illustrated in Figure 8.1b) the average level μ_t increases as x_t increases, the slope β_1 of the straight line is positive. The error term ε_t in the regression model

$$y_t = \beta_0 + \beta_1 x_t + \varepsilon_t$$

measures the effect on y_t (service time) of all factors other than x_t (the number of

TABLE 8.1   *Continued*

(c)   Calculation of the Least Squares Point Estimates

x_t	y_t	$x_t y_t$	x_t^2
1	37	(1)(37) = 37	$(1)^2$ = 1
2	58	(2)(58) = 116	$(2)^2$ = 4
2	68	(2)(68) = 136	$(2)^2$ = 4
3	82	(3)(82) = 246	$(3)^2$ = 9
4	103	(4)(103) = 412	$(4)^2$ = 16
4	109	(4)(109) = 436	$(4)^2$ = 16
4	112	(4)(112) = 448	$(4)^2$ = 16
5	134	(5)(134) = 670	$(5)^2$ = 25
5	138	(5)(138) = 690	$(5)^2$ = 25
6	154	(6)(154) = 924	$(6)^2$ = 36
7	189	(7)(189) = 1323	$(7)^2$ = 49

$$\sum_{t=1}^{11} x_t = 43 \qquad \sum_{t=1}^{11} y_t = 1184 \qquad \sum_{t=1}^{11} x_t y_t = 5438 \qquad \sum_{t=1}^{11} x_t^2 = 201$$

$$b_1 = \frac{11 \sum_{t=1}^{11} x_t y_t - \left(\sum_{t=1}^{11} x_t \right)\left(\sum_{t=1}^{11} y_t \right)}{11 \sum_{t=1}^{11} x_t^2 - \left(\sum_{t=1}^{11} x_t \right)^2} = \frac{11(5438) - (43)(1184)}{11(201) - 1849} = 24.6022$$

and

$$b_0 = \bar{y} - b_1 \bar{x} = 107.6364 - (24.6022)(3.9091) = 11.4641$$

where

$$\bar{y} = \frac{\sum_{t=1}^{11} y_t}{11} = \frac{1184}{11} = 107.6364 \qquad \text{and} \qquad \bar{x} = \frac{\sum_{t=1}^{11} x_t}{11} = \frac{43}{11} = 3.9091$$

microcomputers serviced). One such factor might be the complexity of the problems that must be corrected on a particular service call.

Since we do not know the true values of the parameters β_0 and β_1 describing the line

$$\mu_t = \beta_0 + \beta_1 x_t$$

we must estimate this line. However, we have a problem. How does a person find the "best" line that can be drawn through the points? One approach would be to simply "eyeball" a line through the points. Mathematically, this would really amount to guessing the values of the slope β_1 and the intercept β_0 of the line. For example, we might guess that the slope is 25 and that the intercept is 12. These

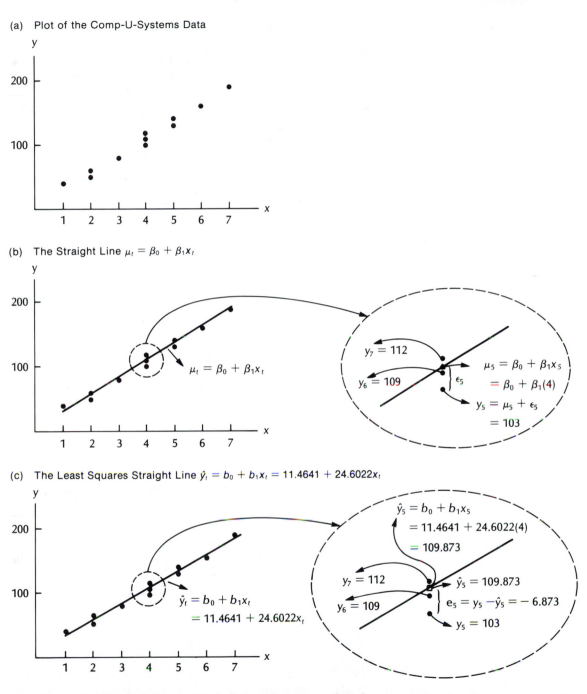

(a)   Plot of the Comp-U-Systems Data

(b)   The Straight Line $\mu_t = \beta_0 + \beta_1 x_t$

$\mu_t = \beta_0 + \beta_1 x_t$

$y_7 = 112$

$y_6 = 109$

$\mu_5 = \beta_0 + \beta_1 x_5$

$= \beta_0 + \beta_1(4)$

$y_5 = \mu_5 + \epsilon_5$

$= 103$

(c)   The Least Squares Straight Line $\hat{y}_t = b_0 + b_1 x_t = 11.4641 + 24.6022 x_t$

$\hat{y}_t = b_0 + b_1 x_t$

$= 11.4641 + 24.6022 x_t$

$\hat{y}_5 = b_0 + b_1 x_5$

$= 11.4641 + 24.6022(4)$

$= 109.873$

$y_7 = 112$

$y_6 = 109$

$\hat{y}_5 = 109.873$

$e_5 = y_5 - \hat{y}_5 = -6.873$

$y_5 = 103$

FIGURE 8.1   *The Plot of the Comp-U-Systems Data, the Line $\mu_t = \beta_0 + \beta_1 x_t$, and the Line $\hat{y}_t = b_0 + b_1 x_t = 11.4641 + 24.6022 x_t$*

guesses would yield a line defined by the equation

$$\hat{y}_t = 12 + 25x_t$$

where $\hat{y}_t$ denotes "predicted service time." Unfortunately, each person guessing would probably come up with a different line. Thus, the following question arises. *Is there one line that we can consider "best"?* Well, there is such a line! Using statistical theory, it can be shown that the best *point estimates* b_1 and b_0 of the slope β_1 and intercept β_0 are calculated as illustrated in Table 8.1c. We call $b_1 = 24.6022$ and $b_0 = 11.4641$ the *least squares point estimates* of β_1 and β_0, and we call the line

$$\hat{y}_t = b_0 + b_1 x_t$$
$$= 11.4641 + 24.6022 x_t$$

the *least squares line* (see Figure 8.1c). To explain the meaning of the term *least squares*, first note that we call $\hat{y}_t$ a "predicted service time" because, since

$$\hat{y}_t = b_0 + b_1 x_t$$
$$= 11.4641 + 24.6022 x_t$$

is the obvious point estimate of the average level

$$\mu_t = \beta_0 + \beta_1 x_t$$

and since (for reasons to be discussed in Section 8.4) we predict the error term ε_t to be zero, it follows that $\hat{y}_t$ is also the point *prediction* of the actual (individual) service time

$$y_t = \mu_t + \varepsilon_t$$
$$= \beta_0 + \beta_1 x_t + \varepsilon_t$$

Now, consider the SAS output in Table 8.2, and recall that we arranged the data in order of increasing values of x (number of microcomputers serviced) when performing the regression analysis. In addition to presenting the least squares point estimates $b_0 = 11.4641$ and $b_1 = 24.6022$, Table 8.2 presents—for each historical (that is, previously observed) service call t—the residual $(y_t - \hat{y}_t)$, which is the difference between the observed service time y_t and the predicted service time

$$\hat{y}_t = b_0 + b_1 x_t$$
$$= 11.4641 + 24.6022 x_t$$

For example, since (see Table 8.2) $y_5 = 103$ and

$$\hat{y}_5 = b_0 + b_1 x_5$$
$$= 11.4641 + 24.6022(4)$$
$$= 109.873$$

TABLE 8.2 *SAS Output of a Regression Analysis of the Comp-U-Systems Data*

DEP VARIABLE: Y

SOURCE	DF	SUM OF SQUARES	MEAN SQUARE	F VALUE	PROB > F	
MODEL	1	19918.844[a]	19918.844	935.149	0.0001	Analysis
ERROR	9	191.702[b]	21.300184[c]			of Variance
C TOTAL	10	20110.545[d]				Table

ROOT MSE	4.615212[e]	R-SQUARE	0.9905[f]	
DEP MEAN	107.636	ADJ R-SQ	0.9894[g]	
C.V.	4.287782			

VARIABLE	DF	PARAMETER[h] ESTIMATE	STANDARD[i] ERROR	T FOR HO:[j] PARAMETER=0	PROB > |T|[k]
INTERCEP	1	11.464088 [l]	3.439026	3.334	0.0087
X	1	24.602210[m]	0.804514	30.580	0.0001

x_t	OBS	ACTUAL[n]	PREDICT[o] VALUE	STD ERR[p] PREDICT	LOWER 95% MEAN	UPPER 95% MEAN	LOWER 95% PREDICT	UPPER 95% PREDICT	RESIDUAL
1	1	37.000	36.066	2.723	29.907	42.226	23.944	48.188	0.933702
2	2	58.000	60.669	2.073	55.980	65.357	49.224	72.113	−2.669
2	3	68.000	60.669	2.073	55.980	65.357	49.224	72.113	7.331
3	4	82.000	85.271	1.572	81.715	88.827	74.241	96.300	−3.271
4	5	{103.000}[q]	{109.873}[r]	{1.393}[s]	[106.721,	113.025][t]	[98.967,	120.779][u]	{−6.873}[v]
4	6	109.000	109.873	1.393	106.721	113.025	98.967	120.779	−.872928
4	7	112.000	109.873	1.393	106.721	113.025	98.967	120.779	2.127
5	8	134.000	134.475	1.645	130.753	138.197	123.391	145.559	−.475138
5	9	138.000	134.475	1.645	130.753	138.197	123.391	145.559	3.525
6	10	154.000	159.077	2.183	154.139	164.016	147.528	170.627	−5.077
7	11	189.000	183.680	2.850	177.233	190.126	171.409	195.950	5.320

SUM OF RESIDUALS −1.06581E−14
SUM OF SQUARED RESIDUALS 191.7017[w]

[a] Explained Variation   [b] SSE [c] s^2 [d] Total Variation   [e] s [f] R^2 [g] $\bar{R}^2$ [h] b_j [i] s_{b_j} [j] t_{b_j} [k] Prob-value   [l] b_0

[m] b_1 [n] y_t [o] $\hat{y}_t$ [p] $s\sqrt{h_{tt}}$ [q] y_5 [r] $\hat{y}_5$ [s] $s\sqrt{h_{55}}$ [t] 95% Confidence Interval for μ_5 [u] 95% Prediction Interval for y_5

[v] $y_5 - \hat{y}_5$ [w] SSE

it follows that the residual for the fifth service call is

$$y_5 - \hat{y}_5 = 103 - 109.873$$

$$= -6.873$$

Note the size of each residual ($y_t - \hat{y}_t$) and hence the size of the *sum of squared residuals* (see Table 8.2)

$$SSE = \sum_{t=1}^{11} (y_t - \hat{y}_t)^2$$

$$= 191.7017$$

depends on the least squares point estimates $b_0 = 11.4641$ and $b_1 = 24.6022$ of the parameters β_0 and β_1. A different set of point estimates of β_0 and β_1 would yield a different sum of squared residuals. The reason we call $b_0 = 11.4641$ and $b_1 = 24.6022$ the least squares point estimates of β_0 and β_1 is that it can be shown (by manipulations we need not go into here) that these point estimates give a value of *SSE* that is smaller than the one any other point estimates would give. Stated *geometrically*, if we calculate the *distance* $(y_t - \hat{y}_t)$ between each point (y_t) on our data plot and the corresponding prediction $\hat{y}_t$ on the least squares line

$$\hat{y}_t = b_0 + b_1 x_t$$

$$= 11.4641 + 24.6022 x_t$$

and if we square each distance and add up all the squares, then we get a smaller sum of squared distances than could be obtained using any other line! In this sense, then, the least squares line is the "best line that can be drawn through the points on the data plot." (See Figure 8.1c).

Now consider a future service call on which we will service $x_t = 4$ microcomputers. It follows that

$$\hat{y}_t = b_0 + b_1 x_t$$

$$= 11.4641 + 24.6022(4)$$

$$= 109.873$$

is the *point estimate* of

$$\mu_t = \beta_0 + \beta_1 x_t = \beta_0 + \beta_1(4)$$

the *average* time required to service four microcomputers, and is the *point prediction* of

$$y_t = \mu_t + \varepsilon_t = \beta_0 + \beta_1(4) + \varepsilon_t$$

the *actual* time required on an (*individual*) future service call to service four microcomputers (note that the SAS output gives the prediction $\hat{y}_t = 109.873$ corresponding to $x_t = 4$). Moreover, it can be shown that (under assumptions to be discussed in Section 8.4) a $100(1 - \alpha)\%$ *confidence interval* for $\mu_t = \beta_0 + \beta_1 x_t$ is

$$[\hat{y}_t \pm t_{[\alpha/2]}^{(n-2)} s\sqrt{h_{tt}}]$$

and a $100(1 - \alpha)\%$ *prediction interval* for $y_t = \mu_t + \varepsilon_t$ is

$$[\hat{y}_t \pm t_{[\alpha/2]}^{(n-2)} s\sqrt{1 + h_{tt}}]$$

Here

1. $t_{[\alpha/2]}^{(n-2)}$ is the point on the scale of the t-distribution having $(n-2)$ degrees of freedom so that the area under the curve of this t-distribution to the right of $t_{[\alpha/2]}^{(n-2)}$ is $\alpha/2$. Since (in the present situation) $n = 11$, it follows that if we wish to calculate 95% confidence and prediction intervals, then we would use $t_{[\alpha/2]}^{(n-2)} = t_{[.05/2]}^{(11-2)} = t_{[.025]}^{(9)} = 2.262$.

2. s is called the *standard error* and is given by the formula (see Table 8.2)

$$ s = \sqrt{\frac{SSE}{n-2}} = \sqrt{\frac{191.7017}{11-2}} = 4.6152 $$

3. h_{tt} is called a *leverage value* and is given by the equation

$$ h_{tt} = \frac{1}{n} + \frac{(x_t - \bar{x})^2}{\displaystyle\sum_{t=1}^{n} x_t^2 - n\bar{x}^2} $$

where

$$ \bar{x} = \frac{\displaystyle\sum_{t=1}^{n} x_t}{n} = \frac{\displaystyle\sum_{t=1}^{11} x_t}{11} $$

$$ = \frac{1+2+2+3+4+4+4+5+5+6+7}{11} $$

$$ = 3.9091 $$

Since $x_t = 4$ microcomputers will be serviced on the future service call, it follows that

$$ h_{tt} = \frac{1}{11} + \frac{(x_t - \bar{x})^2}{\displaystyle\sum_{t=1}^{11} x_t^2 - 11\bar{x}^2} $$

$$ = \frac{1}{11} + \frac{(4 - 3.9091)^2}{201 - 11(3.9091)^2} $$

$$ = .0911 $$

Note that h_{tt} can be obtained from the SAS output in Table 8.2. Although this output does not give h_{tt} directly, it does give $s(h_{tt})^{1/2}$. Looking for the value of $s(h_{tt})^{1/2}$ corresponding to $x_t = 4$, which is

$$ s\sqrt{h_{tt}} = 1.393 $$

which implies that

$$ h_{tt} = \left(\frac{1.393}{s}\right)^2 = \left(\frac{1.393}{4.6152}\right)^2 = .0911 $$

we find that a 95% confidence interval for μ_t is

$$[\hat{y}_t \pm t_{[.025]}^{(11-2)} s\sqrt{h_{tt}}] = [109.873 \pm 2.262(4.6152)\sqrt{.0911}]$$

$$= [106.721, 113.025]$$

Note that this interval is given on the SAS output. This interval says that we are 95% confident that

$$\mu_t = \beta_0 + \beta_1 x_t = \beta_0 + \beta_1(4)$$

the average time required to service four microcomputers, is at least 106.721 minutes and is no more than 113.025 minutes. Furthermore, a 95% prediction interval for $y_t = \mu_t + \varepsilon_t$ is

$$[\hat{y}_t \pm t_{[.025]}^{(11-2)} s\sqrt{1 + h_{tt}}] = [109.873 \pm 2.262(4.6152)\sqrt{1 + .0911}]$$

$$= [98.967, 120.779]$$

This interval is also given on the SAS output. The prediction interval says that we are 95% confident that

$$y_t = \mu_t + \varepsilon_t$$
$$= \beta_0 + \beta_1 x_t + \varepsilon_t = \beta_0 + \beta_1(4) + \varepsilon_t$$

the actual time required to service four microcomputers on a future (individual) service call, will be at least 98.967 minutes and will be no more than 120.779 minutes.

To complete this example, two comments should be made. First, note that the $100(1 - \alpha)\%$ prediction interval for $y_t = \beta_0 + \beta_1 x_t + \varepsilon_t$

$$[\hat{y}_t \pm t_{[\alpha/2]}^{(n-2)} s\sqrt{1 + h_{tt}}]$$

is longer than the $100(1 - \alpha)\%$ confidence interval for $\mu_t = \beta_0 + \beta_1 x_t$

$$[\hat{y}_t \pm t_{[\alpha/2]}^{(n-2)} s\sqrt{h_{tt}}]$$

because, whereas the "error bound" $t_{[\alpha/2]}^{(n-2)} s(h_{tt})^{1/2}$ in the confidence interval measures the uncertainty associated with not knowing the parameters β_0 and β_1, the "error bound" $t_{[\alpha/2]}^{(n-2)} s(1 + h_{tt})^{1/2}$ in the prediction interval measures the uncertainty associated with not knowing β_0 and β_1 and the added uncertainty associated with not knowing the error term ε_t. Second, the farther that the future value of x_t is from $\bar{x}$ (the average of the historical values of x_t), the larger is the leverage value

$$h_{tt} = \frac{1}{n} + \frac{(x_t - \bar{x})^2}{\sum\limits_{t=1}^{n} x_t^2 - n\bar{x}^2}$$

and thus the longer (less accurate) are the confidence and prediction intervals.

(a) If $\beta_1 > 0$, the average level increases as x_t increases

$\mu_t = \beta_0 + \beta_1 x_t$

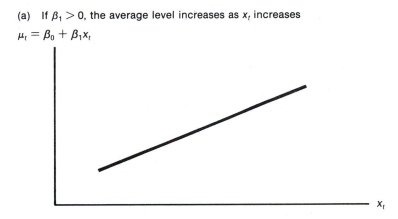

(b) If $\beta_1 < 0$, the average level decreases as x_t increases

$\mu_t = \beta_0 + \beta_1 x_t$

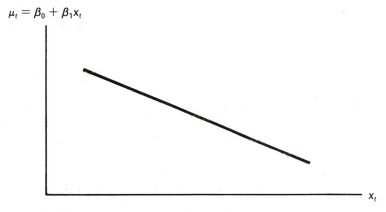

FIGURE 8.2   An Average Level That Changes in a Straight Line Fashion as x_t Increases

The previous example has illustrated the use of the simple linear regression model

$$
\begin{aligned}
y_t &= \mu_t + \varepsilon_t \\
&= \beta_0 + \beta_1 x_t + \varepsilon_t
\end{aligned}
$$

to relate a *dependent variable* y_t (service time) to an *independent variable* x_t (the number of microcomputers serviced). While the simple linear regression model used in Example 8.1 has a positive slope ($\beta_1 > 0$), which implies that μ_t increases as x_t increases (as illustrated in Figures 8.1b and 8.2a), in some situations we use the simple linear regression model having a negative slope ($\beta_1 < 0$), which implies that μ_t decreases as x_t increases (as illustrated in Figure 8.2b).

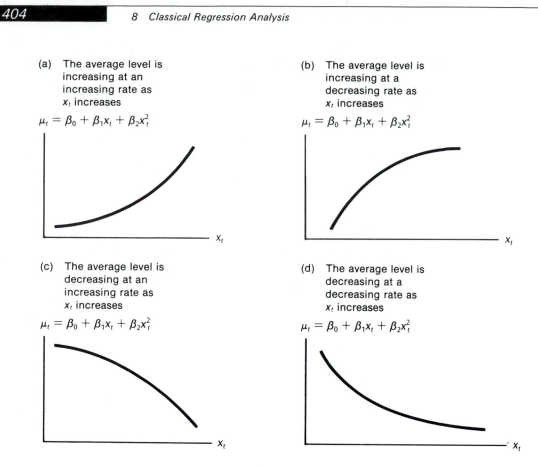

(a) The average level is increasing at an increasing rate as x_t increases

$$\mu_t = \beta_0 + \beta_1 x_t + \beta_2 x_t^2$$

(b) The average level is increasing at a decreasing rate as x_t increases

$$\mu_t = \beta_0 + \beta_1 x_t + \beta_2 x_t^2$$

(c) The average level is decreasing at an increasing rate as x_t increases

$$\mu_t = \beta_0 + \beta_1 x_t + \beta_2 x_t^2$$

(d) The average level is decreasing at a decreasing rate as x_t increases

$$\mu_t = \beta_0 + \beta_1 x_t + \beta_2 x_t^2$$

FIGURE 8.3   An Average Level that Changes in a Quadratic Fashion as x_t Increases

Another useful regression model is the *quadratic regression model*

$$
\begin{aligned}
y_t &= \mu_t + \varepsilon_t \\
&= \beta_0 + \beta_1 x_t + \beta_2 x_t^2 + \varepsilon_t
\end{aligned}
$$

This model says that the values of y_t can be represented by an average level (μ_t) that changes as x_t increases according to the *quadratic curve* defined by the equation

$$\mu_t = \beta_0 + \beta_1 x_t + \beta_2 x_t^2$$

combined with random fluctuations (represented by ε_t) which cause the values of y_t to deviate from the average level. Thus, as illustrated in Figure 8.3, the average level μ_t either is increasing at an increasing or a decreasing rate or is decreasing at an increasing or decreasing rate as x_t increases. The particular numerical values of β_0, β_1, and β_2 determine exactly how the average level changes as x_t increases. We will use the quadratic regression model in the next example.

8.2.2 More Complex Regression Models

We now consider more complicated classical regression models. Again, we begin with an example.

EXAMPLE 8.2

Enterprise Industries produces Fresh, a brand of liquid laundry detergent. The company would like to develop a prediction model that can be used to predict the demand for the extra-large sized bottle of Fresh. With a reliable model, Enterprise Industries can more effectively plan its production schedule, plan its budget, and estimate requirements for producing and storing this product. The demand for Fresh (in hundreds of thousands of bottles) in "sales period t," where each sales period is defined to last four weeks, is denoted by the symbol y_t, is called the *dependent variable*, and is believed (subjectively) to be partially determined by one or more of the *independent variables* x_{t1} = the price (in dollars) of Fresh in period t as offered by Enterprise Industries, x_{t2} = the average industry price (in dollars) in period t of competitors' similar detergents, and x_{t3} = the advertising expenditure (in hundreds of thousands of dollars) in period t of Enterprise Industries to promote Fresh. Assume the data in Table 8.3 has been observed over the past 30 sales periods.

Before using the data to help us determine an appropriate model relating y_t to x_{t1}, x_{t2}, and x_{t3}, consider forming one independent variable from the two independent variables x_{t1} and x_{t2}. That is, consider defining the independent variable

$$x_{t4} = x_{t2} - x_{t1}$$

which is the difference between x_{t2}, the average industry price, and x_{t1}, Enterprise Industries' price for Fresh. Of course, this is not the only way that x_{t1} and x_{t2} can be combined. For example,

$$x_{t5} = \frac{x_{t2}}{x_{t1}}$$

the ratio of the average industry price to Enterprise Industries' price, might be useful in forecasting sales of Fresh. In this example we will concentrate on constructing a regression model that predicts y_t as a function of x_{t4} and x_{t3}. Noting that the observed values of x_{t4}, the *price difference*, are given in Table 8.3, we plot the values of y_t against the increasing values of x_{t4} in Figure 8.4a and we plot the values of y_t against the increasing values of x_{t3} in Figure 8.4b. Since Figure 8.4a indicates that the values of y_t (demand for Fresh) tend to increase in a straight line fashion (with a positive slope) as the values of x_{t4} (price difference) increase, it is reasonable to relate y_t to x_{t4} by the simple linear regression model

$$y_t = \mu_t + \varepsilon_t$$
$$= \beta_0 + \beta_1 x_{t4} + \varepsilon_t$$

TABLE 8.3   *The Historical Observations*

t	Demand y_t	Price x_{t1}	Average Industry Price x_{t2}	Price Difference $x_{t4} = x_{t2} - x_{t1}$	Advertising Expenditure x_{t3}
1	7.38	3.85	3.80	−.05	5.50
2	8.51	3.75	4.00	.25	6.75
3	9.52	3.70	4.30	.60	7.25
4	7.50	3.70	3.70	0	5.50
5	9.33	3.60	3.85	.25	7.00
6	8.28	3.60	3.80	.20	6.50
7	8.75	3.60	3.75	.15	6.75
8	7.87	3.80	3.85	.05	5.25
9	7.10	3.80	3.65	−.15	5.25
10	8.00	3.85	4.00	.15	6.00
11	7.89	3.90	4.10	.20	6.50
12	8.15	3.90	4.00	.10	6.25
13	9.10	3.70	4.10	.40	7.00
14	8.86	3.75	4.20	.45	6.90
15	8.90	3.75	4.10	.35	6.80
16	8.87	3.80	4.10	.30	6.80
17	9.26	3.70	4.20	.50	7.10
18	9.00	3.80	4.30	.50	7.00
19	8.75	3.70	4.10	.40	6.80
20	7.95	3.80	3.75	−.05	6.50
21	7.65	3.80	3.75	−.05	6.25
22	7.27	3.75	3.65	−.10	6.00
23	8.00	3.70	3.90	.20	6.50
24	8.50	3.55	3.65	.10	7.00
25	8.75	3.60	4.10	.50	6.80
26	9.21	3.65	4.25	.60	6.80
27	8.27	3.70	3.65	−.05	6.50
28	7.67	3.75	3.75	0	5.75
29	7.93	3.80	3.85	.05	5.80
30	9.26	3.70	4.25	.55	6.80

Since Figure 8.4b indicates that the values of y_t (demand for Fresh) tend to increase at an increasing rate as the values of x_{t3} (advertising expenditure) increase, it is reasonable to relate y_t to x_{t3} by the quadratic regression model

$$y_t = \mu_t + \varepsilon_t$$
$$= \beta_0 + \beta_1 x_{t3} + \beta_2 x_{t3}^2 + \varepsilon_t$$

Although it is logical to relate y_t to x_{t4} and x_{t3} by combining these models to form

(a)   The Plot of the Values of y_t against the Increasing Values of x_{t4}

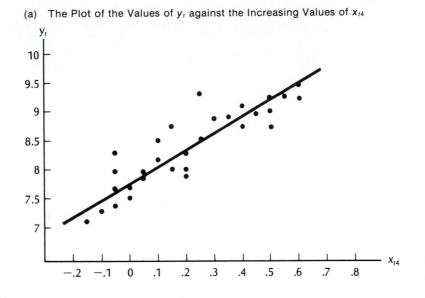

(b)   The Plot of the Values of y_t against the Increasing Values of x_{t3}

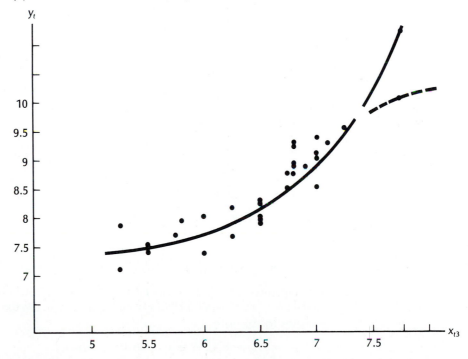

FIGURE 8.4   Data Plots in the Fresh Detergent Problem

the model

$$y_t = \mu_t + \varepsilon_t$$
$$= \beta_0 + \beta_1 x_{t4} + \beta_2 x_{t3} + \beta_3 x_{t3}^2 + \varepsilon_t$$

we will later find that we should also include the *cross-product term* $x_{t4}x_{t3}$ and thus form the model

$$y_t = \mu_t + \varepsilon_t$$
$$= \beta_0 + \beta_1 x_{t4} + \beta_2 x_{t3} + \beta_3 x_{t3}^2 + \beta_4 x_{t4}x_{t3} + \varepsilon_t$$

It can be shown that including $x_{t4}x_{t3}$ in the model implies that *interaction* exists between x_{t4} and x_{t3}, which means that the relationship between y_t and x_{t3} depends upon x_{t4}. Therefore, the effectiveness of additional advertising expenditure depends upon the price difference. We will discuss the exact meaning of "interaction" in a later example.

The regression model

$$y_t = \beta_0 + \beta_1 x_{t4} + \beta_2 x_{t3} + \beta_3 x_{t3}^2 + \beta_4 x_{t4}x_{t3} + \varepsilon_t$$

says that the values of y_t can be represented by an average level (denoted μ_t) that changes over time according to the equation

$$\mu_t = \beta_0 + \beta_1 x_{t4} + \beta_2 x_{t3} + \beta_3 x_{t3}^2 + \beta_4 x_{t4}x_{t3}$$

combined with random fluctuations (represented by the error term ε_t) which cause the values of y_t to deviate from the average level. The error term ε_t represents the effects of all factors that are not explicitly included in the model. One such factor is competitors' advertising expenditures (possibly not explicitly included in the model because Enterprise Industries would have great difficulty determining future values of this variable).

The SAS output of the *least squares point estimates*

$$b_0 = 29.1133$$
$$b_1 = 11.1342$$
$$b_2 = -7.6080$$
$$b_3 = .6712$$

and

$$b_4 = -1.4777$$

of the parameters β_0, β_1, β_2, β_3, and β_4 in the model is given in Table 8.4. To explain the meaning of the term *least squares*, note that we can use the least squares point estimates to calculate

$$\hat{y}_t = b_0 + b_1 x_{t4} + b_2 x_{t3} + b_3 x_{t3}^2 + b_4 x_{t4}x_{t3}$$
$$= 29.1133 + 11.1342 x_{t4} - 7.6080 x_{t3} + .6712 x_{t3}^2 - 1.4777 x_{t4}x_{t3}$$

which is the point estimate of the average level

$$\mu_t = \beta_0 + \beta_1 x_{t4} + \beta_2 x_{t3} + \beta_3 x_{t3}^2 + \beta_4 x_{t4} x_{t3}$$

and is the point prediction of the actual demand for Fresh in period t

$$y_t = \mu_t + \varepsilon_t$$
$$= \beta_0 + \beta_1 x_{t4} + \beta_2 x_{t3} + \beta_3 x_{t3}^2 + \beta_4 x_{t4} x_{t3} + \varepsilon_t$$

since we predict the error term to be zero. Now, note that the SAS output in Table 8.4 presents (for each historical—that is, previously observed—period t) the *residual* $(y_t - \hat{y}_t)$, which is the difference between the observed demand y_t and the demand prediction $\hat{y}_t$ made using the prediction equation. For example, since (see Table 8.4) $y_2 = 8.510$ and

$$\hat{y}_2 = 29.1133 + 11.1342 x_{24} - 7.6080 x_{23} + .6712 x_{23}^2 - 1.4777 x_{24} x_{23}$$

$$= 29.1133 + 11.1342(.25) - 7.6080(6.75)$$

$$\quad + .6712(6.75)^2 - 1.4777(.25)(6.75)$$

$$= 8.633$$

it follows that the residual in period 2 is (see Table 8.4)

$$y_2 - \hat{y}_2 = 8.510 - 8.633 = -.123$$

Note that the size of each residual $(y_t - \hat{y}_t)$ and hence the size of the *sum of squared residuals* (see Table 8.4)

$$SSE = \sum_{t=1}^{30} (y_t - \hat{y}_t)^2 = 1.0644$$

depends upon the least squares estimates $b_0 = 29.1133$, $b_1 = 11.1342$, $b_2 = -7.6080$, $b_3 = .6712$, and $b_4 = -1.4777$ of the parameters β_0, β_1, β_2, β_3, and β_4. A different set of point estimates of β_0, β_1, β_2, β_3, and β_4 would yield a different sum of squared residuals. The reason we call $b_0 = 29.1133$, $b_1 = 11.1342$, $b_2 = -7.6080$, $b_3 = .6712$, and $b_4 = -1.4777$ the *least squares point estimates* is that it can be shown by manipulations we need not go into here that these point estimates give a value of *SSE* that is smaller than the one any other point estimates would give. How the data in Table 8.3 is used to calculate least squares point estimates will not be discussed here. It is sufficient to know that SAS and other computer packages can be used to calculate least squares point estimates.

Next, suppose that in the future sales period $t = 31$, the price for Fresh will be $x_{31,1} = 3.80$, the average industry price will be $x_{31,2} = 3.90$, and the advertising expenditure for Fresh will be $x_{31,3} = 6.80$. Since the price difference will be

$$x_{31,4} = x_{31,2} - x_{31,1} = 3.90 - 3.80 = .10$$

TABLE 8.4 The SAS Output of a Regression Model $y_t = \beta_0 + \beta_1 x_{t4} + \beta_2 x_{t3} + \beta_3 x_{t3}^2 + \beta_4 x_{t4} x_{t3} + \varepsilon_t$

DEP VARIABLE: Y

SOURCE	DF	SUM OF SQUARES	MEAN SQUARE	F VALUE	PROB > F	
MODEL	4^e	12.394190^a	3.098548^b	72.777^c	0.0001^d	Analysis of Variance Table
ERROR	25^f	1.064397^g	0.042576^h			
C TOTAL	29	13.458587^i				

ROOT MSE	0.206339^j	R-SQUARE	0.9209^k	
DEP MEAN	8.382667	ADJ R-SQ	0.9083^l	
C.V.	2.461498			

VARIABLE	DF	PARAMETER ESTIMATE	STANDARD[r] ERROR	T FOR HO:[s] PARAMETER=0	PROB > |T|[t]
INTERCEP	1	29.113287^m	7.483206	3.890	0.0007
X4	1	11.134226^n	4.445854	2.504	0.0192
X3	1	-7.608007^o	2.469109	-3.081	0.0050
X3SQ	1	0.671247^p	0.202701	3.312	0.0028
X43	1	-1.477717^q	0.667165	-2.215	0.0361

OBS	ACTUAL[u]	PREDICT VALUE[v]	STD ERR PREDICT[w]	LOWER 95% MEAN	UPPER 95% MEAN	LOWER 95% PREDICT	UPPER 95% PREDICT	RESIDUAL[x]
1	7.380	7.424	0.081140	7.257	7.591	6.968	7.881	-.044139
2	8.510	8.633	0.049414	8.531	8.735	8.196	9.070	-.122850
3	9.520	9.490	0.127991	9.227	9.754	8.990	9.990	0.029866
4	7.500	7.574	0.079322	7.411	7.738	7.119	8.030	-.074478
5	9.330	8.946	0.086038	8.769	9.123	8.485	9.406	0.384097
6	8.280	8.327	0.056341	8.211	8.443	7.887	8.768	-.047250
7	8.750	8.517	0.063947	8.385	8.649	8.072	8.962	0.233113
8	7.870	7.841	0.163020	7.506	8.177	7.300	8.383	0.028687
9	7.100	7.166	0.144177	6.869	7.463	6.648	7.684	-.066071
10	8.000	7.970	0.080969	7.804	8.137	7.514	8.427	0.029666
11	7.890	8.327	0.056341	8.211	8.443	7.887	8.768	-.437250
12	8.150	7.974	0.062179	7.846	8.102	7.530	8.418	0.176312
13	9.100	9.064	0.060524	8.940	9.189	8.622	9.507	0.035566
14	8.860	8.998	0.055135	8.885	9.112	8.558	9.438	-.138209
15	8.900	8.797	0.047450	8.700	8.895	8.361	9.233	0.102677
16	8.870	8.743	0.047649	8.645	8.841	8.307	9.179	0.126964
17	9.260	9.255	0.080749	9.089	9.422	8.799	9.712	0.004773
18	9.000	9.143	0.067340	9.005	9.282	8.696	9.590	-.143455
19	8.750	8.652	0.051799	8.745	8.958	8.413	9.290	-.101611
20	7.950	7.945	0.082413	7.775	8.115	7.487	8.403	0.005016
21	7.650	7.689	0.082596	7.519	7.859	7.231	8.147	-.038914
22	7.270	7.403	0.103346	7.191	7.616	6.928	7.879	-.133353
23	8.000	8.327	0.056341	8.211	8.443	7.887	8.768	-.327250
24	8.500	8.827	0.131141	8.557	9.097	8.324	9.331	-.327373

25	8.750	8.960	0.069988	8.816	9.104	8.511	9.409	$-.210185$
26	9.210	9.069	0.094423	8.874	9.263	8.601	9.536	0.141240
27	8.270	7.945	0.082413	7.775	8.115	7.487	8.403	0.325016
28	7.670	7.560	0.065802	7.425	7.696	7.114	8.006	0.109641
29	7.930	7.696	0.065635	7.561	7.831	7.250	8.142	0.234223
30	9.260	9.014	0.081742	8.846	9.183	8.557	9.472	0.245527
31		8.526^y	0.082777^z	[8.355,	$8.696]^{aa}$	[8.068,	$8.984]^{bb}$	

SUM OF RESIDUALS 1.00586E$-$13
SUM OF SQUARED RESIDUALS 1.064397^{cc}

[a] Explained Variation   [b] MS(model)   [c] F(model)   [d] Prob-value for F (model)   [e] $n_p - 1$ [f] $n - n_p$ [g] SSE   [h] s^2
[i] Total Variation   [j] s [k] R^2 [l] $\overline{R}^2$ [m] b_0 [n] b_1 [o] b_2 [p] b_3 [q] b_4 [r] s_{b_j} [s] t_{b_j} [t] Prob-value for t_{b_j} [u] y_t
[v] $\hat{y}_t$ [w] $s\sqrt{h_{tt}}$ [x] $y_t - \hat{y}_t$ [y] $\hat{y}_{31}$ [z] $s\sqrt{h_{31,31}}$ [aa] 95% Confidence Interval for μ_{31} [bb] 95% Prediction Interval for y_{31}
[cc] SSE

in sales period 31, it follows that

$$\hat{y}_{31} = b_0 + b_1 x_{31,4} + b_2 x_{31,3} + b_3 x_{31,3}^2 + b_4 x_{31,3} x_{31,4}$$

$$= 29.1133 + 11.1342(.10) - 7.6080(6.80)$$

$$+ .6712(6.80)^2 - 1.4777(.10)(6.80)$$

$$= 8.526 \quad \text{(that is, 852,600 bottles)}$$

is the point estimate of μ_{31} and is the point prediction of $y_{31} = \mu_{31} + \varepsilon_{31}$ (see Table 8.4). Moreover, further examining the SAS output in Table 8.4, we see that

1. A *95% confidence interval* for μ_{31} is [8.355, 8.696] (that is, [835,500 bottles, 869,600 bottles]) which says that Enterprise Industries is 95% confident that μ_{31}, the mean of all potential demands for Fresh that could be observed when the price difference is $x_{31,4} = .10$ and the advertising expenditure for Fresh is $x_{31,3} = 6.80$, is greater than or equal to 835,500 bottles and less than or equal to 869,600 bottles.

2. A *95% prediction interval* for $y_{31} = \mu_{31} + \varepsilon_{31}$ is [8.068, 8.984] (that is, [806,800 bottles, 898,400 bottles]) which (because of the added uncertainty with respect to the error term ε_{31}) is wider than the 95% confidence interval for μ_{31}. The 95% prediction interval for $y_{31} = \mu_{31} + \varepsilon_{31}$ says that Enterprise Industries is 95% confident that $y_{31} = \mu_{31} + \varepsilon_{31}$, the actual demand for Fresh that will occur in future sales period 31, will be greater than or equal to 806,800 bottles and less than or equal to 898,400 bottles. The upper limit of this prediction interval—898,400 bottles—is very important because it says that if Enterprise Industries plans to stock 898,400 extra-large sized bottles of Fresh during the future sales period, it can be very confident that this supply will adequately meet the demand for Fresh (since Enterprise Industries is very confident

that the future demand will be no more than 898,400 bottles). The lower limit of this prediction interval—806,800 bottles—is also very important because it tells Enterprise Industries that it is very confident that it will sell at least 806,800 bottles in the future sales period. Since each bottle will sell for $3.80 in the future sales period, Enterprise Industries is very confident that sales revenue from the extra-large sized bottles of Fresh will be at least 806,800 × $3.80 = $3,065,840 in the future sales period. This minimum revenue figure will help Enterprise Industries to better understand its cash flow situation.

It should be noted that the validity of the confidence and prediction intervals (and the validity of all of the statistical inferences made in this chapter) depend upon several assumptions to be presented in Section 8.4. Following this example, we will show how to calculate the confidence and prediction intervals.

To complete this example, we define the *experimental region* to be the "range" of the combinations of the previously observed (historical) price differences and advertising expenditures (see Table 8.3). Since the combination of the price difference $x_{31,4} = .10$ and the advertising expenditure $x_{31,3} = 6.80$ that will occur in future sales period 31 is in the experimental region (again, see Table 8.3) and since the prediction equation

$$\hat{y}_t = 29.1133 + 11.1342x_{t4} - 7.6080x_{t3} + .6712x_{t3}^2 - 1.4777x_{t4}x_{t3}$$

has been developed by using the historical data that defines the experimental region, it is reasonable to use this prediction equation to predict the demand for Fresh that will occur in future sales period 31. However, since, for example, the combination of the price difference $-.50$ and the advertising expenditure 8.75 is "far outside" of the experimental region, it follows, if this combination were to occur in future sales period 32, that it would not be reasonable to use the prediction equation to predict the demand for Fresh that will occur in that future sales period.

Examples 8.1 and 8.2 illustrate that we can relate a dependent variable y_t to any number of independent variables $x_{t1}, x_{t2}, \ldots, x_{tp}$ as follows

---

The (general) linear regression model is

$$y_t = \mu_t + \varepsilon_t$$
$$= \beta_0 + \beta_1 x_{t1} + \beta_2 x_{t2} + \cdots + \beta_p x_{tp} + \varepsilon_t$$

This model is called a *linear* regression model because

$$\mu_t = \beta_0 + \beta_1 x_{t1} + \beta_2 x_{t2} + \cdots + \beta_p x_{tp}$$

is a linear function of the parameters β_0, β_1, β_2, ..., β_p. For example, the Fresh detergent model

$$y_t = \mu_t + \varepsilon_t$$
$$= \beta_0 + \beta_1 x_{t4} + \beta_2 x_{t3} + \beta_3 x_{t3}^2 + \beta_4 x_{t4} x_{t3} + \varepsilon_t$$

is a *linear* model, because, although this model utilizes x_{t3}^2 and $x_{t4} x_{t3}$, μ_t is a linear function of the *parameters* β_0, β_1, β_2, β_3, and β_4. However, the model

$$y_t = \mu_t + \varepsilon_t$$
$$= \beta_0 + \beta_1 x_t^{\beta_2} + \varepsilon_t$$

is not a linear model, because μ_t is not a linear function of the parameters β_0, β_1, and β_2. We emphasize the concept of a linear regression model because the methods of regression analysis are easiest to use and best developed for such a model.

If b_0, b_1, ..., b_p denote point estimates of the parameters β_0, β_1, ..., β_p in the linear regression model, then

---

We compute *point estimates* and *point predictions* as follows:

$$\hat{y}_t = b_0 + b_1 x_{t1} + \cdots + b_p x_{tp}$$

is the *point estimate* of

$$\mu_t = \beta_0 + \beta_1 x_{t1} + \cdots + \beta_p x_{tp}$$

and is the *point prediction* of

$$y_t = \mu_t + \varepsilon_t$$
$$= \beta_0 + \beta_1 x_{t1} + \cdots + \beta_p x_{tp} + \varepsilon_t$$

since we predict ε_t to be zero.

We define the *least squares point estimates* of β_0, β_1, ..., β_p to be the point estimates b_0, b_1, ..., b_p that minimize the sum of squares residuals

$$\text{SSE} = \sum_{t=1}^{n} (y_t - \hat{y}_t)^2$$

$$= \sum_{t=1}^{n} [y_t - (b_0 + b_1 x_{t1} + \cdots + b_p x_{tp})]^2$$

where n denotes the number of observed combinations of y_t, x_{t1}, ..., x_{tp}. These estimates are calculated as follows.

> The least squares point estimates b_0 and b_1 of the parameters β_0 and β_1 in the simple linear regression model
>
> $$y_t = \beta_0 + \beta_1 x_t + \varepsilon_t$$
>
> are given by the formulas
>
> $$b_1 = \frac{n \sum_{t=1}^{n} x_t y_t - \left(\sum_{t=1}^{n} x_t\right)\left(\sum_{t=1}^{n} y_t\right)}{n \sum_{t=1}^{n} x_t^2 - \left(\sum_{t=1}^{n} x_t\right)^2}$$
>
> and
>
> $$b_0 = \bar{y} - b_1 \bar{x}$$
>
> where
>
> $$\bar{y} = \frac{\sum_{t=1}^{n} y_t}{n} \quad \text{and} \quad \bar{x} = \frac{\sum_{t=1}^{n} x_t}{n}$$

The formula for the least squares point estimates $b_0, b_1, \ldots, b_p$ of the parameters $\beta_0, \beta_1, \ldots, \beta_p$ in the (general) linear regression model

$$y_t = \beta_0 + \beta_1 x_{t1} + \cdots + \beta_p x_{tp} + \varepsilon_t$$

involves a branch of mathematics called *matrix algebra* and is beyond the scope of this text. It is sufficient to know that SAS and other computer packages can be used to calculate least squares point estimates.

Next, letting n_p denote the number of parameters $\beta_0, \beta_1, \ldots, \beta_p$ in the regression model

$$y_t = \mu_t + \varepsilon_t$$
$$= \beta_0 + \beta_1 x_{t1} + \cdots + \beta_p x_{tp} + \varepsilon_t$$

and recalling that

$$\hat{y}_t = b_0 + b_1 x_{t1} + \cdots + b_p x_{tp}$$

is the point estimate of μ_t and the point prediction of $y_t = \mu_t + \varepsilon_t$, it can be proven that

> A $100(1 - \alpha)\%$ confidence interval for μ_t is
>
> $$[\hat{y}_t \pm t_{[\alpha/2]}^{(n-n_p)} s \sqrt{h_{tt}}]$$
>
> and a $100(1 - \alpha)\%$ prediction interval for $y_t = \mu_t + \varepsilon_t$ is
>
> $$[\hat{y}_t \pm t_{[\alpha/2]}^{(n-n_p)} s \sqrt{1 + h_{tt}}]$$

Here

1. $t_{[\alpha/2]}^{(n-n_p)}$ is the point on the scale of the *t*-distribution having $(n - n_p)$ degrees of freedom so that the area under the curve of this *t*-distribution to the right of $t_{[\alpha/2]}^{(n-n_p)}$ is $\alpha/2$.

2.   *s* is called the *standard error* and is computed as follows.

The *standard error* is

$$s = \sqrt{\frac{SSE}{n - n_p}} = \sqrt{\frac{\sum\limits_{t=1}^{n}(y_t - \hat{y}_t)^2}{n - n_p}}$$

3. h_{tt} is called a *leverage value* and is computed as follows.

The leverage value for the simple linear regression model $y_t = \beta_0 + \beta_1 x_t + \varepsilon_t$ is

$$h_{tt} = \frac{1}{n} + \frac{(x_t - \bar{x})^2}{\sum\limits_{t=1}^{n} x_t^2 - n\bar{x}^2}$$

Just as this value is a function of $(x_t - \bar{x})$, the distance between the particular value x_t and the average of all *n* previously observed values of x_t

$$\bar{x} = \frac{\sum\limits_{t=1}^{n} x_t}{n}$$

it can be shown that the leverage value h_{tt} for the (general) linear regression model is a function of the distances between the particular values $x_{t1}, x_{t2}, \ldots, x_{tp}$ and the means

$$\bar{x}_1 = \frac{\sum\limits_{t=1}^{n} x_{t1}}{n} \qquad \bar{x}_2 = \frac{\sum\limits_{t=1}^{n} x_{t2}}{n} \qquad \cdots \qquad \bar{x}_n = \frac{\sum\limits_{t=1}^{n} x_{tp}}{n}$$

Although the formula for h_{tt} for the (general) linear regression model involves matrix algebra and is beyond the scope of this text, SAS and other computer packages can be used to obtain h_{tt} (as we will demonstrate in the following example).

EXAMPLE 8.3   Consider Example 8.2 and the Fresh detergent model

$$y_t = \beta_0 + \beta_1 x_{t4} + \beta_2 x_{t3} + \beta_3 x_{t3}^2 + \beta_4 x_{t4} x_{t3} + \varepsilon_t$$

which contains $n_p = 5$ parameters. Since the SAS output in Table 8.4 tells us that

1. $\hat{y}_{31} = 8.526$ is the point estimate of μ_{31} and the point prediction of $y_{31} = \mu_{31} + \varepsilon_{31}$ (recall that the actual calculation of $\hat{y}_{31}$ has been demonstrated in Example 8.2)

2. $t_{[.025]}^{(n - n_p)} = t_{[.025]}^{(30 - 5)} = t_{[.025]}^{(25)} = 2.06$

   and

   $t_{[.005]}^{(n - n_p)} = t_{[.005]}^{(30 - 5)} = t_{[.005]}^{(25)} = 2.787$

3. $s = \sqrt{\dfrac{SSE}{n - n_p}} = \sqrt{\dfrac{1.0644}{30 - 5}} = .2063$

4. $s\sqrt{h_{31,31}} = .0828$

   which implies that

   $$h_{31,31} = \left(\frac{.0828}{s}\right)^2 = \left(\frac{.0828}{.2063}\right)^2 = .1611$$

it follows that

1. 95% and 99% confidence intervals for μ_{31} are, respectively,

   $[8.526 \pm 2.06(.2063)\sqrt{.1611}] = [8.355, 8.696]$

   and

   $[8.526 \pm 2.787(.2063)\sqrt{.1611}] = [8.295, 8.757]$

2. 95% and 99% prediction intervals for $y_{31} = \mu_{31} + \varepsilon_{31}$ are, respectively,

   $[8.526 \pm 2.06(.2063)\sqrt{1.1611}] = [8.068, 8.984]$

   and

   $[8.526 \pm 2.787(.2063)\sqrt{1.1611}] = [7.906, 9.146]$

Note that the 95% confidence and prediction intervals are given by the SAS output in Table 8.4, and recall that we have interpreted these intervals in Example 8.2.

We next present an example illustrating the meaning of interaction.

EXAMPLE 8.4   In this example we will discuss the exact meaning of the term $x_{t4}x_{t3}$ in the model

$$\begin{aligned}
y_t &= \mu_t + \varepsilon_t \\
&= \beta_0 + \beta_1 x_{t4} + \beta_2 x_{t3} + \beta_3 x_{t3}^2 + \beta_4 x_{t4} x_{t3} + \varepsilon_t
\end{aligned}$$

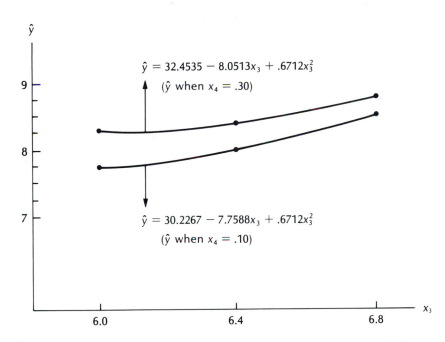

$$\hat{y} = 32.4535 - 8.0513x_3 + .6712x_3^2$$
$$(\hat{y} \text{ when } x_4 = .30)$$

$$\hat{y} = 30.2267 - 7.7588x_3 + .6712x_3^2$$
$$(\hat{y} \text{ when } x_4 = .10)$$

FIGURE 8.5   Two Quadratic Curves Illustrating that, as x_3 (the Advertising Expenditure for Fresh) Increases, μ (the Mean Demand for Fresh) Increases at a Slower Rate When x_4 (the Price Difference) Equals $.30 than When x_4 Equals $.10.

To do this, we will plot the values of the point estimate

$$\hat{y} = b_0 + b_1x_4 + b_2x_3 + b_3x_3^2 + b_4x_4x_3$$
$$= 29.1133 + 11.1342x_4 - 7.6080x_3 + .6712x_3^2 - 1.4777x_4x_3$$

of the mean demand for Fresh

$$\mu = \beta_0 + \beta_1x_4 + \beta_2x_3 + \beta_3x_3^2 + \beta_4x_4x_3$$

against increasing values of x_3, when we fix x_4 at two different values—first at $x_4 = \$.10$ and then at $x_4 = \$.30$. If we fix x_4 at $.10, the point estimate $\hat{y}$ of μ equals

$$\hat{y} = 29.1133 + 11.1342x_4 - 7.6080x_3 + .6712x_3^2 - 1.4777x_4x_3$$
$$= 29.1133 + 11.1342(.10) - 7.6080x_3 + .6712x_3^2 - 1.4777(.10)x_3$$
$$= 30.2267 - 7.7558x_3 + .6712x_3^2$$

In words, this equation may be interpreted as the point estimate of the equation relating μ (the mean demand for Fresh) to x_3 (the advertising expenditure for Fresh) when x_4 (the price difference) is $.10. The quadratic curve defined by this equation is plotted in Figure 8.5 against increasing values of x_3. As illustrated in this figure, note that, for example,

1. When $x_3 = 6$,

    $$\hat{y} = 30.2267 - 7.7558(6) + .6712(6)^2 \approx 7.76$$

2. When $x_3 = 6.4$,

    $$\hat{y} = 30.2267 - 7.7558(6.4) + .6712(6.4)^2 \approx 8.08$$

3. When $x_3 = 6.8$,

    $$\hat{y} = 30.2267 - 7.7558(6.8) + .6712(6.8)^2 \approx 8.52$$

Next, if we fix x_4 at \$.30, the point estimate $\hat{y}$ of μ equals

$$\hat{y} = 29.1133 + 11.1342x_4 - 7.6080x_3 + .6712x_3^2 - 1.4777x_4x_3$$

$$= 29.1133 + 11.1342(.30) - 7.6080x_3 + .6712x_3^2 - 1.4777(.30)x_3$$

$$= 32.4535 - 8.0513x_3 + .6712x_3^2$$

In words, this equation may be interpreted as the point estimate of the equation relating μ (the mean demand for Fresh) to x_3 (the advertising expenditure for Fresh) when x_4 (the price difference) is \$.30. The quadratic curve defined by this equation is plotted in Figure 8.5 against increasing values of x_3. As illustrated in this figure, note that, for example,

1. When $x_3 = 6$,

    $$\hat{y} = 32.4535 - 8.0513(6) + .6712(6)^2 \approx 8.31$$

2. When $x_3 = 6.4$,

    $$\hat{y} = 32.4535 - 8.0513(6.4) + .6712(6.4)^2 \approx 8.42$$

3. When $x_3 = 6.8$,

    $$\hat{y} = 32.4535 - 8.0513(6.8) + .6712(6.8)^2 \approx 8.74$$

Examining Figure 8.5, we note that, as x_3 increases, the quadratic curve defined by the equation

$$\hat{y} = 32.4535 - 8.0513x_3 + .6712x_3^2$$

which is the point estimate of the equation relating μ to x_3 when x_4 is \$.30, increases at a slower rate than the quadratic curve defined by the equation

$$\hat{y} = 30.2267 - 7.7558x_3 + .6712x_3^2$$

which is the point estimate of the equation relating μ to x_3 when x_4 is \$.10. Thus, Figure 8.5 illustrates that there is "less opportunity" for increased advertising expenditure to increase the mean demand for Fresh when x_4 equals \$.30 than when x_4 equals \$.10 (that is, when we are at a larger price advantage, sales are already high, so there is less opportunity for advertising to improve sales). Therefore, we conclude that the larger the price difference, the smaller the increase in the mean demand for Fresh that is caused by an increase in the advertising expenditure for Fresh. This conclusion implies that the relationship between the mean demand for Fresh and x_3 (the advertising expenditure) depends upon x_4 (the price difference). Therefore, we say that there is *interaction* between x_3 and x_4. Including the term $x_{t4}x_{t3}$ in the model

$$
\begin{aligned}
y_t &= \mu_t + \varepsilon_t \\
&= \beta_0 + \beta_1 x_{t4} + \beta_2 x_{t3} + \beta_3 x_{t3}^2 + \beta_4 x_{t4} x_{t3} + \varepsilon_t
\end{aligned}
$$

has taken into account the existence of this interaction. This is because, for example, it is the $x_{t4}x_{t3}$ term that has caused the x_3 terms in the quadratic equations

$$
\hat{y} = 30.2267 - 7.7558x_3 + .6712x_3^2 \quad (x_4 = .10)
$$

and

$$
\hat{y} = 32.4535 - 8.0513x_3 + .6712x_3^2 \quad (x_4 = .30)
$$

of Figure 8.5 to differ (these differing x_3 terms cause the rates at which $\hat{y}$ increases as x_3 increases to be dependent on the level of x_4).

Next, note that Table 8.4 tells us that the least squares point estimate of β_4 is $b_4 = -1.4777$, which is quite "far from zero," and thus indicates that β_4 does not equal zero. This is an indication that we should indeed include the term $\beta_4 x_{t4} x_{t3}$ in the model. If b_4 had been "near zero" (we will discuss in the next section how to determine by hypothesis testing what we really mean by "near zero"), then we might conclude that β_4 equals (or nearly equals) zero and thus remove the term $\beta_4 x_{t4} x_{t3}$ from the previous model to form the model

$$
\begin{aligned}
y_t &= \mu_t + \varepsilon_t \\
&= \beta_0 + \beta_1 x_{t4} + \beta_2 x_{t3} + \beta_3 x_{t3}^2 + \varepsilon_t
\end{aligned}
$$

This model, which does not utilize the term $x_{t4}x_{t3}$, assumes that *no interaction* exists between x_4 and x_3. That is, it assumes that the relationship between the mean demand for Fresh and x_3 does not depend on x_4. This would say that as the advertising expenditure increases, the mean demand for Fresh increases at the same rate for all price differences.

One final comment should be made. Examination of Table 8.3 indicates that most of the price differences that have been observed are positive and that the few negative price differences are not large in absolute value. Since this indicates that Enterprise Industries' price for Fresh is usually smaller than the average industry price, we can say that Enterprise Industries is usually at a price advantage when selling Fresh. Given this pricing strategy, the type of interaction described above is reasonable. This is because

1. In the "worst case" Enterprise Industries is either at a slight price advantage or at a slight price disadvantage, and in such a situation increasing the advertising expenditure for Fresh can still be expected to substantially increase the mean demand for Fresh, and (as previously discussed)
2. The larger the price difference (that is, the larger that Enterprise Industries' price advantage) the smaller the increase in the mean demand for Fresh associated with increasing the advertising expenditure.

If, however, this price situation were to change, the type of interaction we have described might not exist. For example, consider a situation in which (due to frequent changes in pricing strategy) Enterprise Industries is often at a substantial price advantage (that is, the price for Fresh is frequently substantially lower than the average industry price) and is also often at a substantial price disadvantage (that is, the price for Fresh is also frequently substantially higher than the average industry price). In such a situation, in the "worst case" Enterprise Industries is at a substantial price disadvantage, and in such circumstances increasing the advertising expenditure for Fresh would probably not increase the mean demand for Fresh very much. This is because, no matter how much money Enterprise Industries spends advertising Fresh, since customers might not wish to buy a substantially higher priced product, they might not buy Fresh, and thus the demand for Fresh might remain fairly low. Thus, in this situation, the increase in the mean demand for Fresh that is associated with increasing the advertising expenditure for Fresh would be larger when Enterprise Industries is at a substantial price advantage than when Enterprise Industries is at a substantial price disadvantage.

To complete this section, note that, although the subscript "t" in the model

$$y_t = \mu_t + \varepsilon_t$$
$$= \beta_0 + \beta_1 x_{t1} + \beta_2 x_{t2} + \cdots + \beta_p x_{tp} + \varepsilon_t$$

has represented "time" in the Fresh detergent example, this subscript can represent any criterion. For example, if the Fresh detergent data had been observed in the same sales period in 30 different *sales regions* (of the same sales potential), then "t" would represent "sales region." We see, then, that regression models can be used to describe and predict both time series data and cross-sectional data (data observed at essentially the same time). At times we will omit the subscript t and write the model as simply

$$y = \mu + \varepsilon$$
$$= \beta_0 + \beta_1 x_1 + \beta_2 x_2 + \cdots + \beta_p x_p + \varepsilon$$

8.3 MODEL BUILDING

8.3.1 Coefficients of Determination and Correlation

EXAMPLE 8.5

Consider the Fresh detergent model

$$y_t = \mu_t + \varepsilon_t$$

$$= \beta_0 + \beta_1 x_{t4} + \beta_2 x_{t3} + \beta_3 x_{t3}^2 + \beta_4 x_{t4} x_{t3} + \varepsilon_t$$

If we did not have the independent variables x_{t4} and x_{t3} to use to predict y_t, the only reasonable prediction of y_t would be

$$\bar{y} = \frac{\sum_{t=1}^{30} y_t}{30} = 8.38$$

the mean of the 30 observed demand values, and thus the sum of squared residuals would be

$$\sum_{t=1}^{30} (y_t - \bar{y})^2 = 13.4586$$

which is called the *total variation* (see Table 8.4). However, since we do have the regression model, the sum of squared residuals really is

$$SSE = \sum_{t=1}^{30} (y_t - \hat{y}_t)^2 = 1.0644$$

which is also called the *unexplained variation*. Therefore, the regression model has reduced the sum of squared residuals from

$$\sum_{t=1}^{30} (y_t - \bar{y})^2 = 13.4586$$

to

$$SSE = \sum_{t=1}^{30} (y_t - \hat{y}_t)^2 = 1.0644$$

or by an amount equal to

$$\sum_{t=1}^{30} (y_t - \bar{y})^2 - \sum_{t=1}^{30} (y_t - \hat{y}_t)^2 = 13.4586 - 1.0644$$

$$= 12.3942$$

which is called the *explained variation* (see Table 8.4). It can be proven that the explained variation can be calculated by the formula

$$\sum_{t=1}^{30} (\hat{y}_t - \bar{y})^2$$

That is,

$$\sum_{t=1}^{30} (\hat{y}_t - \bar{y})^2 = \sum_{t=1}^{30} (y_t - \bar{y})^2 - \sum_{t=1}^{30} (y_t - \hat{y}_t)^2$$

or

$$\begin{matrix} \text{explained} \\ \text{variation} \end{matrix} = \begin{matrix} \text{total} \\ \text{variation} \end{matrix} - \begin{matrix} \text{unexplained} \\ \text{variation} \end{matrix}$$

Therefore, it follows that

$$\sum_{t=1}^{30} (y_t - \bar{y})^2 = \sum_{t=1}^{30} (\hat{y}_t - \bar{y})^2 + \sum_{t=1}^{30} (y_t - \hat{y}_t)^2$$

or

$$\begin{matrix} \text{total} \\ \text{variation} \end{matrix} = \begin{matrix} \text{explained} \\ \text{variation} \end{matrix} + \begin{matrix} \text{unexplained} \\ \text{variation} \end{matrix}$$

Next, we define R^2, the *multiple coefficient of determination* (see Table 8.4), to be

$$R^2 = \frac{\text{explained variation}}{\text{total variation}} = \frac{12.3942}{13.4586} = .9209$$

which says that the regression model

$$y_t = \beta_0 + \beta_1 x_{t4} + \beta_2 x_{t3} + \beta_3 x_{t3}^2 + \beta_4 x_{t4} x_{t3} + \varepsilon_t$$

explains 92.09% of the total variation in the $n = 30$ observed demands for Fresh.

With this example in mind, and recalling that

$$\hat{y}_t = b_0 + b_1 x_{t1} + \cdots + b_p x_{tp}$$

is the point estimate of μ_t and the point prediction of y_t in the general model

$$\begin{aligned} y_t &= \mu_t + \varepsilon_t \\ &= \beta_0 + \beta_1 x_{t1} + \cdots + \beta_p x_{tp} + \varepsilon_t \end{aligned}$$

we can state that

$$\text{Total variation} = \sum_{t=1}^{n} (y_t - \bar{y})^2 \quad \text{where} \quad \bar{y} = \frac{\sum_{t=1}^{n} y_t}{n}$$

$$Explained\ variation\ =\ \sum_{t=1}^{n} (\hat{y}_t - \bar{y})^2$$

$$Unexplained\ variation\ =\ \sum_{t=1}^{n} (y_t - \hat{y}_t)^2$$

Total Variation = Explained Variation + Unexplained Variation

$$Multiple\ Coefficient\ of\ Determination\ =\ R^2\ =\ \frac{explained\ variation}{total\ variation}$$

$$Multiple\ Correlation\ Coefficient\ =\ R\ =\ \sqrt{R^2}$$

Since the explained variation cannot be greater than the total variation, R^2 is always between 0 and 1. The nearer R^2 is to 1, the greater is the proportion of the total variation than is explained by the regression model under consideration. We will subsequently see how we can use R^2 to help in building a "good" regression model.

Next, note that if we are considering the simple linear regression model

$$y_t\ =\ \beta_0 + \beta_1 x_t + \varepsilon_t$$

we call R^2 the *simple coefficient of determination* and denote this quantity by r^2. It follows that r^2 is the proportion of the total variation in the n observed values of the dependent variable that is explained by the simple linear regression model.

We next define the *simple correlation coefficient*, denoted by r, to be $r = \pm (r^2)^{1/2}$, where we define $r = +(r^2)^{1/2}$ if b_1, the least squares point estimate of the slope β_1, is a positive number, and $r = -(r^2)^{1/2}$ if b_1 is a negative number.

Since for the simple linear regression model the prediction of y_t is

$$\hat{y}_t\ =\ b_0 + b_1 x_t$$

it can be shown (by utilizing the algebraic formulas for the least squares point estimates b_0 and b_1) that

$$r^2\ =\ \frac{\sum_{t=1}^{n} (\hat{y}_t - \bar{y})^2}{\sum_{t=1}^{n} (y_t - \bar{y})^2}$$

is equal to

$$\frac{\left[\sum_{t=1}^{n} (x_t - \bar{x})(y_t - \bar{y}) \right]^2}{\sum_{t=1}^{n} (x_t - \bar{x})^2 \sum_{t=1}^{n} (y_t - \bar{y})^2}$$

where

$$\bar{x} = \frac{\sum_{t=1}^{n} x_t}{n}$$

is the average of the n observed values of the independent variable x, and

$$\bar{y} = \frac{\sum_{t=1}^{n} y_t}{n}$$

is the average of the n observed values of the dependent variable y. Furthermore, taking square roots implies that

---

The *simple correlation coefficient* is

$$r = \frac{\sum_{t=1}^{n} (x_t - \bar{x})(y_t - \bar{y})}{\left[\sum_{t=1}^{n} (x_t - \bar{x})^2 \sum_{t=1}^{n} (y_t - \bar{y})^2 \right]^{1/2}}$$

It can be shown that this formula for r is equivalent to the formula

$$r = b_1 \frac{s_x}{s_y}$$

where

$$s_x = \sqrt{\frac{\sum_{t=1}^{n} (x_t - \bar{x})^2}{n - 1}} \quad \text{and} \quad s_y = \sqrt{\frac{\sum_{t=1}^{n} (y_t - \bar{y})^2}{n - 1}}$$

and b_1 is the least squares point estimate of the slope β_1 in the simple linear regression model.

It follows that this formula for r automatically yields a value of the simple correlation coefficient which is positive if b_1 is positive or negative if b_1 is negative. Since r^2 is always between zero and 1, it follows that r is always between -1 and 1. A value of r close to 1 indicates that the independent variable x and the dependent variable y have a strong tendency to move together in a linear fashion with a positive slope and therefore that x and y are highly related and *positively correlated*. A value of r close to -1 indicates that x and y have a strong tendency to move together in a linear fashion with a negative slope and that the independent variables x and y are highly related and *negatively correlated*.

8.3.2 Testing the Importance of an Independent Variable

If we consider the regression model

$$y_t = \mu_t + \varepsilon_t$$

$$= \beta_0 + \beta_1 x_{t1} + \beta_2 x_{t2} + \cdots + \beta_j x_{tj} + \cdots + \beta_p x_{tp} + \varepsilon_t$$

it would seem logical that we should decide that the independent variable x_{tj} is significantly related to the dependent variable y_t, and thus that we should include x_{tj} in the model if we can be quite certain that we should reject

$$H_0: \beta_j = 0$$

in favor of

$$H_1: \beta_j \neq 0$$

When x_{tj} is significantly related to y_t, we say that x_{tj} is an *important* independent variable for the description and prediction of y_t.

To test H_0 versus H_1 we use the t_{b_j} *statistic*, which is calculated by the equation

$$t_{b_j} = \frac{b_j}{s_{b_j}}$$

where b_j is the least squares point estimate of β_j and s_{b_j} is the *standard error of the point estimate b_j* (note from Table 8.4 that SAS calculates b_j, s_{b_j}, and t_{b_j}). If the absolute value of b_j is "large," then the absolute value of t_{b_j} is "large." It follows that a "large" t_{b_j} statistic implies (1) that we should reject the null hypothesis $H_0: \beta_j = 0$ and (2) that we should include x_{tj} in the regression model. To decide how large (in absolute value) t_{b_j} must be before we reject $H_0: \beta_j = 0$, we consider the errors that can be made in hypothesis testing. A *Type I error* is committed if we reject $H_0: \beta_j = 0$ when $H_0: \beta_j = 0$ is true (which means we would include x_{tj} in the model when x_{tj} should not be included). A *Type II error* is committed if we do not reject $H_0: \beta_j = 0$ when $H_0: \beta_j = 0$ is false (which means we would not include x_{tj} in the model when x_{tj} should be included). We obviously desire that both the *probability of a Type I error*, which we denote by the symbol α, and the *probability of a Type II error*, be *small*. The hypothesis testing procedure that we use in this book assumes that we observe n data points and that we set α equal to a specified value. Here, we usually choose α to be between .05 and .01, with .05 being the most frequent choice. Notice that the lower we set α, the lower is the probability that we will include x_{tj} in a model when x_{tj} should not be included, and therefore, if we can reject $H_0: \beta_j = 0$, the more confident we are that we should include x_{tj}. We usually do not set α lower than .01 because setting α extremely small often leads to a probability of a Type II error (not including x_{tj} when x_{tj} should be included) that is unacceptably large. We now state the following method for testing $H_0: \beta_j = 0$.

Testing H_0: $\beta_j = 0$ versus H_1: $\beta_j \neq 0$

Suppose that the regression model under consideration utilizes n_p parameters, and define the t_{b_j} statistic to be

$$t_{b_j} = \frac{b_j}{s_{b_j}}$$

Then we can reject the null hypothesis H_0: $\beta_j = 0$ in favor of the alternative hypothesis H_1: $\beta_j \neq 0$ by setting the probability of a Type I error equal to α if and only if either of the following equivalent conditions hold:

1. $|t_{b_j}| > t_{[\alpha/2]}^{(n-n_p)}$

   that is,

   $$t_{b_j} > t_{[\alpha/2]}^{(n-n_p)} \qquad \text{or} \qquad t_{b_j} < -t_{[\alpha/2]}^{(n-n_p)}$$

   where the *rejection point* $t_{[\alpha/2]}^{(n-n_p)}$ is the point on the scale of the t-distribution having $(n - n_p)$ degrees of freedom so that the area under the curve of this t-distribution to the right of $t_{[\alpha/2]}^{(n-n_p)}$ is $\alpha/2$. If $(n - n_p)$ is at least 30, $t_{[\alpha/2]}^{(n-n_p)}$ should be approximated by $z_{[\alpha/2]}$, which is the point on the scale of the standard normal curve so that the area under this curve to the right of $z_{[\alpha/2]}$ is $\alpha/2$.

2. *Prob-value* is less than α, where prob-value is twice the area under the curve of the t-distribution having $(n - n_p)$ degrees of freedom to the right of $|t_{b_j}|$. If $(n - n_p)$ is at least 30, then the prob-value is approximately found by calculating twice the area under the standard normal curve to the right of $|t_{b_j}|$.

We will not discuss the precise rationale behind this hypothesis test in this book. However, noting that we can find $z_{[\alpha/2]}$ points in Table A1 (in Appendix A) and that we can find $t_{[\alpha/2]}^{(n-n_p)}$ points in Table A2 (in Appendix A), we fully illustrate and "intuitively" motivate this result in the following example.

EXAMPLE 8.6   Consider the Fresh detergent model

$$y_t = \mu_t + \varepsilon_t$$
$$= \beta_0 + \beta_1 x_{t4} + \beta_2 x_{t3} + \beta_3 x_{t3}^2 + \beta_4 x_{t4} x_{t3} + \varepsilon_t$$

and testing H_0: $\beta_4 = 0$ versus H_1: $\beta_4 \neq 0$. Table 8.4 presents the SAS output of

1. the least squares point estimate of β_4, which is

   $$b_4 = -1.4777$$

2.   the standard error of the point estimate, which is

$$s_{b_4} = .6672$$

and

3.   the associated t-value, which is

$$t_{b_4} = \frac{b_4}{s_{b_4}} = \frac{-1.4777}{.6672} = -2.215$$

Noting that this model uses $n_p = 5$ parameters, it follows that if we wish to test $H_0: \beta_4 = 0$ versus $H_1: \beta_4 \neq 0$ by setting α equal to .05, we could use the rejection point

$$t_{[\alpha/2]}^{(n - n_p)} = t_{[.05/2]}^{(30 - 5)} = t_{[.025]}^{(25)} = 2.06$$

To explain what we mean when we say that α equals .05, note that although the $n = 30$ values of the Fresh detergent demand time series in Table 8.1 yielded the point estimate $b_4 = -1.4777$, standard error $s_{b_4} = 0.6672$, and t-value $t_{b_4} = -2.215$, another realization (that is, another $n = 30$ values) of the Fresh detergent demand time series would yield a somewhat different point estimate b_4, standard error s_{b_4}, and t-value t_{b_4}. It can be proven that, if the null hypothesis $H_0: \beta_4 = 0$ is true, then the population of all possible t_{b_4} statistics that could be observed (by observing all possible realizations of the Fresh detergent demand time series) has (approximately) a t-distribution with $(n - n_p)$ degrees of freedom. It follows that setting α equal to .05 means that, if $H_0: \beta_4 = 0$ is true, then as illustrated in Figure 8.6a:

1.   .95 (that is, 95%) of all possible t_{b_4} statistics would be between $-t_{[.025]}^{(25)} = -2.06$ and $t_{[.025]}^{(25)} = 2.06$ and thus would be close enough to zero to cause us to not reject $H_0: \beta_4 = 0$ (when $H_0: \beta_4 = 0$ is true) —a correct decision.

2.   .025 (that is, 2.5%) of all possible t_{b_4} statistics would be less than $-t_{[.025]}^{(25)} = -2.06$ and .025 (that is, 2.5%) of all possible t_{b_4} statistics would be greater than $t_{[.025]}^{(25)} = 2.06$ and thus—in total—.05 (that is, 5%) of all possible t_{b_4} statistics would be different enough from zero to cause us to reject $H_0: \beta_4 = 0$ (when $H_0: \beta_4 = 0$ is true)—a Type I error.

Recalling that the computed t_{b_4} statistic we have actually observed is $t_{b_4} = -2.215$, it follows, since

$$|t_{b_4}| = 2.215 > 2.06 = t_{[.025]}^{(25)}$$

that we can reject $H_0: \beta_4 = 0$ in favor of $H_1: \beta_4 \neq 0$ by setting α equal to .05.

If we wish to use condition 2, then we must first calculate the prob-value, which is twice the area under the curve of the t-distribution having $n - n_p = 25$ degrees of freedom to the right of $|t_{b_4}| = |-2.215| = 2.215$. Since this area can

(a)   The Rejection Points For Testing H_0: $\beta_4 = 0$ versus H_1: $\beta_4 \neq 0$

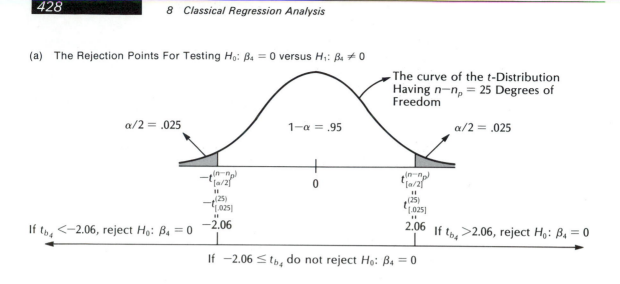

(b)   The Prob-Value for Testing H_0: $\beta_4 = 0$ versus H_1: $\beta_4 \neq 0$:
Reject H_0: $\beta_4 = 0$ if Prob-value is less than α

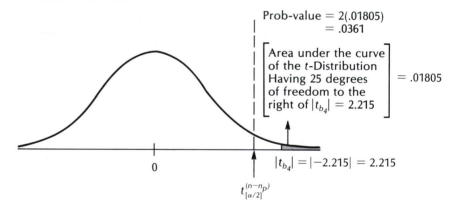

FIGURE 8.6   *Testing H_0: $\beta_4 = 0$ versus H_1: $\beta_4 \neq 0$*

be computer-calculated to be .01805, the prob-value is 2(0.01805) = .0361 (see Figure 8.6b). Since this prob-value is less than $\alpha = .05$, it follows by condition 2 that we can reject H_0: $\beta_4 = 0$ in favor of H_1: $\beta_4 \neq 0$ by setting α equal to .05. To see the logic behind condition 2, note that if, for *any value of α*

$$\text{prob-value} \; = \; 2 \times \left[\begin{array}{c} \text{area under the curve of the} \\ \text{t-distribution having 25 degrees} \\ \text{of freedom to the right of} \\ |t_{b_4}| \; = \; 2.215 \end{array} \right] < \alpha$$

then

$$\left[\begin{array}{c} \text{area under the curve of the} \\ \text{t-distribution having 25 degrees} \\ \text{of freedom to the right of} \\ |t_{b_4}| = 2.215 \end{array} \right] < \alpha/2$$

This implies (as can be seen by comparing Figure 8.6a with Figure 8.6b) that $|t_{b_4}|$ is greater than $t_{[\alpha/2]}^{(n-n_p)}$, which implies by condition 1 that we should reject H_0: $\beta_4 = 0$ in favor of H_1: $\beta_4 \neq 0$. For example, since

prob-value $= .0361$

is less than .05 and less than .04, but not less than 0.01, we can reject H_0: $\beta_4 = 0$ in favor of H_1: $\beta_4 \neq 0$ by setting α equal to .05, or by setting α equal to .04, but not by setting α equal to .01. Although we cannot reject H_0: $\beta_4 = 0$ in favor of H_1: $\beta_4 \neq 0$ by setting α equal to .01, since we can reject H_0 in favor of H_1 by setting α equal to .05, we have substantial evidence that we should include the term $x_{t4}x_{t3}$ in the regression model

$$y_t = \beta_0 + \beta_1 x_{t4} + \beta_2 x_{t3} + \beta_3 x_{t3}^2 + \beta_4 x_{t4} x_{t3} + \varepsilon_t$$

In fact, examining all of the t_{b_j} statistics and prob-values in Table 8.4, we see that, since each of the t_{b_j} statistics is greater than $t_{[.05/2]}^{(25)} = 2.06$, or, equivalently, since each of the prob-values is less than $\alpha = .05$, it follows that we can reject each of the hypotheses H_0: $\beta_0 = 0$, H_0: $\beta_1 = 0$, H_0: $\beta_2 = 0$, H_0: $\beta_3 = 0$, and H_0: $\beta_4 = 0$ by setting $\alpha = .05$. Thus, it seems reasonable to include each of the independent variables x_{t4}, x_{t3}, x_{t3}^2, and $x_{t4}x_{t3}$ in the above model.

As illustrated in Example 8.6, since condition 2 requires calculating the area under the curve of the t-distribution having $n - n_p$ degrees of freedom (or the area under the standard normal curve) to the right of $|t_{b_j}|$, whereas condition 1 requires only that we calculate the t_{b_j} statistic and find the rejection point $t_{[\alpha/2]}^{(n-n_p)}$ in a t-table (or $z_{[\alpha/2]}$ in a normal table), condition 2 is more complicated than condition 1 from a computational stand-point. However, condition 2 has the following advantage over condition 1. If there were several people, all of whom wished to use different values of α, the probability of a Type I error, and if condition 1 were used, then each person would have to find a different rejection point $t_{[\alpha/2]}^{(n-n_p)}$ to decide whether to reject H_0 at his or her particular chosen value of α. However, if condition 2 is used, only the prob-value needs to be calculated, and each person knows that if the prob-value is less than his or her particular chosen value of α, then H_0 should be rejected.

Up to this point we have considered the prob-value as a decision rule for deciding when to reject the null hypothesis H_0: $\beta_j = 0$ in favor of the alternative hypothesis H_1: $\beta_j \neq 0$. However, the prob-value has a more important use than its use as a decision rule. Although setting α at .05 is a frequently used convention, this choice (or any other choice of α) is quite arbitrary. This has led many statisticians to use the prob-value to determine the amount of "probabilistic doubt" cast upon H_0: $\beta_j = 0$ by the t_{b_j} statistic.

Specifically, the prob-value can be interpreted to be the probability, if $H_0: \beta_j = 0$ is true, of observing a t_{b_j} statistic that is at least as far from zero (in the negative or positive direction), and thus at least as contradictory to $H_0: \beta_j = 0$, as the t_{b_j} statistic that we have actually observed. For example, if we consider the Fresh detergent model

$$y_t = \beta_0 + \beta_1 x_{t4} + \beta_2 x_{t3} + \beta_3 x_{t3}^2 + \beta_4 x_{t4} x_{t3} + \varepsilon_t$$

then the prob-value for testing $H_0: \beta_3 = 0$, which is .0028, says that, if $H_0: \beta_3 = 0$ is true, then the t_{b_3} statistic that we have observed ($t_{b_3} = 3.312$) is so rare that only .0028 (that is, 28 in 10,000) of all possible t_{b_3} statistics are at least as far away from zero and thus at least as contradictory to $H_0: \beta_3 = 0$ as this observed t_{b_3} statistic. Therefore, the prob-value of .0028 says that if $H_0: \beta_3 = 0$ is true, then we have observed a t_{b_3} statistic (3.312) that is so rare that its occurrence can be described as a 28 in 10,000 chance. This interpretation helps us to reach one of two possible conclusions: (1) The null hypothesis $H_0: \beta_3 = 0$ is true, and we have observed a t_{b_3} statistic (3.312) so rare that its occurrence can be described as a 28 in 10,000 chance; or (2) The null hypothesis $H_0: \beta_3 = 0$ is false and $H_1: \beta_3 \neq 0$ is true. Since any reasonable person would find it very difficult to believe that a 28 in 10,000 chance has actually occurred, and since we must believe that this has happened if we are to believe that $H_0: \beta_3 = 0$ is true, we would have substantial indication that $H_0: \beta_3 = 0$ is not true. The fact that the prob-value of .0028 is so small casts great doubt on the validity of the null hypothesis $H_0: \beta_3 = 0$ and thus lends great support to the validity of the alternative hypothesis $H_1: \beta_3 \neq 0$. From this example we also see that the smaller the prob-value, the more the observed t_{b_3} statistic contradicts $H_0: \beta_3 = 0$, and hence the more doubt is cast upon the validity of $H_0: \beta_3 = 0$.

Although the t_{b_j} statistic and related prob-value are useful measures of the importance of the independent variable x_{t_j} in the regression model

$$y_t = \beta_0 + \beta_1 x_{t1} + \cdots + \beta_j x_{tj} + \cdots + \beta_p x_{tp} + \varepsilon_t$$

it can be shown that the t_{b_j} statistic and related prob-value measure the *additional importance* of the independent variable x_{tj} over and above the combined importance of the other independent variables in the regression model. For this reason, *multicollinearity* (which exists when independent variables are related to each other and thus [to some extent] contribute redundant information for the description and prediction of the dependent variable) can cause the t_{b_j} statistics to make individual independent variables look unimportant when these independent variables really are important. For example, consider the Fresh detergent model

$$y_t = \beta_0 + \beta_1 x_{t4} + \beta_2 x_{t3} + \beta_3 x_{t3}^2 + \beta_4 x_{t4} x_{t3} + \beta_5 x_{t4} x_{t3}^2 + \varepsilon_t$$
$$(3.5620) \quad (1.1067) \quad (-2.7713) \quad (2.9811) \quad (-.8832) \quad (.7104)$$

which includes both $x_{t4} x_{t3}$ and $x_{t4} x_{t3}^2$. Note that multicollinearity certainly exists in this model, because, for example, the independent variables x_{t4}, $x_{t4} x_{t3}$, and $x_{t4} x_{t3}^2$ are related and the independent variables x_{t3}, x_{t3}^2, $x_{t4} x_{t3}$, and $x_{t4} x_{t3}^2$ are related. Moreover, this multicollinearity causes the t_{b_j} statistics (which are given under the appropriate terms in the model) to make x_{t4}, $x_{t4} x_{t3}$, and $x_{t4} x_{t3}^2$ look unimportant when previous analysis (see

Table 8.4) has indicated that x_{t4} and $x_{t4}x_{t3}$ are important in the model

$$y_t = \beta_0 + \beta_1 x_{t4} + \beta_2 x_{t3} + \beta_3 x_{t3}^2 + \beta_4 x_{t4}x_{t3} + \varepsilon_t$$

This does not imply that we should never include two cross-product terms such as $x_{t4}x_{t3}$ and $x_{t4}x_{t3}^2$ in a regression model (in some situations including two such terms leads to a better model), but it does say that we should always guard against being misled by t_{b_j} statistics when strong multicollinearity exists.

The reason that multicollinearity can cause the t_{b_j} statistic

$$t_{b_j} = \frac{b_j}{s_{b_j}}$$

and the related prob-value to give a misleading impression of the importance of the independent variable x_j in the model

$$y_t = \beta_0 + \beta_1 x_{t1} + \cdots + \beta_j x_{tj} + \cdots + \beta_p x_{tp} + \varepsilon_t$$

is related to the fact that the standard error of the estimate s_{b_j} can be proven to be given by the equation

$$s_{b_j} = s\sqrt{c_{jj}}$$

Here

1. s denotes the standard error and is given by the equation

$$s = \sqrt{\frac{SSE}{n - n_p}}$$

and

2. c_{jj} is given by the equation

$$c_{jj} = \frac{1}{\sum_{t=1}^{n} (x_{tj} - \bar{x}_j)^2 (1 - R_j^2)}$$

where

$$\bar{x}_j = \sum_{t=1}^{n} x_{tj}/n$$

and where R_j^2 is the multiple coefficient of determination that would be calculated by running a regression analysis using the model

$$x_{tj} = \beta_0 + \beta_1 x_{t1} + \cdots + \beta_{j-1} x_{t,j-1} + \beta_{j+1} x_{t,j+1} + \cdots + \beta_p x_{tp}$$

which expresses the independent variable x_j as a function of the remaining independent variables $x_1, \ldots, x_{j-1}, x_{j+1}, \ldots, x_p$.

Since it follows from the discussion of Section 8.3.1 that R_j^2 is the proportion of the total variation in the n observed values of the independent variable x_j that is explained by the overall regression model, it follows that R_j^2 is a measure of the multicollinearity between the n observed values of x_j and the n combinations of observed values of the independent variables $x_1, \ldots, x_{j-1}, x_{j+1}, \ldots, x_p$. The greater this multicollinearity is,

1. the greater R_j^2 is, and thus
2. the larger

$$c_{jj} = \frac{1}{\displaystyle\sum_{t=1}^{n} (x_{tj} - \bar{x}_j)^2 (1 - R_j^2)}$$

is, and thus

3. the larger the denominator of the t_{b_j} statistic

$$t_{b_j} = \frac{b_j}{s\sqrt{c_{jj}}}$$

is (likely to be).

If the t_{b_j} statistic has a large denominator, then the t_{b_j} statistic might, depending on the size of b_j, be small. Thus, strong multicollinearity between the independent variable x_j and the remaining independent variables $x_1, \ldots, x_{j-1}, x_{j+1}, \ldots, x_p$ can cause the t_{b_j} statistic to be small (and thus the related prob-value to be large), which would give the impression that x_j is not important (even if it really is important) in describing and predicting the dependent variable y. Since

$$s_{b_j} = s\sqrt{c_{jj}} = s\sqrt{\frac{1}{\displaystyle\sum_{t=1}^{n} (x_{tj} - \bar{x}_j)^2 (1 - R_j^2)}}$$

$$= s\sqrt{\frac{1}{\displaystyle\sum_{t=1}^{n} (x_{tj} - \bar{x}_j)^2}} \sqrt{\frac{1}{1 - R_j^2}}$$

we call

$$VIF_j = \frac{1}{1 - R_j^2}$$

the *variance inflation factor* for b_j. The larger R_j^2 is, the larger VIF_j is, and thus the more VIF_j is inflating s_{b_j}. Both the largest variance inflation factor and the mean of the variance inflation factors

$$\overline{VIF} = \frac{\displaystyle\sum_{j=1}^{p} VIF_j}{p}$$

related to the independent variables in the model

$$y = \beta_0 + \beta_1 x_1 + \cdots + \beta_p x_p + \varepsilon$$

are used as indicators of the severity of multicollinearity. If the largest variance inflation factor is greater than 10, or if the mean of the variance inflation factors is substantially greater than 1, then multicollinearity may be severe. Whereas we can measure the multicollinearity between the independent variable x_j and the rest of the independent variables $x_1, \ldots, x_{j-1}, x_{j+1}, \ldots, x_p$ in a regression model by calculating the variance inflation factor VIF_j, we can measure the multicollinearity between the independent variable x_j and another independent variable x_k by calculating the *simple correlation coefficient* between x_j and x_k.

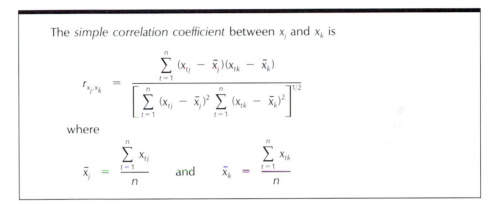

The *simple correlation coefficient* between x_j and x_k is

$$r_{x_j x_k} = \frac{\sum_{t=1}^{n} (x_{tj} - \bar{x}_j)(x_{tk} - \bar{x}_k)}{\left[\sum_{t=1}^{n} (x_{tj} - \bar{x}_j)^2 \sum_{t=1}^{n} (x_{tk} - \bar{x}_k)^2 \right]^{1/2}}$$

where

$$\bar{x}_j = \frac{\sum_{t=1}^{n} x_{tj}}{n} \quad \text{and} \quad \bar{x}_k = \frac{\sum_{t=1}^{n} x_{tk}}{n}$$

The formula for

$$r_{x_j x_k}$$

may be compared with the formula for the simple correlation coefficient between x and y (see Section 8.3.1) and may be seen as a special case of r if we consider x_j to be the independent variable x and x_k to be the dependent variable y. From our interpretation of r in Section 8.3.1, it follows that $r_{x_j x_k}$ is a measure of the linear relationship, or correlation, between x_j and x_k. The larger that $r_{x_j x_k}$ is, the stronger is the linear correlation (and thus the multicollinearity) between x_j and x_k. If $r_{x_j x_k}$ is greater than .9, then the multicollinearity between x_j and x_k may be severe.

If multicollinearity is severe, we must be particularly careful to avoid incorrectly concluding that the independent variables affected by the multicollinearity are unimportant. To lessen the severity of the multicollinearity, we might remove some of the involved independent variables. However, when we remove these variables it becomes difficult to assess their importance.

EXAMPLE 8.7   In Table 8.5 we present data taken from 17 U.S. Navy Hospitals concerning the need for hospital staffing. Here x_1 denotes average daily patient load, x_2 denotes monthly X-ray exposures, x_3 denotes monthly occupied bed days, x_4 denotes eligible population in the area (divided by 1000), x_5 denotes average length of

TABLE 8.5   Hospital Staffing Data*

Hospital	x_1	x_2	x_3	x_4	x_5	y
1	15.57	2463	472.92	18.0	4.45	566.52
2	44.02	2048	1339.75	9.5	6.92	696.82
3	20.42	3940	620.25	12.8	4.28	1033.15
4	18.74	6505	568.33	36.7	3.90	1603.62
5	49.20	5723	1497.60	35.7	5.50	1611.37
6	44.92	11520	1365.83	24.0	4.60	1613.27
7	55.48	5779	1687.00	43.3	5.62	1854.17
8	59.28	5969	1639.92	46.7	5.15	2160.55
9	94.39	8461	2872.33	78.7	6.18	2305.58
10	128.02	20106	3655.08	180.5	6.15	3503.93
11	96.00	13313	2912.00	60.9	5.88	3571.89
12	131.42	10771	3921.00	103.7	4.88	3741.40
13	127.21	15543	3865.67	126.8	5.50	4026.52
14	252.90	36194	7684.10	157.7	7.00	10343.81
15	409.20	34703	12446.33	169.4	10.78	11732.17
16	463.70	39204	14098.40	331.4	7.05	15414.94
17	510.22	86533	15524.00	371.6	6.35	18854.45

*Source: Procedures and Analysis for Staffing Standards Development: Regression Analysis Handbook. San Diego, California: Navy Manpower and Materials Analysis Center, 1979.

patients' stays in days, and y denotes monthly man-hours. We will analyze this data by using the results presented in Myers [1986]. To begin, note that the simple correlation coefficients involving x_1, x_2, x_3, x_4, and x_5 that are greater than .9 are

$$r_{x_1,x_2} = .90738$$

$$r_{x_1,x_3} = .99990$$

$$r_{x_1,x_4} = .93569$$

$$r_{x_2,x_3} = .90715$$

$$r_{x_2,x_4} = .91047$$

and

$$r_{x_3,x_4} = .93317$$

These coefficients, along with the magnitudes of the variance inflation factors for the model involving x_1 $(VIF_1 = 9{,}597.57)$, $x_2(VIF_2 = 7.94)$, $x_3(VIF_3 = 8{,}933.09)$, $x_4(VIF_4 = 23.29)$, and $x_5(VIF_5 = 4.28)$, indicate the presence of very strong multicollinearity. Specifically, the values

$$r_{x_1,x_3} = .99990$$

$$VIF_1 = 9{,}597.57$$

and

$$VIF_3 = 8{,}933.09$$

indicate that x_1 and x_3 contribute very redundant information and thus probably are not needed in the same model. Similarly, the values

$$r_{x_1,x_4} = .93569$$

$$r_{x_3,x_4} = .93317$$

and

$$VIF_4 = 23.29$$

indicate that x_4 might not be needed in a model utilizing x_1 or in a model utilizing x_3. We will further analyze the hospital data in Section 8.3.4.

8.3.3 Comparing Regression Models

Since multicollinearity can hinder the ability of the t_{b_j} statistics and related prob-values to assess the importance of the independent variables in a regression model, it is useful to consider all reasonable regression models and compare them on the basis of several criteria. One such criterion is R^2, the multiple coefficient of determination. We have seen that the larger R^2 is for a particular regression model, the larger is the proportion of the total variation that is explained by the regression model. However, one must balance the magnitude of R^2, or, in general, the "goodness" of any criterion, against the difficulty and expense of using the regression model to describe, predict, and control the dependent variable. Generally speaking, the size of R^2 does not necessarily indicate whether a regression model will accurately describe and predict the dependent variable. For example, although a value of R^2 close to 1 indicates that the regression model under consideration explains a high proportion of the total variation in the n observed values of the dependent variable, a value of R^2 close to 1 does not (necessarily) imply that, if we are interested in predicting future values of the dependent variable, the regression model will produce predictions that are accurate enough for the application at hand. Furthermore, when using R^2 to *compare* overall regression models, one must be aware that it can be proven that *adding any independent variable to a regression model*—even an independent variable which is totally unrelated to the dependent variable—

1. Will decrease the unexplained variation

$$SSE = \sum_{t=1}^{n} (y_t - \hat{y}_t)^2$$

and
2.   Will leave the total variation unchanged, since

$$\text{total variation} \;=\; \sum_{t=1}^{n} (y_t - \bar{y})^2$$

is a function of only the n observed values of the dependent variable (y_1, y_2, . . . , y_n) and thus is independent of the regression model under consideration, and thus

3.   Will increase the

$$\text{explained variation} \;=\; \text{total variation} - \text{unexplained variation}$$

and thus

4.   Will increase

$$R^2 \;=\; \frac{\text{explained variation}}{\text{total variation}}$$

For reasons to be discussed later, two better criteria by which to compare regression models are *corrected R^2* and the *mean square error*, which we now define.

Suppose that the regression model under consideration

$$y \;=\; \beta_0 + \beta_1 x_1 + \cdots + \beta_p x_p + \varepsilon$$

includes $n_p - 1$ independent variables (and thus, because of the intercept β_0, utilizes n_p parameters). Then

1.   The *corrected multiple coefficient of determination (corrected R^2)* is

$$\bar{R}^2 \;=\; \left(R^2 - \frac{n_p - 1}{n - 1} \right) \left(\frac{n - 1}{n - n_p} \right)$$

where R^2 is the multiple coefficient of determination and n is the number of observed values of the dependent variable.
2.   The *mean square error* is

$$s^2 \;=\; \frac{SSE}{n - n_p}$$

To explain the reasoning behind the above definition of $\bar{R}^2$, first note that it can be shown that if the values of the independent variables are completely random (that is, randomly chosen from a population of numbers), they will still explain enough of the total variation in the observed values of the dependent variable to make R^2 equal to, on the average, $(n_p - 1)/(n - 1)$. Therefore, our first step in correcting R^2 is to subtract this random explanation and form the quantity

$$R^2 - \frac{n_p - 1}{n - 1}$$

If the values of the independent variables are completely random, then this corrected version of R^2 equals 0. However, if the values of the independent variables are not completely random, then the quantity

$$R^2 - \frac{n_p - 1}{n - 1}$$

reduces R^2 too much. To see why, note that if R^2 equals 1, then the above corrected version of R^2 does not equal 1, but equals

$$1 - \frac{n_p - 1}{n - 1} = \frac{n - n_p}{n - 1}$$

which is less than 1, since $n - n_p < n - 1$. To define a corrected R^2 which equals 1 if R^2 equals 1, we multiply

$$R^2 - \frac{n_p - 1}{n - 1}$$

by the factor

$$\left(\frac{n - 1}{n - n_p} \right)$$

to form the quantity

$$\left(R^2 - \frac{n_p - 1}{n - 1} \right) \left(\frac{n - 1}{n - n_p} \right)$$

which is the previously defined $\bar{R}^2$.

We next discuss the *mean square error*

$$s^2 = \frac{SSE}{n - n_p}$$

First, since s^2 depends on SSE, we wish s^2 to be reasonably small. Next, note that whereas adding an independent variable to a model will always decrease SSE, if the decrease in SSE caused by the addition of the independent variable is not enough to offset the decrease in the denominator of s^2, $(n - n_p)$, caused by the addition of the independent variable, then s^2 will increase (this will be demonstrated in the next example). If s^2 increases, it can be proven that the width of the prediction interval for y_t will increase. This says that our prediction of y_t has become less precise, which implies that we should not add the independent variable to the regression model. If s^2 decreases fairly substantially, then the width of the prediction interval will probably decrease, which implies that it might be appropriate to add the independent variable to the regression model.

It can be shown that the mean square error, s^2, and $\bar{R}^2$ are related by the equation

$$s^2 = (1 - \bar{R}^2)s_y^2$$

where

$$s_y^2 = \frac{\sum_{t=1}^{n} (y_t - \bar{y})^2}{n - 1}$$

Since s_y^2 is independent of the model under consideration (because s_y^2 depends only on the observed values of the dependent variable $y_1, y_2, \ldots, y_n$), the equation

$$s^2 = (1 - \bar{R}^2)s_y^2$$

implies that s^2 decreases if and only if $\bar{R}^2$ increases. Therefore, both the mean square error, s^2, and adjusted R^2, $\bar{R}^2$, are better criteria for comparing regression models than is R^2 because, while adding an unimportant independent variable to the regression model will increase R^2, adding an unimportant independent variable might well (although it is not guaranteed) increase s^2 and decrease $\bar{R}^2$, which indicates that we should not include the independent variable in a final regression model.

EXAMPLE 8.8

Consider the Fresh detergent model

$$y_t = \beta_0 + \beta_1 x_{t4} + \beta_2 x_{t3} + \beta_3 x_{t3}^2 + \beta_4 x_{t4} x_{t3} + \varepsilon_t$$

for which (see Table 8.4)

1. $SSE = 1.0644$
2. Explained variation $= 12.3942$

3. $R^2 = \dfrac{\text{explained variation}}{\text{total variation}} = \dfrac{12.3942}{13.4586} = .9209$

4. $s^2 = \dfrac{SSE}{n - n_p} = \dfrac{1.0644}{30 - 5} = \dfrac{1.0644}{25} = .0426$

5. $\bar{R}^2 = \left(R^2 - \dfrac{n_p - 1}{n - 1}\right)\left(\dfrac{n - 1}{n - n_p}\right)$

$$= \left(.9029 - \frac{5 - 1}{30 - 1}\right)\left(\frac{30 - 1}{30 - 5}\right) = .9083$$

Adding the independent variable $x_{t4} x_{t3}^2$ to the given model to form the model

$$y_t = \beta_0 + \beta_1 x_{t4} + \beta_2 x_{t3} + \beta_3 x_{t3}^2 + \beta_4 x_{t4} x_{t3} + \beta_5 x_{t4} x_{t3}^2 + \varepsilon_t$$

1. Decreases SSE to 1.0425
2. Increases the explained variation to 12.4161

TABLE 8.6 Several Regression Models That Can be Used to Predict Demand for Fresh Detergent

Model	R^2	s^2
$y_t = \beta_0 + \beta_1 x_{t1} + \varepsilon_t$	.2202	.3784
$y_t = \beta_0 + \beta_1 x_{t1} + \beta_2 x_{t1}^2 + \varepsilon_t$	.2286	.3845
$y_t = \beta_0 + \beta_1 x_{t2} + \varepsilon_t$	.5490	.2168
$y_t = \beta_0 + \beta_1 x_{t2} + \beta_2 x_{t2}^2 + \varepsilon_t$	.5590	.2198
$y_t = \beta_0 + \beta_1 x_{t3} + \varepsilon_t$	.7673	.1119
$y_t = \beta_0 + \beta_1 x_{t3} + \beta_2 x_{t3}^2 + \varepsilon_t$	.8380	.0808
$y_t = \beta_0 + \beta_1 x_{t4} + \varepsilon_t$	.7915	.1002
$y_t = \beta_0 + \beta_1 x_{t4} + \beta_2 x_{t4}^2 + \varepsilon_t$	.8043	.0975
$y_t = \beta_0 + \beta_1 x_{t1} + \beta_2 x_{t2} + \varepsilon_t$	.8288	.0854
$y_t = \beta_0 + \beta_1 x_{t1} + \beta_2 x_{t3} + \varepsilon_t$	.7717	.1138
$y_t = \beta_0 + \beta_1 x_{t2} + \beta_2 x_{t3} + \varepsilon_t$	.8377	.0809
$y_t = \beta_0 + \beta_1 x_{t1} + \beta_2 x_{t2} + \beta_3 x_{t3} + \varepsilon_t$	.8936	.0551
$y_t = \beta_0 + \beta_1 x_{t1} + \beta_2 x_{t2} + \beta_3 x_{t3} + \beta_4 x_{t3}^2 + \varepsilon_t$	.9084	.0493
$y_t = \beta_0 + \beta_1 x_{t1} + \beta_2 x_{t1}^2 + \beta_3 x_{t2} + \beta_4 x_{t2}^2 + \beta_5 x_{t3} + \beta_6 x_{t3}^2 + \varepsilon_t$	.9161	.0491
$y_t = \beta_0 + \beta_1 x_{t4} + \beta_2 x_{t3} + \varepsilon_t$	.8860	.0568
$y_t = \beta_0 + \beta_1 x_{t4} + \beta_2 x_{t3} + \beta_3 x_{t3}^2 + \varepsilon_t$	.9054	.0490
$y_t = \beta_0 + \beta_1 x_{t4} + \beta_2 x_{t4}^2 + \beta_3 x_{t3} + \beta_4 x_{t3}^2 + \varepsilon_t$	.9106	.0481
$y_t = \beta_0 + \beta_1 x_{t4} + \beta_2 x_{t3} + \beta_3 x_{t3}^2 + \beta_4 x_{t4} x_{t3} + \varepsilon_t$	.9209	.0426
$y_t = \beta_0 + \beta_1 x_{t4} + \beta_2 x_{t3} + \beta_3 x_{t3}^2 + \beta_4 x_{t4} x_{t3} + \beta_5 x_{t4} x_{t3}^2 + \varepsilon_t$	.9225	.0434

3. Increases R^2 to

$$\frac{\text{explained variation}}{\text{total variation}} = \frac{12.4161}{13.4586} = .9225$$

4. Increases s^2 to

$$\frac{SSE}{n - n_p} = \frac{1.0425}{30 - 6} = \frac{1.0425}{24} = .0434$$

because the decrease of SSE from 1.0644 to 1.0425 has not been enough to offset the decrease in $n - n_p$ from 25 to 24.

5. Decreases $\bar{R}^2$ to

$$\left(R^2 - \frac{n_p - 1}{n - 1} \right)\left(\frac{n - 1}{n - n_p} \right) = \left(.9225 - \frac{6 - 1}{30 - 1} \right)\left(\frac{30 - 1}{30 - 5} \right)$$

$$= .8701$$

Since adding $x_{t4} x_{t3}^2$ decreases $\bar{R}^2$ and increases s^2 (which would increase the width of the prediction interval for y_t), we should not add $x_{t4} x_{t3}^2$ to the model

$$y_t = \beta_0 + \beta_1 x_{t4} + \beta_2 x_{t3} + \beta_3 x_{t3}^2 + \beta_4 x_{t4} x_{t3} + \varepsilon_t$$

TABLE 8.7   *A Comparison of the 95% Prediction Intervals for y_{31} Given by Three Regression Models in the Fresh Detergent Probem*

Model	95% Prediction Interval for y_{31}
$y_t = \beta_0 + \beta_1 x_{t1} + \beta_2 x_{t2} + \beta_3 x_{t3} + \varepsilon_t$	[7.80, 8.85]
$y_t = \beta_0 + \beta_1 x_{t4} + \beta_2 x_{t3} + \beta_3 x_{t3}^2 + \varepsilon_t$	[7.96, 8.93]
$y_t = \beta_0 + \beta_1 x_{t4} + \beta_2 x_{t3} + \beta_3 x_{t3}^2 + \beta_4 x_{t4} x_{t3} + \varepsilon_t$	[8.068, 8.984]

In fact, recalling that the t_{b_j} statistics and prob-values in Table 8.4 indicate that the independent variables in the model are significant, and noting that this model has the largest R^2 and smallest s^2 of any model in Table 8.6, and also noting that this model provides the shortest prediction interval for y_{31} (when $x_{31,4} = .10$ and $x_{31,3} = 6.80$) of any model in Table 8.7, we conclude that this model is a reasonable final model to use in forecasting demand for Fresh detergent.

To complete this section, we note that, although the model

$$y_t = \beta_0 + \beta_1 x_{t4} + \beta_2 x_{t3} + \beta_3 x_{t3}^2 + \beta_4 x_{t4} x_{t3} + \varepsilon_t$$

seems to be a reasonable model in the Fresh detergent problem, sometimes it is appropriate in time series situations to use lagged (that is, past) values of both the dependent and the independent variables to predict future values of the dependent variable. For example, one possible model involving lagged variables in the Fresh detergent problem is

$$y_t = \beta_0 + \beta_1 y_{t-1} + \beta_2 y_{t-2} + \beta_3 x_{t4} + \beta_4 x_{t3} + \beta_5 x_{t3}^2 + \beta_6 x_{(t-1)3} + \varepsilon_t$$

In Chapter 7 we discussed a precise way to relate a dependent variable to its own past values and to the future and past values of *one* independent variable. When using more than one independent variable, the methods of Chapter 7 become more difficult to apply, and in this case we can use the model-building methods of classical regression analysis as an "intuitive" procedure for developing a forecasting model. For example, considering the second model, we would regard y_{t-1}, y_{t-2}, x_{t4}, x_{t3}, x_{t3}^2, and $x_{(t-1)3}$ as independent variables, and we would use data plots, t_{b_j} statistics (and prob-values), R^2, $\bar{R}^2$, s^2, and prediction interval width to assess the importance of these independent variables and to assess the adequacy of the overall regression model.

8.3.4 Advanced Model Comparison Methods: The C-Statistic and the PRESS Statistic

We begin by defining the C-statistic.

Suppose we are attempting to choose an appropriate set of independent variables (to form a final regression model) from p potential independent variables. Then, if the regression model

$$y = \beta_0 + \beta_1 x_1 + \cdots + \beta_{k-1} x_{k-1} + \varepsilon$$

includes $k - 1$ independent variables (and, because of the intercept β_0, utilizes k parameters), we define the C-*statistic* to be

$$C = \frac{SSE}{s_p^2} - (n - 2k)$$

Here, n is the number of observed values of the dependent variable, SSE is the unexplained variation calculated from the model

$$y = \beta_0 + \beta_1 x_1 + \cdots + \beta_{k-1} x_{k-1} + \varepsilon$$

and s_p^2 is the mean square error calculated from the model

$$y = \beta_0 + \beta_1 x_1 + \cdots + \beta_p x_p + \varepsilon$$

which includes all p independent variables (and thus utilizes $p + 1$ parameters).

Since

$$C = \frac{SSE}{s_p^2} - (n - 2k)$$

is a function of SSE, the unexplained variation calculated from the model

$$y = \beta_0 + \beta_1 x_1 + \cdots + \beta_{k-1} x_{k-1} + \varepsilon$$

and since we want SSE to be small, *we want C to be small*. Although adding even an unimportant independent variable to a regression model will decrease the unexplained variation, adding an unimportant independent variable can increase C. This can happen when the decrease in the unexplained variation caused by the addition of the extra independent variable is not enough to offset the decrease in $n - 2k$ caused by the addition of the extra independent variable (which increases k by 1). It should be noted that although adding an unimportant independent variable to a regression model can increase both s^2 and C, there is no exact relationship between s^2 and C. For example, adding an independent variable to a regression model can decrease s^2 and increase C. We have seen that we want to find a model for which C is small. In addition, it can be shown from the theory behind the C-statistic that *we also wish to find a model for which the C-statistic roughly equals k* (k equals the number of parameters in the model). If a model has a C-statistic substantially greater than k, it can be shown that this model has substantial bias and is undesirable. Thus, although we want to find a model for which

TABLE 8.8   Model Comparison Statistics For All Reasonable Regression Models Describing the Hospital Staffing Data

Variables in Model	s^2	R^2	C	PRESS
x_5	21,940,359	.3348	785.26	455,789,688
x_4	3,816,879	.8843	125.87	70,075,727
x_2	3,517,337	.8934	114.97	158,243,072
x_1	939,990	.9715	21.20	22,243,776
x_3	917,487	.9722	20.38	21,841,500
x_4, x_5	3,165,704	.9104	96.50	71,164,912
x_2, x_5	2,688,542	.9239	80.30	128,703,914
x_2, x_4	2,452,812	.9306	72.29	140,743,693
x_1, x_3	971,399	.9725	21.99	22,564,832
x_1, x_4	914,181	.9741	20.04	35,786,830
x_3, x_4	870,761	.9754	18.57	32,454,824
x_1, x_5	564,291	.9840	8.16	12,977,100
x_3, x_5	538,720	.9848	7.29	12,628,470
x_1, x_2	490,585	.9861	5.66	18,036,834
x_2, x_3	469,456	.9867	4.94	17,853,441
x_2, x_4, x_5	1,816,626	.9523	48.28	107,102,405
x_1, x_3, x_4	818,290	.9785	16.80	34,400,018
x_1, x_4, x_5	583,845	.9846	9.41	17,828,243
x_3, x_4, x_5	570,794	.9850	9.00	16,275,120
x_1, x_3, x_5	569,830	.9850	8.97	13,036,635
x_1, x_2, x_4	528,257	.9861	7.66	32,814,300
x_2, x_3, x_4	504,218	.9868	6.90	30,139,304
x_1, x_2, x_3	482,286	.9873	6.21	22,794,229
x_1, x_2, x_5	403,215	.9894	3.71	18,051,730
x_2, x_3, x_5	377,954	.9901	2.92	17,846,717
x_1, x_3, x_4, x_5	615,741	.9851	10.92	18,621,398
x_1, x_2, x_3, x_4	499,469	.9879	7.54	37,719,883
x_1, x_2, x_3, x_5	389,793	.9905	4.34	22,464,723
x_1, x_2, x_4, x_5	387,001	.9906	4.26	30,255,902
x_2, x_3, x_4, x_5	378,826	.9908	4.03	28,629,419
x_1, x_2, x_3, x_4, x_5	412,277	.9908	6.00	32,195,222

C is as small as possible, if C for such a model is substantially greater than k, we may prefer to choose a different model for which C is slightly larger and more nearly equal to the number of parameters in that (different) model.

To illustrate using the C-statistic, we analyze the hospital staffing data set of Table 8.5. In Table 8.8 we present several statistics related to all reasonable regression models describing the hospital data, and in Table 8.9 we present several statistics related to what

TABLE 8.9 *Model Comparison Statistics for the Six Best Models Describing the Hospital Staffing Data*

Model, Variables, and Variance Inflation Factors	s^2	R^2	C	PRESS
Model 1: $x_1(1.8199)$, $x_5(1.8199)$	564,291	.9840	8.16	12,977,100
Model 2: $x_3(1.8195)$, $x_5(1.8195)$	538,720	.9848	7.29	12,628,470
Model 3: $x_1(5.6605)$, $x_2(5.6605)$	490,585	.9861	5.66	18,036,834
Model 4: $x_2(5.6471)$, $x_3(5.6471)$	469,456	.9867	4.94	17,853,441
Model 5: $x_1(11.3214)$, $x_2(7.7714)$, $x_5(2.4985)$	403,215	.9894	3.71	18,051,730
Model 6: $x_2(7.7373)$, $x_3(11.2693)$, $x_5(2.4929)$	377,954	.9901	2.92	17,846,177

we consider to be the six best models (note that the numbers in parentheses corresponding to the variables are the variance inflation factors for the specified model). First, note that the C-statistics in Tables 8.8 and 8.9 have been computed by using the fact that the mean square error s_p^2 for the model containing all $p = 5$ potential independent variables

$$y = \beta_0 + \beta_1 x_1 + \beta_2 x_2 + \beta_3 x_3 + \beta_4 x_4 + \beta_5 x_5 + \varepsilon$$

is 412,277. For example, then, the C-statistic for

$$\text{Model 6:} \quad y = \beta_0 + \beta_1 x_2 + \beta_2 x_3 + \beta_3 x_5 + \varepsilon$$

which has an unexplained variation of $SSE = 4913402$ and $k = 4$, is

$$C = \frac{SSE}{s_p^2} - (n - 2k)$$

$$= \frac{4913402}{412277} - [17 - 2(4)]$$

$$= 2.92$$

Examining Table 8.9, we conclude that model 6 is the best model with respect to the C-statistic, because model 6 has the smallest C-statistic and because, since $C = 2.92$ is not greater than $k = 4$ (the number of parameters in model 6), the C-statistic indicates that model 6 is not biased.

Since model 6 has not only the smallest C-statistic but also the smallest s^2 of any model in Table 8.9, we might be tempted to conclude that model 6 is the best model to use to predict hospital staffing needs. However, since the hospital staffing data set contains three hospitals (15, 16, and 17) that are substantially larger than the other 14 small to medium-sized hospitals (note that classifying hospital 14 is difficult), it is important to use each regression model under consideration to calculate for hospital t

(that is, for each observation) the *PRESS* (or *deleted*) *residual* $d_t = y_t - \hat{y}_{(t)}$. Here,

$$\hat{y}_{(t)} = b_0^{(t)} + b_1^{(t)} x_{t1} + b_2^{(t)} x_{t2} + \cdots + b_p^{(t)} x_{tp}$$

is the point prediction of y_t calculated by using least squares point estimates $b_0^{(t)}$, $b_1^{(t)}, \ldots, b_p^{(t)}$ that are computed by utilizing all $n = 17$ observations except for observation t. Recalling that the usual residual is $e_t = y_t - \hat{y}_t$, where $\hat{y}_t$ is the point prediction of y_t calculated by using least squares point estimates $b_0, b_1, \ldots, b_p$ that are computed by using all $n = 17$ observations, it can be proven that $d_t = e_t/(1 - h_{tt})$. Here, h_{tt} is the leverage value corresponding to observation t (this leverage value has been previously discussed in Section 8.2 and will be further discussed in Section 8.4.6).

To see how PRESS residuals can help us, note that each of the large hospitals 15, 16, and 17 is somewhat outside of the experimental region defined by the other 16 hospitals, since each of the large hospitals is different from the 14 small to medium-sized hospitals and is different (with respect to at least some of the values of x_1, x_2, x_3, x_4, and x_5) from the other two large hospitals. These facts imply that if we use a particular regression model to calculate for large hospital t the PRESS residual, $d_t = y_t - \hat{y}_{(t)}$, then, since the point prediction $\hat{y}_{(t)}$ of y_t is made by using all 16 hospitals except for large hospital t, it follows that we are predicting y_t by extrapolating the regression model somewhat outside of the experimental region defined by the other hospitals. The smaller the magnitude of d_t, the less dangerous it is to extrapolate the particular regression model to predict y_t. It can be verified that, while the magnitudes of the PRESS residuals corresponding to hospitals 1 through 14 are of roughly the same sizes for the six best models, the magnitudes of the PRESS residuals corresponding to hospitals 15, 16, and 17 and thus the *PRESS statistic* (which is the sum of the squared PRESS residuals) are smallest for model 2. This might be interpreted to mean that it is less dangerous to extrapolate (somewhat) model 2 to predict staffing needs for large hospitals. However, if we use any of the best models to predict the staffing needs for hospital 14, we obtain a point prediction that is much smaller than the observed value $y_{14} = 10\,343.81$ (see Table 8.5), which implies that y_{14} was much greater than would be expected for the values of x_1, x_2, x_3, x_4, and x_5 corresponding to hospital 14. If we remove observation 14 from the data set (which would be appropriate if y_{14} was erroneously recorded or resulted from a situation that would be very unlikely to occur again), and if we perform a regression analysis using the remaining 16 observations, it can be verified that model 6, in addition to giving the smallest values of C, s^2, and PRESS, also yields the smallest (in magnitude) usual and PRESS residuals corresponding to large hospitals 15, 16, and 17. This implies that it might be best to use model 6 (based on hospitals 1 through 13, 15, 16, and 17) to predict staffing needs for large hospitals (particularly for large hospitals having large values of x_2). This is because the magnitudes of the usual and PRESS residuals corresponding to hospital 17—which has a very large value of x_2—are much smaller for model 6 than for model 2 (which does not utilize the independent variable x_2). Finally, if we remove hospital 14 and large hospitals 15, 16, and 17 from the data set, model 2 and model 6 give very similar results, and thus we might choose the simpler model 2 to predict staffing needs for small to medium-sized hospitals.

8.3.5 Stepwise Regression, Forward Selection, Backward Elimination, and Maximum R^2 Improvement

In a regression problem where the number of potential independent variables is not large, we can fairly easily compare all reasonable regression models with respect to various criteria (such as R^2, s^2, and C). However, if we are attempting to choose an appropriate set of independent variables from a large number of potential independent variables, comparing all reasonable regression models can be quite unwieldy. In this case, it is useful to employ a screening procedure that can be used to identify one set (or several sets) of the "most important" independent variables. We now present four such screening procedures.

STEPWISE REGRESSION

Stepwise regression is generally carried out on a computer and is available in most standard regression computer packages. There are slight variations in the way that different computer packages carry out stepwise regression. Assuming that y is the dependent variable and $x_1, x_2, \ldots, x_p$ are the p potential independent variables (where p will generally be large), we explain how most of the computer packages carry out stepwise regression. To make our description as concise as possible, we need to introduce some new terminology. Stepwise regression uses t_{b_j} statistics (and related prob-values) to determine the importance (or significance) of the independent variables in various regression models. In this context, the t_{b_j} *statistic indicates that the independent variable x_j is significant at the α level if and only if the related prob-value is less than α* (which implies that we can reject H_0: $\beta_j = 0$ in favor of H_1: $\beta_j \neq 0$ by allowing the probability of a Type I error to be equal to α). Then stepwise regression is performed as follows.

Before beginning the stepwise procedure, we choose a value of α_{entry}, which we call "the probability of a Type I error related to entering an independent variable into the regression model," and a value of α_{stay}, which we call "the probability of a Type I error related to retaining an independent variable that was previously entered into the model." We discuss the considerations involved in choosing these values after our description of the stepwise procedure. For now, suffice it to say that it is common practice to set both α_{entry} and α_{stay} equal to .05.

Step 1: The stepwise procedure considers the p possible one-independent-variable regression models of the form

$$y = \beta_0 + \beta_1 x_1 + \varepsilon$$

Each different model includes a different potential independent variable. For each model, the t_{b_1} statistic (and prob-value) related to testing H_0: $\beta_1 = 0$ versus H_1: $\beta_1 \neq 0$ is calculated. Denoting the independent variable giving the largest absolute value of the t_{b_1} statistic (and the smallest prob-value) by the symbol $x_{[1]}$, we consider the model

$$y = \beta_0 + \beta_1 x_{[1]} + \varepsilon$$

If the t_{b_1} statistic does not indicate that the independent variable $x_{[1]}$ is significant at the α_{entry} level in the model, then the stepwise procedure terminates by choosing the model

$$y = \beta_0 + \varepsilon$$

If the t_{b_1} statistic indicates that the independent variable $x_{[1]}$ is significant at the α_{entry} level in the model, then $x_{[1]}$ is retained for use in Step 2.

Step 2:   The stepwise procedure considers the $p - 1$ possible two-independent-variable regression models of the form

$$y = \beta_0 + \beta_1 x_{[1]} + \beta_2 x_j + \varepsilon$$

Each different model includes $x_{[1]}$, the independent variable chosen in Step 1, and a different potential independent variable chosen from the remaining $p - 1$ independent variables that were not chosen in Step 1. For each model, the t_{b_2} statistic (and prob-value) related to testing $H_0: \beta_2 = 0$ versus $H_1: \beta_2 \neq 0$ is calculated. Denoting the independent variable giving the largest absolute value of the t_{b_2} statistic (and the smallest prob-value) by the symbol $x_{[2]}$ we consider the model

$$y = \beta_0 + \beta_1 x_{[1]} + \beta_2 x_{[2]} + \varepsilon$$

If the t_{b_2} statistic indicates that the independent variable $x_{[2]}$ is significant at the α_{entry} level in the model, then $x_{[2]}$ is retained in this model, and the stepwise procedure checks to see if the independent variable $x_{[1]}$ should be allowed to stay in the model.

This check should be made, because multicollinearity will probably cause the t_{b_1} statistic calculated for the model

$$y = \beta_0 + \beta_1 x_{[1]} + \beta_2 x_{[2]} + \varepsilon$$

to be different from the t_{b_1} statistic calculated for the model

$$y = \beta_0 + \beta_1 x_{[1]} + \varepsilon$$

If the t_{b_1} statistic does not indicate that the independent variable $x_{[1]}$ is significant at the α_{stay} level in the model

$$y = \beta_0 + \beta_1 x_{[1]} + \beta_2 x_{[2]} + \varepsilon$$

then the stepwise procedure returns to the beginning of Step 2, and starting with a new one-independent-variable model that uses the new significant independent variable $x_{[2]}$, the stepwise procedure attempts to find a new two-independent-variable model

$$y = \beta_0 + \beta_1 x_{[2]} + \beta_2 x_j + \varepsilon$$

If the t_{b_1} statistic indicates that the independent variable $x_{[1]}$ is significant at the α_{stay} level in the model

$$y = \beta_0 + \beta_1 x_{[1]} + \beta_2 x_{[2]} + \varepsilon$$

then both the independent variables $x_{[1]}$ and $x_{[2]}$ are retained for use in Step 3.

Further steps: The stepwise procedure continues by adding independent variables one at a time to the model. At each step an independent variable is added to the model if and only if it has the largest (in absolute value) t_{b_j} statistic of the independent variables not in the model, and if its t_{b_j} statistic indicates that it is significant at the α_{entry} level. After adding an independent variable to the model, the stepwise procedure checks all the independent variables already included in the model and removes any independent variable that does not produce a t_{b_j} statistic indicating that the variable is significant at the α_{stay} level. Only after the necessary removals are made does the stepwise procedure attempt to add another independent variable to the model. The stepwise procedure terminates when all the independent variables not in the model have t_{b_j} statistics indicating that these variables are insignificant at the α_{entry} level or when the variable to be added to the model is the one just removed from it.

Regarding the choice of α_{entry} and α_{stay}, Draper and Smith [1981] state that it is usually best to choose α_{entry} equal to α_{stay}. It is not recommended that α_{stay} be chosen less than α_{entry}, because this makes it too likely that an independent variable that has just been added to the model will (in subsequent steps) be removed from the model. Sometimes, however, it is reasonable to choose α_{stay} to be greater than α_{entry}, because this makes it more likely that an independent variable whose significance decreases as new independent variables are added to the model will be allowed to stay in the model. Draper and Smith go on to suggest that α_{entry} and α_{stay} be set equal to .05 or .10.

However, we should point out that setting α_{entry} and α_{stay} higher than .10 is also reasonable, because this will cause more independent variables to be included in the model and thus will give the analyst an opportunity to consider additional independent variables. Indeed, though the model obtained by the stepwise procedure may be reasonable, it should not necessarily be regarded as the best final regression model. First, since the choices of α_{entry} and α_{stay} are arbitrary, and since the many hypothesis tests performed by the stepwise procedure imply that Type I and Type II errors might be committed, the stepwise procedure might include some unimportant independent variables in the model and exclude some important independent variables from the model. Second, it is sometimes appropriate to include powers (such as squared values) of the independent variables and interaction terms in a final regression model. While such terms can be included in the set of p potential independent variables to be considered by stepwise regression, we often omit them so that the (probably) already large list of potential independent variables is not unduly increased. Thus, if we do omit powers and interaction terms from consideration, and if some of these terms are important, the stepwise procedure will omit some important independent variables. In general, then, stepwise regression should be regarded as a screening procedure that can be used to find at least some of the most important independent variables. Once stepwise regression identifies these independent variables, we should then carefully use the other model-building techniques discussed in this book to arrive at an appropriate final regression model.

TABLE 8.10 The SAS Output of the Stepwise Procedure for the Hospital Staffing Data

STEP 1 VARIABLE X3 ENTERED   R SQUARE = 0.97218120   C(P) = 20.3817958

	DF	SUM OF SQUARES	MEAN SQUARE	F	PROB > F
REGRESSION	1	480950231.62604150	480950231.62604150	524.20	0.0001
ERROR	15	13762308.86295839	917487.25753056		
TOTAL	16	494712540.48899990			

	B VALUE	STD ERROR	PARTIAL REG SS	F	PROB > F
INTERCEPT	−28.12861560				
X3	1.11739237	0.04880403	480950231.62604150	524.20	0.0001

STEP 2 VARIABLE X2 ENTERED   R SQUARE = 0.98671474   C(P) = 4.94164787

	DF	SUM OF SQUARES	MEAN SQUARE	F	PROB > F
REGRESSION	2	488140157.95096330	244070078.97548168	519.90	0.0001
ERROR	14	6572382.53803656	469455.89557404		
TOTAL	16	494712540.48899990			

	B VALUE	STD ERROR	PARTIAL REG SS	F	PROB > F
INTERCEPT	−68.31395896				
X2	0.07486591	0.01913019	7189926.32492182	15.32	0.0016
X3	0.82287456	0.08295986	46187674.54075647	98.39	0.0001

STEP 3 VARIABLE X5 ENTERED   R SQUARE = 0.99006817   C(P) = 2.91769778

	DF	SUM OF SQUARES	MEAN SQUARE	F	PROB > F
REGRESSION	3	489799141.98626880	163266380.66208962	431.97	0.0001
ERROR	13	4913398.50273108	377953.73097931		
TOTAL	16	494712540.48899990			

	B VALUE	STD ERROR	PARTIAL REG SS	F	PROB > F
INTERCEPT	1523.38923568				
X2	0.05298733	0.02009194	2628687.59792946	6.96	0.0205
X3	0.97848162	0.10515362	32726194.93174630	86.59	0.0001
X5	−320.95082518	153.19222065	1658984.03530548	4.39	0.0563

EXAMPLE 8.9   In the hospital staffing problem, we will let x_1, x_2, x_3, x_4, and x_5 (defined in Section 8.3.4) be the $p = 5$ potential independent variables to be considered in stepwise regression. When both α_{entry} and α_{stay} are set equal to .10, the stepwise procedure that we have described (1) adds x_3 on the first step, (2) adds x_2 (and retains x_3) on the second step, (3) adds x_5 (and retains x_2 and x_3 on the third step), and (4) terminates after step 3 when no more independent variables can be added. The SAS output of this stepwise procedure is given in Table 8.10. Note that the stepwise procedure arrives at the final model

$$y = \beta_0 + \beta_1 x_2 + \beta_2 x_3 + \beta_3 x_5 + \varepsilon$$

which is the model that the statistics of Table 8.9 indicate is (probably) best.

FORWARD SELECTION

Forward selection works in the same way as stepwise regression, *except that once an independent variable is entered into the model, it is never removed.* Forward selection is generally considered to be less effective than stepwise regression but to be useful in some problems.

BACKWARD ELIMINATION

In backward elimination, a regression analysis is performed by using a regression model containing all the p potential independent variables. Then, the independent variable having the smallest (in absolute value) t_{b_j} statistic is chosen. If the t_{b_j} statistic indicates that this independent variable is significant at the α_{stay} level (α_{stay} is chosen prior to the beginning of the procedure), then the procedure terminates by choosing the regression model containing all p independent variables. If the t_{b_j} statistic does not indicate that this independent variable is significant at the α_{stay} level, then this independent variable is removed from the model and a regression analysis is performed by using a regression model containing all the remaining independent variables. The procedure continues by removing independent variables one at a time from the model. At each step, an independent variable is removed from the model if it has the smallest (in absolute value) t_{b_j} statistic of the independent variables remaining in the model and if its t_{b_j} statistic does not indicate that it is significant at the α_{stay} level. The procedure terminates when no independent variable remaining in the model can be removed. Backward elimination is generally considered to be a reasonable procedure especially for analysts who like to start with all possible independent variables in the model so that they will not "miss any important variables."

MAXIMUM R^2 IMPROVEMENT

The following description of maximum R^2 improvement is quoted from the *SAS User's Guide, 1982 Edition*, which calls this procedure the MAXR method.

> Unlike the three best techniques above, this method does not settle on a single model. Instead, it looks for the "best" one-variable model, the "best" two-variable model, and so forth.

The MAXR method begins by finding the one-variable model producing the highest R^2. Then another variable, the one that would yield the greatest increase in R^2, is added.

Once the two-variable model is obtained, each of the variables in the model is compared to each variable not in the model. For each comparison, MAXR determines if removing one variable and replacing it with the other variable would increase R^2. After comparing all possible switches, the one that produces the largest increase in R^2 is made.

Comparisons begin again, and the process continues until MAXR finds that no switch could increase R^2. The two-variable model thus achieved is considered the "best" two-variable model the technique can find.

Another variable is then added to the model, and the comparing and switching process is repeated to find the "best" three-variable model, and so forth.

The difference between the stepwise technique and the maximum R^2 improvement method is that all switches are evaluated before any switch is made in the MAXR method. In the stepwise method, the "worst" variable may be removed without considering what adding the "best" remaining variable might accomplish.

The MAXR method is becoming increasingly popular and is generally considered to be superior to stepwise regression, forward selection, and backward elimination.

8.3.6   An F-Test for the Overall Model

In this section we will present an *F*-test related to the utility of the overall regression model

$$y = \beta_0 + \beta_1 x_1 + \cdots + \beta_p x_p + \varepsilon$$

Specifically, we will consider testing the null hypothesis

$$H_0: \beta_1 = \beta_2 = \cdots = \beta_p = 0$$

which says that none of the independent variables $x_1, x_2, \ldots, x_p$ affect y, versus the alternative hypothesis

$$H_1: \quad \text{At least one of} \quad \beta_1, \beta_2, \ldots, \beta_p \quad \text{does not equal zero}$$

which says that at least one of the independent variables $x_1, x_2, \ldots, x_p$ affects y. If we can reject H_0 in favor of H_1 by specifying a small probability of a Type I error, then it is reasonable to conclude that at least one of $x_1, x_2, \ldots, x_p$ *significantly* affects y. In this case, we should use t_{b_j} statistics and other techniques to determine which of $x_1, x_2, \ldots, x_p$ significantly affect y. Recalling that n_p denotes the number of parameters in the overall regression model, and letting the *explained variation* also be called the *sum of squares due to the overall model* (denoted SS_{model}), we summarize the *overall F-test* as follows.

Testing $H_0: \beta_1 = \beta_2 = \cdots = \beta_p = 0$ versus H_1: At least one of $\beta_1, \beta_2, \ldots, \beta_p$ does not equal zero

Define the *overall F-statistic* to be

$$F(\text{model}) = \frac{MS_{\text{model}}}{MSE}$$

where

$$MS_{\text{model}} = \frac{SS_{\text{model}}}{n_p - 1}$$

$$= \frac{\text{explained variation}}{n_p - 1}$$

and

$$MSE = \frac{SSE}{n - n_p}$$

(Note here that *MS* denotes "mean square" and *E* denotes "error.")

$$= s^2$$

Also define prob-value to be the area to the right of $F(\text{model})$ under the curve of the F-distribution having $(n_p - 1)$ and $(n - n_p)$ degrees of freedom (see Figure 8.7, b and c).

Then, we can reject H_0 in favor of H_1 by setting the probability of a Type I error equal to α if and only if either of the following equivalent conditions hold:

1. $F(\text{model}) > F_{[\alpha]}^{(n_p - 1, n - n_p)}$

    where

    $F_{[\alpha]}^{(n_p - 1, n - n_p)}$

    is the point on the scale of the F-distribution having $n_p - 1$ and $n - n_p$ degrees of freedom so that the area under this curve to the right of $F_{[\alpha]}^{(n_p - 1, n - n_p)}$ is α (see Figure 8.7a), or

2. Prob-value $< \alpha$ (see Figure 8.7, b and c)

We will first consider condition 1. This condition—which says that we should reject H_0 in favor of H_1 if

$$F(\text{model}) = \frac{MS_{\text{model}}}{MSE} = \frac{(\text{explained variation})/(n_p - 1)}{SSE/(n - n_p)}$$

is "large"—is reasonable, because a large value of $F(\text{model})$ would be caused by an explained variation that is large relative to the unexplained variation (SSE), which would occur if at least one of the independent variables in the regression model "significantly affects" the dependent variable. This would imply that $H_0: \beta_1 = \beta_2 = \cdots = \beta_p = 0$ is false and that H_1: At least one of $\beta_1, \beta_2, \ldots, \beta_p$ does not equal zero is true. Moreover, the equivalence of conditions 1 and 2 follows by carefully examining Figure 8.7, a, b, and c.

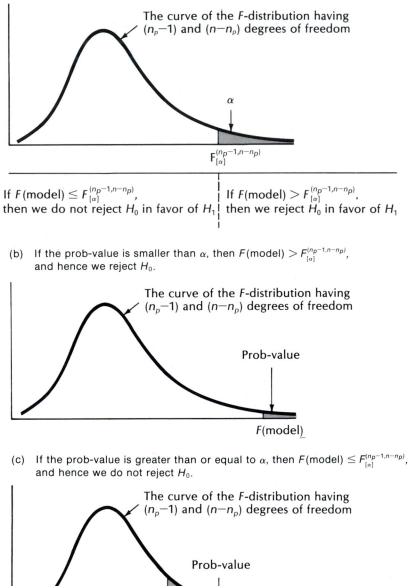

(a)   The rejection point $F_{[\alpha]}^{(n_p-1, n-n_p)}$ based on setting the probability of a Type 1 error equal to α

The curve of the F-distribution having (n_p-1) and $(n-n_p)$ degrees of freedom

α

$F_{[\alpha]}^{(n_p-1, n-n_p)}$

If $F(\text{model}) \leq F_{[\alpha]}^{(n_p-1, n-n_p)}$, then we do not reject H_0 in favor of H_1 | If $F(\text{model}) > F_{[\alpha]}^{(n_p-1, n-n_p)}$, then we reject H_0 in favor of H_1

(b)   If the prob-value is smaller than α, then $F(\text{model}) > F_{[\alpha]}^{(n_p-1, n-n_p)}$, and hence we reject H_0.

The curve of the F-distribution having (n_p-1) and $(n-n_p)$ degrees of freedom

Prob-value

$F(\text{model})$

(c)   If the prob-value is greater than or equal to α, then $F(\text{model}) \leq F_{[\alpha]}^{(n_p-1, n-n_p)}$, and hence we do not reject H_0.

The curve of the F-distribution having (n_p-1) and $(n-n_p)$ degrees of freedom

Prob-value

$F(\text{model})$

FIGURE 8.7   *Testing $H_0 \colon \beta_1 = \beta_2 = \cdots = \beta_p = 0$ Versus $H_1 \colon$ At Least One of $\beta_1, \beta_2, \ldots, \beta_p$ Does Not Equal Zero*

EXAMPLE 8.10

Consider the Fresh detergent model

$$y_t = \beta_0 + \beta_1 x_{t4} + \beta_2 x_{t3} + \beta_3 x_{t3}^2 + \beta_4 x_{t4} x_{t3} + \varepsilon_t$$

In Table 8.4 we present the SAS output of F(model), which is

$$F(model) = \frac{MS_{model}}{MSE} = \frac{SS_{model}/(n_p - 1)}{s^2}$$

$$= \frac{\text{explained variation}/(n_p - 1)}{SSE/(n - n_p)}$$

$$= \frac{12.3942/(5 - 1)}{1.0644/(30 - 5)}$$

$$= \frac{3.0985}{.0426}$$

$$= 72.777$$

and the *prob-value* related to F(model), which is the area to the right of F(model) under the curve of the F-distribution having $n_p - 1 = 5 - 1 = 4$ and $n - n_p = 30 - 5 = 25$ degrees of freedom, and which can be computer-calculated to be less than or equal to .0001 (that is, when the SAS output reads .0001, the prob-value is less than or equal to .0001). Noting from Table A4 that

$$F_{[.05]}^{(n_p - 1, n - n_p)} = F_{[.05]}^{(4, 25)} = 2.76$$

is the point on the scale of the F-distribution having $n_p - 1 = 4$ and $n - n_p = 25$ degrees of freedom so that the area to the right of this point under the curve of this F-distribution is .05, it follows, since

$$F(model) = 72.777 > 2.76 = F_{[.05]}^{(n_p - 1, n - n_p)}$$

or (equivalently) since

$$\text{prob-value} = .0001 < .05 = \alpha$$

that we can reject

$$H_0: \beta_1 = \beta_2 = \beta_3 = \beta_4 = \beta_5 = 0$$

in favor of

$$H_1: \quad \text{At least one of} \quad \beta_1, \beta_2, \beta_3, \beta_4, \text{ and } \beta_5 \quad \text{does not equal zero}$$

by setting α equal to .05. Thus, we conclude that *at least* one of the independent variables in the model

$$y_t = \beta_0 + \beta_1 x_{t4} + \beta_2 x_{t3} + \beta_3 x_{t3}^2 + \beta_4 x_{t4} x_{t3} + \varepsilon_t$$

is important.

The importance of the overall *F*-test is that this test (for the significance of at least one independent variable) is based on an *overall* probability of a Type I error, whereas the individual *t*-tests (in Section 8.3.2) of significance for individual independent variables are *each* based on the probability of a Type I error, meaning that the probability that *at least* one of the *t*-tests would declare an unimportant variable significant is difficult to determine. In spite of this, if the overall *F*-test declares that at least one independent variable is significant, it is common practice to use the individual *t*-tests of Section 8.3.2 (and the comparison methods of Section 8.3.3) to determine *which* of the independent variables are significant.

8.3.7   An F-Test for Portions of a Model

In this section we will present an *F*-test related to the utility of a portion of a regression model. For example, if we consider the Fresh detergent model

$$y = \beta_0 + \beta_1 x_4 + \beta_2 x_3 + \beta_3 x_3^2 + \beta_4 x_4 x_3 + \varepsilon$$

it might be useful to test the null hypothesis

$$H_0: \beta_3 = \beta_4 = 0$$

which says that neither of the "higher order terms" x_3^2 and $x_4 x_3$ affect *y*, versus the alternative hypothesis

$$H_1: \quad \text{At least one of} \quad \beta_3 \text{ and } \beta_4 \quad \text{does not equal zero}$$

which says that at least one of the higher order terms x_3^2 and $x_4 x_3$ affects *y*. In general, consider the regression model

$$y = \beta_0 + \beta_1 x_1 + \cdots + \beta_g x_g + \beta_{g+1} x_{g+1} + \cdots + \beta_p x_p + \varepsilon$$

and consider testing the null hypothesis

$$H_0: \beta_{g+1} = \beta_{g+2} = \cdots = \beta_p = 0$$

which says that none of the independent variables $x_{g+1}, x_{g+2}, \ldots, x_p$ affects *y*, versus the alternative hypothesis

$$H_1: \quad \text{At least one of} \quad \beta_{g+1}, \beta_{g+2}, \ldots, \beta_p \quad \text{does not equal zero}$$

which says that at least one of the independent variables $x_{g+1}, x_{g+2}, \ldots, x_p$ affects *y*. If we can reject H_0 in favor of H_1 by specifying a *small* probability of a Type I error, then it is reasonable to conclude that at least one of $x_{g+1}, x_{g+2}, \ldots, x_p$ *significantly* affects *y*. In this case, we should use t_{b_j} statistics and other techniques to determine which of $x_{g+1}, x_{g+2}, \ldots, x_p$ significantly affect *y*. In order to test H_0 versus H_1, consider the following two models:

Complete model: $y = \beta_0 + \beta_1 x_1 + \cdots + \beta_g x_g + \beta_{g+1} x_{g+1} + \cdots + \beta_p x_p + \varepsilon$

Reduced model: $y = \beta_0 + \beta_1 x_1 + \cdots + \beta_g x_g + \varepsilon$

Here, the complete model is assumed to have n_p parameters, the reduced model is the complete model under the assumption that H_0 is true, and $(p - g)$ denotes the number of regression parameters we have set equal to zero in the statement of the null hypothesis H_0. Using regression analysis, we calculate SSE_C, the unexplained variation for the complete model, and SSE_R, the unexplained variation for the reduced model. Then we consider the difference

$$SS_{drop} = SSE_R - SSE_C$$

which we call the *drop in the unexplained variation attributable to the independent variables* $x_{g+1}, x_{g+2}, \ldots, x_p$. Now, it can be shown that the "extra" independent variables $x_{g+1}, x_{g+2}, \ldots, x_p$ will always make SSE_C somewhat smaller than SSE_R, and, hence, will always make SS_{drop} positive. The question is whether this difference is large enough to conclude that at least one of the independent variables $x_{g+1}, x_{g+2}, \ldots, x_p$ significantly affects y. Since the value of SS_{drop} depends on the units in which the observed values of the dependent variable are measured, we need to modify SS_{drop} and formulate a unitless measure of the additional importance of the set of variables $x_{g+1}, x_{g+2}, \ldots, x_p$. Such a statistic is called the *partial F-statistic* and is denoted by the symbol $F(x_{g+1}, \ldots, x_p | x_1, x_2, \ldots, x_g)$. Below we define this statistic and show how it is used to test H_0 versus H_1.

Testing H_0: $\beta_{g+1} = \beta_{g+2} = \cdots = \beta_p = 0$ versus H_1: At least one of $\beta_{g+1}, \beta_{g+2}, \ldots, \beta_p$ does not equal zero

Define the partial *F*-statistic to be

$$F(x_{g+1}, \ldots, x_p | x_1, \ldots, x_g) = \frac{MS_{drop}}{MSE_C}$$

where

$$MS_{drop} = \frac{SS_{drop}}{p - g} = \frac{SSE_R - SSE_C}{p - g}$$

and

$$MSE_C = \frac{SSE_C}{n - n_p} \qquad \text{(Note here that } MSE_C \text{ equals } s^2 \text{ for the complete model.)}$$

Also define prob-value to be the area to the right of $F(x_{g+1}, \ldots, x_p | x_1, \ldots, x_g)$ under the curve of the *F*-distribution having $(p - g)$ and $(n - n_p)$ degrees of freedom.

Then we can reject H_0 in favor of H_1 by setting the probability of a Type I error equal to α if and only if either of the following equivalent conditions hold:

1. $F(x_{g+1}, \ldots, x_p | x_1, \ldots, x_g) > F_{[\alpha]}^{(p-g, n-n_p)}$

   where

   $F_{[\alpha]}^{(p-g, n-n_p)}$

is the point on the scale of the *F*-distribution having $(p - g)$ and $(n - n_p)$ degrees of freedom so that the area under this curve to the right of $F_{[\alpha]}^{(p-g, n-n_p)}$ is α, or

2. Prob-value $< \alpha$

Before presenting an example, note that condition 1—which says that we should reject H_0 in favor of H_1 if

$$F(x_{g+1}, \ldots, x_p | x_1, \ldots, x_g) = \frac{(SSE_R - SSE_C)/(p - g)}{SSE_C/(n - n_p)}$$

is "large"—is reasonable, because a large value of $F(x_{g+1}, \ldots, x_p | x_1, \ldots, x_g)$ would be obtained when $(SSE_R - SSE_C)$ is large. A large value of $(SSE_R - SSE_C)$ would be obtained if at least one of the independent variables $x_{g+1}, x_{g+2}, \ldots, x_p$ makes SSE_C substantially smaller than SSE_R, which would indicate that $H_0: \beta_{g+1} = \beta_{g+2} = \cdots = \beta_p = 0$ is false and that H_1: At least one of $\beta_{g+1}, \beta_{g+2}, \ldots, \beta_p$ does not equal zero is true.

Also note that there is a relationship between the t_{b_j} statistic and the partial *F*-statistic. It can be proven that

$$(t_{b_j})^2 = F(x_j | x_1, \ldots, x_{j-1}, x_{j+1}, \ldots, x_p)$$

and that

$$(t_{[\alpha/2]}^{(n-n_p)})^2 = F_{[\alpha]}^{(1, n-n_p)}$$

Hence, the conclusion that

$$|t_{b_j}| > t_{[\alpha/2]}^{(n-n_p)}$$

which leads us to reject $H_0: \beta_j = 0$ in favor of $H_1: \beta_j \neq 0$, will be made if and only if

$$F(x_j | x_1, \ldots, x_{j-1}, x_{j+1}, \ldots, x_p) > F_{[\alpha]}^{(1, n-n_p)}$$

which again leads us to reject $H_0: \beta_j = 0$ in favor of $H_1: \beta_j \neq 0$. Thus, the conditions for rejecting $H_0: \beta_j = 0$ in favor of $H_1: \beta_j \neq 0$ using the t_{b_j} statistic and partial *F*-statistic are equivalent.

EXAMPLE 8.11

When we use the Fresh detergent model

Complete model: $y_t = \beta_0 + \beta_1 x_{t4} + \beta_2 x_{t3} + \beta_3 x_{t3}^2 + \beta_4 x_{t4} x_{t3} + \varepsilon_t$

which has $n_p = 5$ parameters, to carry out a regression analysis of the data in Table 8.3 we obtain an unexplained variation of $SSE_C = 1.0644$. In order to test

$$H_0: \beta_3 = \beta_4 = 0$$

versus

H_1: At least one of β_3 and β_4 does not equal zero

note that $(p - g) = 2$, since two parameters (β_3 and β_4) are set equal to zero in the statement of the null hypothesis H_0. Also, note that, under the assumption that H_0 is true, the complete model becomes the following

Reduced model: $y_t = \beta_0 + \beta_1 x_{t4} + \beta_2 x_{t3} + \varepsilon_t$

for which the unexplained variation is $SSE_R = 1.5337$. Thus, in order to test H_0 versus H_1 we use the following partial F-statistic and prob-value:

1. $F(x_{t3}^2, x_{t4}x_{t3} \mid x_{t4}, x_{t3}) = \dfrac{MS_{drop}}{MSE_C} = \dfrac{.2347}{.0426} = 5.5094$

since

$$MS_{drop} = \frac{SS_{drop}}{p - g} = \frac{SSE_R - SSE_C}{p - g} = \frac{1.5337 - 1.0644}{2}$$

$$= \frac{.4693}{2} = .2347$$

and

$$MSE_C = \frac{SSE_C}{n - n_p} = \frac{1.0644}{30 - 5} = \frac{1.0644}{25} = .0426$$

2.   prob-value = .0111, which is the (computer-calculated) area to the right of 5.5094 under the curve of the F-distribution having 2 and 25 degrees of freedom.

If we wish to use condition 1 to determine whether we can reject H_0 in favor of H_1 by setting α, the probability of a Type I error, equal to .05, then we would use the rejection point

$$F_{[\alpha]}^{(p-g, n-n_p)} = F_{[.05]}^{(2, 25)} = 3.39$$

Since $F(x_{t3}^2, x_{t4}x_{t3} \mid x_{t4}, x_{t3}) = 5.5094 > 3.39 = F_{[.05]}^{(2, 25)}$, we can reject H_0: $\beta_3 = \beta_4 = 0$ in favor of H_1: At least one of β_3 and β_4 does not equal zero by setting α equal to .05. Alternatively, since prob-value = .0111 is less than .05 and .02, we can reject H_0 in favor of H_1 by setting α equal to .05 or .02. However, since prob-value = .0111 is not less than .01, we cannot reject H_0 in favor of H_1 by setting α equal to .01. In summary, the smallness of the prob-value provides substantial evidence that H_0: $\beta_3 = \beta_4 = 0$ is false and that H_1: At least one of β_3 and β_4 does not equal zero is true. Therefore, we have substantial evidence that at least one of the higher order terms x_3^2 and $x_4 x_3$ significantly affects y (demand for Fresh). Thus, we should use t_{b_j} statistics and other techniques to determine which of these higher order terms significantly affect demand for Fresh (see Examples 8.6 and 8.7).

Before leaving this section, we will make two comments. First, it can be shown that multicollinearity can hinder the ability of the partial F-statistic to indicate that at least one of the independent variables $x_{g+1}, x_{g+2}, \ldots, x_p$ is significant (although, in many cases, multicollinearity affects the partial F-statistic related to $x_{g+1}, x_{g+2}, \ldots, x_p$ less than it affects the individual t_{b_j} statistics related to $x_{g+1}, x_{g+2}, \ldots, x_p$). Second, we will show (in Section 8.5) that the partial F-statistic can be used to help compare population means.

8.4 ASSUMPTIONS AND RESIDUAL ANALYSIS

8.4.1   The Inference Assumptions

The validity of the statistical inferences (hypothesis tests, confidence intervals, and prediction intervals) obtained by using the classical regression model

$$y_t = \mu_t + \varepsilon_t$$
$$= \beta_0 + \beta_1 x_{t1} + \cdots + \beta_p x_{tp} + \varepsilon_t$$

depends on three assumptions called the *inference assumptions*. Recall that, although we express these assumptions by referring to the time period t in which y_t has been observed, the "t" can refer to any criterion (such as "sales region").

The Inference Assumptions

  Inference Assumption 1 (Normality):   For each and every time period t, the error term ε_t follows a normal probability distribution (with mean zero).

  Inference Assumption 2 (Constant Variances):   The variance of $\varepsilon_t = y_t - \mu_t$, which measures the spread of all of the potential values of y_t around the average level μ_t, is the same for each and every time period t.

  Inference Assumption 3 (Independence):   The error terms $\varepsilon_1, \varepsilon_2, \ldots$ are statistically independent.

8.4.2   Validation of the Normality Assumption

Recalling that the error term

$$\varepsilon_t = y_t - \mu_t$$
$$= y_t - (\beta_0 + \beta_1 x_{t1} + \cdots + \beta_p x_{tp})$$

measures the effects of the factors causing y_t to deviate from μ_t (that is, of all factors other than the independent variables $x_{t1}, x_{t2}, \ldots, x_{tp}$), Inference Assumption 1 (Normality) says that all of the values of y_t that could possibly be observed in time period t are distributed in a normal curve fashion around μ_t. We now present an example that illustrates checking for violations of the normality assumption.

EXAMPLE 8.12

Considering the Fresh detergent model

$$y_t = \mu_t + \varepsilon_t$$
$$= \beta_0 + \beta_1 x_{t4} + \beta_2 x_{t3} + \beta_3 x_{t3}^2 + \beta_4 x_{t4} x_{t3} + \varepsilon_t$$

the point prediction of the tth error term

$$\varepsilon_t = y_t - \mu_t$$
$$= y_t - (\beta_0 + \beta_1 x_{t4} + \beta_2 x_{t3} + \beta_3 x_{t3}^2 + \beta_4 x_{t4} x_{t3})$$

is the tth residual

$$e_t = y_t - \hat{y}_t$$
$$= y_t - (b_0 + b_1 x_{t4} + b_2 x_{t3} + b_3 x_{t3}^2 + b_4 x_{t4} x_{t3})$$
$$= y_t - (29.1133 + 11.1342 x_{t4} - 7.6080 x_{t3} + .6712 x_{t3}^2 - 1.4777 x_{t4} x_{t3})$$

Noting that the $n = 30$ residuals are given in the SAS output of Table 8.4, the most obvious way to check the validity of the normality assumption is to construct a frequency distribution and histogram of the residuals. A frequency distribution of the residuals is given in Table 8.11. This frequency distribution is depicted graphically in the form of a histogram in Figure 8.8a. We see that the

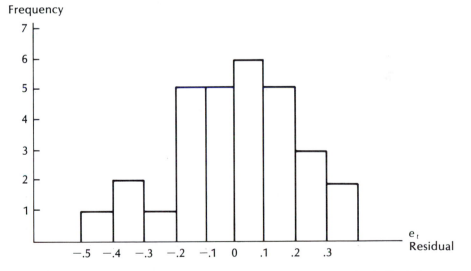

(a)   Histogram of the Residuals

FIGURE 8.8   *Graphical Methods for Checking the Validity of the Normality Assumption for the Fresh Detergent Model*

$$y_t = \beta_0 + \beta_1 x_{t4} + \beta_2 x_{t3} + \beta_3 x_{t3}^2 + \beta_4 x_{t4} x_{t3} + \varepsilon_t$$

(Part (b), Normal Plot of the Residuals, follows on page 460.)

FIGURE 8.8   Continued

(b)   Normal Plot of the Residuals

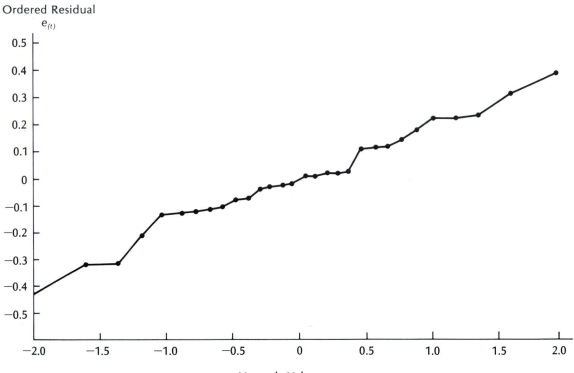

TABLE 8.11   Frequency Distribution of the Residuals for the Fresh Detergent Model $y_t = \beta_0 + \beta_1 x_{t4} + \beta_2 x_{t3} + \beta_3 x_{t3}^2 + \beta_4 x_{t4} x_{t3} + \varepsilon_t$

Subinterval	Frequency
$-.5$ to $-.4$	1
$-.4$ to $-.3$	2
$-.3$ to $-.2$	1
$-.2$ to $-.1$	5
$-.1$ to 0	5
0 to $.1$	6
$.1$ to $.2$	5
$.2$ to $.3$	3
$.3$ to $.4$	2

TABLE 8.12   *Ordered Residuals and Normal Plot Calculations for the Fresh Detergent Model* $y_t = \beta_0 + \beta_1 x_{t4} + \beta_2 x_{t3} + \beta_3 x_{t3}^2 + \beta_4 x_{t4} x_{t3} + \varepsilon_t$

(a)   Ordered residuals

$e_{(t)}$
$-.437250$
$-.327373$
$-.327250$
$-.210185$
$-.143455$
$-.138209$
$-.133353$
$-.122850$
$-.101611$
$-.074478$
$-.066071$
$-.047250$
$-.044139$
$-.038914$
.004773
.005016
.028687
.029666
.029866
.035566
.102677
.109641
.126964
.141240
.176312
.233113
.234223
.245527
.325016
.384097

(Part (b), Normal Plot Calculations, follows on page 462.)

histogram looks reasonably bell-shaped and symmetric. Therefore, the histogram does not suggest a serious violation of the normality assumption (note that only pronounced departures from the normality assumption are considered to be worrisome).

Another way to check the normality assumption is to construct a *normal plot* of the residuals. In order to construct this plot, we must first arrange the residuals in order from smallest to largest. These ordered residuals are given in Table 8.12a.

TABLE 8.12   Continued

(b)   Normal Plot Calculations

t	$\dfrac{3t-1}{3n+1}$	$z_{(t)}$	t	$\dfrac{3t-1}{3n+1}$	$z_{(t)}$
1	$\dfrac{3(1)-1}{3(30)+1} = \dfrac{2}{91} = .0220$	-2.01	16	.5165	.04
2	$\dfrac{3(2)-1}{3(30)+1} = \dfrac{5}{91} = .0549$	-1.60	17	.5495	.12
3	$\dfrac{3(3)-1}{3(30)+1} = \dfrac{8}{91} = .0879$	-1.35	18	.5824	.21
4	.1209	-1.17	19	.6154	.29
5	.1538	-1.02	20	.6484	.38
6	.1868	$-.89$	21	.6813	.47
7	.2198	$-.77$	22	.7143	.57
8	.2527	$-.67$	23	.7473	.67
9	.2857	$-.57$	24	.7802	.77
10	.3187	$-.47$	25	.8132	.89
11	.3516	$-.38$	26	.8462	1.02
12	.3846	$-.29$	27	.8791	1.17
13	.4176	$-.21$	28	.9121	1.35
14	.4505	$-.12$	29	.9451	1.60
15	.4835	$-.04$	30	.9780	2.02

Denoting the tth ordered residual as $e_{(t)}$ ($t = 1, 2, \ldots, 30$), we next compute for each value of t the point $z_{(t)}$ such that the area under the standard normal curve to the left of $z_{(t)}$ is

$$\frac{3t-1}{3n+1} = \frac{3t-1}{3(30)+1} = \frac{3t-1}{91}$$

These computations are summarized in Table 8.12b. We now plot the ordered residuals in Table 8.12a on the vertical axis against the values of $z_{(t)}$ in Table 8.12b on the horizontal axis. The resulting normal plot is depicted in Figure 8.8b. Looking at this figure, we see that it has a *straight line appearance*, which (it can be shown) implies that the normality assumption is not violated. Substantial departures from a straight line appearance (admittedly a subjective decision) indicate a violation of the normality assumption.

The methods illustrated here are not the only ways to check the normality assumption. For example, several statistical hypothesis tests can be used to check this assumption. The interested reader is referred to Mendenhall and Reinmuth [1982].

8.4.3 Validation of the Constant Variance Assumption

Inference Assumption 2 (Constant Variances) says that the variance of the normal curve describing all of the values of y_t that could possibly be observed in one time period equals the variance of the normal curve describing all of the values of y_t that could possibly be observed in any other time period.

To check the assumption that the variance of ε_t is the same for each and every time period t, we plot the residuals against the following criteria:

1. Increasing values of each of the independent variables
2. Increasing values of $\hat{y}_t$
3. The time order in which the data was observed.

In general, a residual plot with the appearance of Figure 8.9a—in which the residuals tend to fan out with increasing values of a particular criterion—indicates that the residuals $e_1, e_2, \ldots, e_n$ and, hence, the error terms $\varepsilon_1, \varepsilon_2, \ldots, \varepsilon_n$, are increasing in absolute value with increasing values of the criterion. This would suggest that the variance of ε_t is increasing with increasing values of the criterion. A plot with the appearance of Figure 8.9b—in which the residuals tend to "funnel in" with increasing values of a particular criterion—indicates that the residuals $e_1, e_2, \ldots, e_n$ and, hence, the error terms $\varepsilon_1, \varepsilon_2, \ldots, \varepsilon_n$, are decreasing in absolute value with increasing values of the criterion. Such a plot suggests that the variance of ε_t is decreasing with increasing values of the criterion. However, a plot with the appearance of Figure 8.9c—in which the residuals tend to form a horizontal band—indicates that the residuals $e_1, e_2, \ldots, e_n$ and, hence, the error terms $\varepsilon_1, \varepsilon_2, \ldots, \varepsilon_n$, are remaining relatively constant in absolute value with increasing values of a particular criterion. This suggests that the variance of ε_t is not changing with increasing values of the criterion. Therefore, a plot with the appearance of Figure 8.9c does not provide evidence to suggest that the constant variance assumption is violated. We now present an example.

EXAMPLE 8.13   Again consider the Fresh detergent model

$$y_t = \mu_t + \varepsilon_t$$

$$= \beta_0 + \beta_1 x_{t4} + \beta_2 x_{t3} + \beta_3 x_{t3}^2 + \beta_4 x_{t4} x_{t3} + \varepsilon_t$$

In order to check the validity of the constant variance assumption, we plot the residuals for this model against the following criteria:

1. Increasing values of each of the independent variables x_{t4} and x_{t3} (in Figures 8.10 and 8.11),
2. Increasing values of $\hat{y}_t$ (in Figure 8.12), and
3. The time order in which the data was observed (in Figure 8.13).

Since the residual plots in these figures all have a horizontal band appearance, we

(a) Non-Constant Error Variance: Error Variance Increases With Increasing Values of the
     Criterion

$e_t = y_t - \hat{y}_t$

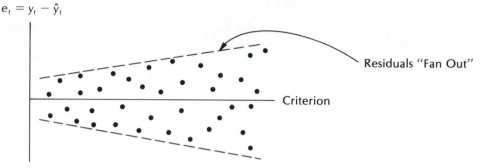

Residuals "Fan Out"

Criterion

(b) Non-Constant Error Variance: Error Variance Decreases With Increasing Values of the
     Criterion

$e_t = y_t - \hat{y}_t$

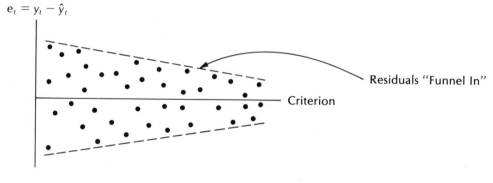

Residuals "Funnel In"

Criterion

(c) Constant Error Variance With Increasing Values of the Criterion

$e_t = y_t - \hat{y}_t$

Residuals form a
"Horizontal Band"

Criterion

FIGURE 8.9 *Possible Implications of a Residual Plot Against Increasing Values of a Particular Criterion (such as One of the Independent Variables, the Predicted Value of the Dependent Variable, or Time)*

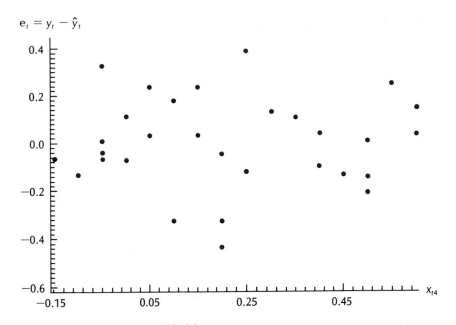

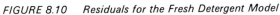

FIGURE 8.10   Residuals for the Fresh Detergent Model

$$y_t = \beta_0 + \beta_1 x_{t4} + \beta_2 x_{t3} + \beta_3 x_{t3}^2 + \beta_4 x_{t4} x_{t3} + \varepsilon_t$$

Plotted Against Increasing Values of x_{t4}, the Price Difference

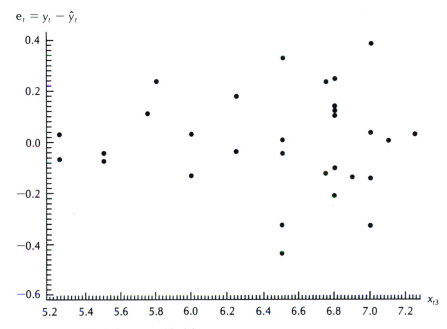

FIGURE 8.11   Residuals for the Fresh Detergent Model

$$y_t = \beta_0 + \beta_1 x_{t4} + \beta_2 x_{t3} + \beta_3 x_{t3}^2 + \beta_4 x_{t4} x_{t3} + \varepsilon_t$$

Plotted Against Increasing Values of x_{t3}, the Advertising Expenditure for Fresh

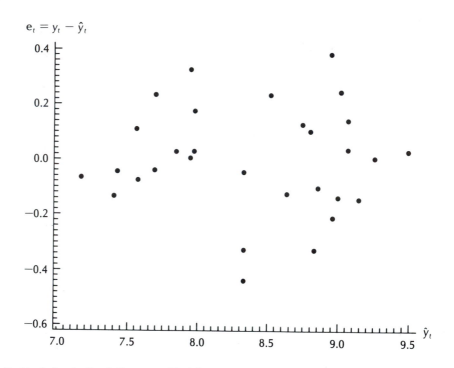

FIGURE 8.12   *Residuals for the Fresh Detergent Model*

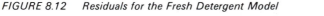

Plotted Against the Increasing Values of $\hat{y}_t$, Predicted Demand for Fresh

conclude that there is no evidence of a violation of the constant variance assumption for the Fresh detergent model.

8.4.4   Validation of the Independence Assumption Using Residual Plots and the Durbin-Watson Statistic

To understand Inference Assumption 3 (Independence), we note that the error terms ε_1, ε_2, . . . can be *autocorrelated*. We say that error terms occurring over time have *positive autocorrelation* if a positive error term in time period t tends to produce, or be followed by, another positive error term in time period $t + k$ (a later time period), and if a negative error term in time period t tends to produce, or be followed by, another negative error term in time period $t + k$. Said even more "intuitively," positive autocorrelation exists when positive error terms tend to be followed over time by positive error terms, and when negative error terms tend to be followed over time by negative

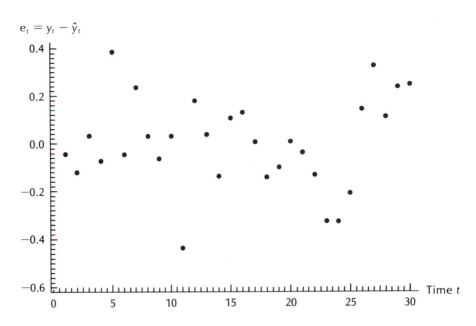

FIGURE 8.13   *Residuals for the Fresh Detergent Model*

$$y_t = \beta_0 + \beta_1 x_{t4} + \beta_2 x_{t3} + \beta_3 x_{t3}^2 + \beta_4 x_{t4} x_{t3} + \varepsilon_t$$

Plotted In Time Order

error terms. An example of positive autocorrelation in the error terms is depicted in Figure 8.14a, which illustrates that positive autocorrelation in the error terms can produce a cyclical pattern over time.

   If we consider μ_t, the average level of the dependent variable in time period t, note that a positive error term produces a value of the dependent variable that is greater than μ_t, and a negative error term produces a value of the dependent variable that is smaller than μ_t. This says that positive autocorrelation in the error terms means that greater than average values of the dependent variable tend to be followed by greater than average values of the dependent variable, and smaller than average values of the dependent variable tend to be followed by smaller than average values of the dependent variable.

EXAMPLE 8.14   In the Fresh detergent problem, the historical demand data given in Table 8.3 was collected in time sequence over 30 sales periods. If we consider the model

$$y_t = \mu_t + \varepsilon_t$$
$$= \beta_0 + \beta_1 x_{t4} + \beta_2 x_{t3} + \beta_3 x_{t3}^2 + \beta_4 x_{t4} x_{t3} + \varepsilon_t$$

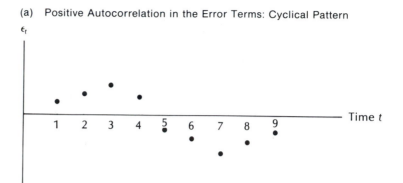

(a) Positive Autocorrelation in the Error Terms: Cyclical Pattern

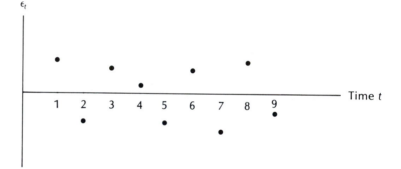

(b) Negative Autocorrelation in the Error Terms: Alternating Pattern

FIGURE 8.14   *Positive and Negative Autocorrelation*

then the effect of competitors' average advertising expenditure is included in ε_t, the error term. If, for the moment, we assume that competitors' average advertising expenditure significantly affects the demand for Fresh, then a "higher than average" competitors' average advertising expenditure probably causes demand for Fresh to be lower than average and, hence, probably causes a negative error term to occur. On the other hand, a "lower than average" competitors' average advertising expenditure probably causes demand for Fresh to be higher than average and, hence, probably causes a positive error term to occur. If, then, Enterprise Industries' competitors tend to spend money on advertising in a cyclical fashion—spending large amounts for several consecutive sales periods (during an advertising campaign) and then spending lesser amounts for several consecutive sales periods—and if the Fresh detergent demand data is collected in successive sales periods, a negative error term in one sales period will tend to be followed by a negative error term in the next sales period, and a positive error term in one sales period will tend to be followed by a positive error term in the next sales period. In this case, the error terms would display positive auto-

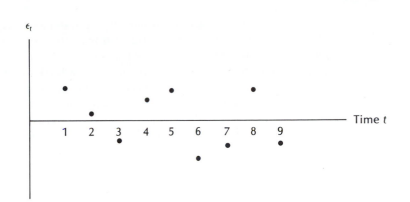

FIGURE 8.15   Little or No Autocorrelation in the Error Terms: Random Pattern

correlation, and thus these error terms would not be statistically independent. However, in spite of this possibility, we will soon use a residual plot to verify that the independence assumption does in fact hold for the Fresh detergent model.

Error terms occurring over time have *negative autocorrelation* if a positive error term in time period t tends to produce, or be followed by, a negative error term in time period $t + k$, and if a negative error term in time period t tends to produce, or be followed by, a positive error term in time period $t + k$. Said even more "intuitively," negative autocorrelation exists when positive error terms tend to be followed over time by negative error terms and when negative error terms tend to be followed over time by positive error terms. An example of negative autocorrelation in the error terms is depicted in Figure 8.14b, which illustrates that negative autocorrelation in the error terms can produce an alternating pattern over time. Since a positive error term produces a greater than average value of the dependent variable, and a negative error term produces a smaller than average value of the dependent variable, negative autocorrelation in the error terms means that greater than average values of the dependent variable tend to be followed by smaller than average values of the dependent variable, and smaller than average values of the dependent variable tend to be followed by greater than average values of the dependent variable. An example of negative autocorrelation might be provided by a retailer's weekly stock orders. A larger than average stock order one week might result in an oversupply and, hence, a smaller than average order the next week.

Although our discussion here has been quite intuitive, the ideas we have presented will allow us to give a relatively simple interpretation of Inference Assumption 3. In essence, this independence assumption says that the error terms $\varepsilon_1, \varepsilon_2, \ldots$ display no positive or negative autocorrelation. That is, the independence assumption says that the error terms display a random pattern, as illustrated in Figure 8.15.

Since the residuals $e_1, e_2, \ldots, e_n$ are point estimates of the error terms $\varepsilon_1, \varepsilon_2, \ldots, \varepsilon_n$, a residual plot against time can be used to detect violations of Inference

Assumption 3. If a residual plot against the time sequence in which the data has been collected has the appearance of Figure 8.14a—that is, if the plot displays a cyclical pattern—this suggests that the error terms $\varepsilon_1, \varepsilon_2, \ldots, \varepsilon_n$ are positively autocorrelated and that the independence assumption does not hold. Another way to detect positive autocorrelation is to look at the signs of the time ordered residuals. Letting "+" denote a positive residual and "−" denote a negative residual, we call a sequence of residuals with the same sign (for instance, $+ + +$) a *run*. If positive autocorrelation exists, the signs of the residuals should display relatively few runs, which will be of fairly long duration. For instance, the pattern

$$+ + + + - - - + + + - - - - + + + + - - -$$

in the time-ordered residuals would indicate that positive autocorrelation exists and that the independence assumption is violated.

A plot of the time-ordered residuals having the appearance of Figure 8.14b—that is, the plot displays an alternating pattern—suggests that the error terms $\varepsilon_1, \varepsilon_2, \ldots, \varepsilon_n$ are negatively autocorrelated and that the independence assumption does not hold. If we look at the signs of the time-ordered residuals, negative autocorrelation is characterized by many runs of relatively short duration. For example, the pattern

$$+ - + - + + - + - + - - + -$$

in the time-ordered residuals would indicate that negative autocorrelation exists and that the independence assumption is violated.

However, a plot of the time-ordered residuals displaying a random pattern, as illustrated in Figure 8.15, says that the error terms $\varepsilon_1, \varepsilon_2, \ldots, \varepsilon_n$ have little or no autocorrelation, which suggests that these error terms are independent, and that Inference Assumption 3 holds.

EXAMPLE 8.15   The residuals for the Fresh detergent model

$$y_t = \beta_0 + \beta_1 x_{t4} + \beta_2 x_{t3} + \beta_3 x_{t3}^2 + \beta_4 x_{t4} x_{t3} + \varepsilon_t$$

are plotted in time order in Figure 8.13. Looking at this figure, we see that these residuals probably display a random pattern as shown in Figure 8.15. This implies that there is probably little or no autocorrelation in the error terms for the Fresh detergent model, and that, therefore, these error terms are probably independent. Hence, it is reasonable to believe that Inference Assumption 3 holds for this model.

When a residual plot suggests that the error terms may be autocorrelated, one might wish to use a formal statistical test for autocorrelation. Two tests that can be used are the *runs test* and the *Durbin-Watson test*. For a description of the runs test, see Draper and Smith [1981]. Here, we will discuss the Durbin-Watson test for autocorrelation.

The Durbin-Watson test for autocorrelation is based on what is called the *Durbin-Watson statistic*. This statistic is defined as follows.

The *Durbin-Watson statistic* is

$$d = \frac{\sum_{t=2}^{n} (e_t - e_{t-1})^2}{\sum_{t=1}^{n} e_t^2}$$

where $e_1, e_2, \ldots, e_n$ are the time-ordered residuals.

If we consider testing the null hypothesis

H_0:   The error terms are not autocorrelated

versus the alternative hypothesis

H_1:   The error terms are positively autocorrelated

then Durbin and Watson have shown that there are points (denoted $d_{L,\alpha}$ and $d_{U,\alpha}$) such that, if α is the probability of a Type I error, then

1.   If $d < d_{L,\alpha}$, we reject H_0,
2.   If $d > d_{U,\alpha}$, we do not reject H_0, and
3.   If $d_{L,\alpha} \leqslant d \leqslant d_{U,\alpha}$, the test is inconclusive.

Here small values of d lead to the conclusion of positive autocorrelation because, d's being small indicates that the differences $(e_t - e_{t-1})$ are small, which indicates that the adjacent residuals e_t and e_{t-1} are of the same magnitude, which in turn says that the adjacent error terms ε_t and ε_{t-1} are positively correlated.

So that the Durbin-Watson test may be easily done, tables containing the points $d_{L,\alpha}$ and $d_{U,\alpha}$ have been constructed. These tables give the appropriate $d_{L,\alpha}$ and $d_{U,\alpha}$ points for various values of (1) α, the probability of a Type I error, (2) the number, $n_p - 1$, of independent variables in the regression model, and (3) the number, n, of observations. The tables for $\alpha = .05$ and $\alpha = .01$ are given as Tables A5 and A6 (in Appendix A).

The Durbin-Watson test can also be used to test for negative autocorrelation. If we consider testing the null hypothesis

H_0:   The error terms are not autocorrelated

versus the alternative hypothesis

H_1:   The error terms are negatively autocorrelated

Durbin and Watson have shown that, based on setting the probability of a Type I error equal to α, the points $d_{L,\alpha}$ and $d_{U,\alpha}$ are such that

1.   If $(4 - d) < d_{L,\alpha}$, we reject H_0,
2.   If $(4 - d) > d_{U,\alpha}$, we do not reject H_0, and
3.   If $d_{L,\alpha} \leqslant (4 - d) \leqslant d_{U,\alpha}$, the test is inconclusive.

Here large values of d (and, hence, small values of $4 - d$) lead to the conclusion of negative autocorrelation because, if d is large, this indicates that the differences $(e_t - e_{t-1})$ are large, which says that the adjacent error terms ε_t and ε_{t-1} are negatively autocorrelated.

Finally, if we wish to test the null hypothesis

H_0:   The error terms are not autocorrelated

versus the alternative hypothesis

H_1:   The error terms are positively or negatively autocorrelated

Durbin and Watson have shown that, based on setting the probability of a Type I error equal to α,

1.   If $d < d_{L,\alpha/2}$ or if $(4 - d) < d_{L,\alpha/2}$, we reject H_0,
2.   If $d > d_{U,\alpha/2}$ or if $(4 - d) > d_{U,\alpha/2}$, we do not reject H_0, and
3.   If $d_{L,\alpha/2} \leqslant d \leqslant d_{U,\alpha/2}$ and $d_{L,\alpha/2} \leqslant (4 - d) \leqslant d_{U,\alpha/2}$, the test is inconclusive.

In practice, positive autocorrelation is found more often that negative autocorrelation. Therefore, the first test we have presented (the test for positive autocorrelation) is used more often than the others. For example, for the Fresh detergent model

$$y_t = \beta_0 + \beta_1 x_{t4} + \beta_2 x_{t3} + \beta_3 x_{t3}^2 + \beta_4 x_{t4} x_{t3} + \varepsilon_t$$

the test for positive autocorrelation is inconclusive at $\alpha = .05$, because $d = 1.59$ is between $d_{L,.05} = 1.14$ and $d_{U,.05} = 1.74$ (see Table A5). However, the Durbin-Watson statistic does not measure all types of autocorrelations. In the next subsection (which can be understood only by those who have read Chapters 2, 3, 4, and 5), we present a superior way to check for autocorrelated error terms.

Three final comments should be made. First, one good way to insure that Inference Assumptions 1, 2, and 3 hold (at least approximately) is to explicitly include in the regression model describing y_t all of the factors (or independent variables) that significantly affect y_t (then the error term ε_t describes the effects of only unimportant factors). Second, only pronounced departures from the Inference Assumptions are considered worrisome. Indeed, the Inference Assumptions will probably not exactly hold in any real regression situation. Third, a remedy for a violation of Inference Assumption 2 is discussed in Section 8.4.6, and remedies for violations of Inference Assumption 3 are discussed in Chapters 2 through 5 and Chapter 7. In Bowerman, O'Connell, and Dickey [1986], a discussion of remedies for violations of Inference Assumption 1 and a more complete discussion of remedies for violations of Inference Assumption 2 are given.

8.4.5   *Using the RSAC and RSPAC to Check for Autocorrelated Error Terms (Requires Chapters 2, 3, 4, and 5)*

Whereas the SAS output in Table 8.4 results from using a *SAS regression procedure*, Figure 8.16 presents the SAS output of estimation, diagnostic checking, and forecasting

FIGURE 8.16 SAS Output of Estimation, Diagnostic Checking, and Forecasting for the Model $y_t = \beta_0 + \beta_1 x_{t4} + \beta_2 x_{t3} + \beta_3 x_{t3}^2 + \beta_4 x_{t4} x_{t3} + \varepsilon_t$

(a) Estimation

SAS

ARIMA: PRELIMINARY ESTIMATION

CONSTANT TERM ESTIMATE = 8.38267
WHITE NOISE VARIANCE EST = 0.44862

ARIMA: LEAST SQUARES ESTIMATION

ITERATION	SSE	MU	NUM1	NUM2	NUM3	NUM4	CONSTANT	LAMBDA
0	3994.5	8.38267	2.66521	1.04345	.0846481	0.388116	8.38267	1.0E-05
1	1.31015	11.1531	3.63798	-1.67605	0.18464	-.324815	11.1531	1.0E-06
2	1.07572	25.2579	9.52539	-6.33464	0.566792	-1.23028	25.2579	1.0E-07
3	1.0644	29.0107	11.0914	-7.57413	0.668468	-1.47113	29.0107	1.0E-08
4	1.0644	29.113	11.1341	-7.60791	0.67124	-1.4777	29.113	1.0E-09
5	1.0644	29.1133	11.1342	-7.60801	0.671247	-1.47772	29.1133	1.0E-10

PARAMETER	ESTIMATE	STD ERROR	T RATIO	LAG	VARIABLE
MU	29.1133	7.48219	3.89	0	Y
NUM1	11.1342	4.44555	2.50	0	X4
NUM2	-7.60801	2.46877	-3.08	0	X3
NUM3	0.671247	0.202673	3.31	0	X3SQ
NUM4	-1.47772	0.667117	-2.22	0	X43

CONSTANT ESTIMATE = 29.1133
VARIANCE ESTIMATE = 0.0425759
STD ERROR ESTIMATE = 0.206339
NUMBER OF RESIDUALS = 30

CORRELATIONS OF THE ESTIMATES

	MU	NUM1	NUM2	NUM3	NUM4
MU	1.000	0.695	-0.999	0.996	-0.711
NUM1	0.695	1.000	-0.698	0.699	-0.998
NUM2	-0.999	-0.698	1.000	-0.999	0.715
NUM3	0.996	0.699	-0.999	1.000	-0.719
NUM4	-0.711	-0.998	0.715	-0.719	1.000

(Parts (b), the RSAC, (c), the RSPAC, and (d), the Forecast, follow on pages 474 and 475.)

FIGURE 8.16 Continued (b) The RSAC

AUTOCORRELATION CHECK OF RESIDUALS

TO LAG	CHI SQUARE	DF	PROB	AUTOCORRELATIONS					
6	10.04	5	0.074	0.179	0.215	-0.014	-0.250	-0.112	-0.340
12	10.81	11	0.459	-0.087	-0.034	-0.057	0.031	0.002	0.060
18	18.43	17	0.362	0.112	0.051	-0.030	-0.198	-0.088	-0.207
24	28.66	23	0.192	0.015	-0.147	0.100	0.206	0.102	0.067

AUTOCORRELATION PLOT OF RESIDUALS

LAG	COVARIANCE	CORRELATION	STD
0	0.0425759	1.00000	0
1	0.00913337	0.21452	0.182574
2	0.0076108	0.17876	0.190791
3	-0.0005798	-0.01362	0.196295
4	-0.0106257	-0.24957	0.196326
5	-.00476252	-0.11186	0.206631
6	-0.0144964	-0.34048	0.208639
7	-.00144134	-0.03385	0.226404
8	-.00368491	-0.08655	0.226573
9	-.00244218	-0.05736	0.227673
10	0.0013204	0.03101	0.228154
11	.000101269	0.00238	0.228294
12	0.00257063	0.06038	0.228295
13	0.00216542	0.05086	0.228827
14	0.00475517	0.11169	0.229203
15	-.00127747	-0.03000	0.23101
16	-0.0084251	-0.19788	0.23114
17	-.00375366	-0.08816	0.23672
18	-0.0087952	-0.20658	0.237812
19	-.00625064	-0.14681	0.24372
20	.000640903	0.01505	0.24665
21	0.00427656	0.10045	0.246681
22	0.00876666	0.20591	0.24804
23	0.00434258	0.10200	0.253674
24	0.00287321	0.06748	0.255037

'.' MARKS TWO STANDARD ERRORS

(c) The RSPAC

PARTIAL AUTOCORRELATIONS

LAG	CORRELATION	-1 ... 0 ... 1
1	0.21452	
2	0.13914	
3	-0.08175	
4	-0.27649	
5	-0.00193	
6	-0.25938	
7	0.09397	
8	-0.08311	
9	-0.08221	
10	-0.08262	
11	0.02697	
12	-0.07157	
13	0.05124	
14	0.04579	
15	-0.13118	
16	-0.26950	
17	0.02450	
18	-0.16007	
19	-0.13840	
20	0.03656	
21	0.04241	
22	-0.05165	
23	-0.01472	
24	-0.11937	

'.' MARKS TWO STANDARD ERRORS

(d) Forecasting of y_{31}

SAS

FORECASTS FOR VARIABLE Y

OBS	FORECAST	STD ERROR	LOWER 95%	UPPER 95%
- - - - - - - - - FORECAST BEGINS - - - - - - - -				
31	8.5259	0.2063	8.1215	8.9303

475

for the model

$$y_t = \beta_0 + \beta_1 x_{t4} + \beta_2 x_{t3} + \beta_3 x_{t3}^2 + \beta_4 x_{t4} x_{t3} + \varepsilon_t$$

that is generated using a *SAS time series procedure*. Since the prob-values related to the values of Q^* for $K = 6, 12, 18$, and 24 are fairly large (see Figure 8.16b) we conclude that the error terms are *statistically independent*. Because the error terms are statistically independent, although we can use the prediction interval for y_{31} given in Figure 8.16d

[8.1215, 8.9303]

a *more correct* (and *longer*) prediction interval for y_{31} results from using the SAS regression procedure. This interval (given in Table 8.4) is

[8.068, 8.984]

However, if a regression model has dependent error terms $\varepsilon_t, \varepsilon_{t-1}, \ldots$, it is best to use the SAS time series procedure to calculate prediction intervals.

8.4.6   *Outlying and Influential Observations**

An observation that is well separated from the rest of the data is called an *outlier*, and an observation that causes the least squares point estimates to be substantially different from what they would be if the observation were removed from the data set is called *influential*. An observation may be an outlier with respect to its y value and/or its x values, but an outlier may or may not be influential. We illustrate these ideas by considering Figure 8.17 which is a hypothetical plot of the values of a dependent variable y against the increasing values of an independent variable x (we assume that the regression analysis involves only one independent variable). Observation 1 in Figure 8.17 is outlying with respect to its y value but not with respect to its x value (since its x value is near the middle of the other x values). Moreover, observation 1 may not be influential, because there are several observations with similar x values and nonoutlying y values, which will keep the least squares point estimates from being excessively influenced by observation 1. Observation 2 in Figure 8.17 is outlying with respect to its x value, but since its y value is consistent with the regression relationship displayed by the non-outlying observations, it probably is not influential. Observation 3, however, is probably influential, because it is outlying with respect to its x value, and because its y value is not consistent with the regression relationship displayed by the other observations.

In order to identify outliers with respect to their x values, consider the regression model

$$y_t = \beta_0 + \beta_1 x_{t1} + \beta_2 x_{t2} + \cdots + \beta_p x_{tp} + \varepsilon_t$$

In Section 8.2 we stated that the leverage value h_{tt} is a function of the distance between the x-values of the tth observation

$$x_{t1}, x_{t2}, \ldots, x_{tp}$$

*Much of the discussion of this section is based on a similar discussion by Neter, Wasserman, and Kutner [1985].

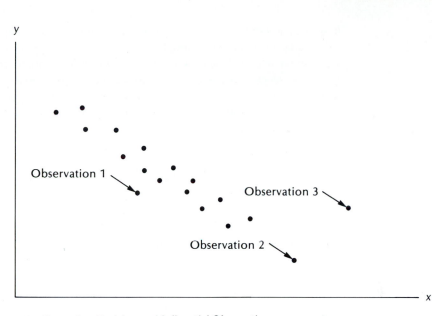

FIGURE 8.17   *Data Plot Illustrating Outlying and Influential Observations*

and the means

$$\bar{x}_1 = \frac{\sum_{t=1}^{n} x_{t1}}{n} \qquad \bar{x}_2 = \frac{\sum_{t=1}^{n} x_{t2}}{n} \quad \cdots \quad \bar{x}_p = \frac{\sum_{t=1}^{n} x_{tp}}{n}$$

of the x values that have occurred in all n previously observed periods. If the leverage value h_{tt} is large, the tth observation is outlying with respect to its x values, and thus the tth observation will exert substantial leverage in determining the values of the least squares point estimates. (To see why this is true, note that since observation 2 and observation 3 in Figure 8.17 are outlying with respect to their x values, they will exert substantial leverage in determining the values of the least squares point estimates.) A leverage value h_{tt} is generally considered to be large if it is substantially greater than most of the other leverage values or if it is greater than twice the average leverage value —that is, if it is greater than

$$2\bar{h} = 2 \frac{\sum_{t=1}^{n} h_{tt}}{n} = 2 \frac{n_p}{n}$$

Note that it can be proven that any particular leverage value h_{tt} is between 0 and 1 and that

$$\sum_{t=1}^{n} h_{tt} = n_p$$

It should be noted, however, that an observation with a large leverage value is not necessarily influential. For example, both observation 2 and observation 3 in Figure 8.17 would have large leverage values, because both observations are outliers with respect to their x values. However, as previously stated, whereas observation 3 probably is very influential (because its y value is not consistent with the regression relationship displayed by the other observations), observation 2 probably is not influential (because its y value is consistent with the regression relationship displayed by the nonoutlying observations).

Next, note that residual plots against time, increasing values of the independent variables, or increasing values of the predicted values of the dependent variable can indicate the presence of outliers with respect to their y values. Any residual that is substantially different from the others is suspect. As a rule of thumb, if the absolute value of the *standardized residual*

$$\frac{e_t}{s} = \frac{y_t - \hat{y}_t}{s}$$

or the (more preferred) *studentized residual*

$$\frac{e_t}{s\sqrt{1 - h_{tt}}}$$

is larger than 3 or 4, an outlier with respect to its y value is suspected. Perhaps the most preferred way to identify an outlier with respect to its y value is to make use of the *deleted residual*

$$d_t = y_t - \hat{y}_{(t)}$$

where

$$\hat{y}_{(t)} = b_0^{(t)} + b_1^{(t)}x_{t1} + b_2^{(t)}x_{t2} + \cdots + b_p^{(t)}x_{tp}$$

is the point prediction of y_t calculated by using least squares point estimates $b_0^{(t)}$, $b_1^{(t)}$, $b_2^{(t)}$, . . . , $b_p^{(t)}$. These estimates are calculated by using all n observations except for the tth observation (that is, the value of the dependent variable and the values of the independent variables that have occurred in the tth period). We do not use the tth observation because, if we do not eliminate the tth observation, and if the tth observation is an outlier with respect to y_t, its y value, then the tth observation might cause $e_t = y_t - \hat{y}_t$ to be small, which would falsely imply that the tth observation is not an outlier with respect to its y value. If we let s_{d_t} denote the standard error of d_t, then it can be shown that *the studentized deleted residual is*

$$\frac{d_t}{s_{d_t}} = e_t \left[\frac{n - n_p - 1}{SSE(1 - h_{tt}) - e_t^2} \right]^{1/2}$$

where $e_t = y_t - \hat{y}_t$, and the population of all possible values of d_t/s_{d_t} has a t-distribution with $n - n_p - 1$ degrees of freedom. If the absolute value of the studentized deleted residual d_t/s_{d_t} is greater than

$$t_{[.025]}^{(n - n_p - 1)}$$

this residual is considered to be large, and the tth observation is considered to be an outlier with respect to its y value.

If we have concluded that the tth observation is an outlier with respect to its x or y value, we can determine whether the tth observation is influential by calculating *Cook's distance measure.*

$$D_t = \frac{e_t^2}{n_p s^2} \left[\frac{h_{tt}}{(1 - h_{tt})^2} \right]$$

Although the following cannot be understood by examining the given formula, it can be shown that D_t is a composite measure of the magnitudes of the following differences:

$$(b_0 - b_0^{(t)}) \quad (b_1 - b_1^{(t)}) \quad (b_2 - b_2^{(t)}) \quad \ldots \quad (b_p - b_p^{(t)})$$

If D_t is large, the least squares point estimates $b_0, b_1, b_2, \ldots, b_p$ calculated by using all n observations differ substantially from the least squares point estimates $b_0^{(t)}, b_1^{(t)}, b_2^{(t)}, \ldots, b_p^{(t)}$ calculated by using all n observations except for the tth observation, and thus the tth observation is influential. To understand what we mean by a large D_t value, note that although the population of all possible values of D_t does not have an F-distribution, practice has shown that

1. If D_t is less than

$$F_{[.80]}^{(n_p, n - n_p)}$$

(the 20th percentile of the F-distribution having n_p and $n - n_p$ degrees of freedom), then the tth observation should not be considered influential.

2. If D_t is greater than

$$F_{[.50]}^{(n_p, n - n_p)}$$

(the 50th percentile of the F-distribution having n_p and $n - n_p$ degrees of freedom), then the tth observation should be considered influential.

3. If

$$F_{[.80]}^{(n_p, n - n_p)} \leq D_t \leq F_{[.50]}^{(n_p, n - n_p)}$$

then the nearer D_t is to

$$F_{[.50]}^{(n_p, n - n_p)}$$

the greater the extent of the influence of the tth observation.

Once we have identified influential outlying observations, we must decide what to do about these observations. If an influential outlying observation has been caused by incorrect measurement (perhaps resulting from a faulty instrument) or erroneous recording (for example, an incorrect decimal point), the observation should be corrected (if it can be corrected), and the regression analysis should be rerun. If the observation cannot be corrected, then it should probably be dropped from the data set. If the influential outlying observation is accurate, it is possible that the regression model under consideration is inadequate in that it does not contain an important independent variable that would explain this observation or does not have the correct functional form. This possibility should be investigated and if need be, the model should be improved. Finally, if no explanation can be found for an influential outlying observation, it might be appropriate to drop this observation from the data set. As an alternative, instead of calculating point estimates $b_0, b_1, b_2, \ldots, b_p$ that minimize the sum of *squared* residuals, we could dampen the effect of the influential outlying observation by calculating point estimates $b_0, b_1, b_2, \ldots, b_p$ that minimize the sum of the *absolute values* of the residuals. The reader interested in this approach should see Kennedy and Gentle [1980].

EXAMPLE 8.16   Home Real Estate Company has used the data in Table 8.13 to develop the model

$$y_t = \beta_0 + \beta_1 x_{t1} + \beta_2 x_{t2} + \beta_3 x_{t2}^2 + \varepsilon_t$$

relating y_t (the selling price of a house) to x_{t1} (the number of square feet in the house) and x_{t2} (the age of the house). Examining the column labeled HAT DIAG H in Table 8.13, we see that the leverage value $h_{66} = .9587$ is greater than $2(n_p/n) = 2(4/16) = .5$ and indicates that observation 6 is an outlier with respect to its x values (intuitively, this is because house 6 is somewhat smaller and much older than the other houses). The column labeled RSTUDENT shows the fact that the studentized deleted residual $d_6/s_{d_6} = -2.5079$ is (in absolute value) greater than

$$t_{[.025]}^{(n - n_p - 1)} = 2.201$$

and indicates that observation 6 is an outlier with respect to its y value. Next, we let $f_6 = \hat{y}_6 - \hat{y}_{(6)}$ denote the difference between the point predictions of y_6 made with and without using observation 6, and we let s_{f_6} denote the standard error of this difference. Then, examining the column labeled DFFITS, the fact that the

TABLE 8.13 SAS Output of the Data in the Home Real Estate Company Problem and of Diagnostics for Detecting Outlying and Influential Observations for the Model

$$y_t = \beta_0 + \beta_1 x_{t1} + \beta_2 x_{t2} + \beta_3 x_{t2}^2 + \varepsilon_t$$

OBS[a]	Y[b]	X1[c]	X2[d]	HAT DIAG[e] H	RESIDUAL[g]	RSTUDENT[i]	DFFITS[k]
1	68.7	2.05	3.43	0.2728	−1.63733	−0.6669	−0.4085
2	54.9	1.70	11.61	0.1603	0.716355	0.2671	0.1167
3	51.5	1.47	8.31	0.1980	0.226113	0.0860	0.0427
4	71.6	1.75	0.00	0.1781	3.85264	1.6082	0.7487
5	58.4	1.94	7.41	0.2040	−4.96169	−2.3066	−1.1676
6	40.7	1.19	31.70	0.9587[f]	−1.19347[h]	−2.5079[j]	−12.0874[l]
7	51.7	1.56	16.10	0.2042	3.23091	1.3285	0.6730
8	71.9	1.95	2.05	0.1879	2.2038	0.8603	0.4138
9	57.1	1.60	1.74	0.1449	−4.62767	−1.9862	−0.8175
10	58.3	1.49	2.76	0.1982	0.535255	0.2039	0.1014
11	73.5	1.91	0.00	0.2307	1.92183	0.7657	0.4193
12	58.5	1.38	0.00	0.4231	−.388631	−0.1745	−0.1494
13	49.1	1.55	12.61	0.1829	−.849241	−0.3214	−0.1520
14	67.5	1.88	2.80	0.1286	0.448381	0.1638	0.0629
15	53.7	1.60	7.08	0.1111	−1.84092	−0.6786	−0.2399
16	50.0	1.55	18.00	0.2166	2.36366	0.9456	0.4972

OBS	COOK'S[m] D	DFBETAS[o] INTERCEP	DFBETAS[q] X1	DFBETAS[s] X2	DFBETAS[u] X2SQ	STUDENT[w] RESIDUAL
1	0.044	0.3069	−0.3432	0.0314	−0.0858	−0.683
2	0.004	−0.0235	0.0197	0.0896	−0.0758	0.278
3	0.000	0.0288	−0.0289	0.0184	−0.0264	0.090
4	0.124	0.1118	0.0190	−0.5538	0.4262	1.511
5	0.251	0.8156	−0.8515	−0.4547	0.2591	−1.978
6	25.351[n]	−0.0030[p]	−0.2241[r]	3.9352[t]	−7.4644[v]	−2.089
7	0.106	−0.0014	−0.0392	0.5225	−0.4174	1.288
8	0.044	−0.2252	0.2797	−0.1347	0.1532	0.870
9	0.134	−0.5151	0.4020	0.3884	−0.1503	−1.780
10	0.003	0.0847	−0.0755	−0.0256	−0.0065	0.213
11	0.046	−0.1330	0.2003	−0.2658	0.2539	0.779
12	0.006	−0.1289	0.1137	0.0758	−0.0283	−0.182
13	0.006	−0.0350	0.0420	−0.1140	0.1111	−0.334
14	0.001	−0.0269	0.0354	−0.0160	0.0162	0.171
15	0.015	−0.1069	0.1005	−0.1040	0.1415	−0.694
16	0.062	−0.0407	0.0100	0.3627	−0.2839	0.950

[a] House [b] Selling price (thousands of dollars) [c] Square footage (thousands of square feet) [d] Age (years) [e] h_{tt} [f] h_{66}

[g] $e_t = y_t - \hat{y}_t$ [h] $e_6 = y_6 - \hat{y}_6$ [i] d_t/s_{d_t} [j] d_6/s_{d_6} [k] $t/s_{t_t} = (\hat{y}_t - \hat{y}_{(t)})/s_{t_t}$ [l] $f_6/s_{f_6} = (\hat{y}_6 - \hat{y}_{(6)})/s_{f_6}$ [m] d_t [n] d_6 [o] $g_0^{(t)}/s_{g_0^{(t)}}$

[p] $g_0^{(6)}/s_{g_0^{(6)}}$ [q] $g_1^{(t)}/s_{g_1^{(t)}}$ [r] $g_1^{(6)}/s_{g_1^{(6)}}$ [s] $g_2^{(t)}/s_{g_2^{(t)}}$ [t] $g_2^{(6)}/s_{g_2^{(6)}}$ [u] $g_3^{(t)}/s_{g_3^{(t)}}$ [v] $g_3^{(6)}/s_{g_3^{(6)}}$ [w] $e_t/s\sqrt{1 - h_{tt}}$

absolute value of $f_6/s_{f_6} = -12.0874$ is greater than 2 (a commonly used critical value for DFFITS) indicates that removing observation 6 from the data set would substantially change the point prediction of y_6. Examining the column labeled COOK'S D, the fact that $D_6 = 25.351$ is greater than

$$F_{[.05]}^{(n_p, n - n_p)} = 5.41$$

which is itself greater than

$$F_{[.50]}^{(n_p, n - n_p)}$$

implies that removing observation 6 from the data set would substantially change (as a group) the least squares point estimates of the parameters β_0, β_1, β_2, and β_3. To determine whether the least squares point estimate of a particular parameter β_j would substantially change, we let $g_j^{(6)} = b_j - b_j^{(6)}$ denote the difference between the point estimates of β_j calculated with and without using observation 6, we let $s_{g_j^{(6)}}$ denote the standard error of this difference, and we calculate $g_j^{(6)}/s_{g_j^{(6)}}$. Specifically, examining the columns labeled DFBETAS, the fact that the absolute values of $g_0^{(6)}/s_{g_0^{(6)}} = -.0030$ and $g_1^{(6)}/s_{g_1^{(6)}} = -.2241$ are less than 2 (a commonly used critical value for DFBETAS) indicates that the least squares point estimates of β_0 and β_1 probably would not change substantially, while the fact that the absolute values of $g_2^{(6)}/s_{g_2^{(6)}} = 3.9352$ and $g_3^{(6)}/s_{g_3^{(6)}} = -7.4644$ are greater than 2 indicates that the least squares point estimates of β_2 and β_3 probably would change substantially. The functional form of the model can be verified to be correct. Therefore, although we will lose information concerning smaller and older homes, we will remove outlying, influential observation 6 from the data set. It can be verified that doing this (1) changes the least squares point estimates of β_0, β_1, β_2, and β_3 from 25.8480, 23.9425, −1.4538, and .0335 to 25.8650, 24.6601, −2.2297, and .0847; (2) reduces the standard error for the model from 2.8117 to 2.3425; and (3) changes the point prediction of and 95% prediction interval for the selling price of a five-year-old house containing 1700 square feet from 60.118 and [53.762, 66.474] to 58.756 and [53.274, 64.237] (which is a shorter interval).

8.4.7   Handling Nonconstant Error Variances

Consider the regression model

$$y_t = \mu_t + \varepsilon_t$$
$$= \beta_0 + \beta_1 x_{t1} + \cdots + \beta_j x_{tj} + \cdots + \beta_p x_{tp} + \varepsilon_t$$

If a residual plot against increasing values of x_{tj} fans out, it implies that the error variance σ_t^2 (the variance of ε_t) and standard deviation σ_t increase as x_{tj} increases. In such a situation it might be reasonable to conclude that $\sigma_t = x_{tj}^c \sigma$, which says that σ_t is proportional to some power, x_{tj}^c, of x_{tj}. Frequently, either $c = 1/2$, in which case $\sigma_t = x_{tj}^{1/2} \sigma$, or $c = 1$, in which case $\sigma_t = x_{tj} \sigma$. If $\sigma_t = x_{tj}^c \sigma$, then it can be proven that we can equalize the error variance by using the *transformed* (or *weighted*) regression model

$$\frac{y_t}{x_{tj}^c} = \beta_0\left(\frac{1}{x_{tj}^c}\right) + \beta_1\left(\frac{x_{t1}}{x_{tj}^c}\right) + \cdots + \beta_j\left(\frac{x_{tj}}{x_{tj}^c}\right) + \cdots + \beta_p\left(\frac{x_{tp}}{x_{tj}^c}\right) + \frac{\varepsilon_t}{x_{tj}^c}$$

That is, if we use this model, the variance of the transformed error term

$$\eta_t = \frac{\varepsilon_t}{x_{tj}^c}$$

is the same for all values of x_{tj}. Then, if $b_0, b_1, \ldots, b_j, \ldots, b_p$ are the least squares point estimates of the parameters in the transformed model, it follows that

$$\hat{y}_t = b_0 + b_1 x_{t1} + \cdots + b_j x_{tj} + \cdots + b_p x_{tp}$$

is the point estimate of μ_t and the point prediction of $y_t = \mu_t + \varepsilon_t$. Moreover, if s is the standard error for the transformed model, and h_{tt}^* is a special leverage value that can be computed by using SAS (see Section 8.6), it follows that

1. A $100(1 - \alpha)\%$ confidence interval for μ_t is

    $$[\hat{y}_t \pm t_{[\alpha/2]}^{(n-n_p)} s \sqrt{h_{tt}^*}]$$

    and

2. A $100(1 - \alpha)\%$ prediction interval for $y_t = \mu_t + \varepsilon_t$ is

    $$[\hat{y}_t \pm t_{[\alpha/2]}^{(n-n_p)} s \sqrt{x_{tj}^{2c} + h_{tt}^*}]$$

EXAMPLE 8.17

The National Association of Retail Hardware Stores (NARHS), a nationally known trade association, wishes to investigate the relationship between x, home value (in thousands of dollars), and y, yearly expenditure on upkeep like lawn care, painting, repairs (in dollars). A random sample of 40 homeowners is taken; the results are given in Table 8.14. Figure 8.18 gives a plot of y versus x. From this plot it appears that y is increasing (probably in a quadratic fashion) as x increases and that the variance of the y values is also increasing as x increases. Increasing variance makes some intuitive sense here, since people with more expensive homes generally have higher incomes and can afford to pay to have upkeep done or perform upkeep chores themselves if they wish, thus causing a relatively large variation in upkeep expenses.

Since y appears to increase in a quadratic fashion as x increases, we begin by trying the model

$$y_t = \beta_0 + \beta_1 x_t + \beta_2 x_t^2 + \varepsilon_t$$

When we estimate β_0, β_1, and β_2 using the data in Table 8.14, we obtain the prediction equation

$$\hat{y}_t = -14.4436 + 3.0585 x_t + .0246 x_t^2$$

TABLE 8.14   *NARHS Upkeep Expenditure Data*

House	Value of House (thousands of dollars), x	Expenditure on Upkeep (dollars), y
1	118.50	706.04
2	76.54	398.60
3	92.43	436.24
4	111.03	501.71
5	80.34	426.45
6	49.84	144.24
7	114.52	644.23
8	50.89	211.54
9	128.93	675.87
10	48.14	189.02
11	85.50	459.04
12	115.51	813.62
13	114.16	602.39
14	102.95	428.52
15	92.86	387.50
16	84.39	434.63
17	123.53	698.00
18	77.77	355.75
19	112.10	737.59
20	101.02	706.66
21	76.52	424.57
22	116.09	656.92
23	62.72	301.03
24	84.91	321.07
25	88.64	519.40
26	81.41	348.50
27	60.22	162.17
28	95.55	482.55
29	79.39	460.07
30	89.25	475.45
31	136.10	835.16
32	24.45	62.70
33	52.28	239.89
34	143.09	1005.32
35	41.86	184.18
36	43.10	212.80
37	66.79	313.45
38	106.43	658.47
39	61.01	195.08
40	99.01	545.42

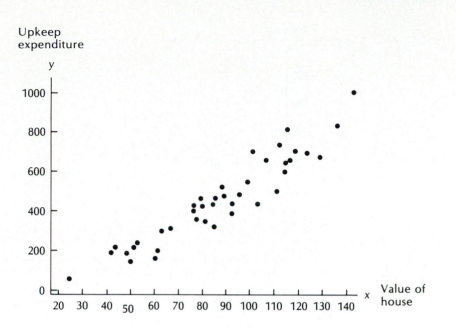

FIGURE 8.18   *Upkeep Expenditure Versus Value of House for NARHS Data*

Figure 8.19 shows a plot of the residuals for this model against increasing values of x. We see that the residuals appear to fan out as x increases, indicating the existence of a nonconstant error variance. Therefore, we will assume that $\sigma_t = x_t^c \sigma$. Although the value of c is not at all obvious, we will start by arbitrarily assuming that $c = 1$, which implies that $\sigma_t = x_t \sigma$. Then the transformed model becomes

$$\frac{y_t}{x_t} = \beta_0 \left(\frac{1}{x_t}\right) + \beta_1 \left(\frac{x_t}{x_t}\right) + \beta_2 \left(\frac{x_t^2}{x_t}\right) + \frac{\varepsilon_t}{x_t}$$

$$= \beta_0 \left(\frac{1}{x_t}\right) + \beta_1(1) + \beta_2 x_t + \eta_t$$

For example, for periods 1 and 30, the values of the dependent variable y_t/x_t and the independent variables $1/x_t$, 1, and x_t are:

Period 1: $\dfrac{y_1}{x_1} = \dfrac{706.04}{118.50} = 5.9581,$

$\dfrac{1}{x_1} = \dfrac{1}{118.5} = .008439,$ 1, $x_1 = 118.5$

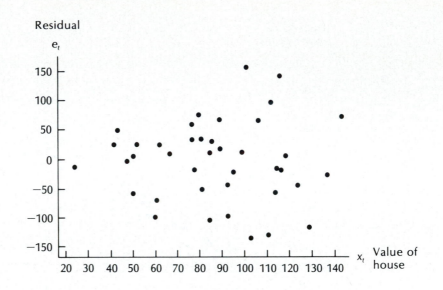

FIGURE 8.19 *Residuals for the Quadratic NARHS Model Plotted Against Increasing Values of x_t*

Period 30: $\dfrac{y_{30}}{x_{30}} = \dfrac{545.42}{99.01} = 5.5087,$

$$\frac{1}{x_{30}} = \frac{1}{99.01} = .0101, \quad 1, \quad x_{30} = 99.01$$

Using the transformed model, the least squares point estimates of β_0, β_1, and β_2 are $b_0 = -26.7501$, $b_1 = 3.4089$, and $b_2 = .0224$. Calculating residual t for the transformed model by the equation

$$\frac{y_t}{x_t} - \left(b_0 \left(\frac{1}{x_t} \right) + b_1(1) + b_2 x_t \right)$$

$$= \frac{y_t}{x_t} - \left(-26.7501 \left(\frac{1}{x_t} \right) + 3.4089 + .0224 x_t \right)$$

it follows that a plot of these residuals in Figure 8.20 has a horizontal band appearance. This indicates that the assumption that $\sigma_t = x_t \sigma$, and the resulting transformed model, have equalized the error variances. Therefore, consider a house worth, say, \$110,000 (that is, $x_t = 110$). It follows that

$$\hat{y}_t = b_0 + b_1 x_t + b_2 x_t^2$$

$$= -26.7501 + 3.4089(110) + .0224(110)^2$$

$$= 619.27$$

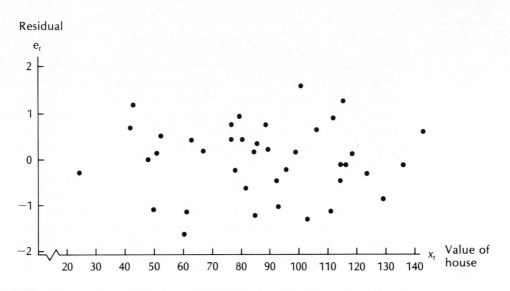

Residual
e_t

FIGURE 8.20   Residuals for the Transformed NARHS Model Plotted Against Increasing Values of x_t

is the point estimate of μ_t, the mean yearly upkeep expenditure for all houses worth \$110,000, and is a point prediction of $y_t = \mu_t + \varepsilon_t$, the upkeep expenditure in a future year for an individual house worth \$110,000. Furthermore, since the standard error for the transformed model is $s = .7935$, and the special leverage value corresponding to $x_t = 110$ is $h_{tt}^* = 506.354$ (see Section 8.6), it follows that a 95% confidence interval for μ_t is

$$[\hat{y}_t \pm t_{[.025]}^{(40-3)} \, s \, \sqrt{h_{tt}^*}] = [619.27 \pm 2.026(.7935)\sqrt{506.354}]$$

$$= [583.10, \ 655.44]$$

and that a 95% prediction interval for $y_t = \mu_t + \varepsilon_t$ is

$$[\hat{y}_t \pm t_{[.025]}^{(40-3)} \, s \sqrt{x_t^2 + h_{tt}^*}] = [619.27 \pm 2.026(.7935)\sqrt{(110)^2 + 506.354}]$$

$$= [438.77, \ 799.77]$$

8.5   USING DUMMY VARIABLES

An important application of classical regression analysis involves *using dummy variables to compare several population means* in what we refer to as *analysis of variance* situations. In this section we consider an example of carrying out this type of analysis.

EXAMPLE 8.18

Part 1: The Problem and Data

To increase the demand for Fresh, Enterprise Industries' marketing department is conducting a study in which it wishes to compare the effectiveness of three different advertising campaigns: campaigns A, B, and C. Campaign A consists entirely of television commercials, campaign B consists of a balanced mixture of television and radio commercials, and campaign C consists of a balanced mixture of television, radio, newspaper, and magazine ads. In order to conduct this study, Enterprise Industries randomly selected an advertising campaign to be used during each of the past 30 sales periods. Table 8.15 lists the campaigns (A, B, or C) used during these sales periods. After a campaign had been selected for a (four-week) sales period, Enterprise Industries employed the campaign for the first three weeks of the sales period. No campaign was used during the last week so that the effect of the advertising campaign used in the current sales period would not carry over to the next sales period.

Part 2: The Model and Interpretation of the Model Parameters

In this example, we will see how Enterprise Industries can use *dummy variables* to study and compare the effectiveness of advertising campaigns A, B, and C. In order to do this, we will define two dummy variables as follows:

$$D_{t,B} = \begin{cases} 1 \text{ if campaign B is used in sales period } t \\ 0 \text{ otherwise, and} \end{cases}$$

$$D_{t,C} = \begin{cases} 1 \text{ if campaign C is used in sales period } t \\ 0 \text{ otherwise} \end{cases}$$

Using these dummy variables, we now consider a regression model relating y_t to x_{t3}, x_{t4}, $D_{t,B}$, and $D_{t,C}$. This model is

$$y_t = \mu_t + \varepsilon_t$$

$$= \beta_0 + \beta_1 x_{t4} + \beta_2 x_{t3} + \beta_3 x_{t3}^2 + \beta_4 x_{t4} x_{t3} + \beta_5 D_{t,B} + \beta_6 D_{t,C} + \varepsilon_t$$

To see what the dummy variables do for us, we wish to interpret the meanings of the parameters β_5 and β_6 in this model (and also the meaning of the difference $\beta_6 - \beta_5$). To do this, consider a sales period in which (1) the price difference (x_{t4}) is d, (2) Enterprise Industries' advertising expenditure (x_{t3}) is a, and (3) advertising campaign A is used (which implies that both $D_{t,B}$ and $D_{t,C}$ equal zero). The mean demand for Fresh for all such sales periods, which we denote as $\mu_{[d,a,A]}$, is (employing the model given above)

$$\mu_{[d,a,A]} = \beta_0 + \beta_1 d + \beta_2 a + \beta_3 a^2 + \beta_4 da + \beta_5(0) + \beta_6(0)$$

$$= \beta_0 + \beta_1 d + \beta_2 a + \beta_3 a^2 + \beta_4 da$$

TABLE 8.15   *Advertising   Campaigns   Used   by   Enterprise Industries*

t	Advertising Campaign Used in Sales Period t
1	B
2	B
3	B
4	A
5	C
6	A
7	C
8	C
9	B
10	C
11	A
12	C
13	C
14	A
15	B
16	B
17	B
18	A
19	B
20	B
21	C
22	A
23	A
24	A
25	A
26	B
27	C
28	B
29	C
30	C

Next, consider a sales period in which (1) the price difference (x_{t4}) is d, (2) Enterprise Industries' advertising expenditure (x_{t3}) is a, and (3) advertising campaign B is used (which implies that $D_{t,B} = 1$ and $D_{t,C} = 0$). The mean demand for all such sales periods, denoted $\mu_{[d,a,B]}$, is

$$\mu_{[d,a,B]} = \beta_0 + \beta_1 d + \beta_2 a + \beta_3 a^2 + \beta_4 da + \beta_5(1) + \beta_6(0)$$
$$= \beta_0 + \beta_1 d + \beta_2 a + \beta_3 a^2 + \beta_4 da + \beta_5$$

Last, consider a sales period in which (1) the price difference (x_{t4}) is d, (2) Enterprise Industries' advertising expenditure (x_{t3}) is a, and (3) advertising campaign C is used (which implies that $D_{t,B} = 0$ and $D_{t,C} = 1$). The mean demand for all such sales periods, denoted $\mu_{[d,a,C]}$, is

$$\mu_{[d,a,C]} = \beta_0 + \beta_1 d + \beta_2 a + \beta_3 a^2 + \beta_4 da + \beta_5(0) + \beta_6(1)$$
$$= \beta_0 + \beta_1 d + \beta_2 a + \beta_3 a^2 + \beta_4 da + \beta_6$$

In order to interpret the meaning of β_5, consider the difference between $\mu_{[d,a,B]}$ and $\mu_{[d,a,A]}$, which is

$$\mu_{[d,a,B]} - \mu_{[d,a,A]} = (\beta_0 + \beta_1 d + \beta_2 a + \beta_3 a^2 + \beta_4 da + \beta_5)$$
$$- (\beta_0 + \beta_1 d + \beta_2 a + \beta_3 a^2 + \beta_4 da)$$
$$= \beta_5$$

Since this difference equals β_5, we can interpret β_5 as follows. The parameter β_5 is the change in mean demand for Fresh associated with changing from using advertising campaign A to using advertising campaign B, while the price difference and advertising expenditure remain constant. Thus, intuitively, β_5 measures the effect on mean demand for Fresh of changing from advertising campaign A to advertising campaign B.

We can interpret the meaning of β_6 by considering the difference between $\mu_{[d,a,C]}$ and $\mu_{[d,a,A]}$, which is

$$\mu_{[d,a,C]} - \mu_{[d,a,A]} = (\beta_0 + \beta_1 d + \beta_2 a + \beta_3 a^2 + \beta_4 da + \beta_6)$$
$$- (\beta_0 + \beta_1 d + \beta_2 a + \beta_3 a^2 + \beta_4 da)$$
$$= \beta_6$$

Since this difference equals β_6, the parameter β_6 can be interpreted as the change in mean demand for Fresh associated with changing from using advertising campaign A to using advertising campaign C, while the price difference and advertising expenditure remain constant. So, intuitively, β_6 measures the effect on mean demand for Fresh of changing from advertising campaign A to advertising campaign C.

Finally, we can interpret the meaning of the difference $\beta_6 - \beta_5$ by considering the difference between $\mu_{[d,a,C]}$ and $\mu_{[d,a,B]}$, which is

$$\mu_{[d,a,C]} - \mu_{[d,a,B]} = (\beta_0 + \beta_1 d + \beta_2 a + \beta_3 a^2 + \beta_4 da + \beta_6)$$
$$- (\beta_0 + \beta_1 d + \beta_2 a + \beta_3 a^2 + \beta_4 da + \beta_5)$$
$$= \beta_6 - \beta_5$$

This says that $\beta_6 - \beta_5$ can be interpreted as the change in mean demand for Fresh associated with changing from using advertising campaign B to using advertising campaign C, while the price difference and advertising expenditure remain constant. Thus, intuitively, $\beta_6 - \beta_5$ measures the effect on mean demand for Fresh of changing from advertising campaign B to advertising campaign C.

In summary, then, when we employ the dummy variables $D_{t,B}$ and $D_{t,C}$, the parameters

$$\beta_5 = \mu_{[d,a,B]} - \mu_{[d,a,A]}$$

$$\beta_6 = \mu_{[d,a,C]} - \mu_{[d,a,A]}$$

and

$$\beta_6 - \beta_5 = \mu_{[d,a,C]} - \mu_{[d,a,B]}$$

measure, respectively, the effects on mean demand for Fresh of

1. changing from advertising campaign A to advertising campaign B,
2. changing from advertising campaign A to advertising campaign C, and
3. changing from advertising campaign B to advertising campaign C.

Before continuing, note that the dummy variables in the current example are used to model the effects of three advertising campaigns, which arise from an experiment designed to determine the best advertising campaign. In some situations, dummy variables are used to measure the effects of qualitative (that is, nonnumerical) independent variables not related to a designed experiment. For example, although Enterprise Industries might not know exactly how much its competitors spend on advertising in different sales periods, it might qualitatively classify these expenditures as high (A), medium (B), and low (C) and use the dummy variables of the model to assess the effects of these expenditure levels.

Part 3 *Using the Partial F-Statistic to Test for Differences Between Advertising Campaigns A, B, and C*

Enterprise Industries first wishes to see whether or not there are any differences between the effects of advertising campaigns A, B, and C. In order to make this determination, we can employ a *partial F-test*. To do this, we consider the following complete model

$$\text{Complete model: } y_t = \beta_0 + \beta_1 x_{t4} + \beta_2 x_{t3} + \beta_3 x_{t3}^2 + \beta_4 x_{t4} x_{t3} + \beta_5 D_{t,B}$$
$$+ \beta_6 D_{t,C} + \varepsilon_t$$

which has $n_p = 7$ parameters, and for which the unexplained variation is $SSE_C = .3936$. Since we have previously seen that

$$\beta_5 = \mu_{[d,a,B]} - \mu_{[d,a,A]}$$

and

$$\beta_6 = \mu_{[d,a,C]} - \mu_{[d,a,A]}$$

the null hypothesis

$$H_0: \beta_5 = \beta_6 = 0 \quad \text{or} \quad H_0: \beta_5 = 0 \quad \text{and} \quad \beta_6 = 0$$

is equivalent to

$$H_0: \mu_{[d,a,B]} - \mu_{[d,a,A]} = 0 \quad \text{and} \quad \mu_{[d,a,C]} - \mu_{[d,a,A]} = 0$$

which is equivalent to

$$H_0: \mu_{[d,a,A]} = \mu_{[d,a,B]} = \mu_{[d,a,C]}$$

To test

$$H_0: \beta_5 = \beta_6 = 0 \quad \text{or, equivalently,} \quad \mu_{[d,a,A]} = \mu_{[d,a,B]} = \mu_{[d,a,C]}$$

which says that advertising campaigns A, B, and C have the same effects on the mean demand for Fresh versus

$$H_1: \text{At least one of } \beta_5 \text{ and } \beta_6 \text{ does not equal zero}$$

or, equivalently,

at least two of $\mu_{[d,a,A]}, \mu_{[d,a,B]},$ and $\mu_{[d,a,C]}$ differ from each other

which says that at least two of advertising campaigns A, B, and C have different effects on the mean demand for Fresh, first note that $(p - g) = 2$, since two parameters (β_5 and β_6) are set equal to zero in the statement of the null hypothesis H_0. Also note that, under the assumption that H_0 is true, the complete model becomes the following reduced model:

$$\text{Reduced model:} \quad y_t = \beta_0 + \beta_1 x_{t4} + \beta_2 x_{t3} + \beta_3 x_{t3}^2 + \beta_4 x_{t4} x_{t3} + \varepsilon_t$$

for which the unexplained variation is $SSE_R = 1.0644$. Thus, in order to test H_0 versus H_1 we make the following computations:

$$
\begin{aligned}
MS_{\text{drop}} &= \frac{SS_{\text{drop}}}{p - g} \\[2mm]
&= \frac{SSE_R - SSE_C}{p - g} \\[2mm]
&= \frac{1.0644 - .3936}{2} \\[2mm]
&= \frac{.6708}{2} \\[2mm]
&= .3354
\end{aligned}
$$

and

$$MSE_C = \frac{SSE_C}{n - n_p}$$

$$= \frac{.3936}{30 - 7}$$

$$= \frac{.3936}{23}$$

$$= .0171$$

1. The appropriate partial F-statistic is

$$F(D_{t,B}, D_{t,C} \mid x_{t4}, x_{t3}, x_{t3}^2, x_{t4}x_{t3}) = \frac{MS_{drop}}{MSE_C}$$

$$= \frac{.3354}{.0171} = 19.614$$

2. The prob-value related to this partial F-statistic is the area to the right of 19.614 under the curve of the F-distribution having 2 and 23 degrees of freedom, which can be computer-calculated to be less than .0001 (for the purposes of subsequent discussion, we will assume that the prob-value equals .0001).

If we wish to use condition (1) in Section 8.3.5 to determine whether we can reject H_0 in favor of H_1 by setting α equal to .05, then we use the rejection point

$$F_{[\alpha]}^{(p-g,n-n_p)} = F_{[.05]}^{(2,23)} = 3.42$$

Since $F(D_{t,B}, D_{t,C} \mid x_{t4}, x_{t3}, x_{t3}^2, x_{t4}x_{t3}) = 19.614 > 3.42 = F_{[.05]}^{(2,23)}$, we can reject H_0 in favor of H_1 by setting α equal to .05. Alternatively, since prob-value $= .0001$ is less than .05 and .01, it follows by condition 2 of Section 8.3.5 that we can reject H_0 in favor of H_1 by setting α equal to .05 or .01. These results provide substantial evidence that at least two of advertising campaigns A, B, and C have different effects on the mean demand for Fresh.

It should be noted that using the partial F-test has an advantage over determining whether there are differences in the effects of advertising campaigns A, B, and C by using t-statistics (and related prob-values) to perform individual tests of

$$H_0 : \beta_5 = \mu_{[d,a,B]} - \mu_{[d,a,A]} = 0$$

versus

$$H_1 : \beta_5 = \mu_{[d,a,B]} - \mu_{[d,a,A]} \neq 0$$

$$H_0 : \beta_6 = \mu_{[d,a,C]} - \mu_{[d,a,A]} = 0$$

versus

$$H_1 : \beta_6 = \mu_{[d,a,C]} - \mu_{[d,a,A]} \neq 0$$

and

$$H_0 : \beta_6 - \beta_5 = \mu_{[d,a,C]} - \mu_{[d,a,B]} = 0$$

versus

$$H_1 : \beta_6 - \beta_5 = \mu_{[d,a,C]} - \mu_{[d,a,B]} \neq 0$$

This is because, although we can set the probability of a Type I error equal to .05 for each individual test, the probability of falsely rejecting H_0 in at least one of these tests can be shown to be greater than .05. On the other hand, we can use the partial F-statistic to determine whether there are differences in the effects of advertising campaigns A, B, and C by testing the single hypothesis $H_0 : \mu_{[d,a,A]} = \mu_{[d,a,B]} = \mu_{[d,a,C]}$, and we can do so by setting the (overall) probability a Type I error equal to .05.

Part 4   Pairwise Comparisons of Advertising Campaigns A, B, and C

Since the partial F-test provides strong evidence to suggest that at least two of advertising campaigns A, B, and C have different effects on the mean demand for Fresh, it makes sense to use t-statistics and confidence intervals to investigate the exact nature of these differences. That is, Enterprise Industries now wishes to determine exactly which advertising campaigns differ and also wishes to estimate the magnitude of the differences by making pairwise comparisons of the campaigns.

In Table 8.16 we present the SAS output of the least squares point estimates of the parameters $\beta_0, \beta_1, \beta_2, \beta_3, \beta_4, \beta_5, \beta_6$, and $\beta_6 - \beta_5$ in the regression model

$$y_t = \beta_0 + \beta_1 x_{t4} + \beta_2 x_{t3} + \beta_3 x_{t3}^2 + \beta_4 x_{t4} x_{t3} + \beta_5 D_{t,B} + \beta_6 D_{t,C} + \varepsilon_t$$

Moreover, letting θ denote any one of the parameters, Table 8.16 also presents the $t_{\hat{\theta}}$ statistic and related prob-value for testing $H_0 : \theta = 0$. Although SAS does not calculate confidence intervals, it follows that we can calculate a 95% confidence interval for θ by the formula

$$[\hat{\theta} \pm t_{[.025]}^{(n - n_p)} s_{\hat{\theta}}] = [\hat{\theta} \pm t_{[.025]}^{(30 - 7)} s_{\hat{\theta}}]$$

$$= [\hat{\theta} \pm 2.069 s_{\hat{\theta}}]$$

In Table 8.16 we present 95% confidence intervals for β_5, β_6, and $\beta_6 - \beta_5$. It should also be noted that another portion (not given) of the SAS output tells us that the standard error for the model is $s = .1308$.

We will now use this information to study the effects of advertising campaigns A, B, and C. We see that:

TABLE 8.16 SAS Output of the t_θ Statistics and Prob-Values for Testing the Hypotheses $H_0: \theta = 0$, and the 95% Confidence Intervals for β_5, β_6, and $\beta_6 - \beta_5$

	PARAMETER[a]	ESTIMATE[b]	T FOR HO:[c] PARAMETER=0	PR > |T|[d]	STD ERROR OF[e] ESTIMATE	[$\hat{\theta} \pm 2.069 s_{\hat{\theta}}$] 95% Confidence Interval for θ
$\beta_0 =$	INTERCEPT	25.61269602	5.34	0.0001	4.79378249	
$\beta_1 =$	X4	9.05868432	2.99	0.0066	3.03170457	
$\beta_2 =$	X3	−6.53767133	−4.13	0.0004	1.58136655	
$\beta_3 =$	X3SQ	0.58444394	4.50	0.0002	0.12987222	
$\beta_4 =$	X43	−1.15648054	−2.54	0.0184	0.45573648	
$\beta_5 =$	DB	0.21368626	3.44	0.0023	0.06215362	[.0851, .3423]
$\beta_6 =$	DC	0.38177617	6.23	0.0001	0.06125253	[.2550, .5085]
$\beta_6 - \beta_5 =$	P1	0.1680899	2.64	0.0147	0.06370381	[.0363, .2999]

[a] θ [b] $\hat{\theta}$ [c] $t_{\hat{\theta}}$ [d] Prob-value [e] $s_{\hat{\theta}}$

1. Since $b_5 = .2137$ is the least squares point estimate of

$$\beta_5 = \mu_{[d,a,B]} - \mu_{[d,a,A]}$$

which equals the change in mean demand for Fresh associated with changing from advertising campaign A to advertising campaign B, while the price difference and advertising expenditure remain constant, Enterprise Industries estimates that the effect of changing from advertising campaign A to advertising campaign B is to increase (since b_5 is positive) mean demand for Fresh by .2137 (or 21,370 bottles). Furthermore, the 95% confidence interval for β_5

[.0851, .3423]

makes Enterprise Industries 95% confident that the effect of changing from advertising campaign A to advertising campaign B is to increase mean demand for Fresh by between .0851 (8,510 bottles) and .3423 (34,230 bottles). In addition, the prob-values of .0023 associated with the null hypothesis

$$H_0 : \beta_5 = \mu_{[d,a,B]} - \mu_{[d,a,A]} = 0$$

says that we have very strong evidence that

$$\beta_5 = \mu_{[d,a,B]} - \mu_{[d,a,A]} > 0$$

or that $\mu_{[d,a,B]} > \mu_{[d,a,A]}$. That is, Enterprise Industries has very strong evidence that advertising campaign B is more effective than advertising campaign A.

2. Since $b_6 = .3818$ is the least squares point estimate of

$$\beta_6 = \mu_{[d,a,C]} - \mu_{[d,a,A]}$$

which equals the change in mean demand for Fresh associated with changing from advertising campaign A to advertising campaign C, while the price difference and advertising expenditure remain constant, Enterprise Industries estimates that the effect of changing from advertising campaign A to advertising campaign C is to increase (since b_6 is positive) mean demand for Fresh by .3818 (or 38,180 bottles). Moreover, the 95% confidence interval for β_6

[.2550, .5085]

makes Enterprise Industries 95% confident that the effect of changing from advertising campaign A to advertising campaign C is to increase mean demand for Fresh by between .2550 (25,500 bottles) and .5085 (50,850 bottles). In addition, since the prob-value associated with the null hypothesis

$$H_0 : \beta_6 = \mu_{[d,a,C]} - \mu_{[d,a,A]} = 0$$

is less than .0001, we have overwhelming evidence that

$$\beta_6 = \mu_{[d,a,C]} - \mu_{[d,a,A]} > 0$$

or that $\mu_{[d,a,C]} > \mu_{[d,a,A]}$. That is, Enterprise Industries has overwhelming evidence that advertising campaign C is more effective than advertising campaign A.

3. Since $b_6 - b_5 = .3818 - .2137 = .1681$ is the point estimate of

$$\beta_6 - \beta_5 = \mu_{[d,a,C]} - \mu_{[d,a,B]}$$

which is the change in mean demand for Fresh associated with changing from advertising campaign B to advertising campaign C, while the price difference and advertising expenditure remain constant, Enterprise Industries estimates that the effect of changing from advertising campaign B to advertising campaign C is to increase (since $b_6 - b_5$ is positive) mean demand for Fresh by .1681 (or 16,810 bottles). Furthermore, the 95% confidence interval for $\beta_6 - \beta_5$

[.0363, .2999]

makes Enterprise Industries 95% confident that the effect of changing from advertising campaign B to advertising campaign C is to increase mean demand for Fresh by between .0363 (3,630 bottles) and .2999 (29,990 bottles).

In addition, since the prob-value for testing the null hypothesis

$$H_0: \beta_6 - \beta_5 = \mu_{[d,a,C]} - \mu_{[d,a,B]} = 0$$

is .0147, we have strong evidence that

$$\beta_6 - \beta_5 = \mu_{[d,a,C]} - \mu_{[d,a,B]} > 0$$

or that $\mu_{[d,a,C]} > \mu_{[d,a,B]}$. That is, Enterprise Industries has strong evidence that advertising campaign C is more effective than advertising campaign B.

Part 5: Using the Model to Forecast Demand for Fresh

Suppose that, on the basis of the analysis in 2 and 3 above, Enterprise Industries concludes that campaign C is the most effective advertising campaign, and, on the basis of that conclusion, decides to employ it in future sales periods. Then, a point prediction of the actual demand for Fresh in (future) sales period 31 (when the price difference will be $x_{31,4} = x_{31,2} - x_{31,1} = 3.90 - 3.80 = .10$, and the advertising expenditure will be $x_{31,3} = 6.80$) is

$$\hat{y}_{31} = b_0 + b_1 x_{31,4} + b_2 x_{31,3} + b_3 x_{31,3}^2 + b_4 x_{31,4} x_{31,3} + b_5 D_{31,B} + b_6 D_{31,C}$$

$$= 25.6127 + 9.0587(.10) - 6.5377(6.80) + .5844(6.80)^2$$

$$- 1.1565(.10)(6.80) + .2137(0) + .3818(1)$$

$$= 8.6825 \quad (\text{or } 868{,}250 \text{ bottles})$$

and a 95% prediction interval for y_{31} is (using SAS)

$$[8.385, \ 8.980]$$

or [838,500 bottles, 898,000 bottles].

Part 6: Analyzing Some Other Potential Models

Notice that the regression model we have used

$$
\begin{aligned}
y_t &= \mu_t + \varepsilon_t \\
&= \beta_0 + \beta_1 x_{t4} + \beta_2 x_{t3} + \beta_3 x_{t3}^2 + \beta_4 x_{t4} x_{t3} + \beta_5 D_{t,B} + \beta_6 D_{t,C} + \varepsilon_t
\end{aligned}
$$

does not contain any interaction terms involving the dummy variables $D_{t,B}$ and $D_{t,C}$. If such interaction terms are included, however, these terms do not have much importance over and above the combined importance of the other independent variables. For example, consider the model

$$
\begin{aligned}
y_t &= \mu_t + \varepsilon_t \\
&= \beta_0 + \beta_1 x_{t4} + \beta_2 x_{t3} + \beta_3 x_{t3}^2 + \beta_4 x_{t4} x_{t3} + \beta_5 D_{t,B} + \beta_6 D_{t,C} \\
&\quad + \beta_7 x_{t4} D_{t,B} + \beta_8 x_{t4} D_{t,C} + \beta_9 x_{t3} D_{t,B} + \beta_{10} x_{t3} D_{t,C} \\
&\quad + \beta_{11} x_{t4} x_{t3} D_{t,B} + \beta_{12} x_{t4} x_{t3} D_{t,C} + \varepsilon_t
\end{aligned}
$$

This model contains the "two-factor" interaction terms $x_{t4} D_{t,B}$, $x_{t4} D_{t,C}$, $x_{t3} D_{t,B}$, and $x_{t3} D_{t,C}$, as well as the "three-factor" interaction terms $x_{t4} x_{t3} D_{t,B}$ and $x_{t4} x_{t3} D_{t,C}$, in addition to the terms in the model we have previously discussed. Although "three-factor" interaction terms can be very difficult to interpret, intuitively the term $x_{t4} x_{t3} D_{t,B}$, for example, measures the amount of interaction between the interaction variable $x_{t4} x_{t3}$ and the dummy variable $D_{t,B}$, which represents the effect of advertising campaign B. Table 8.17 gives the SAS output of the least squares point estimates of the parameters in the "expanded" model (these estimates are obtained by using the data in Tables 8.3 and 8.15), along with the t_{b_j} statistics and associated prob-values for testing $H_0: \beta_j = 0$ ($j = 0, 1, \ldots, 12$). Looking at Table 8.17, we see that the prob-values for all of the interaction terms in the expanded model are quite large (most are substantially greater than .05). In addition, the standard error for this expanded model is $s = .1311$, which is larger than the standard error (.1308) for our previous model. This larger standard error, along with the large prob-values associated with the interaction terms in the expanded model, indicate that the inclusion of the additional interaction terms is not warranted. In order to reduce the degree of multicollinearity, one might wish to try some other models with fewer interaction terms. For example, we might consider the model

TABLE 8.17 SAS Output of the Least Squares Point Estimates, t_{b_j} Statistics, and Prob-Values for the Model

$$y_t = \beta_0 + \beta_1 x_{t4} + \beta_2 x_{t3} + \beta_3 x_{t3}^2 + \beta_4 x_{t4} x_{t3} + \beta_5 D_{t,B} + \beta_6 D_{t,C} + \beta_7 x_{t4} D_{t,B} + \beta_8 x_{t4} D_{t,C}$$
$$+ \beta_9 x_{t3} D_{t,B} + \beta_{10} x_{t3} D_{t,C} + \beta_{11} x_{t4} x_{t3} D_{t,B} + \beta_{12} x_{t4} x_{t3} D_{t,C} + \varepsilon_t$$

VARIABLE	DF	PARAMETER ESTIMATE	STANDARD ERROR	T FOR H0: PARAMETER=0	PROB > |T|
INTERCEP	1	32.590707	5.740052	5.678	0.0001
X4	1	5.887626	5.658572	1.040	0.3127
X3	1	−8.603161	1.838238	−4.680	0.0002
X3SQ	1	0.735327	0.147080	5.000	0.0001
X43	1	−0.665556	0.831766	−0.800	0.4347
DB	1	−0.761000	1.169902	−0.650	0.5241
DC	1	−2.415693	1.398314	−1.728	0.1022
X4DB	1	8.277369	6.093695	1.358	0.1921
X4DC	1	11.961760	9.746570	1.227	0.2364
X3DB	1	0.156574	0.192191	0.815	0.4265
X3DC	1	0.447779	0.224349	1.996	0.0622
X43DB	1	−1.231316	0.887758	−1.387	0.1834
X43DC	1	−1.836136	1.445220	−1.270	0.2210

TABLE 8.18 SAS Output of the Least Squares Point Estimates, t_{b_j} Statistics, and Prob-Values for the Model

$$y_t = \beta_0 + \beta_1 x_{t4} + \beta_2 x_{t3} + \beta_3 x_{t3}^2 + \beta_4 x_{t4} x_{t3} + \beta_5 D_{t,B} + \beta_6 D_{t,C} + \beta_7 x_{t4} D_{t,B} + \beta_8 x_{t4} D_{t,C}$$
$$+ \beta_9 x_{t3} D_{t,B} + \beta_{10} x_{t3} D_{t,C} + \varepsilon_t$$

VARIABLE	DF	PARAMETER ESTIMATE	STANDARD ERROR	T FOR H0: PARAMETER=0	PROB > |T|
INTERCEP	1	30.727132	5.611330	5.476	0.0001
X4	1	12.100134	3.600116	3.361	0.0033
X3	1	−8.016542	1.797726	−4.459	0.0003
X3SQ	1	0.689466	0.143943	4.790	0.0001
X43	1	−1.575823	0.531068	−2.967	0.0079
DB	1	−0.722749	1.176255	−0.614	0.5462
DC	1	−1.680605	1.139724	−1.475	0.1567
X4DB	1	−0.132349	0.454120	−0.291	0.7739
X4DC	1	−0.443757	0.442450	−1.003	0.3285
X3DB	1	0.149000	0.193337	0.771	0.4504
X3DC	1	0.331765	0.184911	1.794	0.0887

$$y_t = \beta_0 + \beta_1 x_{t4} + \beta_2 x_{t3} + \beta_3 x_{t3}^2 + \beta_4 x_{t4} x_{t3} + \beta_5 D_{t,B} + \beta_6 D_{t,C}$$
$$+ \beta_7 x_{t4} D_{t,B} + \beta_8 x_{t4} D_{t,C} + \beta_9 x_{t3} D_{t,B} + \beta_{10} x_{t3} D_{t,C} + \varepsilon_t$$

However, examination of Table 8.18, which presents the SAS output of key quantities related to this model, indicates that inclusion of the additional interaction terms in this model is not warranted. Moreover, when other models are analyzed, we find that the inclusion of interaction terms other than $x_{t4} x_{t3}$ is not warranted.

8.6 USING SAS

Below we present the SAS statements that generate the output in Table 8.2.

```
DATA COMP;
INPUT X Y ;
CARDS;
1    37
2    58
2    68
3    82
4   103
4   109
4   112
5   134
5   138
6   154
7   189
PROC PRINT;
PROC REG DATA=COMP;
MODEL Y=X / P CLM CLI;    } Specifies model y_t = β_0 + β_1 x_t + ε_t
              Note:   P = residuals desired.
                      CLM = 95% confidence intervals desired.
                      CLI = 95% prediction intervals desired.
```

Below we present the SAS statements that generate the output in Table 8.4 and Figure 8.16.

```
DATA DETR;
INPUT Y X4 X3;
X3SQ = X3 · X3;    } Transformation to obtain x_{t3}^2 term.
X43 = X4 · X3;    } Transformation to obtain x_{t4} x_{t3} term.
CARDS;
```

```
7.38   −0.05   5.50
8.51    0.25   6.75
         ⋮
9.26    0.55   6.80
 ·      0.10   6.80
```

PROC PRINT;
PROC REG DATA = DETR; } PROC REG is the regression procedure.
MODEL Y = X4 X3 X3SQ X43/ P CLM CLI;   } Gives output in Table 8.4.

Specifies model:
$$y_t = \beta_0 + \beta_1 x_{t4} + \beta_2 x_{t3} + \beta_3 x_{t3}^2 + \beta_4 x_{t4} x_{t3} + \varepsilon_t$$

PROC ARIMA DATA=DETR;
IDENTIFY VAR=Y NOPRINT CROSSCOR=(X4 X3 X3SQ X43);
ESTIMATE INPUT = (X4 X3 X3SQ X43) PRINTALL PLOT;   Gives output in Figure 8.15.
FORECAST LEAD = 1;

Below we present the SAS statements that generate the output in Table 8.13.

DATA HOUSE;
INPUT Y X1 X2;
X2SQ = X2 · X2;   } Transformation to obtain x_{t2}^2 term.
CARDS;

```
68.7   2.05    3.43
54.9   1.70   11.61
        ⋮
50.0   1.55   18.00
 ·     1.70    5.00
```
} Decimal point denotes a missing value. Used to obtain a prediction when $x_{t1} = 1.70$ and $x_{t2} = 5.00$

PROC PRINT;
PROC REG DATA = HOUSE;
MODEL Y = X1 X2 X2SQ / P R INFLUENCE CLM CLI;

Specifies model $y_t = \beta_0 + \beta_1 x_{t1} + \beta_2 x_{t2} + \beta_3 x_{t2}^2 + \varepsilon_t$

Note: P = residuals desired.
R = diagnostics for outliers desired.
INFLUENCE = diagnostics for influential outliers desired.
CLM = 95% confidence intervals desired.
CLI = 95% prediction intervals desired.

Below we present the SAS statements that generate the output in Table 8.16.

```
DATA DETR;
INPUT Y X4 X3 DB DC;        }   DB and DC are names assigned to the dummy variables for advertising
                                campaigns B and C.
X3SQ = X3 * X3;
X43 = X4 * X3;
CARDS;
7.38  −0.05  5.50  1  0  ⎫
8.51   0.25  6.75  1  0  ⎪
9.52   0.60  7.25  1  0  ⎪
7.50   0.00  5.50  0  0  ⎬   Data (See Tables 8.3 and 8.15) including dummy variables.
       ⋮                  ⎪
9.26   0.55  6.80  0  1  ⎭
```

$$\cdot \quad 0.10 \quad 6.80 \quad 0 \quad 1 \left.\right\}$$ Generates prediction when $x_{t4} = .10$, $x_{t3} = 6.80$, and advertising campaign C is used.

PROC GLM DATA = DETR;   } PROC GLM is the general linear models procedure.
MODEL Y = <u>X4 X3 X3SQ X43 DB DC</u>/ CLI;

Specifies model: $y_t = \beta_0 + \beta_1 x_{t4} + \beta_2 x_{t3} + \beta_3 x_{t3}^2 + \beta_4 x_{t4} x_{t3} + \beta_5 D_{t,B} + \beta_6 D_{t,C} + \varepsilon_t$

ESTIMATE 'P1' DB −1 DC 1;

This statement gives $\hat{\theta}$, $s_{\hat{\theta}}$, and $t_{\hat{\theta}}$ for $\theta = \beta_6 - \beta_5$.

DB −1 Specifies the −1 coefficient multiplied by β_5 (the parameter that corresponds to DB) in the expression $\beta_6 - \beta_5$.

DC 1 Specifies the 1 coefficient multiplied by β_6 (the parameter that corresponds to DC) in the expression $\beta_6 - \beta_5$.

P1 Specifies the name assigned to the parameter $\theta = \beta_6 - \beta_5$.

Note: The ESTIMATE statement can only be used with PROC GLM.

Finally, we present the SAS program to fit the NARHS-transformed model of Example 8.17:

$$\frac{y_t}{x_t} = \beta_0 \left(\frac{1}{x_t}\right) + \beta_1(1) + \beta_2 x_t + \eta_t$$

```
DATA NARHS;
INPUT TY INX ONE X;
CARDS;
5.9581  .008439  1  118.5  ⎫
        ⋮                  ⎬   See Example 8.17 for the data.
5.5087  .0101    1  99.01  ⎭
```

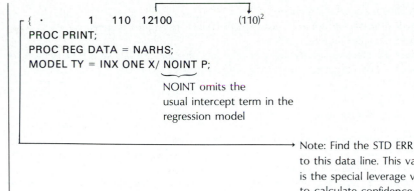

PROC PRINT;
PROC REG DATA = NARHS;
MODEL TY = INX ONE X/ NOINT P;

NOINT omits the
usual intercept term in the
regression model

→ Note: Find the STD ERR PREDICT corresponding
to this data line. This value is $s(h_{tt}^*)^{1/2}$, where h_{tt}^*
is the special leverage value used in Example 8.17
to calculate confidence and prediction intervals. Then,
since the SAS output gives us the standard error
s, we can solve for h_{tt}^*.

EXERCISES

8.1   State University wishes to predict the amount of fuel (in tons of coal) that will be used to
heat University buildings during the next week. Fuel consumption in week t (denoted y_t) is
thought to depend on two independent variables:
1. x_{t1} = the average hourly temperature in week t
2. x_{t2} = the chill index in week t (an index from 0 to 30 which measures effects other than
temperature, such as wind velocity, cloud cover, etc.)
The following data has been observed over the past eight weeks.

The tth Week	Average Hourly Temperature x_{t1}	Chill Index x_{t2}	Weekly Fuel Consumption y_t
1st week	$x_{11} = 28$	$x_{12} = 18$	$y_1 = 12.4$
2nd week	$x_{21} = 32.5$	$x_{22} = 24$	$y_2 = 12.3$
3rd week	$x_{31} = 28$	$x_{32} = 14$	$y_3 = 11.7$
4th week	$x_{41} = 39$	$x_{42} = 22$	$y_4 = 11.2$
5th week	$x_{51} = 57.8$	$x_{52} = 16$	$y_5 = 9.5$
6th week	$x_{61} = 45.9$	$x_{62} = 8$	$y_6 = 9.4$
7th week	$x_{71} = 58.1$	$x_{72} = 1$	$y_7 = 8.0$
8th week	$x_{81} = 62.5$	$x_{82} = 0$	$y_8 = 7.5$

Suppose that a weather forecasting service tells us that the average hourly temperature will
be 40°F and the chill index will be 10 in week 9. We wish to predict y_9, the fuel consumption
in week 9.
Using the model $y_t = \mu_t + \varepsilon_t = \beta_0 + \beta_1 x_{t1} + \varepsilon_t$, hand calculate the following:
a.   The least squares point estimates of β_0 and β_1.
b.   A point estimate of μ_9, mean fuel consumption when the average hourly temperature
will be 40° F.

    c. A 95% confidence interval for μ_9.

    d. A point prediction for $y_9 = \mu_9 + \varepsilon_9$, the fuel consumption in week 9.

    e. A 95% prediction interval for y_9.

8.2 Consider the State University fuel consumption situation of Exercise 8.1. In Table 1 we present the SAS output resulting from using the model

$$y_t = \mu_t + \varepsilon_t$$
$$= \beta_0 + \beta_1 x_{t1} + \beta_2 x_{t2} + \varepsilon_t$$

to perform a regression analysis of the data in Exercise 8.1.

    a. Identify b_0, s_{b_0}, t_{b_0}, and the related prob-value. What do t_{b_0} and the related prob-value say about the importance of the intercept β_0?

    b. Identify b_1, s_{b_1}, t_{b_1}, and the related prob-value. What do t_{b_1} and the related prob-value say about the importance of the independent variable x_1, average hourly temperature?

    c. Identify b_2, s_{b_2}, t_{b_2}, and the related prob-value. What do t_{b_2} and the related prob-value say about the importance of the independent variable x_2, the chill index?

    d. Identify the total variation, the explained variation, the unexplained variation, and R^2. Interpret R^2.

    e. Identify s^2, s, F(model), and the related prob-value. What do F(model) and the related prob-value say about the overall significance of the model?

    f. Identify $\hat{y}_9$, the point estimate of μ_9 and point prediction of y_9. Show how $\hat{y}_9$ was calculated.

    g. Identify and interpret the 95% confidence interval for μ_9.

    h. Identify and interpret the 95% prediction interval for y_9.

    i. Write a SAS program that will generate the output in Table 1.

8.3 Consider the State University fuel consumption situation of Exercises 8.1 and 8.2. In Table 2 we present the SAS output resulting from using the model

$$y_t = \mu_t + \varepsilon_t$$
$$= \beta_0 + \beta_1 x_{t1} + \beta_2 x_{t2} + \beta_3 x_{t1} x_{t2} + \varepsilon_t$$

to perform a regression analysis of the fuel consumption data in Exercise 8.1.

    a. What do the t_{b_j} statistics and related prob-values say about the importance of the intercept β_0 and the independent variables x_{t1}, x_{t2}, and $x_{t1} x_{t2}$ in the model.

    b. Identify and interpret R^2 for the model.

    c. Identify s^2, s, F(model), and the related prob-value for the model.

    d. What do F(model) and the related prob-value say about the overall significance of the model?

    e. Identify and interpret the point estimate of and 95% confidence interval for μ_9 obtained using the model.

    f. Identify and interpret the point prediction of and 95% prediction interval for y_9 obtained using the model.

    g. Compare the model of this exercise with the model of Exercise 8.2 on the basis of R^2.

    h. Compare the model of this exercise with the model of Exercise 8.2 on the basis of the standard error, s.

    i. Compare the prediction intervals obtained using the model of this exercise and the model of Exercise 8.2.

TABLE 1

DEP VARIABLE: Y

SOURCE	DF	SUM OF SQUARES	MEAN SQUARE	F VALUE	PROB > F
MODEL	2	25.462472	12.731236	229.368	0.0001
ERROR	5	0.277528	0.055506		
C TOTAL	7	25.740000			

ROOT MSE	0.235596		R-SQUARE	0.9892
DEP MEAN	10.250000		ADJ R-SQ	0.9849
C.V.	2.298501			

VARIABLE	DF	PARAMETER ESTIMATE	STANDARD ERROR	T FOR H0: PARAMETER=0	PROB > |T|
INTERCEP	1	12.917034	0.549200	23.520	0.0001
X1	1	-0.087064	0.009035063	-9.636	0.0002
X2	1	0.090221	0.014122	6.389	0.0014

OBS	ACTUAL	PREDICT VALUE	STD ERR PREDICT	LOWER 95% MEAN	UPPER 95% MEAN	LOWER 95% PREDICT	UPPER 95% PREDICT	RESIDUAL
1	12.400	12.103	0.134178	11.758	12.448	11.406	12.800	0.296776
2	11.700	11.742	0.157256	11.338	12.147	11.014	12.470	-.042341
3	12.300	12.253	0.137739	11.899	12.607	11.551	12.954	0.047239
4	11.200	11.506	0.131314	11.169	11.844	10.813	12.200	-.306406
5	9.400	9.643	0.101292	9.382	9.903	8.983	10.302	-.242576
6	9.500	9.328	0.180017	8.866	9.791	8.566	10.090	0.171716
7	8.000	7.949	0.143545	7.580	8.318	7.240	8.658	0.051145
8	7.500	7.476	0.155875	7.075	7.876	6.749	8.202	0.024446
9	.	10.337	0.109411	10.055	10.618	9.669	11.004	.

SUM OF RESIDUALS   3.33067E-15
SUM OF SQUARED RESIDUALS 0.2775282

TABLE 2

DEP VARIABLE: Y

SOURCE	DF	SUM OF SQUARES	MEAN SQUARE	F VALUE	PROB > F
MODEL	3	25.471752	8.490584	126.608	0.0002
ERROR	4	0.268248	0.067062		
C TOTAL	7	25.740000			

ROOT MSE	0.258963	R-SQUARE	0.9896
DEP MEAN	10.250000	ADJ R-SQ	0.9818
C.V.	2.526473		

VARIABLE	DF	PARAMETER ESTIMATE	STANDARD ERROR	T FOR H0: PARAMETER=0	PROB > |T|
INTERCEP	1	12.576028	1.097621	11.458	0.0003
X1	1	-0.080924	0.019263	-4.201	0.0137
X2	1	0.112914	0.062948	1.794	0.1473
X12	1	-0.000455525	0.00122456	-0.372	0.7288

OBS	ACTUAL	PREDICT VALUE	STD ERR PREDICT	LOWER 95% MEAN	UPPER 95% MEAN	LOWER 95% PREDICT	UPPER 95% PREDICT	RESIDUAL
1	12.400	12.113	0.149820	11.697	12.529	11.282	12.944	0.286979
2	11.700	11.712	0.190691	11.183	12.242	10.819	12.605	-.012385
3	12.300	12.301	0.198682	11.749	12.852	11.394	13.207	-6.2E-04
4	11.200	11.513	0.145508	11.109	11.917	10.689	12.338	-.313257
5	9.400	9.598	0.164241	9.142	10.054	8.746	10.449	-.197661
6	9.500	9.284	0.230956	8.643	9.925	8.321	10.247	0.216024
7	8.000	7.961	0.161015	7.514	8.408	7.114	8.807	0.039203
8	7.500	7.518	0.206278	6.946	8.091	6.599	8.437	-.018284
9	.	10.286	0.181757	9.781	10.791	9.408	11.164	.

SUM OF RESIDUALS   4.44089E-15
SUM OF SQUARED RESIDUALS   0.2682483

j. Which model—the model of this exercise or the model of Exercise 8.2—seems best? Fully explain your answer.

k. Why does x_{t2} seem of dubious importance in the model of this exercise (t_{b_2} = 1.794 and prob-value = .1473) when x_{t2} is important (t_{b_2} = 6.389 and prob-value = .0041) in the model of Exercise 8.2?

8.4 A study* was undertaken to examine the profit y per sales dollar earned by a construction company and its relationship to the size x_1 of the construction contract (in hundreds of thousands of dollars) and the number x_2 of years of experience of the construction supervisor. Data were obtained from a sample of n = 18 construction projects undertaken by the construction company over the past two years. These data are shown in the table below.

Profit y_t	Contract Size (in $100,000s) x_{t1}	Supervisor Experience (in years) x_{t2}
2.0	5.1	4
3.5	3.5	4
8.5	2.4	2
4.5	4.0	6
7.0	1.7	2
7.0	2.0	2
2.0	5.0	4
5.0	3.2	2
8.0	5.2	6
5.0	4.3	6
6.0	2.9	2
7.5	1.1	2
4.0	2.6	4
4.0	4.0	6
1.0	5.3	4
5.0	4.9	6
6.5	5.0	6
1.5	3.9	4

a. Plot the values of y_t against the increasing values of x_{t1}. Explain why the plot indicates that the quadratic regression model

$$y_t = \mu_t + \varepsilon_t$$

$$= \beta_0 + \beta_1 x_{t1} + \beta_2 x_{t1}^2 + \varepsilon_t$$

might be an appropriate regression model relating y_t to x_{t1}.

*The idea for and data in this problem have been taken from an example in Mendenhall and Reinmuth [1982]. We thank these authors and Duxbury Press for permission to use the idea and data.

b. Plot the values of y_t against the increasing values of x_{t2}. This plot suggests that the relationship between y_t and x_{t2} might be quadratic. However, it is hard to tell because there are only three levels of x_{t2} in the data (2, 4, and 6).

c. Further analysis (which we will discuss later) indicates that the regression model

$$y_t = \mu_t + \varepsilon_t$$

$$= \beta_0 + \beta_1 x_{t1} + \beta_2 x_{t2} + \beta_3 x_{t1}^2 + \beta_4 x_{t1} x_{t2} + \varepsilon_t$$

is a reasonable regression model relating y_t to x_{t1} and x_{t2}. Note that this model includes the independent variables x_{t1} and x_{t1}^2 suggested by the plot made in part a but that it includes only the linear term x_{t2} (and not the quadratic term x_{t2}^2 that is possibly suggested by the plot made in part b). One reason for not including x_{t2}^2 is that interaction between x_{t1} and x_{t2} is partially causing the plot made in part b to suggest that the relationship between y_t and x_{t2} might be quadratic. The regression model takes this interaction into account by utilizing the term $x_{t1} x_{t2}$. Graphical analysis can be used to suggest the need for the interaction term $x_{t1} x_{t2}$. To carry out this analysis:

1. Plot y_t against x_{t1} when $x_{t2} = 2$.
2. Plot y_t against x_{t1} when $x_{t2} = 4$.
3. Plot y_t against x_{t1} when $x_{t2} = 6$.
4. Combine the plots made in (1), (2), and (3) by making these plots on the same set of axes. Use a different color for each level of x_{t2} ($= 2, 4, 6$).
5. Using the plot made in (4) as the basis for your answer, does the relationship between y_t and x_{t1} depend on the level of x_{t2} ($= 2$, 4, 6)? That is, does the relationship between y_t and x_{t1} change as the level of x_{t2} changes? Explain your answer in terms of the plot made in (4).
6. What do the plots made in (4) say about whether interaction exists between x_{t1} and x_{t2}?
7. Discuss what interaction between x_{t1} and x_{t2} means in the construction company problem. Intuitively, why does interaction between x_{t1} and x_{t2} make sense?

d. If the least squares point estimates of the parameters in the model are calculated, we find that

$$b_0 = 19.3050 \quad b_1 = -1.4866 \quad b_2 = -6.3715 \quad b_3 = -.7522 \quad b_4 = 1.7171$$

By using the least squares point estimates and plugging values of x_{t1} and x_{t2} into the prediction equation

$$\hat{y}_t = b_0 + b_1 x_{t1} + b_2 x_{t2} + b_3 x_{t1}^2 + b_4 x_{t1} x_{t2}$$

1. Plot $\hat{y}_t$ against x_{t1} (for $x_{t1} = 3, 4$, and 5) when $x_{t2} = 2$.
2. Plot $\hat{y}_t$ against x_{t1} (for $x_{t1} = 3, 4$, and 5) when $x_{t2} = 4$.
3. Plot $\hat{y}_t$ against x_{t1} (for $x_{t1} = 3, 4$, and 5) when $x_{t2} = 6$.
4. Combine the plots made in (1), (2), and (3) by making these three plots on the same set of axes. Use a different color for each level of x_{t2} ($= 2, 4, 6$).
5. What do the plots made in (4) suggest concerning the policy the company should follow in assigning supervisors to construction projects? Explain your answer.

e. The SAS output of the t_{b_j} statistics and prob-values that can be used to test $H_0: \beta_0 = 0$, $H_0: \beta_1 = 0$, $H_0: \beta_2 = 0$, $H_0: \beta_3 = 0$, and $H_0: \beta_4 = 0$ is as follows:

VARIABLE	DF	PARAMETER ESTIMATE	STANDARD ERROR	T FOR H0: PARAMETER=0	PROB > |T|
INTERCEP	1	19.304957	2.052057	9.408	0.0001
X1	1	−1.486602	1.177734	−1.262	0.2290
X2	1	−6.371452	1.042308	−6.113	0.0001
X1SQ	1	−0.752248	0.225237	−3.340	0.0053
X12	1	1.717053	0.253779	6.766	0.0001

1. What do the t_{b_j} statistics and related prob-values say about the importance of the intercept β_0 and the independent variables x_{t1}, x_{t2}, x_{t1}^2, and $x_{t1}x_{t2}$? Explain your answer.

2. Explain why we might wish to include x_{t1} in the model when the prob-value for testing $H_0: \beta_1 = 0$ does not cast substantial doubt on the validity of this hypothesis.

f. Write a SAS program that will carry out the regression analysis of the construction company data using the model given in (c) above.

g. Two other regression models which might be used to relate y_t to x_{t1} and x_{t2} are

$$y_t = \mu_t + \varepsilon_t$$

$$= \beta_0 + \beta_1 x_{t1} + \beta_2 x_{t2} + \beta_3 x_{t1}^2 + \beta_4 x_{t2}^2 + \beta_5 x_{t1}x_{t2} + \varepsilon_t$$

$$\quad (.0001) \quad (.2355) \quad (.0002) \quad (.0903) \quad (.6865) \quad (.0135)$$

and

$$y_t = \mu_t + \varepsilon_t$$

$$= \beta_0 + \beta_1 x_{t1} + \beta_2 x_{t2} + \beta_3 x_{t1}^2 + \beta_4 x_{t1}x_{t2} + \beta_5 x_{t1}^2 x_{t2} + \varepsilon_t$$

$$\quad (.2696) \quad (.1306) \quad (.7845) \quad (.0032) \quad (.4056) \quad (.0471)$$

1. The prob-values related to the importance of the independent variables in the models are placed under these variables. Discuss what these prob-values say about adding x_{t2}^2 or $x_{t1}^2 x_{t2}$ to the model

$$y_t = \mu_t + \varepsilon_t$$

$$= \beta_0 + \beta_1 x_{t1} + \beta_2 x_{t2} + \beta_3 x_{t1}^2 + \beta_4 x_{t1}x_{t2} + \varepsilon_t$$

2. Note that when x_{t2}^2 or $x_{t1}^2 x_{t2}$ is added to the model, the prob-values related to some of the other independent variables in this model make these variables look less important than do the prob-values in the SAS output of part (e). Explain this phenomenon.

h. The following table summarizes the values of R^2, $\bar{R}^2$, and s^2, and the 95% prediction intervals for y_t (for the future combination $x_{t1} = 4.8$ and $x_{t2} = 6$) that result when several models are used to perform regression analyses of the construction company data.

Model	R^2	$\overline{R}^2$	s^2	Prediction Interval
$y = \beta_0 + \beta_1 x_1 + \beta_2 x_2 + \varepsilon$	.3284	.2389	3.9971	$[-.2618, 8.834]$
$y = \beta_0 + \beta_1 x_1 + \beta_2 x_2 + \beta_3 x_1^2 + \varepsilon$	.3797	.2468	3.9556	$[-.1358, 8.986]$
$y = \beta_0 + \beta_1 x_1 + \beta_2 x_2 + \beta_3 x_1^2 + \beta_4 x_1 x_2 + \varepsilon$	.8628	.8206	.9422	$[3.758, 8.362]$
$y = \beta_0 + \beta_1 x_1 + \beta_2 x_2 + \beta_3 x_1^2 + \beta_4 x_1 x_2$ $+ \beta_5 x_1^2 x_2 + \varepsilon$	.9025	.8619	.7251	$[3.968, 8.043]$
$y = \beta_0 + \beta_1 x_1 + \beta_2 x_2 + \beta_3 x_1^2 + \beta_4 x_1 x_2$ $+ \beta_5 x_2^2 + \varepsilon$	.8647	.8084	1.0063	$[3.620, 8.432]$

1. Compare these models on the basis of R^2.
2. Compare these models on the basis of $\overline{R}^2$.
3. Compare these models on the basis of s^2.
4. Compare the 95% prediction intervals obtained using these models. Which model yields the most accurate interval?
5. On the basis of the comparisons in (1), (2), (3), and (4), which model seems best? Explain your answer.
6. Considering parts e and g, does the best model have the most significant prob-values? Why or why not?
7. Using the third model above and the least squares point estimates given in part d, calculate a point prediction of profit for a $480,000 construction project which is handled by a supervisor with six years of experience.

i. The residuals for the model

$$y_t = \beta_0 + \beta_1 x_{t1} + \beta_2 x_{t2} + \beta_3 x_{t1}^2 + \beta_4 x_{t1} x_{t2} + \varepsilon_t$$

are given below.

OBS	y_t	x_{t1}	x_{t2}	$\hat{y}_t$	$e_t = y_t - \hat{y}_t$
1	2.000	5.1	4	1.699	0.300627
2	3.500	3.5	4	3.440	0.060264
3	8.500	2.4	2	6.903	1.597
4	4.500	4.0	6	4.303	0.196875
5	7.000	1.7	2	7.699	-.698810
6	7.000	2.0	2	7.448	-.448066
7	2.000	5.0	4	1.921	0.079017
8	5.000	3.2	2	5.091	-.091041
9	8.000	5.2	6	6.577	1.423
10	5.000	4.3	6	5.075	-.074741
11	6.000	2.9	2	5.883	0.116596
12	7.500	1.1	2	7.794	-.294086
13	4.000	2.6	4	2.726	1.274
14	4.000	4.0	6	4.303	-.303125
15	1.000	5.3	4	1.211	-.211019
16	5.000	4.9	6	6.212	-1.212
17	6.500	5.0	6	6.349	0.151395
18	1.500	3.9	4	3.366	-1.866

1. Construct a histogram of the residuals. Does the normality assumption appear to be violated? Explain your answer.
2. Construct a normal plot of the residuals. Does the normality assumption appear to be violated. Explain your answer.
3. Plot the residuals against the increasing values of x_{t1}. Does the constant variance assumption appear to be violated? Explain your answer.
4. Plot the residuals against the increasing values of x_{t2}. Does the constant variance assumption appear to be violated? Explain your answer.
5. Plot the residuals against the increasing values of $\hat{y}_t$. Does the constant variance assumption appear to be violated? Explain your answer.
6. Assuming that the data and thus residuals in the table are listed in the time order in which they have been observed, plot the residuals against time. Does this residual plot suggest that the independence assumption is violated? Explain your answer.
7. Calculate the Durbin-Watson statistic for the residuals.
8. Use the Durbin-Watson statistic of part (7) to test for positive autocorrelation at $\alpha = .05$.
9. Use the Durbin-Watson statistic of part (7) to test for negative autocorrelation at $\alpha = .05$.

8.5 Suppose that a regression model yields the following residual plot against time.

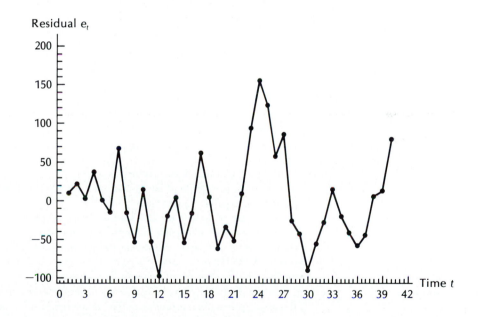

Moreover the plotted residuals have these signs:

+ + + + + − + − − + − − − + − − + + − − − − + + + + + + − − − − −

+ − − − − + + +

   a.   Discuss why the residual plot suggests the existence of positive autocorrelation.
   b.   Assuming that the Durbin-Watson statistic equals .8397, test for positive autocorrelation at $\alpha = .05$ and at $\alpha = .01$.

8.6   Market Planning, Inc., a marketing research firm, has obtained the following prescription sales data for $n = 20$ independent pharmacies.*

Pharmacy	Sales y	Floor Space x_1	Prescription % x_2	Parking x_3	Income x_4
1	22	4900	9	40	18
2	19	5800	10	50	20
3	24	5000	11	55	17
4	28	4400	12	30	19
5	18	3850	13	42	10
6	21	5300	15	20	22
7	29	4100	20	25	8
8	15	4700	22	60	15
9	12	5600	24	45	16
10	14	4900	27	82	14
11	18	3700	28	56	12
12	19	3800	31	38	8
13	15	2400	36	35	6
14	22	1800	37	28	4
15	13	3100	40	43	6
16	16	2300	41	20	5
17	8	4400	42	46	7
18	6	3300	42	15	4
19	7	2900	45	30	9
20	17	2400	46	16	3

These variables can be described precisely as follows:

   y = average weekly prescription sales over the past year (in units of $1,000)
   x_1 = floor space (in square feet)
   x_2 = percent of floor space allocated to the prescription department
   x_3 = number of parking spaces available for the store
   x_4 = monthly per capita income for the surrounding community (in units of $100)

   a.   By plotting y versus x_1, y versus x_2, y versus x_3, and y versus x_4, specify one or more regression models that might be useful in relating y to x_1, x_2, x_3, and x_4.
   b.   Suppose that $y = \beta_0 + \beta_1 x_1 + \beta_2 x_2 + \varepsilon$ is a reasonable regression model describing y. If the least squares point estimates of the parameters in this model are calculated, we find that $b_0 = 48.291$, $b_1 = -.004$, and $b_2 = -.582$. Predict the average weekly prescription sales for a pharmacy having 4,000 sq ft, with 20% of this space allocated to the prescription department.

*The idea for and the data in this problem were taken from an example in Ott (1984). We thank Dr. Ott and Duxbury Press for permission to use the idea and the data.

8.7 Consider Exercise 8.6. Table 3 gives the SAS output of the values of R^2 and the C-statistic that result from regression models representing all possible combinations of the independent variables defined in Exercise 8.6, along with the dummy variable SHOPCNTR (which equals 1 for the pharmacies—1, 2, 3, 6, 8, 9, 10, 17, and 19—located in shopping centers, and equals 0 otherwise). Based on the C-statistic, which model seems best? Why?

8.8 Consider Exercise 8.6. Following is the SAS output of a stepwise regression of the data in Exercise 8.6. Here, the potential independent variables are all the independent variables defined in Exercise 8.6, plus the dummy variable SHOPCNTR defined in Exercise 8.7. Moreover, both α_{entry} and α_{stay} have been set equal to .15.

STEPWISE REGRESSION PROCEDURE FOR DEPENDENT VARIABLE VOLUME

STEP 1   VARIABLE PRESC_RX
       ENTERED   R SQUARE = 0.43933184   C(P) = 10.17094219

	DF	SUM OF SQUARES	MEAN SQUARE	F	PROB > F
REGRESSION	1	329.74051403	329.74051403	14.10	0.0014
ERROR	18	420.80948597	23.37830478		
TOTAL	19	750.55000000			

	B VALUE	STD ERROR	TYPE II SS	F	PROB > F
INTERCEPT	25.98133346				
PRESC_RX	−0.32055657	0.08535423	329.74051403	14.10	0.0014

- -

STEP 2   VARIABLE FLOOR_SP
       ENTERED   R SQUARE = 0.66566267   C(P) = 1.60624219

	DF	SUM OF SQUARES	MEAN SQUARE	F	PROB > F
REGRESSION	2	499.61311336	249.80655668	16.92	0.0001
ERROR	17	250.93688664	14.76099333		
TOTAL	19	750.55000000			

	B VALUE	STD ERROR	TYPE II SS	F	PROB > F
INTERCEPT	48.29085530				
FLOOR_SP	−0.00384228	0.00113262	169.87259933	11.51	0.0035
PRESC_RX	−0.58189034	0.10263739	474.44587802	32.14	0.0001

- -

NO OTHER VARIABLES MET THE 0.1500 SIGNIFICANCE LEVEL
FOR ENTRY INTO THE MODEL

---

a. What is the first independent variable entered?

b. What is the second independent variable entered? Has the first independent variable entered been retained? Why?

c. What is the final model arrived at by stepwise regression? Is this model the same model that you chose in Exercise 8.7?

TABLE 3

REGRESSION ANALYSES
PROC RSQUARE—ALL POSSIBLE SUBSETS ANALYSIS

N = 20   REGRESSION MODELS FOR DEPENDENT VARIABLE VOLUME

NUMBER IN MODEL	R-SQUARE	C(P)	VARIABLES IN MODEL
1	0.00480421	30.45388047	PARKING
1	0.03353172	29.11293360	FLOOR_SP
1	0.04105340	28.76183600	SHOPCNTR
1	0.14798995	23.77023759	INCOME
1	0.43933184	10.17094219	PRESC_RX
2	0.04210776	30.71262010	PARKING SHOPCNTR
2	0.06855667	29.47803470	FLOOR_SP PARKING
2	0.20543099	23.08899693	PARKING INCOME
2	0.23487329	21.71468547	FLOOR_SP INCOME
2	0.25653635	20.70349407	FLOOR_SP SHOPCNTR
2	0.49576794	9.53661080	SHOPCNTR INCOME
2	0.53142435	7.87223587	PRESC_RX PARKING
2	0.54748785	7.12242198	PRESC_RX INCOME
2	0.64706473	2.47435928	PRESC_RX SHOPCNTR
2	0.66566267	1.60624219	FLOOR_SP PRESC_RX
3	0.25569607	22.74271718	FLOOR_SP PARKING INCOME
3	0.26507110	22.30510820	FLOOR_SP PARKING SHOPCNTR
3	0.49828073	11.41931841	PARKING SHOPCNTR INCOME
3	0.50012580	11.33319388	FLOOR_SP SHOPCNTR INCOME
3	0.60243233	6.55771633	PRESC_RX PARKING INCOME
3	0.64711563	4.47198330	PRESC_RX SHOPCNTR INCOME
3	0.66259120	3.74961255	PRESC_RX PARKING SHOPCNTR
3	0.66641145	3.57129027	FLOOR_SP PRESC_RX INCOME
3	0.67943313	2.96346249	FLOOR_SP PRESC_RX PARKING
3	0.69072432	2.43641080	FLOOR_SP PRESC_RX SHOPCNTR
4	0.50128901	13.27889728	FLOOR_SP PARKING SHOPCNTR INCOME
4	0.66300855	5.73013127	PRESC_RX PARKING SHOPCNTR INCOME
4	0.68058567	4.90966443	FLOOR_SP PRESC_RX PARKING INCOME
4	0.69326657	4.31774327	FLOOR_SP PRESC_RX SHOPCNTR INCOME
4	0.69873952	4.06227626	FLOOR_SP PRESC_RX PARKING SHOPCNTR
5	0.70007369	6.00000000	FLOOR_SP PRESC_RX PARKING SHOPCNTR INCOME

8.9 An advertising firm has decided to use six groups of 50 people in an experiment designed to estimate $p(x)$, the probability that a person viewing a 60-second commercial in which a product name is mentioned x times will remember the product name one week later. For $t = 1, 2, \ldots, 6$, the advertising firm shows the $n_t = 50$ people in group t a 60-second commercial in which the product name is mentioned x_t times and a week later records the number y_t, and the proportion $\hat{p}_t = y_t/50$, of the 50 people who recall the product name. The following data is obtained.

t	x_t	y_t	$\hat{p}_t$	$l_t = \ln [\hat{p}_t/(1 - \hat{p}_t)]$	$w_t = 1/[n_t \hat{p}_t(1 - \hat{p}_t)]^{1/2}$
1	1	4	.08	-2.4423	.5212
2	2	7	.14	-1.8153	.4076
3	3	20	.40	-0.4055	.2886
4	4	35	.70	.8473	.3085
5	5	44	.88	1.9924	.4352
6	6	46	.92	2.4423	.5212

If we fit to this data the *weighted logistic* regression model

$$\frac{l_t}{w_t} = \beta_0 \left(\frac{1}{w_t}\right) + \beta_1 \left(\frac{x_t}{w_t}\right) + \frac{\varepsilon_t}{w_t}$$

where w_t is the approximate estimated standard deviation of l_t, we find that $b_0 = -3.6895$, $b_1 = 1.0934$, and $s^2 = 0.4687$. Then, for any function $g(x)$ of x, let

$$T[g(x)] = \exp [g(x)]/\{1 + \exp [g(x)]\}$$

and let $B(x) = z_{[\alpha/2]} s(h_{xx}^\star)^{1/2}$, where $h_{xx}^\star$ is a special leverage value (calculated by the SAS program below) and equals $.0292$ for $x = 4$. It can be proven that a point estimate of $p(x)$ is $T(b_0 + b_1 x)$, and that a $100(1 - \alpha)\%$ confidence interval for $p(x)$ is $\{T[b_0 + b_1 x - B(x)], T[b_0 + b_1 x + B(x)]\}$. Use these facts to find a point estimate of $p(4)$ and a 95% confidence interval for $p(4)$. Also, to see the shape of the logistic curve, plot the values of $T(b_0 + b_1 x)$ versus the values of x for $x = 1, 2, \ldots, 6$.

Now, the SAS program to fit the weighted logistic regression model

$$\frac{l_t}{w_t} = \beta_0 \left(\frac{1}{w_t}\right) + \beta_1 \left(\frac{x_t}{w_t}\right) + \frac{\varepsilon_t}{w_t}$$

uses dependent variable l_t/w_t and independent variables $1/w_t$ and x_t/w_t. For example, for $t = 1$ and 6, we have the following values.

For $t = 1$: $\quad \dfrac{l_1}{w_1} = \dfrac{-2.4423}{.5212} = -4.6859, \quad \dfrac{1}{w_1} = \dfrac{1}{.5212} = 1.9186, \quad \dfrac{x_1}{w_1} = \dfrac{1}{.5212} = 1.9186$

For $t = 6$: $\quad \dfrac{l_6}{w_6} = \dfrac{2.4423}{.5212} = 4.6859, \quad \dfrac{1}{w_6} = \dfrac{1}{.5212} = 1.9186, \quad \dfrac{x_6}{w_6} = \dfrac{6}{.5212} = 11.5119$

Therefore, the SAS program is

```
DATA COM;
INPUT TL INW INX;
CARDS;
 −4.6859  1.9186   1.9186
                :
                :
   4.6859  1.9186  11.5119
```
{ ·   1   4

```
PROC PRINT;
PROC REG DATA = COM;
MODEL TL = INW INX/NOINT P;
```

↳ Note: We find the STD ERR PREDICT corresponding to this data line. This value is $s(h^*_{xx})^{1/2}$, where h^*_{xx} is the special leverage value corresponding to $x = 4$ that we use to calculate the confidence interval for $p(4)$. Then, since the SAS output gives us the standard error s, we can solve for h^*_{xx}.

8.10   Consider analyzing the Fresh detergent data of Tables 8.3 and 8.15 using the model

$$y_t = \mu_t + \varepsilon_t$$

$$= \beta_0 + \beta_1 x_{t4} + \beta_2 x_{t3} + \beta_3 x_{t3}^2 + \beta_4 x_{t4} x_{t3} + \beta_5 D_{t,A} + \beta_6 D_{t,C} + \varepsilon_t$$

where $D_{t,A}$ and $D_{t,C}$ are dummy variables defined as follows:

$$D_{t,A} = \begin{cases} 1 & \text{if campaign A is used in sales period } t \\ 0 & \text{otherwise} \end{cases}$$

$$D_{t,C} = \begin{cases} 1 & \text{if campaign C is used in sales period } t \\ 0 & \text{otherwise} \end{cases}$$

a. Use SAS to run a regression analysis using the model and the data in Tables 8.3 and 8.15.

b. We can employ the overall F-test to test the adequacy of the dummy variable model. Set up the appropriate null and alternative hypotheses for the overall F-test.

c. Using the overall F-statistic and the associated prob-value on your SAS output, perform the overall F-test. What do the results say about the adequacy of the model? Explain in detail. Use $\alpha = .05$.

d. Using the explained variation (SS_{model}) and unexplained variation (SSE) on your SAS output, demonstrate the calculation of (that is, hand calculate) the overall F-statistic for the model.

e. Using the fact that

$$\mu_t = \beta_0 + \beta_1 x_{t4} + \beta_2 x_{t3} + \beta_3 x_{t3}^2 + \beta_4 x_{t4} x_{t3} + \beta_5 D_{t,A} + \beta_6 D_{t,C}$$

show that

$$\mu_{[d,a,A]} - \mu_{[d,a,B]} = \beta_5$$

$$\mu_{(d,a,C)} - \mu_{(d,a,B)} = \beta_6$$

$$\mu_{(d,a,C)} - \mu_{(d,a,A)} = \beta_6 - \beta_5$$

Explain the practical interpretations of the three differences.

f. We can test to see whether or not there are any differences between the effects of advertising campaigns A, B, and C by using a partial *F*-test. Set up the appropriate null and alternative hypotheses for this partial *F*-test. Hint: The null hypothesis should state that there are no differences between the effects of campaigns A, B, and C.

g. Define the appropriate complete model and reduced model for the partial *F*-test.

h. By running regression analyses for the appropriate complete and reduced models, use SAS to determine SSE_C and SSE_R.

i. Using SSE_C and SSE_R, calculate the partial *F*-statistic for the partial *F*-test in (f) above.

j. Using the partial *F*-statistic calculated in (i), carry out the partial *F*-test. Can the null hypothesis you set up in (f) be rejected? What do the results of this test say about differences between the effects of campaigns A, B, and C? Use $\alpha = .05$.

k. Set up the appropriate null and alterative hypotheses needed to test to see whether or not there is any difference between the effects of campaigns A and B.

l. Use the appropriate t_{b_j} statistic and associated prob-value to test the hypotheses in (k). Can the null hypothesis be rejected? What do the results of this test say about whether or not there is a difference between campaigns A and B? Use $\alpha = .05$.

m. Calculate a 95% confidence interval for $\mu_{(d,a,A)} - \mu_{(d,a,B)}$. Interpret the practical meaning of this interval.

n. Set up the appropriate null and alternative hypotheses needed to test to see whether or not there is any difference between the effects of campaigns C and B.

o. Use the appropriate t_{b_j} statistic and associated prob-value to test the hypotheses in (n) above. Can the null hypothesis be rejected? What do the results of this test say about whether or not there is a difference between campaigns C and B? Use $\alpha = .05$.

p. Calculate a 95% confidence interval for $\mu_{(d,a,C)} - \mu_{(d,a,B)}$. Interpret the practical meaning of this interval.

q. Set up the appropriate null and alternative hypotheses needed to test to see whether or not there is any difference between the effects of campaigns C and A.

r. Use PROC GLM in SAS to calculate the appropriate $t_{\hat\theta}$ statistic and associated prob-value needed to test the hypotheses in (q). Use these quantities to test these hypotheses. Can the null hypotheses be rejected? What do the results of this test say about whether or not there is a difference between campaigns C and A? Use $\alpha = .05$.

s. Calculate a 95% confidence interval for $\mu_{(d,a,C)} - \mu_{(d,a,A)}$. Interpret the practical meaning of this interval.

APPENDIX A

STATISTICAL TABLES

TABLE A1 Normal Curve Areas

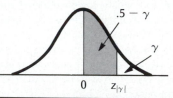

$z_{[\gamma]}$	.00	.01	.02	.03	.04	.05	.06	.07	.08	.09
0.0	.0000	.0040	.0080	.0120	.0160	.0199	.0239	.0279	.0319	.0359
0.1	.0398	.0438	.0478	.0517	.0557	.0596	.0636	.0675	.0714	.0753
0.2	.0793	.0832	.0871	.0910	.0948	.0987	.1026	.1064	.1103	.1141
0.3	.1179	.1217	.1255	.1293	.1331	.1368	.1406	.1443	.1480	.1517
0.4	.1554	.1591	.1628	.1664	.1700	.1736	.1772	.1808	.1844	.1879
0.5	.1915	.1950	.1985	.2019	.2054	.2088	.2123	.2157	.2190	.2224
0.6	.2257	.2291	.2324	.2357	.2389	.2422	.2454	.2486	.2517	.2549
0.7	.2580	.2611	.2642	.2673	.2704	.2734	.2764	.2794	.2823	.2852
0.8	.2881	.2910	.2939	.2967	.2995	.3023	.3051	.3078	.3106	.3133
0.9	.3159	.3186	.3212	.3238	.3264	.3289	.3315	.3340	.3365	.3389
1.0	.3413	.3438	.3461	.3485	.3508	.3531	.3554	.3577	.3599	.3621
1.1	.3643	.3665	.3686	.3708	.3729	.3749	.3770	.3790	.3810	.3830
1.2	.3849	.3869	.3888	.3907	.3925	.3944	.3962	.3980	.3997	.4015
1.3	.4032	.4049	.4066	.4082	.4099	.4115	.4131	.4147	.4162	.4177
1.4	.4192	.4207	.4222	.4236	.4251	.4265	.4279	.4292	.4306	.4319
1.5	.4332	.4345	.4357	.4370	.4382	.4394	.4406	.4418	.4429	.4441
1.6	.4452	.4463	.4474	.4484	.4495	.4505	.4515	.4525	.4535	.4545
1.7	.4554	.4564	.4573	.4582	.4591	.4599	.4608	.4616	.4625	.4633
1.8	.4641	.4649	.4656	.4664	.4671	.4678	.4686	.4693	.4699	.4706
1.9	.4713	.4719	.4726	.4732	.4738	.4744	.4750	.4756	.4761	.4767
2.0	.4772	.4778	.4783	.4788	.4793	.4798	.4803	.4808	.4812	.4817
2.1	.4821	.4826	.4830	.4834	.4838	.4842	.4846	.4850	.4854	.4857
2.2	.4861	.4864	.4868	.4871	.4875	.4878	.4881	.4884	.4887	.4890
2.3	.4893	.4896	.4898	.4901	.4904	.4906	.4909	.4911	.4913	.4916
2.4	.4918	.4920	.4922	.4925	.4927	.4929	.4931	.4932	.4934	.4936
2.5	.4938	.4940	.4941	.4943	.4945	.4946	.4948	.4949	.4951	.4952
2.6	.4953	.4955	.4956	.4957	.4959	.4960	.4961	.4962	.4963	.4964
2.7	.4965	.4966	.4967	.4968	.4969	.4970	.4971	.4972	.4973	.4974
2.8	.4974	.4975	.4976	.4977	.4977	.4978	.4979	.4979	.4980	.4981
2.9	.4981	.4982	.4982	.4983	.4984	.4984	.4985	.4985	.4986	.4986
3.0	.4987	.4987	.4987	.4988	.4988	.4989	.4989	.4989	.4990	.4990

Source: A Hald, *Statistical Tables and Formulas* (New York: Wiley, 1952), abridged from Table 1. Reproduced by permission of the publisher.

TABLE A2   Critical Values of t

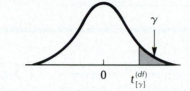

$$0 \qquad t_{[\gamma]}^{(df)}$$

df	$t_{[.10]}^{(df)}$	$t_{[.05]}^{(df)}$	$t_{[.025]}^{(df)}$	$t_{[.01]}^{(df)}$	$t_{[.005]}^{(df)}$
1	3.078	6.314	12.706	31.821	63.657
2	1.886	2.920	4.303	6.965	9.925
3	1.638	2.353	3.182	4.541	5.841
4	1.533	2.132	2.776	3.747	4.604
5	1.476	2.015	2.571	3.365	4.032
6	1.440	1.943	2.447	3.143	3.707
7	1.415	1.895	2.365	2.998	3.499
8	1.397	1.860	2.306	2.896	3.355
9	1.383	1.833	2.262	2.821	3.250
10	1.372	1.812	2.228	2.764	3.169
11	1.363	1.796	2.201	2.718	3.106
12	1.356	1.782	2.179	2.681	3.055
13	1.350	1.771	2.160	2.650	3.012
14	1.345	1.761	2.145	2.624	2.977
15	1.341	1.753	2.131	2.602	2.947
16	1.337	1.746	2.120	2.583	2.921
17	1.333	1.740	2.110	2.567	2.898
18	1.330	1.734	2.101	2.552	2.878
19	1.328	1.729	2.093	2.539	2.861
20	1.325	1.725	2.086	2.528	2.845
21	1.323	1.721	2.080	2.518	2.831
22	1.321	1.717	2.074	2.508	2.819
23	1.319	1.714	2.069	2.500	2.807
24	1.318	1.711	2.064	2.492	2.797
25	1.316	1.708	2.060	2.485	2.787
26	1.315	1.706	2.056	2.479	2.779
27	1.314	1.703	2.052	2.473	2.771
28	1.313	1.701	2.048	2.467	2.763
29	1.311	1.699	2.045	2.462	2.756
inf.	1.282	1.645	1.960	2.326	2.576

Source: From "Table of Percentage Points of the t-Distribution," by Maxine Merrington, *Biometrika* 32 (1941), 300. Reproduced by permission of the *Biometrika* Trustees.

TABLE A3   *Critical Values of Chi-Square*

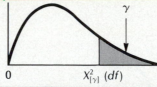

df	$\chi^2_{[.995]}(df)$	$\chi^2_{[.99]}(df)$	$\chi^2_{[.975]}(df)$	$\chi^2_{[.95]}(df)$	$\chi^2_{[.90]}(df)$
1	0.0000393	0.0001571	0.0009821	0.0039321	0.0157908
2	0.0100251	0.0201007	0.0506356	0.102587	0.210720
3	0.0717212	0.114832	0.215795	0.341846	0.584375
4	0.206990	0.297110	0.484419	0.710721	0.063623
5	0.411740	0.554300	0.831211	1.145476	1.61031
6	0.675727	0.872085	1.237347	1.63539	2.20413
7	0.989265	1.239043	1.68987	2.16735	2.83311
8	1.344419	1.646482	2.17973	2.73264	3.48954
9	1.734926	2.087912	2.70039	3.32511	4.16816
10	2.15585	2.55821	3.24697	3.94030	4.86518
11	2.60321	3.05347	3.81575	4.57481	5.57779
12	3.07382	3.57056	4.40379	5.22603	6.30380
13	3.56503	4.10691	5.00874	5.89186	7.04150
14	4.07468	4.66043	5.62872	6.57063	7.78953
15	4.60094	5.22935	6.26214	7.26094	8.54675
16	5.14224	5.81221	6.90766	7.96164	9.31223
17	5.69724	6.40776	7.56418	8.67176	10.0852
18	6.26481	7.01491	8.23075	9.39046	10.8649
19	6.84398	7.63273	8.90655	10.1170	11.6509
20	7.43386	8.26040	9.59083	10.8508	12.4426
21	8.03366	8.89720	10.28293	11.5913	13.2396
22	8.64272	9.54249	10.9823	12.3380	14.0415
23	9.26042	10.19567	11.6885	13.0905	14.8479
24	9.88623	10.8564	12.4011	13.8484	15.6587
25	10.5197	11.5240	13.1197	14.6114	16.4734
26	11.1603	12.1981	13.8439	15.3791	17.2919
27	11.8076	12.8786	14.5733	16.1513	18.1138
28	12.4613	13.5648	15.3079	16.9279	18.9392
29	13.1211	14.2565	16.0471	17.7083	19.7677
30	13.7867	14.9535	16.7908	18.4926	20.5992
40	20.7065	22.1643	24.4331	26.5093	29.0505
50	27.9907	29.7067	32.3574	34.7642	37.6886
60	35.5346	37.4848	40.4817	43.1879	46.4589
70	43.2752	45.4418	48.7576	51.7393	55.3290
80	51.1720	53.5400	57.1532	60.3915	64.2778
90	59.1963	61.7541	65.6466	69.1260	73.2912
100	67.3276	70.0648	74.2219	77.9295	82.3581

Source: From "Tables of the Percentage Points of the χ^2-Distribution," by Catherine M. Thompson, *Biometrika* 32 (1941), 188–189. Reproduced by permission of the *Biometrika* Trustees.

TABLE A3   (Continued)

$\chi^2_{[.10]}(df)$	$\chi^2_{[.05]}(df)$	$\chi^2_{[.025]}(df)$	$\chi^2_{[.01]}(df)$	$\chi^2_{[.005]}(df)$	df
2.70554	3.84146	5.02389	6.63490	7.87944	1
4.60517	5.99147	7.37776	9.21034	10.5966	2
6.25139	7.81473	9.34840	11.3449	12.8381	3
7.77944	9.48773	11.1433	13.2767	14.8602	4
9.23635	11.0705	12.8325	15.0863	16.7496	5
10.6446	12.5916	14.4494	16.8119	18.5476	6
12.0170	14.0671	16.0128	18.4753	20.2777	7
13.3616	15.5073	17.5346	20.0902	21.9550	8
14.6837	16.9190	19.0228	21.6660	23.5893	9
15.9871	18.3070	20.4831	23.2093	25.1882	10
17.2750	19.6751	21.9200	24.7250	26.7569	11
18.5494	21.0261	23.3367	26.2170	28.2995	12
19.8119	22.3621	24.7356	27.6883	29.8194	13
21.0642	23.6848	26.1190	29.1413	31.3193	14
22.3072	24.9958	27.4884	30.5779	32.8013	15
23.5418	26.2962	28.8454	31.9999	34.2672	16
24.7690	27.5871	30.1910	33.4087	35.7185	17
25.9894	28.8693	31.5264	34.8053	37.1564	18
27.2036	30.1435	32.8523	36.1908	38.5822	19
28.4120	31.4104	34.1696	37.5662	39.9968	20
29.6151	32.6705	35.4789	38.9321	41.4010	21
30.8133	33.9244	36.7807	40.2894	42.7956	22
32.0069	35.1725	38.0757	41.6384	44.1813	23
33.1963	36.4151	39.3641	42.9798	45.5585	24
34.3816	37.6525	40.6465	44.3141	46.9278	25
35.5631	38.8852	41.9232	45.6417	48.2899	26
36.7412	40.1133	43.1944	46.9630	49.6449	27
37.9159	41.3372	44.4607	48.2782	50.9933	28
39.0875	42.5569	45.7222	49.5879	52.3356	29
40.2560	43.7729	46.9792	50.8922	53.6720	30
51.8050	55.7585	59.3417	63.6907	66.7659	40
63.1671	67.5048	71.4202	76.1539	79.4900	50
74.3970	79.0819	83.2976	88.3794	91.9517	60
85.5271	90.5312	95.0231	100.425	104.215	70
96.5782	101.879	106.629	112.329	116.321	80
107.565	113.145	118.136	124.116	128.299	90
118.498	124.342	129.561	135.807	140.169	100

TABLE A4   Percentage Points of the F-Distribution ($\gamma = .05$)

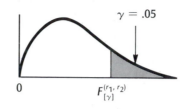

Denominator Degrees of Freedom (r_2)	Numerator Degrees of Freedom (r_1)								
	1	2	3	4	5	6	7	8	9
1	161.4	199.5	215.7	224.6	230.2	234.0	236.8	238.9	240.5
2	18.51	19.00	19.16	19.25	19.30	19.33	19.35	19.37	19.38
3	10.13	9.55	9.28	9.12	9.01	8.94	8.89	8.85	8.81
4	7.71	6.94	6.59	6.39	6.26	6.16	6.09	6.04	6.00
5	6.61	5.79	5.41	5.19	5.05	4.95	4.88	4.82	4.77
6	5.99	5.14	4.76	4.53	4.39	4.28	4.21	4.15	4.10
7	5.59	4.74	4.35	4.12	3.97	3.87	3.79	3.73	3.68
8	5.32	4.46	4.07	3.84	3.69	3.58	3.50	3.44	3.39
9	5.12	4.26	3.86	3.63	3.48	3.37	3.29	3.23	3.18
10	4.96	4.10	3.71	3.48	3.33	3.22	3.14	3.07	3.02
11	4.84	3.98	3.59	3.36	3.20	3.09	3.01	2.95	2.90
12	4.75	3.89	3.49	3.26	3.11	3.00	2.91	2.85	2.80
13	4.67	3.81	3.41	3.18	3.03	2.92	2.83	2.77	2.71
14	4.60	3.74	3.34	3.11	2.96	2.85	2.76	2.70	2.65
15	4.54	3.68	3.29	3.06	2.90	2.79	2.71	2.64	2.59
16	4.49	3.63	3.24	3.01	2.85	2.74	2.66	2.59	2.54
17	4.45	3.59	3.20	2.96	2.81	2.70	2.61	2.55	2.49
18	4.41	3.55	3.16	2.93	2.77	2.66	2.58	2.51	2.46
19	4.38	3.52	3.13	2.90	2.74	2.63	2.54	2.48	2.42
20	4.35	3.49	3.10	2.87	2.71	2.60	2.51	2.45	2.39
21	4.32	3.47	3.07	2.84	2.68	2.57	2.49	2.42	2.37
22	4.30	3.44	3.05	2.82	2.66	2.55	2.46	2.40	2.34
23	4.28	3.42	3.03	2.80	2.64	2.53	2.44	2.37	2.32
24	4.26	3.40	3.01	2.78	2.62	2.51	2.42	2.36	2.30
25	4.24	3.39	2.99	2.76	2.60	2.49	2.40	2.34	2.28
26	4.23	3.37	2.98	2.74	2.59	2.47	2.39	2.32	2.27
27	4.21	3.35	2.96	2.73	2.57	2.46	2.37	2.31	2.25
28	4.20	3.34	2.95	2.71	2.56	2.45	2.36	2.29	2.24
29	4.18	3.33	2.93	2.70	2.55	2.43	2.35	2.28	2.22
30	4.17	3.32	2.92	2.69	2.53	2.42	2.33	2.27	2.21
40	4.08	3.23	2.84	2.61	2.45	2.34	2.25	2.18	2.12
60	4.00	3.15	2.76	2.53	2.37	2.25	2.17	2.10	2.04
120	3.92	3.07	2.68	2.45	2.29	2.17	2.09	2.02	1.96
∞	3.84	3.00	2.60	2.37	2.21	2.10	2.01	1.94	1.88

Source: From "Tables of Percentage Points of the Inverted Beta (F)-Distribution," By Maxine Merrington and Catherine M. Thompson, *Biometrika* 33 (1943), 73–88. Reproduced by permission of the *Biometrika* Trustees.

TABLE A4   Percentage Points of the F-Distribution ($\gamma = .05$) (Continued)

Denominator Degrees of Freedom (r_2)	Numerator Degrees of Freedom (r_1)									
	10	12	15	20	24	30	40	60	120	∞
1	6056	6106	6157	6209	6235	6261	6287	6313	6339	6366
2	99.40	99.42	99.43	99.45	99.46	99.47	99.47	99.48	99.49	99.50
3	27.23	27.05	26.87	26.69	26.60	26.50	26.41	26.32	26.22	26.13
4	14.55	14.37	14.20	14.02	13.93	13.84	13.75	13.65	13.56	13.46
5	10.05	9.89	9.72	9.55	9.47	9.38	9.29	9.20	9.11	9.02
6	7.87	7.72	7.56	7.40	7.31	7.23	7.14	7.06	6.97	6.88
7	6.62	6.47	6.31	6.16	6.07	5.99	5.91	5.82	5.74	5.65
8	5.81	5.67	5.52	5.36	5.28	5.20	5.12	5.03	4.95	4.86
9	5.26	5.11	4.96	4.81	4.73	4.65	4.57	4.48	4.40	4.31
10	4.85	4.71	4.56	4.41	4.33	4.25	4.17	4.08	4.00	3.91
11	4.54	4.40	4.25	4.10	4.02	3.94	3.86	3.78	3.69	3.60
12	4.30	4.16	4.01	3.86	3.78	3.70	3.62	3.54	3.45	3.36
13	4.10	3.96	3.82	3.66	3.59	3.51	3.43	3.34	3.25	3.17
14	3.94	3.80	3.66	3.51	3.43	3.35	3.27	3.18	3.09	3.00
15	3.80	3.67	3.52	3.37	3.29	3.21	3.13	3.05	2.96	2.87
16	3.69	3.55	3.41	3.26	3.18	3.10	3.02	2.93	2.84	2.75
17	3.59	3.46	3.31	3.16	3.08	3.00	2.92	2.83	2.75	2.65
18	3.51	3.37	3.23	3.08	3.00	2.92	2.84	2.75	2.66	2.57
19	3.43	3.30	3.15	3.00	2.92	2.84	2.76	2.67	2.58	2.49
20	3.37	3.23	3.09	2.94	2.86	2.78	2.69	2.61	2.52	2.42
21	3.31	3.17	3.03	2.88	2.80	2.72	2.64	2.55	2.46	2.36
22	3.26	3.12	2.98	2.83	2.75	2.67	2.58	2.50	2.40	2.31
23	3.21	3.07	2.93	2.78	2.70	2.62	2.54	2.45	2.35	2.26
24	3.17	3.03	2.89	2.74	2.66	2.58	2.49	2.40	2.31	2.21
25	3.13	2.99	2.85	2.70	2.62	2.54	2.45	2.36	2.27	2.17
26	3.09	2.96	2.81	2.66	2.58	2.50	2.42	2.33	2.23	2.13
27	3.06	2.93	2.78	2.63	2.55	2.47	2.38	2.29	2.20	2.10
28	3.03	2.90	2.75	2.60	2.52	2.44	2.35	2.26	2.17	2.06
29	3.00	2.87	2.73	2.57	2.49	2.41	2.33	2.23	2.14	2.03
30	2.98	2.84	2.70	2.55	2.47	2.39	2.30	2.21	2.11	2.01
40	2.80	2.66	2.52	2.37	2.29	2.20	2.11	2.02	1.92	1.80
60	2.63	2.50	2.35	2.20	2.12	2.03	1.94	1.84	1.73	1.60
120	2.47	2.34	2.19	2.03	1.95	1.86	1.76	1.66	1.53	1.38
∞	2.32	2.18	2.04	1.88	1.79	1.70	1.59	1.47	1.32	1.00

TABLE A5   Critical Values for the Durbin-Watson d Statistic (α = .05)

n	$n_p - 1 = 1$ $d_{L,.05}$	$d_{U,.05}$	$n_p - 1 = 2$ $d_{L,.05}$	$d_{U,.05}$	$n_p - 1 = 3$ $d_{L,.05}$	$d_{U,.05}$	$n_p - 1 = 4$ $d_{L,.05}$	$d_{U,.05}$	$n_p - 1 = 5$ $d_{L,.05}$	$d_{U,.05}$
15	1.08	1.36	0.95	1.54	0.82	1.75	0.69	1.97	0.56	2.21
16	1.10	1.37	0.98	1.54	0.86	1.73	0.74	1.93	0.62	2.15
17	1.13	1.38	1.02	1.54	0.90	1.71	0.78	1.90	0.67	2.10
18	1.16	1.39	1.05	1.53	0.93	1.69	0.82	1.87	0.71	2.06
19	1.18	1.40	1.08	1.53	0.97	1.68	0.86	1.85	0.75	2.02
20	1.20	1.41	1.10	1.54	1.00	1.68	0.90	1.83	0.79	1.99
21	1.22	1.42	1.13	1.54	1.03	1.67	0.93	1.81	0.83	1.96
22	1.24	1.43	1.15	1.54	1.05	1.66	0.96	1.80	0.86	1.94
23	1.26	1.44	1.17	1.54	1.08	1.66	0.99	1.79	0.90	1.92
24	1.27	1.45	1.19	1.55	1.10	1.66	1.01	1.78	0.93	1.90
25	1.29	1.45	1.21	1.55	1.12	1.66	1.04	1.77	0.95	1.89
26	1.30	1.46	1.22	1.55	1.14	1.65	1.06	1.76	0.98	1.88
27	1.32	1.47	1.24	1.56	1.16	1.65	1.08	1.76	1.01	1.86
28	1.33	1.48	1.26	1.56	1.18	1.65	1.10	1.75	1.03	1.85
29	1.34	1.48	1.27	1.56	1.20	1.65	1.12	1.74	1.05	1.84
30	1.35	1.49	1.28	1.57	1.21	1.65	1.14	1.74	1.07	1.83
31	1.36	1.50	1.30	1.57	1.23	1.65	1.16	1.74	1.09	1.83
32	1.37	1.50	1.31	1.57	1.24	1.65	1.18	1.73	1.11	1.82
33	1.38	1.51	1.32	1.58	1.26	1.65	1.19	1.73	1.13	1.81
34	1.39	1.51	1.33	1.58	1.27	1.65	1.21	1.73	1.15	1.81
35	1.40	1.52	1.34	1.58	1.28	1.65	1.22	1.73	1.16	1.80
36	1.41	1.52	1.35	1.59	1.29	1.65	1.24	1.73	1.18	1.80
37	1.42	1.53	1.36	1.59	1.31	1.66	1.25	1.72	1.19	1.80
38	1.43	1.54	1.37	1.59	1.32	1.66	1.26	1.72	1.21	1.79
39	1.43	1.54	1.38	1.60	1.33	1.66	1.27	1.72	1.22	1.79
40	1.44	1.54	1.39	1.60	1.34	1.66	1.29	1.72	1.23	1.79
45	1.48	1.57	1.43	1.62	1.38	1.67	1.34	1.72	1.29	1.78
50	1.50	1.59	1.46	1.63	1.42	1.67	1.38	1.72	1.34	1.77
55	1.53	1.60	1.49	1.64	1.45	1.68	1.41	1.72	1.38	1.77
60	1.55	1.62	1.51	1.65	1.48	1.69	1.44	1.73	1.41	1.77
65	1.57	1.63	1.54	1.66	1.50	1.70	1.47	1.73	1.44	1.77
70	1.58	1.64	1.55	1.67	1.52	1.70	1.49	1.74	1.46	1.77
75	1.60	1.65	1.57	1.68	1.54	1.71	1.51	1.74	1.49	1.77
80	1.61	1.66	1.59	1.69	1.56	1.72	1.53	1.74	1.51	177
85	1.62	1.67	1.60	1.70	1.57	1.72	1.55	1.75	1.52	1.77
90	1.63	1.68	1.61	1.70	1.59	1.73	1.57	1.75	1.54	1.78
95	1.64	1.69	1.62	1.71	1.60	1.73	1.58	1.75	1.56	1.78
100	1.65	1.69	1.63	1.72	1.61	1.74	1.59	1.76	1.57	1.78

TABLE A6   Critical Values for the Durbin-Watson d Statistic ($\alpha = .01$)

n	$n_p - 1 = 1$ $d_{L,.01}$	$d_{U,.01}$	$n_p - 1 = 2$ $d_{L,.01}$	$d_{U,.01}$	$n_p - 1 = 3$ $d_{L,.01}$	$d_{U,.01}$	$n_p - 1 = 4$ $d_{L,.01}$	$d_{U,.01}$	$n_p - 1 = 5$ $d_{L,.01}$	$d_{U,.01}$
15	0.81	1.07	0.70	1.25	0.59	1.46	0.49	1.70	0.39	1.96
16	0.84	1.09	0.74	1.25	0.63	1.44	0.53	1.66	0.44	1.90
17	0.87	1.10	0.77	1.25	0.67	1.43	0.57	1.63	0.48	1.85
18	0.90	1.12	0.80	1.26	0.71	1.42	0.61	1.60	0.52	1.80
19	0.93	1.13	0.83	1.26	0.74	1.41	0.65	1.58	0.56	1.77
20	0.95	1.15	0.86	1.27	0.77	1.41	0.68	1.57	0.60	1.74
21	0.97	1.16	0.89	1.27	0.80	1.41	0.72	1.55	0.63	1.71
22	1.00	1.17	0.91	1.28	0.83	1.40	0.75	1.54	0.66	1.69
23	1.02	1.19	0.94	1.29	0.86	1.40	0.77	1.53	0.70	1.67
24	1.04	1.20	0.96	1.30	0.88	1.41	0.80	1.53	0.72	1.66
25	1.05	1.21	0.98	1.30	0.90	1.41	0.83	1.52	0.75	1.65
26	1.07	1.22	1.00	1.31	0.93	1.41	0.85	1.52	0.78	1.64
27	1.09	1.23	1.02	1.32	0.95	1.41	0.88	1.51	0.81	1.63
28	1.10	1.24	1.04	1.32	0.97	1.41	0.90	1.51	0.83	1.62
29	1.12	1.25	1.05	1.33	0.99	1.42	0.92	1.51	0.85	1.61
30	1.13	1.26	1.07	1.34	1.01	1.42	0.94	1.51	0.88	1.61
31	1.15	1.27	1.08	1.34	1.02	1.42	0.96	1.51	0.90	1.60
32	1.16	1.28	1.10	1.35	1.04	1.43	0.98	1.51	0.92	1.60
33	1.17	1.29	1.11	1.36	1.05	1.43	1.00	1.51	0.94	1.59
34	1.18	1.30	1.13	1.36	1.07	1.43	1.01	1.51	0.95	1.59
35	1.19	1.31	1.14	1.37	1.08	1.44	1.03	1.51	0.97	1.59
36	1.21	1.32	1.15	1.38	1.10	1.44	1.04	1.51	0.99	1.59
37	1.22	1.32	1.16	1.38	1.11	1.45	1.06	1.51	1.00	1.59
38	1.23	1.33	1.18	1.39	1.12	1.45	1.07	1.52	1.02	1.58
39	1.24	1.34	1.19	1.39	1.14	1.45	1.09	1.52	1.03	1.58
40	1.25	1.34	1.20	1.40	1.15	1.46	1.10	1.52	1.05	1.58
45	1.29	1.38	1.24	1.42	1.20	1.48	1.16	1.53	1.11	1.58
50	1.32	1.40	1.28	1.45	1.24	1.49	1.20	1.54	1.16	1.59
55	1.36	1.43	1.32	1.47	1.28	1.51	1.25	1.55	1.21	1.59
60	1.38	1.45	1.35	1.48	1.32	1.52	1.28	1.56	1.25	1.60
65	1.41	1.47	1.38	1.50	1.35	1.53	1.31	1.57	1.28	1.61
70	1.43	1.49	1.40	1.52	1.37	1.55	1.34	1.58	1.31	1.61
75	1.45	1.50	1.42	1.53	1.39	1.56	1.37	1.59	1.34	1.62
80	1.47	1.52	1.44	1.54	1.42	1.57	1.39	1.60	1.36	1.62
85	1.48	1.53	1.46	1.55	1.43	1.58	1.41	1.60	1.39	1.63
90	1.50	1.54	1.47	1.56	1.45	1.59	1.43	1.61	1.41	1.64
95	1.51	1.55	1.49	1.57	1.47	1.60	1.45	1.62	1.42	1.64
100	1.52	1.56	1.50	1.58	1.48	1.60	1.46	1.63	1.44	1.65

Source: From J. Durbin and G. S. Watson, "Testing for Serial Correlation in Least Squares Regression, II," *Biometrika* 30 (1951), 159–178. Reproduced by permission of the *Biometrika* Trustees.

APPENDIX B

TIME SERIES DATA

TABLE B1   (a) "Historical" Airline Passenger Data—Monthly Total International Airline Passengers (1000's of passengers) 1949–1959

	Jan.	Feb.	Mar.	Apr.	May	June	July	Aug.	Sept.	Oct.	Nov.	Dec.
1949	112	118	132	129	121	135	148	148	136	119	104	118
1950	115	126	141	135	125	149	170	170	158	133	114	140
1951	145	150	178	163	172	178	199	199	184	162	146	166
1952	171	180	193	181	183	218	230	242	209	191	172	194
1953	196	196	236	235	229	243	264	272	237	211	180	201
1954	204	188	235	227	234	264	302	293	259	229	203	229
1955	242	233	267	269	270	315	364	347	312	274	237	278
1956	284	277	317	313	318	374	413	405	355	306	271	306
1957	315	301	356	348	355	422	465	467	404	347	305	336
1958	340	318	362	348	363	435	491	505	404	359	310	337
1959	360	342	406	396	420	472	548	559	463	407	362	405

(b) "Future" Airline Passenger Data—Monthly Total International Airline Passengers (1000's of passengers) 1960

	Jan.	Feb.	Mar.	Apr.	May	June	July	Aug.	Sept	Oct.	Nov.	Dec.
1960	417	391	419	461	472	535	622	606	508	461	390	432

Source: FAA Statistical Handbook of Civil Aviation (several annual issues)

TABLE B2   Number of Employees in Wholesale and Retail Trade in Wisconsin 1961–1975 (thousands of employees)

Year	Jan.	Feb.	Mar.	Apr.	May	June	July	Aug.	Sept.	Oct.	Nov.	Dec.
1961	239.6	236.4	236.8	241.5	243.7	246.1	244.1	244.2	244.8	246.6	250.9	261.4
1962	237.6	235.7	236.1	242.6	244.5	246.6	245.7	247.7	248.9	251.4	255.9	263.7
1963	242.2	239.3	239.7	247.2	249.3	252.3	252.8	253.6	254.2	256.1	260.3	268.8
1964	250.1	247.9	249.0	253.8	258.3	261.3	261.3	261.9	263.3	267.3	270.6	281.5
1965	261.9	258.6	259.7	266.0	271.1	274.4	274.0	273.8	274.9	280.0	285.4	295.9
1966	275.4	273.6	275.9	281.1	285.2	289.1	289.2	288.9	291.1	295.3	300.2	310.9
1967	286.9	283.0	286.2	291.5	295.4	299.7	297.9	298.1	300.2	304.8	311.9	320.9
1968	298.3	295.5	297.2	302.7	306.7	309.1	308.7	309.9	310.8	314.7	321.2	329.0
1969	307.6	305.5	308.0	314.4	320.5	323.4	323.0	324.4	326.1	329.3	335.0	341.9
1970	321.8	317.3	318.6	323.4	327.1	327.9	325.3	325.7	330.0	333.5	337.1	341.3
1971	321.6	318.2	319.6	326.2	332.3	334.2	334.5	335.5	335.1	338.2	341.9	347.9
1972	329.5	326.4	329.1	337.2	344.9	349.6	351.0	353.8	354.5	357.4	362.1	367.5
1973	347.9	345.0	348.9	355.3	362.4	366.6	366.0	370.2	370.9	374.5	380.2	384.6
1974	360.6	354.4	357.4	367.0	375.7	381.0	381.2	383.0	384.3	387.0	391.7	396.0
1975	374.0	370.4	373.2	381.1	389.9	394.6	394.0	397.0	397.2	399.4		

Source: State of Wisconsin Department of Industry, Labor, and Human Relations, Bureau of Research and Statistics.

TABLE B3   Number of Employees in the Fabricated Metals Industry in Wisconsin 1961–1975 (thousands of employees)

Year	Jan.	Feb.	Mar.	Apr.	May	June	July	Aug.	Sept.	Oct.	Nov.	Dec.
1961	31.1	29.5	31.0	31.5	32.0	32.8	32.8	33.1	33.6	33.4	33.5	33.5
1962	33.1	33.1	33.4	33.9	34.4	35.2	35.0	35.3	34.6	34.1	33.9	33.8
1963	33.6	33.8	34.0	34.4	34.7	35.4	35.5	35.5	35.2	34.5	34.4	34.3
1964	33.9	34.1	34.5	34.3	35.0	36.0	35.0	37.2	37.4	36.9	37.2	37.3
1965	37.2	37.6	36.0	38.8	38.8	39.5	39.6	40.1	39.4	38.9	39.5	39.2
1966	38.7	39.1	39.5	39.7	40.0	41.4	41.3	39.8	38.5	40.2	41.0	40.9
1967	40.5	40.3	40.4	40.3	40.4	41.8	41.1	42.1	40.8	40.5	40.6	40.7
1968	40.6	40.7	40.7	40.7	40.9	41.6	41.2	42.0	41.8	42.3	43.1	42.9
1969	42.6	43.1	43.3	43.5	43.9	45.7	45.6	46.5	45.3	45.2	45.3	45.0
1970	44.2	44.3	44.4	43.4	42.8	44.3	44.4	44.8	44.4	43.1	42.6	42.4
1971	42.2	41.8	40.1	42.0	42.4	43.1	42.4	43.1	43.2	42.8	43.0	42.8
1972	42.5	42.6	42.3	42.9	43.6	44.7	44.5	45.0	44.8	44.9	45.2	45.2
1973	45.0	45.5	46.2	46.8	47.5	48.3	48.3	49.1	48.9	49.4	50.0	50.0
1974	49.6	49.9	49.6	50.7	50.7	50.9	50.5	51.2	50.7	50.3	49.2	48.1
1975	46.6	45.3	44.6	44.0	43.7	43.8	43.0	43.6	44.0	45.0		

Source: State of Wisconsin Department of Industry, Labor, and Human Relations, Bureau of Research and Statistics.

TABLE B4 Monthly Values of Average Weekly Total Investments at Large New York City Banks 1965–1974 (millions of dollars)

Year \ Month	Jan.	Feb.	Mar.	Apr.	May	June	July	Aug.	Sept.	Oct.	Nov.	Dec.
1965	5466.1	4845.9	4671.2	4528.4	4418.1	4461.0	4643.7	4460.0	4260.0	4627.9	4612.4	4759.6
1966	4565.7	4193.9	3822.4	4144.9	3823.7	3653.6	3710.4	3705.8	3999.6	3694.6	3855.4	4323.9
1967	4475.4	4785.4	4882.0	4648.7	4656.4	4775.3	5017.9	4859.6	5287.4	5794.2	5832.6	5597.6
1968	5313.2	5192.7	4807.5	4765.6	4631.0	4727.8	5773.2	5298.9	5909.9	5647.8	5645.1	5965.1
1969	5165.4	4580.8	4144.9	4509.0	3984.8	4030.4	4275.2	4291.7	4324.2	4164.4	4615.9	4962.4
1970	4550.2	4136.5	4232.1	4936.8	4507.1	4329.7	4500.2	5169.4	4993.0	4879.9	5059.0	5459.8
1971	5544.6	5342.9	5106.0	5439.2	4863.7	4657.2	4988.7	4427.2	4562.6	4528.4	5330.7	5263.6
1972	5112.2	5031.1	5551.8	5223.9	5004.4	4901.1	4602.1	4746.2	5247.3	4715.1	4862.8	5039.9
1973	4931.2	4362.8	4137.6	4316.1	3760.8	4247.1	3831.9	3768.8	4154.2	4257.6	4867.9	5787.9
1974	5435.0	5117.0	4984.0	4781.0	3793.0	3679.0	3373.0	4118.0	4679.0	4144.0	4625.0	4932.0

Source: Federal Reserve Bulletin, published by the Division of Administrative Services, Board of Governors of the Federal Reserve System.

528

t	Advertising Input x_t	Sales Output y_t	t	Advertising Input x_t	Sales Output y_t
0	116.44	202.66	50	123.90	266.69
1	119.58	232.91	51	122.45	253.07
2	125.74	272.07	52	122.85	249.12
3	124.55	290.97	53	129.28	253.59
4	122.35	299.09	54	129.77	262.13
5	120.44	296.95	55	127.78	279.66
6	123.24	279.49	56	134.29	302.92
7	127.99	255.75	57	140.61	310.77
8	121.19	242.78	58	133.64	307.83
9	118.00	255.34	59	135.45	313.19
10	121.81	271.58	60	130.93	312.80
11	126.54	268.27	61	118.65	301.23
12	129.85	260.51	62	120.34	286.64
13	122.65	266.34	63	120.35	257.17
14	121.64	281.24	64	117.09	229.60
15	127.24	286.19	65	117.56	227.62
16	132.35	271.97	66	121.69	238.21
17	130.86	265.01	67	128.19	252.07
18	122.90	274.44	68	134.79	269.86
19	117.15	291.81	69	128.93	291.62
20	109.47	290.91	70	121.63	314.06
21	114.34	264.95	71	125.43	318.56
22	123.72	228.40	72	126.80	289.11
23	130.33	209.33	73	131.56	255.88
24	133.17	231.69	74	126.43	249.81
25	134.25	281.56	75	116.19	268.82
26	129.75	327.16	76	112.72	288.24
27	130.05	344.24	77	109.53	281.26
28	133.42	324.74	78	110.38	250.92
29	135.16	289.36	79	107.31	222.26
30	130.89	262.92	80	93.59	209.94
31	123.48	263.65	81	89.80	213.30
32	118.46	276.38	82	88.70	207.19
33	122.11	276.34	83	86.64	186.13
34	128.75	258.27	84	89.26	171.20
35	127.09	242.89	85	96.51	170.33
36	114.55	255.98	86	107.35	183.69
37	113.26	278.53	87	110.35	211.30
38	111.51	273.21	88	102.66	252.66
39	111.73	246.37	89	97.56	286.20
40	114.08	221.10	90	98.06	279.45
41	114.32	210.41	91	103.93	237.06
42	115.03	222.19	92	115.66	193.40
43	124.28	245.27	93	112.91	180.79
44	132.69	262.58	94	116.89	215.73
45	134.64	283.25	95	116.84	264.98
46	133.28	311.12	96	109.55	294.07
47	128.00	326.28	97	110.63	299.08
48	129.97	322.04	98	111.32	271.10
49	128.35	295.37	99	117.09	230.56

Source: *Forecasting: Methods and Applications* by S. Makridakis, S.C. Wheelwright, and V.E. McGee, Wiley, New York, 1983.

t	Leading Indicator x_t	Sales y_t	t	Leading Indicator x_t	Sales y_t	t	Leading Indicator x_t	Sales y_t
1	10.01	200.1	51	10.77	220.0	101	12.90	249.4
2	10.07	199.5	52	10.88	218.7	102	13.12	249.0
3	10.32	199.4	53	10.49	217.0	103	12.47	249.9
4	9.75	198.9	54	10.50	215.9	104	12.47	250.5
5	10.33	199.0	55	11.00	215.8	105	12.94	251.5
6	10.13	200.2	56	10.98	214.1	106	13.10	249.0
7	10.36	198.6	57	10.61	212.3	107	12.91	247.6
8	10.32	200.0	58	10.48	213.9	108	13.39	248.8
9	10.13	200.3	59	10.53	214.6	109	13.13	250.4
10	10.16	201.2	60	11.07	213.6	110	13.34	250.7
11	10.58	201.6	61	10.61	212.1	111	13.34	253.0
12	10.62	201.5	62	10.86	211.4	112	13.14	253.7
13	10.86	201.5	63	10.34	213.1	113	13.49	255.0
14	11.20	203.5	64	10.78	212.9	114	13.87	256.2
15	10.74	204.9	65	10.80	213.3	115	13.39	256.0
16	10.56	207.1	66	10.33	211.5	116	13.59	257.4
17	10.48	210.5	67	10.44	212.3	117	13.27	260.4
18	10.77	210.5	68	10.50	213.0	118	13.70	260.0
19	11.33	209.8	69	10.75	211.0	119	13.20	261.3
20	10.96	208.8	70	10.40	210.7	120	13.32	260.4
21	11.16	209.5	71	10.40	210.1	121	13.15	261.6
22	11.70	213.2	72	10.34	211.4	122	13.30	260.8
23	11.39	213.7	73	10.55	210.0	123	12.94	259.8
24	11.42	215.1	74	10.46	209.7	124	13.29	259.0
25	11.94	218.7	75	10.82	208.8	125	13.26	258.9
26	11.24	219.8	76	10.91	208.8	126	13.08	257.4
27	11.59	220.5	77	10.87	208.8	127	13.24	257.7
28	10.96	223.8	78	10.67	210.6	128	13.31	257.9
29	11.40	222.8	79	11.11	211.9	129	13.52	257.4
30	11.02	223.8	80	10.88	212.8	130	13.02	257.3
31	11.01	221.7	81	11.28	212.5	131	13.25	257.6
32	11.23	222.3	82	11.27	214.8	132	13.12	258.9
33	11.33	220.8	83	11.44	215.3	133	13.26	257.8
34	10.83	219.4	84	11.52	217.5	134	13.11	257.7
35	10.84	220.1	85	12.10	218.8	135	13.30	257.2
36	11.14	220.6	86	11.83	220.7	136	13.06	257.5
37	10.38	218.9	87	12.62	222.2	137	13.32	256.8
38	10.90	217.8	88	12.41	226.7	138	13.10	257.5
39	11.05	217.7	89	12.43	228.4	139	13.27	257.0
40	11.11	215.0	90	12.73	233.2	140	13.64	257.6
41	11.01	215.3	91	13.01	235.7	141	13.58	257.3
42	11.22	215.9	92	12.74	237.1	142	13.87	257.5
43	11.21	216.7	93	12.73	240.6	143	13.53	259.6
44	11.91	216.7	94	12.76	243.8	144	13.41	261.1
45	11.69	217.7	95	12.92	245.3	145	13.25	262.9
46	10.93	218.7	96	12.64	246.0	146	13.50	263.3
47	10.99	222.9	97	12.79	246.3	147	13.58	262.8
48	11.01	224.9	98	13.05	247.7	148	13.51	261.8
49	10.84	222.2	99	12.69	247.6	149	13.77	262.2
50	10.76	220.7	100	13.01	247.8	150	13.40	262.7

Source: Time Series Analysis: Forecasting and Control, 2nd ed., by G.E.P. Box and G.M. Jenkins, Holden Day, San Francisco, 1976.

Abraham, B., and J. Ledolter. *Statistical Methods for Forecasting*. New York: Wiley, 1983.

Anderson, T.W. *The Statistical Analysis of Time Series*. New York: Wiley, 1971.

Anscombe, F.J., and J.W. Tukey. "The Examination and Analysis of Residuals." *Technometrics* 5 (1963): 141–160.

Bowerman, B.L., and R.T. O'Connell. *Time Series and Forecasting: An Applied Approach*. Boston: Duxbury Press, 1979.

——— and D.A. Dickey. *Linear Statistical Models: An Applied Approach*. Boston: Duxbury Press, 1986.

Box, G.E.P., and D.R. Cox. "An Analysis of Transformations." *Journal of Royal Statistical Society B* 26 (1964): 211–243.

Box, G.E.P., and G.M. Jenkins. *Time Series Analysis: Forecasting and Control*. 2d ed. San Francisco: Holden-Day, 1976.

Brown, Bernice B. *Delphi Process: A Methodology Used for the Elicitation of Opinion of Experts*. P-3925, RAND Corporation, Santa Monica, California, September 1968.

Brown, R.G. *Statistical Forecasting for Inventory Control*. New York: McGraw-Hill, 1959.

———. *Smoothing, Forecasting and Prediction of Discrete Time Series*. Englewood Cliffs, N.J.: Prentice-Hall, Inc., 1962.

———. *Decision Rules for Inventory Management*. New York: Holt, Rinehart and Winston, 1967.

Chatfield, C., and D.L. Prothero, "Box-Jenkins Seasonal Forecasting: Problems in a Case Study," (with discussion). *Journal of the Royal Statistical Society*. A136 (1973).

Chow, W.M. "Adaptive Control of the Exponential Smoothing Constant." *Journal of Industrial Engineering* 16, no. 5 (1965): 314–317.

Cochran, G.W., and G.M. Cox. *Experimental Designs*. 2d ed. New York: Wiley, 1957.

Dalkey, Norman C. *Delphi*. P-3704, RAND Corporation, Santa Monica, California, October 1967.

———. *The Delphi Method: An Experimental Study of Group Opinion*. RM-5888-PR, RAND Corporation, Santa Monica, California, June 1969.

Davis, O.L. *The Design and Analysis of Industrial Experiments*. New York: Hafner, 1956.

Draper, N., and H. Smith. *Applied Regression Analysis*. 2d ed. New York: Wiley, 1981.

Durbin, J., and G.S. Watson. "Testing for Serial Correlation in Least Squares Regression, I." *Biometrika* 37 (1950): 409–428.

———. "Testing for Serial Correlation in Least Squares Regression, II." *Biometrika* 38 (1951): 159–179.

Fuller, W.A. *Introduction to Statistical Time Series*. New York: Wiley, 1976.

Gerstenfeld, Arthur. "Technological Forecasting." *Journal of Business* 44, no. 1 (1971).

Gordon, T.J., and H. Hayward. "Initial Experiments with the Cross-Impact Method of Forecasting." *Futures* 1, no. 2 (1968).

Graybill, F.A. *Theory and Application of the Linear Model*. Boston: Duxbury Press, 1976.

Hillmer, S.C., and G.C. Tiao. "Likelihood Function of Stationary Multiple Autoregressive Moving Average Models." *Journal of the American Statistical Association* 74 (1979): 652–660.

Johnson, L.A., and D.C. Montogomery. *Forecasting and Time Series Analysis*. New York: McGraw-Hill, 1976.

Kennedy, W.J., Jr., and J.E. Gentle. *Statistical Computing*. New York: Dekker, 1980.

Kleinbaum, D., and L. Kupper. *Applied Regression Analysis and Other Multivariable Methods*. Boston: Duxbury Press, 1978.

Mabert, V.A. *An Introduction to Short Term Forecasting Using the Box-Jenkins Methodology*. Publication No. 2 in the American Institute of Industrial Engineers, Inc., Monograph Series, 1976.

Makridakis, S., S.C. Wheelwright, and V.E. McGee, *Forecasting: Methods and Applications*. New York: Wiley, 1983.

McKenzie, E. "General Exponential Smoothing and The Equivalent ARMA Process." *Journal of Forecasting* 3 (1984): 333–444.

Meeker, W.Q. "TSERIES—A User-Oriented Computer Program for Time Series Analysis." *The American Statistician* 32, no. 3 (1978).

Mendenhall, W. *Introduction to Linear Models and the Design and Analysis of Experiments*. Belmont, Mass.: Wadsworth, 1968.

———— and J. Reinmuth. *Statistics for Management and Economics*. 4th ed. Boston: Duxbury Press, 1982.

Miller, R.B., and D.W. Wichern. *Intermediate Business Statistics: Analysis of Variance, Regression, and Time Series*. New York: Holt, Rinehart, and Winston, 1977.

Myers, Raymond H. *Classical and Modern Regression with Applications*. Boston: Duxbury Press, 1986.

Nelson, C.R., *Applied Time Series Analysis for Managerial Forecasting*. San Francisco: Holden-Day, 1973.

Neter, John, and William Wasserman. *Applied Linear Statistical Models*. Homewood, IL: R.D. Irwin, 1974.

———— and G.A. Whitmore. *Applied Statistics*. 2d ed. Boston: Allyn and Bacon, 1982.

———— and M.H. Kutner. *Applied Linear Statistical Models*. 2d ed. Homewood, IL: Richard Irwin, 1985.

Ott, Lyman. *An Introduction to Statistical Methods and Data Analysis*. 2d ed. Boston: Duxbury Press, 1984.

Pankratz, A. *Forecasting with Univariate Box-Jenkins Models: Concepts and Cases*. New York: Wiley, 1983.

SAS User's Guide, 1982 Edition. Cary, NC: SAS Institute, 1982.

Scheffé, H. *The Analysis of Variance*. New York: Wiley, 1959.

Searle, S.R. *Linear Models*. New York: Wiley, 1971.

Sigford, J.V., and R.H. Parvin. "Project PATTERN: A Methodology for Determining Relevance in

Complex Decision Making." *IEEE Transactions on Engineering Management* 12, no. 1 (1965).

Wheelwright, S.C. and S. Makridakis. *Forecasting Methods for Management.* New York: Wiley and Sons, 1973.

Winer, B.J. *Statistical Principles in Experimental Design.* New York: McGraw-Hill, 1962.

Winters, P.R. "Forecasting Sales by Exponentially Weighted Moving Averages." *Management Science* 6, no. 3 (1960): 324–342.

Wonnacott, T.H., and R.J. Wonnacott. *Introductory Statistics for Business and Economics.* 2d ed. New York: Wiley, 1977.

―――. *Regression: A Second Course in Statistics.* New York: Wiley, 1981.

Younger, M.S. *A First Course in Linear Regression.* 2d ed. Boston: Duxbury Press, 1985.

Zwicky, Fritz. "Morphology of Propulsive Power." *Monographs on Morphological Research No. 1.* Society of Morphological Research, Pasadena, California, 1962.

Index

Abraham, B., 300, 328, 370

Absolute deviations, 14

Adaptive control procedures, 21, 270–272

Additive (Box-Jenkins) model, 119

Additive (constant) seasonal variation, 208, 231, 289

Additive Winters' method, 289–290, 308

Additive Winters' model, 325

Analysis of variance, 487

ARIMA models, 20

Autocorrelated error terms, 466

Autoregressive models, 20

Backshift operator B, 56

Backward elimination, 449

Bowerman, B. L., 252, 326, 472

Box, G. E. P., 20, 59, 112, 137, 144, 316–317, 349, 376, 378

Box-Pierce statistic, 148–151

Broida, M., 252

Brown, B. B., 10

Brown, R. G., 300

Business cycle, 5

Causal models, 10–11

Centered moving average, 241–242, 244–245

Choice of a forecasting method, 17–19, 324–326

Chow, W. M., 271–272

Complete model, 454–455

Confidence interval for μ_t, in regression, 400, 414

Constant term δ, 56–61, 100

Constant variance assumption, 47–48, 56, 101, 208, 458, 463–466

Cook's distance measure, 479–480

Corrected multiple coefficient of determination (corrected R^2), 436–438

Correlation matrix, 147

Critical t-values for identifying spikes, 164–165

Cross-impact method, 10

C-statistic, 440–443

Cut off after lag k (in the SAC)
  at the seasonal level, 90
  nonseasonal data, 37–38

Cut off after lag k (in the SPAC)
  at the seasonal level, 90
  nonseasonal data, 43–44

Cycle, 4–6

Cyclical component (factor), 239

Cyclical factor in multiplicative decomposition, 243, 247–248

Dalkey, N. C., 10

DECOMP (computer program), 252

Decomposition, 239–251

Deleted residual, 478

Delphi method, 9–10

Dependent variable, 11, 147, 393

Deseasonalized observations, 242, 245–247

Deterministic trend, 59, 308

DFBETAS, 481–482

DFFITS, 480–482

Diagnostic checking, 20, 25, 148

Dickey, D. A., 472

Double exponential smoothing, 290–293, 325

Draper, N., 447, 470

Dummy variable
  comparing population means, 487–500
  modeling seasonal variation, 206, 208
  related to other models, 326

Durbin-Watson
  statistic, 470–471
  test for autocorrelation, 470–472

Dying down behavior
  in the SAC, 37–39, 90
  in the SPAC, 44, 90